THE MATHEMATICAL HERITAGE
of
HENRI POINCARÉ

PROCEEDINGS OF SYMPOSIA
IN PURE MATHEMATICS
Volume 39, Part 2

THE MATHEMATICAL HERITAGE
of
HENRI POINCARÉ

AMERICAN MATHEMATICAL SOCIETY
PROVIDENCE, RHODE ISLAND

PROCEEDINGS OF SYMPOSIA IN PURE MATHEMATICS
OF THE AMERICAN MATHEMATICAL SOCIETY
VOLUME 39

PROCEEDINGS OF THE SYMPOSIUM
ON THE MATHEMATICAL HERITAGE OF HENRI POINCARÉ
HELD AT INDIANA UNIVERSITY
BLOOMINGTON, INDIANA
APRIL 7–10, 1980

EDITED BY
FELIX E. BROWDER

Prepared by the American Mathematical Society
with partial support from National Science Foundation grant MCS 79-22916

1980 *Mathematics Subject Classification*. Primary 01-XX, 14-XX, 22-XX, 30-XX, 32-XX, 34-XX, 35-XX, 47-XX, 53-XX, 55-XX, 57-XX, 58-XX, 70-XX, 76-XX, 83-XX.

Library of Congress Cataloging in Publication Data
Main entry under title:

The Mathematical Heritage of Henri Poincaré.

(Proceedings of symposia in pure mathematics; v. 39, pt. 1–)
Bibliography: p.
1. Mathematics–Congresses. 2. Poincaré, Henri, 1854–1912–Congresses.
I. Browder, Felix E. II. Series: Proceedings of symposia in pure mathematics; v. 39, pt. 1, etc.
QA1.M4266   1983   510   83-2774
ISBN 0-8218-1442-7 (set)   ISBN 0-8218-1449-4 (part 2)
ISBN 0-8218-1448-6 (part 1)   ISSN 0082-0717

Table of Contents

PART 1

PART 2

Section 5. Topological methods in nonlinear problems

Section 6. Mechanics and dynamical systems

Section 7. Ergodic theory and recurrence

Section 8. Historical material

Section 5

TOPOLOGICAL METHODS
IN NONLINEAR PROBLEMS

Proceedings of Symposia in Pure Mathematics
Volume 39 (1983), Part 2

Lectures on Morse Theory, Old and New

RAOUL BOTT[1]

Morse Theory is a beautiful and natural extension of the minimum principle for a continuous function on a compact space. In these lectures I would like to discuss it in the context of two problems in analysis which have self-evident geometric interest as well as physical origins.

The first question is simply this. Let M be a compact connected C^∞ manifold endowed with a fixed Riemannian structure. For instance you might think of the two-sphere S^2 with the Riemann structure inherited from an imbedding of S^2 in $\mathbf{R}^3$.

Question. Does such an M always carry a nontrivial closed geodesic?

Recall here first of all that on a compact manifold any two points P and Q can be joined by a geodesic which minimizes the length of all piecewise smooth curves joining P to Q in M. In one way or another this is then an application of the minimum principle, and conceptually you should think of pulling a string confined to M and joining P and Q as tight as possible. When the string has assumed a position in which it cannot be tightened any more, then it describes a geodesic joining P to Q. If it cannot be tightened further even after a "jiggling", then it describes the minimal geodesic in question.

This "pulling tight" principle works also for finding closed geodesics, provided only that we have some constraint to pull against.

Thus if α is a piecewise smooth map of the circle

$$\alpha\colon S^1 \to M$$

which *cannot* be deformed to a point in M, then shortening α in its homotopy class will indeed produce a closed geodesic.

Put differently, let ΛM, denote the space of continuous maps from S^1 to M:

$$\Lambda M = \mathrm{Map}(S^1, M),$$

in the compact open topology.

Also let $\Lambda_* M$ denote the component of the constant maps of S^1 to M. Then a classical theorem going back to Hadamard, Cartan, etc., asserts that

THEOREM. *Every component of ΛM other than $\Lambda_* M$ contains a bona fide closed geodesic.*

Reprinted from Bulletin Amer. Math. Soc. (N.S.) 7 (1982), 331–358.

1980 *Mathematics Subject Classification.* Primary 58E05.

[1]The following is the text of a series of four lectures which I delivered in Peking during the summer of 1980. However the content is an expanded and somewhat altered version of my lecture at the Poincaré Symposium. Rather than giving a survey I tried in both instances to produce as self-contained an account as possible of two specific problems in the subject.

Let me indicate a proof, once you grant me the following fundamental existence theorem of Riemannian geometry.

LEMMA. *There exists a constant $\varepsilon(M) = \varepsilon > 0$ such that any two points p, q on M with distance $\rho(p, q) < \varepsilon$ are joined by a unique minimizing geodesic segment $s(p, q)$ of length $\rho(p, q)$. Furthermore $s^2(p, q)$ varies smoothly with (p, q) in the region $\rho(p, q) < \varepsilon$ of $M \times M$.*

Armed with this fact, which in turn follows directly from the existence theorems governing elliptic ordinary differential equations, one may argue as follows to establish our theorem.

Let $\alpha\colon S^1 \to M$, be some point in ΛS, not in the component $\Lambda_* S$. From the continuity of α it follows that we can subdivide the circle S^1 into a finite number of intervals Δ_i, $i = 1,\ldots,n$, such that for $p, q \in \Delta_i$, $\alpha(p)$ and $\alpha(q)$ are within ε of each other. Now let $P_0, P_1,\ldots,P_{n-1}, P_0$, denote the endpoints of the Δ_i, cyclicly arranged on S^1, and let $s(P_0,\ldots,P_{n-1}, P_0)$ be the *geodesic polygon* spanned by the geodesic segments $s(P_i, P_{i+1})$—whose existence follows from our lemma—parametrized proportionally to arc length, and in proportion to the length of Δ_i. Then it should be clear from the picture below that we can deform α in Λ into $s(P_0,\ldots,P_0)$.

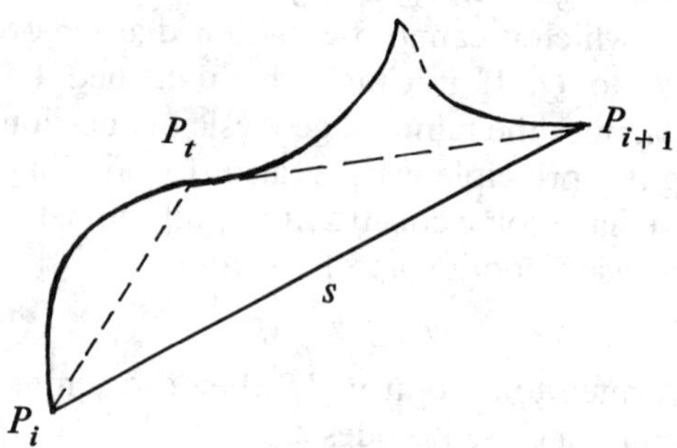

FIGURE 1

Here think of t as a deformation parameter which controls a point P_t on Δ_i moving from P_{i+1} to P_i as t goes from 0 to 1. Now let α_t be the curve which follows α until P_t and then replaces the rest of the curve by $s(P_t, P_{i+1})$.

This is Morse's basic deformation principle and can be used to deform all geodesic problems into finite dimensional ones. In any case at this stage we have seen that:

Each component of ΛM contains a geodesic polygon.

To proceed further choose $0 < \varepsilon < \varepsilon(M)$ and let

$$P_n M \subset M \times M \times \cdots \times M \quad (n \text{ copies})$$

be the subset of n-tuples $(P_1,\ldots,P_n)$ with the property that

$$(1.1) \qquad \rho(P_1, P_2)^2 + \rho(P_2, P_3)^2 + \cdots + \rho(P_n, P_1)^2 \leqslant \varepsilon.$$

Then $P_n M$ is a compact subset of $M^{(n)}$. Further (1.1) implies that each term on the left is $< \varepsilon$, so that every point of $P_n M$ determines a closed n-sided geodesic polygon with vertices at the P_i. If we parametrize the polygons proportionally to arc length, starting at P_1 say, we finally obtain a natural inclusion

$$\iota\colon P_n M \hookrightarrow \Lambda M,$$

which is clearly continuous.

At first sight it might seem that P_n contains only "short" polygons. However observe that by subdividing a polygon, say by introducing new vertices at the midpoints of the edges, the expression on the left of (1.1) is reduced because each term $\rho_i^2 = \rho(P_i, P_{i+1})^2$ is replaced by $(\rho_i/2)^2 + (\rho_i/2)^2 = \rho_i^2/2$. It follows that *any geodesic polygon in ΛM occurs as the image of a point in P_n for n large enough.*

At this stage it is clear that we may confine our search for closed geodesics among the geodesic polygons of P_n in each component of ΛM. For this purpose let

$$E\colon P_n M \to \mathbf{R}$$

be the *energy function*

$$(1.2) \qquad E(P_1,\ldots,P_n) = \sum_{\iota=1}^{n} \rho(P_n, P_{i+1})^2; \qquad P_{n+1} \equiv P_1,$$

given by the L.H.S. of (1.1). This energy function is clearly smooth in a vicinity of $P_n \subset M^{(n)}$. Hence E must assume a minimum in each component. Further by increasing n, if necessary, we can arrange it that E takes on this minimum at an interior point, i.e. one with $E < \varepsilon$.

At such a point dE, the differential of E, must therefore vanish. It remains to establish the following assertion: *A critical point of E on $P_n M$ gives rise to a polygon without corners and all of whose edges have equal length. In short, to a closed geodesic.*

This comes about by virtue of the first variation formula for our function ρ^2 in the vicinity of the diagonal in $M \times M$. Indeed in the region $\rho(P, Q)^2 < \varepsilon^2$, one has the following.

LEMMA. (a) *The diagonal $M \subset M \times M$ is a critical submanifold for ρ^2, whose Hessian is nondegenerate in the normal direction to M.*

(b) *At a point (P, Q) of the diagonal in our region, $d\rho^2$ is given by the formula*

$$(1.2) \qquad d\rho^2(Y_P, Y_Q) = \rho\{(X^+, Y_Q) - (X^-, Y_P)\}.$$

Here, X^+, X^- denote the tangents of unit length to $s(P, Q)$ at Q and P respectively, the Y's are tangent vectors at P and Q and $(\ ,\ )$ denotes the inner product.

Summing this expression at the vertices $(P_1,\ldots,P_n)$ of a point in $P_n M$ yields

$$(1.3) \qquad dE(Y_1,\ldots,Y_n) = \sum_{2}^{n} (Y_i, |S_{i-1}| X_{i-1}^+ - |S_i| X_i^-),$$

where the index $n + 1$ is again to be taken as equal to 1.

At a critical point, therefore, we must have

$$(1.4) \qquad |S_{i-1}| \, X_{i-1}^{+} = |S_i| \, X_i^{-}, \qquad i = 2, \dots, n+1,$$

which precisely expresses the no corner, equal length condition.   Q.E.D.

This completely elementary argument therefore establishes the classical Theorem I. An analogous argument could be used to prove the existence of a minimizing geodesic joining two points on M, or the existence of a geodesic joining two submanifolds N_1 and N_2 in M with minimal length.

But consider now the case of a compact simply connected manifold M, for example S^2. Then ΛM has only one component on which the minimum principle only yields the trivial "point paths" of ΛM.

Note by the way if $e: \Lambda M \to M$ denotes the evaluation map $\alpha \mapsto \alpha(0)$ then these point paths furnish us with section $\eta: M \to \Lambda M$ to e. Technically *e is a fibration in the sense of Serre, with fiber the space of loops* ΩM, that is, the subspace of ΛM consisting of maps α with $\alpha(0)$ some fixed point p of M.

From these two remarks it follows by quite elementary homotopy theory, and Serre's form of the Hurewicz theorem, that the *homotopy groups of* ΛM *cannot all be trivial*. Indeed, from the homotopy exact sequence of a fibering and the existence of a section to e, it follows that

$$(1.5) \qquad \pi_q(\Lambda M) = \pi_q(M) \oplus \pi_q(\Omega M).$$

Next, from the near tautologous isomorphism $\pi_{q+1}(M) \simeq \pi_q(\Omega M)$, $q \geqslant 1$, it follows that

$$(1.6) \qquad \pi_q(\Lambda M) = \pi_q(M) \oplus \pi_{q+1}(M).$$

Finally the $\pi_q(M)$ cannot all be trivial, by Serre's Hurewicz theorem and Poincaré duality.   Q.E.D.

At this stage it suggests itself that one should be able to use the fact that $\pi_q(\Lambda M) \neq 0$ for some q, as a constraint against which one could again minimize and so produce a new extremum. This plan can indeed be carried out and the guiding principle for it was formulated already by G. B. Birkhoff before 1920. It is known as his minimax principle.

To illustrate its application in our present context, let us first simplify matters by once again replacing ΛM by $P_n M$ for n large enough. Indeed the same retraction described earlier, but now done with a compact set of parameters, easily leads to the following [see [**B1**] for details].

LEMMA. *For any fixed* q, *there exists an* n_q *such that*

$$(1.7) \qquad \pi_k(P_n M) \simeq \pi_k(\Lambda M) \quad \textit{for all } k \leqslant q \textit{ and } n \geqslant n_q.$$

In short, the P_n approximate ΛM arbitrarily well in homotopy, and therefore in homology as well.

To prove the existence of a classical geodesic in ΛM we now argue as follows. Let $\xi \in \pi_q(M)$ be a nontrivial element of *lowest dimension*. Then according to (1.6), ξ gives rise to a nontrivial element $T\xi$ in $\pi_{q-1}(\Lambda M)$.

Next choose $n > n_q$, so that $P = P_n M$ approximates ΛM to dimension q. Then $T\xi \in \pi_{q-1}(P)$ is also nontrivial.

On P we now again consider our energy function E, whose critical points yield closed geodesics. Hence we will be done once we find a *critical point of E on P other than a point path*, i.e. one with $E > 0$. These point paths of course constitute a submanifold $M \subset P$, on which the energy function assumes an absolute minimum. Assume then—we are out to find a contradiction—that E has no other critical points on P. Then the negative gradient of E, that is, the vector field X on P, defined by the formula

$$(1.8) \qquad -(X, Y) = dE(Y),$$

is nonvanishing on $P - M$, and always points downwards. Hence following the flow generated by X will eventually deform P into a tubular neighborhood of M, which in turn can be retracted to M. It follows that under our assumption *all homotopy elements of P come from M*. But this is manifestly not the case for $T\xi$. Indeed, by construction $T\xi \in \pi_q(P) \neq 0$ while $\pi_q(M) \equiv 0$.   Q.E.D.

This argument therefore establishes the beautiful theorem of Lyusternik and Fet [**L-F**]:

THEOREM. *Let M be compact and simply connected. Then M carries at least one closed geodesic.*

Let me now explain how this argument is related to the "minimax principle". For that purpose consider the set of maps η: $S_q \to P$ representing $T\xi$, and try to push η as far down, relative to E, as possible. In short consider the real number

$$(1.9) \qquad \kappa = \inf_{\eta} \mathrm{Max}(E, \eta), \qquad [\eta] \in T\xi.$$

As we just saw $\kappa > 0$. The minimax principle simply asserts, *that this κ must be a critical value of E*. The proof is again a quite elementary consequence of pushing down in the direction of steepest descent—i.e. along the negative gradient—and I think of it usually as a corollary of what one might call the *first theorem of Morse Theory*. To formulate it and to deduce the minimax principle from it, let us abstract the situation though, so that from now on in this lecture, P will just denote some arbitrary smooth manifold, and E a smooth function on P, whose "half-spaces" $P_a = \{ p \in P \mid E(p) < a \}$ *however are assumed to be compact.*

This understood let $a \leqslant b$ be real numbers and consider the inclusion of half-spaces $P_a \subset P_b$.

THEOREM A. *If there is no critical point of E in the region $a \leqslant E \leqslant b$, then*

$$(1.10) \qquad P_a \simeq P_b$$

in the sense that they are diffeomorphic.

PROOF. Consider a trajectory of our negative gradient as it leaves the set $E = b$ at time 0. At time $(b - a)$ it is intersecting $E = b$ transversally. Hence by compactness all of them intersect $E = a - \varepsilon$ for some fixed $\varepsilon > 0$. Pictorially each trajectory thus has the three singled out points (see diagram) of intersection with these three level surfaces. Now simply deform the interval $[0, 2]$ into $[1, 2]$ by pushing downwards, but all the time keeping some vicinity of 2 pointwise fixed.

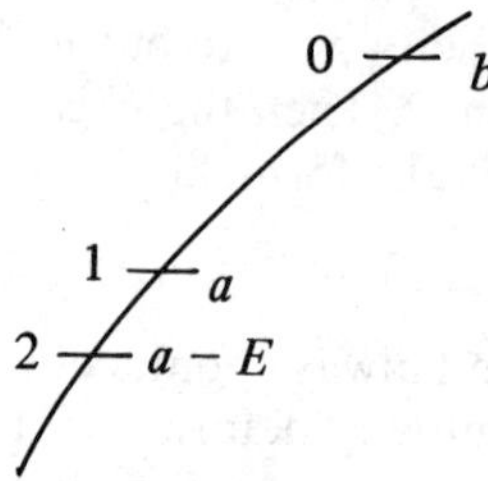

Performing this simultaneously for all of these trajectories, yields the desired diffeomorphism. This argument simultaneously shows that

COROLLARY 1. *Under the conditions of Theorem* I *the inclusion* $P_a \hookrightarrow P_b$ *is a homotopy equivalence.*

COROLLARY 2. *The minimax principle is valid.*

PROOF. Suppose η_n is a sequence of maps with Max $E \mid \eta_n$ tending to κ. Then if κ is not critical, pushing down a fixed ε along the trajectories of X produces a new sequence η_n^1 still representing the same element but with Max $E \mid \eta_n \to \kappa - \varepsilon$. Thus κ is not the inf.   Q.E.D.

We have carried through this discussion in terms of the homotopy functor, but notice that any homotopy invariant functor would do just as well in both these corollaries. Thus singular theory, or in the equivariant situation, equivariant singular theory, or K theory, etc. could clearly also be used to predict critical points of a function. On the other hand Theorem A furnishes us with no overall estimate of just how *many* critical points to expect, and in my second lecture I will indicate two quite different steps in this direction, one due to Morse and the other due to Lyusternik and Schnirelmann, both these ideas therefore stemming from the 20's.

Finally a word about the course I steered in this lecture. The polygonal approximation principle is Morse's and otherwise I have followed the account given, say in [K], where the reader will also find a very thorough bibliography. My only contribution is the observation that P_n, defined simply as the half-space $E < \varepsilon$ already approximates ΛM. For the explicit homotopy equivalences the reader is referred to [B1]—where they are carried out for the fixed endpoint case. But the argument transparently carries over to our situation. Following Palais and Smale, Klingenberg of course carries out everything in the infinite dimensional context of Hilbert-manifolds. That is, his ΛM is defined as the space of H^1-maps of S^1 to M, and the gradient deformations are then carried out directly in this context.

I know of no aspect of the geodesic question where this approach is essential; however it clearly has some aesthetic advantages, and points the way for situations where finite dimensional approximations are not possible—for instance in the Yang-Mills situation, to be discussed in my third lecture.

Lecture 2. In the last lecture we saw how any change in homotopy type of half-spaces $W_a \subset W_b$ of a smooth function f on a compact manifold W predicts a critical point in the range $a < f < b$, and how pushing a nontrivial homotopy and homology class of P_b—which is not in P_a—down, will lead one to a critical value of the function f. This procedure however lacks any quantitative information, for it is quite possible that two, say, nonhomologous classes "get stuck" at the same critical point.

In Morse Theory this lack is redressed in the following manner: First of all one studies what happens for the "generic function" on M and then refers all other cases to the generic one, by some limiting procedure.

Let me now describe this development in some detail.

Consider then a critical point p of f on W, and let $x_1 \cdots x_n$ be local coordinates on W centered at p. The fact that p is a critical point expresses itself in the vanishing of $df = \Sigma(\partial f/\partial x_i)\, dx_i$ at p. That is

$$(2.1) \qquad \frac{\partial f}{\partial x_i}\Big|_p = 0.$$

Consider next the Hessian matrix

$$(2.2) \qquad Hf_p = \frac{\partial^2 f}{\partial x^i \partial x^j}\Big|_p.$$

This Hessian of course depends on the local coordinates, but the *rank* of Hf and the number of negative *eigenvalues* of Hf is seen to be invariant under coordinate changes. Morse introduces the terms

$$(2.3) \qquad \textit{nullity of } p \;(\text{rel } f) - \dim W - \text{rank } Hf\big|_p,$$

$$\textit{index of } p \;(\text{rel } f) = \text{number of negative eigenvalues of } Hf\big|_p$$

and calls a function f *nondegenerate* if all its critical points have nullity 0.

These are the generic functions in the sense that *in the vicinity of every function* one may find a generic one. In any case, for a generic f Morse introduces the quantity

$$(2.4) \qquad \mathfrak{M}_t(f) = \sum_p t^{\lambda(p)}, \qquad p \in C(f),$$

where the sum is extended over the critical points $C(f)$ of f, and $\lambda(p) = $ index of p relative to f.

This sum turns out to be *finite* because nondegeneracy easily implies the discreteness of the critical points and W was assumed compact. This polynomial, which I will call the *Morse polynomial* (or *series*) of f is then Morse's *quantitative* measure of the critical behavior of f, and he shows that the homology of W sets a definite *lower bound* on it. Precisely let

$$(2.5) \qquad P_t(W) = \sum t^k \dim H_k(W; K)$$

be the Poincaré series of W with homology taken relative to some fixed coefficient field K. Then the following inequalities hold.

Morse inequalities. For every nondegenerate f there exists a polynomial $Q_t(f) = q_0 + q_1 t + \cdots$ with *nonnegative coefficients* such that

(2.6) $$\mathfrak{M}_t(f) - P_t(W) = (1 + t)Q_t(f).$$

We often write $\mathfrak{M}_t(f) \geqslant P_t(W)$ for (2.6). Clearly this inequality implies that $\mathfrak{M}_t(f)$ majorizes $P_t(W)$ *coefficient by coefficient.* Thus (2.6) predicts at *least* $P_1(M)$ critical points for *any nondegenerate f* on M. However (2.6) is much stronger than this estimate, namely the $(1 + t)$ factor on the right implies a feedback relationship between critical points of various indices. The power of this feedback is maybe best illustrated by the following corollary of the Morse inequalities.

MORSE'S LACUNARY PRINCIPLE. *Suppose that no consecutive powers of t occur in $\mathfrak{M}_t(f)$. Then $Q_t(f) \equiv 0$ so that*

(2.7) $$\mathfrak{M}_t(f) = P_t(W)$$

for every coefficient field K. In particular, W is then free of torsion.

PROOF. The first nonvanishing power of t on the left of (2.6) clearly implies that the next power also occurs on the right and hence by (2.6) must also occur on the left in $\mathfrak{M}_t(f)$.   Q.E.D.

The power of this principle is that it sometimes allows one to compute the complete additive homology structure of W, from purely local computations near the critical points of f.

A favorite example of mine is the following. Consider the unit sphere S^{2n+1}

$$\sum_0^n |z_i|^2 = 1, \qquad i = 0, \ldots, n,$$

in $\mathbf{C}^{n+1}$, and on it the function $\varphi(z) = \sum_0^n \lambda_i |z_i|^2$ where $\lambda_0 < \lambda_1 < \cdots < \lambda_n$ are a sequence of distinct real numbers. It is clear that φ is invariant under the action

$$e^{i\theta} \colon (z_0, \ldots, z_n) \to \left(e^{i\theta} z_0, \ldots, e^{i\theta} z_n\right)$$

of S^1 on S^{2n+1}, and hence descends to CP_n the projective space. Now, by the principle of Lagrange multipliers, if you wish, the extrema of φ correspond to the coordinate axes, and the eigenvalues of the Hessian of φ along the ith-axis are easily seen to be the set

$$\lambda_0 - \lambda_i, \lambda_1 - \lambda_i, \ldots, \lambda_n - \lambda_i$$

with λ_i excluded. Over the reals their multiplicity is 2, and so the index of the ith critical point is $2i$. Thus

(2.1) $$\mathfrak{M}_t(\varphi) = 1 + t^2 + \cdots + t^{2n}.$$

The lacunary principle applies and we conclude that $P_t(CP_n) = 1 + t^2 + \cdots + t^{2n}$.   Q.E.D.

There are two rather different approaches to proving the Morse inequalities, and as both are very instructive, I will say a few words about each of them.

The level surface method 1. Consider a nondegenerate critical point p of our nondegenerate f on W, and assume that it is the only critical point at its level. The "Morse Lemma" now asserts that there is a coordinate system $x_1, \ldots, x_n$ about p such that near p

$$f = f(p) - x_1^2 - x_2^2 - \cdots - x_\lambda^2 + x_{\lambda+1}^2 + \cdots + x_n^2$$

with $\lambda = \lambda_p$ the index of p rel f.

Using this explicit description of f near p one proves what I call Theorem B of the Morse theory.

THEOREM B. *Let $a < f(p) < b$ be such that f has no critical points in the range $a < f < b$ other than p.*

Then the diffeomorphism type of M_b differs from that of M_a by the attachment of a thickened λ-cell

$$(2.2) \qquad M_b \simeq M_b \bigcup_\alpha e_\lambda \times e_{n-\lambda},$$

and the homotopy type of M_b is therefore that of M_a with λ-cell attached

$$(2.3) \qquad M_b \approx M_a \cup e_\lambda.$$

PROOF BY PICTURE. Consider the case of a critical point p of index 1 on a surface so that near p, f can be taken to be $f = -x^2 + y^2$. Then near p the level surfaces of f take the form:

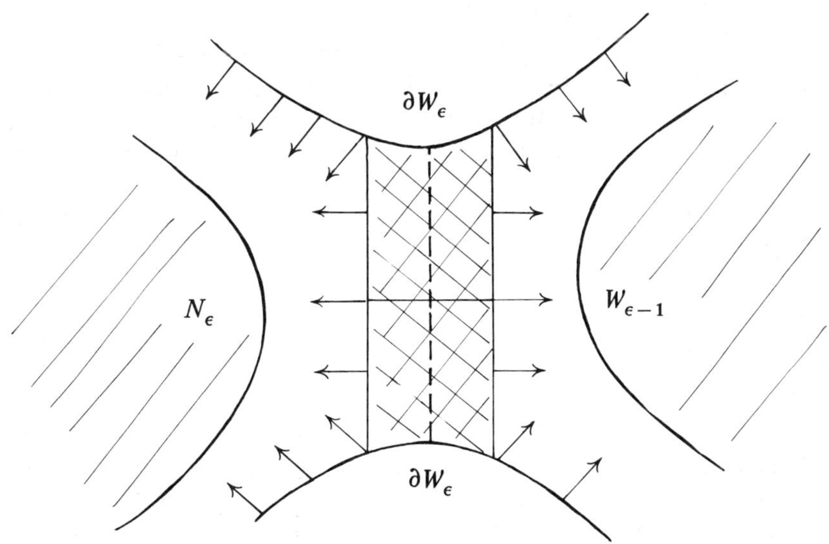

FIGURE 2

Hence if one deletes the cross-hatched region Y from W_ε, then following the gradient lines will deform $W_\varepsilon - Y$ to $W_{-\varepsilon}$. But Y is simply a "thickened 1-cell". The thickening direction is here the y-direction, and the homotopically essential part of Y is already the 1-cell given by its intersection with the X-axis: Thus the homotopy type of W_ε is also described by $(W_\varepsilon - Y) \cup e_1$ as indicated below.

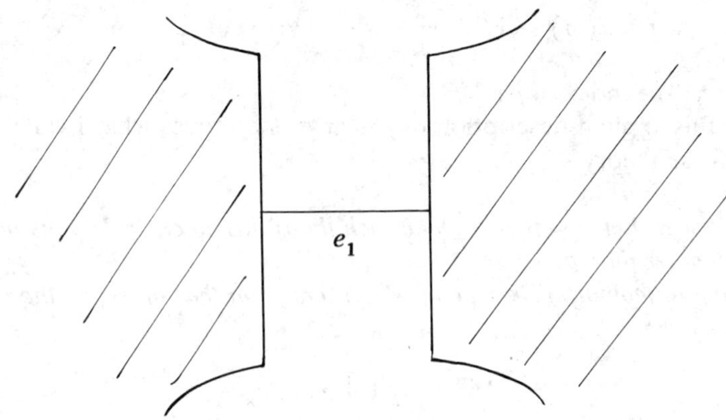

FIGURE 3

So much for a pictorial explanation of Theorem B. Finally a short explanation of how the Morse inequalities follow from Theorems A and B. Consider then the step from $W_{-\varepsilon}$ to W_ε with p the only critical point of f in the range $-\varepsilon < f < \varepsilon$. Assume also that p is nondegenerate of index λ. Let $\mathfrak{M}_t^a(f) = \sum_p t^{\lambda_p}$, with $p \in W_a$, $p \in C(f)$, be the Morse polynomial of a half-space W_a, and let P_t be the corresponding Poincaré Polynomial of W_a

$$P_t(W_a) = \sum t^k \dim H_k(W_a).$$

We will actually refine our earlier formulation of the inequalities to the statement that $\mathfrak{M}_t^a(f) \geqslant P_t(W^a)$ for each regular value a, and then proceed from $W_{-\varepsilon}$ to $W_{+\varepsilon}$ by induction. Now the change in $\mathfrak{M}_t$ from $-\varepsilon$ to ε is clearly t^λ.

On the other hand the change from ΔP_t from $P_t(W^{-\varepsilon})$ to $P_t(W^{+\varepsilon})$ can be two fold. Either,

(1) $\Delta P_t = t^\lambda$ or,
(2) $\Delta P_t = -t^{\lambda - 1}$.

Once this is granted the inequalities at ε follow from these at $-\varepsilon$. Indeed,

$$\Delta(\mathfrak{M}_t - P_t) = 0 \quad \text{or} \quad t^{\lambda - 1}(1 + t)$$

depending on the two cases. In either case the Q term of the inequality is augmented by a polynomial with nonnegative coefficients.   Q.E.D.

The crucial step is therefore the alternative for ΔP_t above, and this is a standard result in homology theory. It is also a very intuitive one. Consider the boundary ∂e_λ of the attaching cell. It is a $\lambda - 1$ sphere $S_p^{\lambda - 1}$ in $W_{-\varepsilon}$. The cycle carried by this sphere either bounds a chain in $W_{-\varepsilon}$ or not. In the first case we cap the chain bounded by $S_p^{\lambda-1}$ with e_λ to create a new nontrivial homology class in W_ε. This corresponds to the alternative: $\Delta P_t = t^\lambda$. In the second case e_λ manifestly has as boundary the nontrivial cycle $S^{\lambda-1}$ in $W_{-\varepsilon}$. Hence in this situation P_t decreases by a $t^{\lambda-1}$. Q.E.D.

Note that to ascertain *which* alternative is valid involves a *global analysis of* $W_{-\varepsilon}$. Note also that if for all critical points the first alternative holds, i.e. $\Delta P_t = t^\lambda$, then $\mathfrak{M}_t(f) = P_t(W)$, that is f has precisely the minimal number of critical points which the topology permits. We often refer to a critical point with $\Delta P_t = t^\lambda$ as *completable*, and to a function for which $\mathfrak{M}_t(f) = P_t(W)$ as a *perfect Morse function* on W.

Note also that a given f can be perfect for one coefficient field and not perfect for another.

Let me conclude this line of proof with the statement of the Morse inequalities in relative form. The proof is the same.

The relative inequalities. Let f be nondegenerate and let $a < b$ be two regular values of f. Then if

$$\mathfrak{M}_t(f)_a^b = \sum t^{\lambda_p}, \qquad p \in C(f) \cap (W_a - W_b),$$

and

$$P_t(W^b, W^a) = \sum \dim H_k(W^b, W^a) t^k$$

the Morse inequalities still hold, that is, $\mathfrak{M}_t(f)_a^b \geqslant P_t(W^b, W^a)$.

The dynamical systems method. Here we pass from f to its gradient $X = \overrightarrow{df}$, relative to some Riemann structure on W, and the action of $\mathbf{R}^1$ on W induced by flowing down, that is along $-X$.

Consider a point $p \in W - C(f)$. Its orbit under this action imbeds $\mathbf{R}^1$ in M, and the closure of the orbit is a segment joining some critical point p to a lower one g. Now, again using the Morse lemma, say, one sees that the trajectories of $\mathbf{R}^1$ which "start" at a fixed critical point p, constitute a cell, W_p, of dimension λ_1, while those which *end* at p constitute a cell W_p^*, of *codimension* λ_p. Furthermore these two cells intersect transversally at p. Indeed their tangent planes at p are precisely the directories of *steepest descent* and *ascent* respectively. In this way, then, one obtains two "stratifications" of W into the "stable" and "unstable" cells;

$$(2.4) \qquad \begin{aligned} W &= \coprod W_p, & \dim W_p &= \lambda_p, p \in C(f), \\ W &= \coprod W_p^*, & \dim W_p^* &= \dim W - \lambda_p, \end{aligned}$$

each indexed by the critical points $C(f)$ of f.

Although the closures of these cells can be badly behaved, this decomposition still has sufficient properties to enable one to deduce the Morse inequalities, and I indicate Smale's argument in this direction. By deforming the gradient, X, of f a trifle, if necessary, in the class of vector fields for which

$$Xf \geqslant 0 \quad \text{and} \quad Xf_p > 0 \quad \text{if } p \text{ is not critical,}$$

he shows that these two cell decompositions induced by X will be brought into normal form, in the sense that any two cells W_i and W_j^* will intersect normally at any $p \in W_i \cap W_j^*$. That is, at such a p

$$\dim W_i + \dim W_j^* - \dim W_i \cap W_j^* = n.$$

It follows immediately that if a trajectory starts at p and ends at q then $\dim W_p > \dim W_q$. Indeed, the interior of this trajectory must be in $W_p \cap W_q^*$ and $\dim W_p \cap W_q^* \geqslant 1$. Hence the above formula reads

$$\dim W_p + \left(n - \dim W_q\right) \geqslant n + 1 \Rightarrow \dim W_p \geqslant \dim W_q + 1. \quad \text{Q.E.D.}$$

It follows that if K_p is the union of all the cells of dim $\leqslant p$, then the $\cdots K_p \subset K_{p+1} \subset \cdots$ defines a finite filtration of W by closed sets, and

$$K^p - K^{p-1} = \coprod W_p, \qquad \dim W_p = p,$$

is the union of disjoint open sets.

Since for Cech theory

$$H^*(K^p, K^{p-1}) = \sum H_C^*(K^p - K^{p-1}),$$

where C denotes compact carriers, this implies that,

$$\dim H^q(K^p, K^{p-1}) = \begin{cases} \text{number of } p \text{ cells in 2.3 if } q = p, \\ 0 \quad \text{otherwise.} \end{cases}$$

Now the inequalities follow by quite standard arguments.

This method is less elementary than the one outlined before, but has several advantages. First of all it points the way to extending the Morse inequalities to arbitrary flows, i.e. vector fields X, satisfying certain generic conditions. This led Smale to the so-called Morse-Smale flows and diffeomorphisms (see [**SS**]).

Secondly a slight modification of our discussion leads us naturally to the Lyusternik-Schnirelmann estimate on the *number* of critical points of a function f on W. Indeed suppose now that p is an arbitrary isolated critical point of f. We still have the set W_p of trajectories of X leaving p, and the corresponding partition (2.3). What we do not know any more is whether W_p is a cell or not. *However W_p will still be contractible to p*, and we can furthermore "thicken" W_p a little so as to preserve this property. It follows that under our assumption on f, W admits a cover $\{\tilde{W}_p\}$ indexed by $C(f)$, by *open contractible* sets. By the very definition of the concept of category, of a space, this implies that

$$\text{Cat}(W) \leqslant \textit{number of critical points of } f.$$

From this in turn we have the cohomological criterion: *Suppose* $\omega_1, \ldots, \omega_m$ *are cohomology classes on* W *with*

$$\omega_1 \wedge \cdots \wedge \omega_m \neq 0; \qquad \dim \omega_i > 0.$$

Then any function with isolated critical points on W *must have at least* $(m + 1)$ *critical points.*

PROOF. If it had fewer, we say only m of them, we could cover W by open sets $\{\tilde{W}_i\}$, $i = 1, \ldots, m$. From the contractability of these and $\dim \omega_i > 0$ it follows that we can choose representations of our ω_i which come from $\tilde{\omega}_i \in H^*(W, \tilde{W}_i)$. But then their product comes from $H^*(W; \cup_1^m \tilde{W}_i) = 0$. Q.E.D.

Finally let me remark that this principle of Lyusternik-Schnirelmann dually leads to the following refinement of the minimax principle. Suppose again that f has isolated critical points only, and now let z_1 and z_2 be two homology classes. In pushing them down, rel f, they each must get stuck at a critical level say K_1 and K_2, but these might well be equal. The *Lyusternik-Schnirelmann principle* however asserts.

Suppose $z_1 = z_2 \cap \omega$ where ω is a cohomology class of dim > 0 and $\cap$ denotes the Cap product. *Then under our assumptions $K_1 < K_2$.*

One calls the relation $z_1 = z_2 \cap W$, among homology classes *subordination*: z_1 is subordinated to z_2, written $z_1 < z_2$. The L.-S. principle clearly implies that the number of critical points of a smooth function f on W is bounded by 1 plus the cardinality of the longest chain of subordinated classes $z_1 < z_2 < \cdots < z_m$ on W.

This corollary is of course the same as our previous estimate as follows from Poincaré duality.

To sum up, I hope we have learned the following, which in some sense comprises the *Elementary aspects* of critical point theory:

(1) A nondegenerate smooth function on a compact manifold has at least $P_1(W; K) = \Sigma_q \dim H^q(W; K)$ critical points.

(2) A smooth function on W has at least $\mathrm{Cat}(W)$ critical points.

(3) If all the critical points of a nondegenerate function are completable, relative to the coefficient field K then f is K-perfect, that is, $\mathfrak{M}_t(f) = P_t(M; K)$, and

(4) if $\mathfrak{M}_t(f)$ is lacunary, then $\mathfrak{M}_t(f) = P_t(M; K)$ for all K.

(5) The gradient flow gives rise to a stratification M, from which one can also deduce the Morse inequalities.

The dates of these various concepts are roughly these: Morse inequalities—1924, Lyusternik-Schnirelmann relations 1929.

The stratification goes back to Thom 1949, Theorem B as stated is due to Smale in the late 50's and in its homotopy version appears in the early 50's in papers of Thom, Pitcher and myself.

Lecture 3. In the last lecture we saw how in the Morse theory a nondegenerate critical point, p, is counted by t^λ, λ being its index. From Theorem B it then also follows that this t^λ is the Poincaré series of the *local relative* homology

$$H^*\left(W_c \cap U_p, W_c^- \cap U_p\right)$$

where W_c^- is the open half-space $f < f(p)$ and U_p is an open neighborhood of p. Such a local Poincaré series, can then be used to define a Morse series for quite general critical sets, and in some sense the Morse inequalities will then still be valid.

On the other hand the actual evaluation of the local contribution becomes quite difficult. There is however one extension of nondegeneracy where the local computation carries over practically word for word from our earlier one, and our first aim in this lecture will be to describe this extension.

Accordingly we define a connected submanifold $N \subset W$ to be a *nondegenerate critical manifold of W* if the following conditions are satisfied.

(3.1)   *Each point $p \in N$ is a critical point of f.*

(3.2)   *The Hessian of f is nondegenerate in the normal direction to N.*

Spelled out this last condition takes this form: Let $p \in N$ and let $(x_1, \ldots, x_k, x_{k+1}, \ldots, x_n)$ be a system of local coordinates in W centered at p, such that near p, N is given by the $n - k$ equations $N: x_{k+1} = 0, \ldots, x_n = 0_x$. Then,

$$(3.3) \qquad \det\left(\frac{\partial^2 f}{\partial x^i \partial x^j} \right) \bigg|_p \neq 0 \quad \text{for } i, j = k+1, \ldots, n.$$

Alternatively, consider a small tubular ε-neighborhood $W_\varepsilon(N)$ of N, which is fibered over N by the normal discs swept out by geodesics of length $\leqslant \varepsilon$ in the normal direction to N, relative to some Riemann structure on W.

Then (3.3) is equivalent to the assumption: *f restricted to each normal disc is nondegenerate*.

Note that this structure automatically gives us a way of decomposing the normal bundle νN into a *positive* and *negative* part

$$(3.4) \qquad\qquad \nu N = \nu^+ N \oplus \nu^- N,$$

where $\nu_p^+ N$ and $\nu_p^- N$ are respectively spanned by the *positive and negative Eigen-directions of the Hessian of f*.

The fiber dimension of $\nu^- N$ will be denoted by λ^N and referred to as the index of N rel f. Finally if θ^- denotes orientation bundle of $\nu^- N$, we assert that the proper way to "count" N in the Morse theory is by the polynomial $t^N P_t(N; \theta^-)$.

Precisely, one has the following extension of the Morse inequalities. Suppose f is nondegenerate in the *extended sense that all its critical sets are nondegenerate critical manifolds. Assume also that W is compact. Then if we define the Morse series of f relative to a coefficient field K by*

$$(3.5) \qquad \mathfrak{M}_t(f) = \sum t^{\lambda_N} P_t(N; \theta^- \otimes K), \qquad N \subset C(f),$$

the Morse inequalities hold:

$$(3.6) \qquad\qquad \mathfrak{M}_t(f) \geqslant P_t(W; K).$$

I have time for only two observations to explain this extension. In the context of Theorem B, what happens now, is that as we pass a critical level, we attach *a thickened version of the negative disc-bundle over M to W_a to obtain W_b.*

On the other hand in the context of the flow engendered by the gradient, the cells of the stable and unstable stratifications:

$$W = \bigcup_N W_N, \qquad W^* = \bigcup_N W_N^*, \qquad N \subset C(f),$$

are now replaced by the negative and positive bundles of N:

$$W_N^- = \nu^- N, \qquad W_N^+ = \nu^+ N.$$

In short, everywhere the cells are replaced by the corresponding cell-bundles, and each W_N is counted by its *cohomology with compact support*. Indeed, the Thom-isomorphism

$$(3.7) \qquad H_C^*(\nu^- N) \simeq H^{*-\lambda_N}(N; \theta^-)$$

converts this counting procedure into (3.5).

REMARKS. (1) One of the advantages of this extended notion of nondegenerateness is, that the pullback of a function f under a fiber projection $W' \overset{\pi}{\to} W$ *stays nondegenerate*. Note also that under such a pullback a critical manifold N goes over to $\pi^{-1}N$ while its index is preserved:

$$(3.8) \qquad \lambda_N = \lambda_{\pi^{-1}N}.$$

On the other hand it is now a trivial statement to say that W admits a perfect nondegenerate Morse function; the constant function is now trivially perfect!

(2) A very simple illustration of this situation is furnished by the function induced by the z-coordinate on the torus obtained by rotating the circle indicated below about the z-axis.

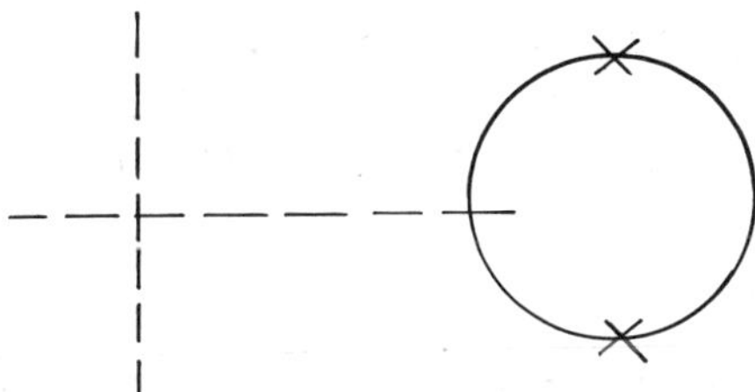

Clearly here the minima and the maxima are nondegenerate circles, with indices 0 and 1. Hence

$$\mathfrak{M}_t(z) = (1 + t) + t(1 + t).$$

Thus z here induces a perfect nondegenerate Morse function.

Actually the example which first motivated me to introduce and exploit this concept, was precisely the closed geodesic problem of my first lecture. For consider the critical point of our energy function E on one of our polygonal space P_n.

Clearly if $(P) = (P_1, \ldots, P_n)$ represents a bona fide closed geodesic, i.e. not a point path, then every *rotation* and *reflection* of the corresponding map

$$\alpha_P \colon S^1 \to M$$

is still in P_n and still critical.

Thus in the closed geodesic problem every bona fide geodesic gives rise to a critical set consisting of *two disjoint circles*. The best one can hope for, then, is that E be nondegenerate in our extended sense. Generically this also turns out to be true, as follows already from general position arguments in Morse's work. Precisely, the following holds.

PROPOSITION. *For a generic Riemann structure on M, and a generic $0 < \varepsilon < \varepsilon(M)$, the energy functional E is nondegenerate on all the spaces $P_n \subset \Lambda M$.*

Actually E is nondegenerate also in highly symmetric situations—such as for instance the spheres in their usual metric. In fact we have the following theorem.

THEOREM. *Let S^n, $n \geqslant 2$, be the standard n-sphere. Then the bona fide closed geodesics are represented by nondegenerate critical manifolds in any P_m in which they occur (ε generic).*

Furthermore these manifolds are of two types:

$$N_0 = S^n \text{ corresponding to the point paths,}$$

and

$$N_k = T_1 S^n \text{ the unit tangent bundle to } S^n,$$

corresponding to the great circle starting at a unit vector $p \in T_1 S^n$ and traversing that circle k times.

Finally, their indices are given by

$$\text{index } N_k = (2k - 1)(n - 1).$$

COROLLARY. *The* mod 2 *Morse series of E on P_m is given by*

$$(3.9) \quad \mathfrak{M}_t(E; P_m) = (1 + t^n) + \sum_{k=1}^{p} (1 + t^n)(1 + t^{n-1}) t^{(2k-1)(n-1)}$$

where $p \to \infty$ as $m \to \infty$.

Actually the techniques of [S-B] show that all these critical manifolds are also of "completable type" and hence that (3.9) also computes the Poincaré polynomial of P_m. Hence finally in view of Proposition (2.6) we obtain the

COROLLARY. *The* mod 2 *Poincaré series of ΛS^n is given by*

$$(3.10) \quad P_t(\Lambda S^n) = (1 + t^n) + \frac{t^{n-1}}{1 - t^{2(n-1)}} \cdot (1 + t^n)(1 + t^{n-1}).$$

Indeed here we have just let $p \to \infty$.

REMARKS. (1) Actually one can show more, the function E is *perfect* for *all coefficient systems*. For n even ΛS^n therefore inherits the 2-torsion of $T_1 S^n$.

(2) Nowadays the formula (3.10) can also be easily computed from the fibering $\Lambda \xrightarrow{e} S^n$ discussed earlier, however the derivation sketched above moves easily within the Morse theory, and would have been quite accessible to Morse in the 1930's.

(3) Finally let me discuss a consequence of (3.9). Assume then that $M = S^n$ in some Riemannian structure which is generic in the sense of our earlier

proposition. Then the Morse inequalities clearly imply an infinite number of bona fide closed geodesics. Indeed each such contributes $2(1 + t)t^\lambda$ and so we need an infinite number of them to dominate $P_t(\Lambda S^n)$.

Unfortunately one has to admit that this result is not very satisfying, because of the possibility of *iterating* a closed geodesic.

Thus if $\alpha\colon S^1 \to M$ is bona fide closed geodesic then the composition

$$S^1 \overset{[n]}{\to} S^1 \overset{\alpha}{\to} M$$

where $[n]$ denotes the n-fold covering $z \to z^n$, $S^1 = \{|z| = 1\}$, is a new one, called the nth iterate of α and denoted by α^n. Geometrically the iterates of course seem redundant, however they must be counted in the Morse calculus. In short if one defines α to be a *prime* closed geodesic if it is not the iterate of any other closed geodesic, then the natural and as it seems, very difficult question is to estimate the number of these prime geodesics on M. For some reason the large literature on this subject teems with mistakes. Morse in his attack introduced the circular connectivities, to be discussed later, and made a mistake which I myself followed later on.

The later results of Alber, for instance as explained in Klingenberg's book [K], are false. The ambitious main result of that book to the effect that there are infinitely many closed prime geodesics on any compact manifold is now also conceded to have gaps in its proof, etc., etc.

Let me conclude this lecture therefore by discussing the most modest but hopefully correct implication of the Morse inequalities to this question.

A first step in this direction is clearly to find some estimate of the behavior of the index $\lambda(\alpha^n)$ under iteration. In 1956 I investigated this question in [B1] and came up with the following really quite elementary estimate. To formulate it we must first recall the "Index Theorem of Morse" in the present context.

Suppose then that α is a bona fide closed geodesic, and think of α as a periodic map

$$\alpha\colon \mathbf{R} \to M, \qquad \alpha(x + 1) = \alpha.$$

The normal bundle to α is then a vector bundle ν over $\mathbf{R}$, and the Jacobi equations of α are naturally to be considered as a second order differential equation L on the space of smooth sections of ν. Explicitly L takes the form

$$(3.11) \qquad LY \equiv -\nabla^2_X Y + R(X, Y)X,$$

where X is the tangent field along α, Y a normal field, that is $Y \in \Gamma(\nu_\alpha)$ and R is the curvature tensor.

In any case (3.11) is an elliptic ordinary differential equation, which relative to a parallel frame along α takes the form

$$L(\eta) \equiv -\frac{d^2\eta_i}{dt^2} + k_{ij}(t)\eta_j, \qquad i, j = 1, \ldots, (n - 1),$$

with $k_{ij}(t)$ a periodic function of t.

For instance on the sphere the matrix $\|k_{ij}\|$ reduces to $k \cdot$ (identity where $1/k$ is the constant curvature of S^n).

Consider now the eigenvalue problem

$$(3.12) \hspace{4cm} LY = \lambda Y$$

subject to the *periodic boundary condition*

$$(3.13) \hspace{4cm} Y(t + 1) = Y(t).$$

This is then a well-posed Sturm-Liouville type problem, which in view of the positivity of the operator d^2/dt^2, has only a finite number of algebraic eigenvalues.

Morse defines the index and nullity of α by the formula

$$(3.14) \quad \begin{aligned} &\text{index } \alpha = \text{number of negative eigenvalues of (3.9) subject to (3.10)} \\ &\text{nullity } \alpha = \text{multiplicity of 0 as an eigenvalue in (3.9).} \end{aligned}$$

This definition is pertinent for our purposes, as it is not difficult to show that (see for instance [**B1**])

PROPOSITION. *If $p \in P_n M$ is any polygonal representative of the closed geodesic α, then the index and nullity of p rel E are given by (3.14).*

REMARK. The definitions (3.11) are the natural candidates for—and turn out to be—the nullity and index of the energy function in the Hilbert-manifold approach to ΛM. Hence our proposition just expresses the compatability of the finite dimensional approximations to the Hilbert-manifold approach.

Using this equivalence, one now proceeds as follows. First of all, let

$$(3.15) \hspace{3cm} J_\alpha = \text{Space of solutions of } LY = 0.$$

This is then the $2(n - 1)$-dimensional vector space of *Jacobi-fields* along α.

The periodicity of the equation $LY = 0$ now implies that the transformation $t \to t + 1$ induces a linear map $T_\alpha : J_\alpha \to J_\alpha$ called the *linear Poincaré map* of α. This transformation plays a fundamental role in all questions concerning α, and also enters our question concerning the index of the iterates of α.

Indeed as is shown in [**B3**], (3.15) determines a *nonnegative integer valued function* Λ on the circle $|z| = 1$, which *jumps only at the points $\{\xi_i\}$ of $|z| = 1$ in the spectrum of T_α—and the jumps of Λ are there bounded by the multiplicity of ξ_i —such that*

$$(3.16) \hspace{3cm} \lambda(\alpha^n) = \sum \Lambda(\omega),$$

where the ω ranges over the nth roots of $+1$ or -1 depending on whether α is orientable or not.

Here of course α is called orientable if the parallel transport of a frame along α preserves the orientation.

Immediate consequences of (3.16) are:

If the spectrum of T_α is off the unit circle then

$$(3.17) \hspace{3cm} \lambda(\alpha^n) = n\lambda(\alpha).$$

If the spectrum intersects the unit circle at the points $e^{2\pi i \rho_k}$, with ρ_k irrational, then

$$(3.18) \qquad \lambda(\alpha^n) = a_{-1} + a_0 n + \sum a_i [n\rho_k]$$

where the a_i are integers and $[\ \ ]$ denotes the greatest integer less than.

In any case we have

$$(3.19) \qquad \lambda^*(\alpha) = \frac{\lim(\alpha^n)}{n} = \frac{1}{2\pi} \int_0^{2\pi} \Lambda(\omega)\, d\omega = a_0 + \sum a_k \rho_k$$

with the $\{\rho_k\}$ as above.

A less immediate corollary, due to Ziller, is that

$$(3.20) \qquad na - b \leqslant \lambda(\alpha^n) \leqslant na + b$$

where a and b are numbers determined by the intervals into which the spectrum of T_α partitions the unit circle.

These matters are discussed in great detail in Klingenberg's book, where he also brings independent proofs and relates these questions with the Maslov cycle, etc. In that context T_α of course arises as the interesting part of the differential of the map

$$G_\alpha: T_1 M \to T_1 M,$$

given by applying the geodesic flow with time equal to the length of α. Then every point p on α is clearly a fixed point of G_α whose differential has normal and tangential components given by

$$(3.21) \qquad dG_{\alpha,p} = T_\alpha \oplus 1.$$

In this connection let me mention the classification of α as hyperbolic, elliptic, and parabolic, according to whether the spectrum of T_α, is (i) off the unit circle, (ii) on the unit circle but not at ± 1, with dG_α equivalent to a rotation, or (iii) concentrated at ± 1.

Returning to our main concern let us try and estimate the number of prime geodesics in the nondegenerate case. The following now follows trivially.

THEOREM. *Let M be a compact simply connected manifold and assume that the Betti numbers* $\dim H_k(\Lambda M; K)$ *tend to ∞ with k, for some coefficient field K. Then in any generic Riemann structure, M will have an infinite number of prime geodesics.*

PROOF. Generic implies that all bona fide closed geodesics correspond to two nondegenerate critical circles in any P_n approximating ΛM. It follows that the combination of any α and all its iterates α^n, is given by an expression of the form

$$2(1 + t) \sum_{n=1}^{\infty} t^{\lambda(\alpha^n)},$$

and in view of (3.21) the coefficients of this series remain bounded.

Hence if M had only a finite number of prime closed geodesics the Morse series of E on P_n, n large, would also, contradicting the Morse inequalities. Q.E.D.

Note that if one drops the nondegeneracy condition, then the Morse inequalities by themselves will not yield an infinite number of prime geodesics. One needs a bound on the number of nondegenerate critical points into which the iterates of a degenerate one bifurcates. Such a bound was obtained by Gromoll and Meyer, in 1964 [G-M] and led them to the beautiful result that Theorem 5 is valid, *for all Riemann structures on M*. I will not have time, unfortunately, to comment on their theorem here if I am to speak about the *equivariant* theory in some detail next time.

Let me therefore stick to the nondegenerate situation and see what, if anything, the Morse inequalities predict for the spheres.

In view of our preceeding computations, we already know that the Morse series

$$\mathfrak{M}_t = \lim_{n \to \infty} \mathfrak{M}_t(E \text{ on } P_n \text{ wth } E > 0)$$

will take the form $\mathfrak{M}_t = 2(1 + t)\overline{\mathfrak{M}}_t$ in *any nondegenerate* situation. For instance, with $M = S^{n+1}$, $n \geqslant 2$, the Morse inequalities therefore take the form

$$(3.22) \quad 2(1 + t)\overline{\mathfrak{M}}_t - \frac{t^n}{1 - t^{2n}}(1 + t^n)(1 + t^{n+1}) \geqslant (1 + t)Q(t).$$

Unfortunately, already the iterates of a single prime geodesic, could satisfy these inequalities so that (3.22) does not advance this cause at all.

In my next lecture I would like to discuss an equivariant version of the Morse theory and explain its connection, both to this problem and the Yang-Mills theory.

Lecture 4. The equivariant case. The most telling criticism of our results so far is the following.

If one deforms the n-sphere into an ellipsoid

$$(4.1) \qquad \sum_{i=1}^{n+1} a_i^2 x_i^2 = 1$$

with $a_1 < a_2 < \cdots < a_{n+1}$, the first critical manifold, $T_1 S^n$, decomposes into the $n(n + 1)/2$ geodesics given by the intersection of the coordinate planes with (4.1). On the other hand $T_1 S^n$ contributes

$$t^{n-1}P_t(T_1 S^n) = t^{n-1}(1 + t^{n-1})(1 + t^n)$$

to the Morse series of ΛS^n. Hence under small perturbations $T_1 S^n$ should contribute no more than $P_1(T_1 S^n) = 4$ critical points.

The correct diagnosis of this ailment is that our energy function has a built-in *symmetry* which has to be taken into account before the proper correspondence between geometry and topology is realized.

The symmetry in question is of course due to the fact, that the energy-integral

$$(4.2) \qquad E = \int_0^1 |\dot{\alpha}|^2 \, dt$$

is *invariant under rotations and reflections of S^1.*

Thus if we consider a model for ΛM, on which E is well defined—for instance the Hilbert-manifold of H^1-maps of S^1 to M, or even the simpler model of piecewise smooth maps of S^1 to M, parametrized proportionately to arc length, then E is *not* an *arbitrary* function on ΛM—it is a priori invariant under the natural action of $O(2)$ on ΛM, induced by the action of $O(2)$ on S^1.

Correspondingly in the context of our polygonal approximations $P_n(M)$, E is seen to be a priori invariant under the finite group generated by cyclic permutations and reversal, of the vertices of our polygons.

In both cases one is therefore led to the general question of *how the Morse theory is to be altered to take into account a priori symmetries of a function f under the action of a compact Lie group G on a manifold W.*

There are two cases to be considered.

Case 1. *The action of G on W is free.*

In this case the quotient space W/G is itself a manifold, and the function f naturally descends to a smooth function

$$(4.3) \qquad\qquad f/G\colon M/G \to \mathbf{R}.$$

It is clear therefore, that in this situation the appropriate theory is simply the old or usual Morse theory of f/G.

Case 2. *The action is not free.*

In this case W/G fails to be a manifold, so that to apply the Morse theory to f/G, one would first of all have to extend it intelligently to nonmanifolds.

Actually this can be done to a certain extent; and this procedure has been the main tool in the past especially for the closed geodesic problem.

I would like to champion a quite different approach here, which is a natural extension of Case 1 from the topologists point of view even though it might seem bizarre from an analysts point of view.

We have already seen how Theorems A and B yield the Morse inequalities via the standard properties of the homology functor. In the present context one should therefore be able to obtain the appropriate extensions of these inequalities via these same theorems, by simply replacing the functor H_*, by the *equivariant homology functor H_*^G*. This program works very nicely and yields the following result.

In accordance with the principle $H_* \mapsto H_*^G$, define the *equivariant Poincaré series* by

$$(4.4) \qquad\qquad P_t(W) = \sum t^k \dim H_k^G(W),$$

and similarly the *equivariant Morse series*, for a *nondegenerate f, by*

$$(4.5) \qquad \mathfrak{M}_t^G(f) = \sum_N t^{\lambda_N} P_t^G(N; \theta^-), \qquad N \in C(f).$$

With this understood, we have the following consequences of Theorems A and B.

THEOREM V. *The equivariant Morse inequalities: For any nondegenerate G-invariant f on the compact manifold W, the Morse inequalities hold also in the equivariant sense*

$$(4.6) \qquad\qquad \mathfrak{M}_t^G(f) - P_t^G(W) = (1 + t)Q_t^G(f).$$

To apply this principle, one of course has to learn to compute with the equivariant homology, and I would like to say a few introductory words for the nonspecialists on this score. In homotopy theory it was realized a long time ago, that the quotient construction $W \to W/G$ works properly *only* when the action is free. Now, starting with this premise what is to be done about nonfree actions? Well first of all it is a triviality that if *U is any space on which G acts freely, then the diagonal action of G on $W \times U$ is free.* This suggests that as far as homotopy is concerned, one should find a space U on which (1) G acts freely, and (2) whose homotopy is trivial, (i.e., U is contractible).

Such spaces turned out to *exist, be essentially unique,* and play an absolutely essential role in all of modern topology.

In any case granting the existence of such a U for G, the *homotopy quotient W_G of any action* is defined by

$$(4.7) \qquad W_G = U \times W/G,$$

where here G of course acts *diagonally* on the product.

Elementary properties of this construction are: *W_G projects naturally on W/G and U/G; and*

(4.8) *If G acts freely on W, then the projection*

$$W_G \to W/G$$

is a homotopy equivalence.

(4.9) *The projection*

$$W_G \to U/G$$

is always a fibering with fiber W.

Note by the way, that in this calculus U/G plays the role of the *homotopy quotient of the trivial action of G on a point.* This space is again of fundamental importance in topology, is usually denoted by BG, and is referred to as the *classifying space of G.* It is a topological space which somehow reflects both the algebraic and the topological properties of G.

In any case, all this granted, the *equivariant version, F^G, of any functor F on spaces is now simply defined by*

$$(4.10) \qquad F^G(W) \equiv F(W_G).$$

In particular note that

$$(4.11) \qquad H_*^G(\text{point}) = H_*(BG),$$

so that *in the equivariant Morse series, a nondegenerate critical point, contributes the expression*

$$(4.12) \qquad P_t^G(\text{point}) = P_t(BG).$$

More generally one proves that: *If N is a nondegenerate critical manifold consisting of a single orbit $N = G/H$ then*

$$(4.13) \qquad P_t^G(G/H) = P_t^H(H) = P_t(BH).$$

In short these classifying spaces and their *ordinary homology,* play an essential role in the equivariant theory.

Below I list a few examples, which are of course absolutely standard in modern topology.

EXAMPLES

G	U	$U/G = BG$	$P_t(BG)$
$\mathbf{Z}$	$\mathbf{R}$	$\mathbf{R}/\mathbf{Z} = S^1$	$(1 + t)$
$\mathbf{Z}^n$	$\mathbf{R}^n$	$\mathbf{R}^n/\mathbf{Z}^n = S^1 \times \cdots \times S^1$ $\underset{(n)}{}$	$(1 + t)^n$
$\mathbf{Z}_2$	$S(\mathfrak{H})$	$\mathbf{R}P_\infty$	$1/(1 - t)$
$U(1) = S^1$	$S(\mathfrak{H})$	$\mathbf{C}P_\infty$	$1/(1 - t^2)$
$U(2)$	2-frames in H	$\mathbf{G}_2(H)$	$\dfrac{1}{(1 - t^2)(1 - t^4)}$
$U(n)$	n frames in H	$\mathbf{G}_n(H)$	$\dfrac{1}{(1 - t^2) \cdots (1 - t^{2n})}$

(4.14)

Here $\mathbf{Z}$ denotes the integers, $\mathbf{Z}^n$ the direct product of $\mathbf{Z}$ with itself n times, $\mathbf{Z}_2$ the group $\{\pm 1\}$, and $U(n)$ of course the unitary group. The first two U's come to mind immediately, on the other hand the rest may strike nonspecialists as surprising. In all of these $\mathfrak{H}$ denotes a complex infinite-dimensional Hilbert space, and $S(\mathfrak{H})$ its unit sphere. The space of n-frames on $\mathfrak{H}$ is then the space of n-tuples $\{x_1, \ldots, x_n\}$ of elements in $S(\mathfrak{H})$ which are mutually orthogonal. $U(n)$ clearly acts on these, i.e.

$$\{x_i\} \to \left\{ \sum_j U_{ij} x_j \right\}$$

and the quotient gives precisely the Grassmannian $\mathbf{G}_n(\mathfrak{H})$ of all n-dimensional subspaces of $\mathfrak{H}$. For $n = 1$, this is simply the projective space.

All these examples then rely on the beautiful fact that the unit sphere in an infinite-dimensional Hilbert Space is contractible! In the final column, P_t is computed with any field K, except for the $\mathbf{Z}_2$ case. There P_t is given for the field $\mathbf{Z}_2$.

With these basics out of the way we can test this new calculus in some simple examples.

Consider then the action of the circle S^1 on the two sphere S^2, given by rotation about the z-axis in $\mathbf{R}^3$,

$$S^2 \colon x^2 + y^2 + z^2 = 1.$$

Also, let f be the height function z, on S^2. Note that in this case S^1 does not act freely at the points $z = \pm 1$, i.e., precisely at the critical points f. Note also that the s of interval $-1 \leqslant z \leqslant 1$ now parametrizes the quotient S^2/S^1, so that

(4.15)
$$P_t(S^2/S^1) = 1.$$

Thus the quotient has a quite *inappropriate homology structure*. On the other hand

(4.16)
$$P_t^G(S^2) = \frac{1 + t^2}{1 - t^2},$$

as follows easily from (4.9).

Correspondingly note that

$$\mathfrak{M}_t^G(z) = \frac{1}{1 - t^2} + \frac{t^2}{1 - t^2}, \qquad G = S^1,$$

corresponding to the fixed points $z = -1$ and $z = +1$ respectively. Thus, our function z is perfect, both, equivariantly and in the normal sense.

If we pass to the function z^2 on S^2,

$$\mathfrak{M}_t^G(z^2) = 1 + \frac{2t^2}{1 - t^2}, \qquad G = S^1,$$

with one corresponding to the minimum circle orbit at $z = 0$. Because S^1 acts *freely here*, this orbit counts in terms of its index and the ordinary Poincaré polynomial of $N/G = pt$.

This function is, strangely enough, still perfect equivariantly, however it is not in the normal sense; indeed

$$\mathfrak{M}_t(z^2) = (1 + t) + 2t^2.$$

Let us next apply this principle to the closed geodesic question. For instance what is the contribution to $\mathfrak{M}_t^G(E)$ of the first critical manifold $T_1 S^n$. Clearly $O(2)$ acts freely in this instance, so that, by (4.8),

$$P_t^G(T_1 S^n) = P_t(T_1 S^n / O(2)).$$

Now $(T_1 S^n)/O(2)$ is simply the Grassmannian $G_2(n + 1)$ of 2-planes in $\mathbf{R}^{n+1}$. In particular,

$$P_1\{G_2(n + 1)\} = \frac{n(n + 1)}{2}.$$

Thus the equivariant theory predicts the right number of nondegenerate closed geodesics into which $T_1 S^n$ splits under any small deformation. Notice on the other hand that the higher critical sets, i.e. the kth iterate of the geodesics in $T^1 S_n$, contribute by

$$(4.17) \qquad t^{(2k-1)(n-1)} \cdot P_t^G(T_1 S^n), \qquad G = O(2),$$

where however $O(2)$ *now acts* with $\mathbf{Z}/k\mathbf{Z}$ in the kernel! Over the rationals Q, this has *no effect* but mod k, say, (4.17) therefore introduces all manner of torsion into ΛS^n, and in one way or another it was this torsion which was improperly accounted for in Morse's attempts on this question long ago.

Indeed Morse's *circular connectivities* of S^n were given by the series

$$(4.18) \qquad \frac{t^{n-1} P_t(G_2(n + 1))}{1 - t^{2(n-1)}},$$

with $P_t(G_2, n)$ taken over the field $\mathbf{Z}_2$. On the other hand, the rational equivariant Morse series, in our present sense, is precisely given by the above expression, with $P_t(G_2, n)$ *computed over the rational* $\mathbf{Q}$.

The equivariant approach to this problem has been worked out in detail by my student Nancy Hingston, and it seems to us that it yields all correct known results in this direction more directly than any other method.

But let me now finally turn to the case where this equivariant theory really works per excellence and where Michael Atiyah and I were first led to apply it. This is the case of the Yang-Mills theory in dimension 2.

Although this theory is not directly pertinent to physics, it is of great interest in the theory of "stable bundles" in algebraic geometry. In particular it recaptures and refines results on the topology of the space of moduli of these bundles due to Harder [H]. Furthermore Harder's results originated in deep theorems in number theory, so that our contribution can at least be thought of as one more testimonial for the unity of mathematics. Let me therefore quickly sketch the barest outline of this application.

Recall first, that the Yang-Mills functional $A \to S(A)$ is defined on the space $\mathfrak{A}(P)$ of connections on some principal bundle P over a compact manifold M. More precisely:

$$(4.19) \qquad S(A) = \int_M \| F_A \|^2 \, dv$$

where F_A is the curvature of A and the norm is taken relative to a fixed Riemann structure on M, and an Ad-invariant positive quadratic form on the Lie algebra of the structure group G of P. We only consider the case G compact, so that such a form always exists.

Our theory is especially appropriate here, as $\mathfrak{a}(P)$ is contractible(!) and S has a larger group of symmetries. Indeed S is invariant under the group

$$\mathcal{G}(P) = \operatorname{Aut} P \text{ over the identity on } M,$$

of Gauge transformations, which acts naturally on $\mathfrak{a}(P)$. From our point of view, the appropriate topological invariant is therefore

$$(4.20) \qquad P_t^{\mathcal{G}}(\mathfrak{A}) \equiv P_t(B\mathcal{G}).$$

Note that this equality follows from the contractability of $\mathfrak{a}(P)$ and (4.10).

The Poincaré series on the right is computable when M is a compact Riemann surface of genus g, and the G is the unitary group $U(n)$.

THEOREM.

$$(4.21) \quad \mathscr{P}_t(B\mathcal{G}; K) = \frac{\{(1+t)(1+t^3) \cdots (1+t^{2n-1})\}^{2g}}{[(1-t^2)(1-t^4) \cdots (1-t^{2n-2})]^2 (1-t^{2n})}$$

when $\mathcal{G}$ is the group of Gauge transformations of any principal bundle P over M, with structure group $U(n)$, and K is any coefficient field.

We can now finally state and apply the main assertion of Atiyah's and mine.

THEOREM. *In the situation envisaged above, the Yang-Mills functional is perfect in the equivariant sense.*

In short,

$$(4.22) \qquad \mathscr{M}_t(S) = P_t^{G}(\mathfrak{A}) = P_t(B\mathcal{G}).$$

I have time here for only a hint at the proof. The perfection of this functional is in the present case closely related to a principle we call "self-completion", which I will explain here briefly in the context of the previous sections. Let then G act on the manifold W as before and let f be G-invariant, with N a nondegenerate critical manifold which we assume to consist of the G-orbit of $p \in W$,

$$(4.23) \qquad N = G/H, \qquad H \text{ the stabilizer of } p.$$

Then H acts on the normal space $\nu_p(N)$, and also on the negative normal space $\nu_p^-(N)$, relative to some H-invariant metric on W. We call this the *negative isotropy representation*, and denote it by λ_N. This representation is the proper analogue of the integer λ_N—the index of N—in the standard theory. Consider now the vector bundle over BH associated to the universal bundle U, by λ_H, and let $e_N \in H^*(BH)$ be its Euler class if it is orientable. (Note that if H is connected, this will automatically be the case indeed then BH is simply connected.) With this understood, we call e_N *K-injective if the multiplication by* e_N

$$(4.24) \qquad Ue_N \colon H^*(BH; K) \to H^*(BH; K)$$

is injective.

THEOREM. *Suppose $C(f)$ consists only of orbits $N = G/H$ as above, with H connected and e_N κ-injective for every $N \subset C(f)$. Then the nondegenerate G-invariant function f on the compact manifold W is K-perfect:*

$$(4.25) \qquad \mathfrak{M}_t^G(f) = P_t^G(W; K).$$

For example consider the S^1 action on S^2 of our earlier example and the function z on S^2. At the minimum ν_N^- is the *zero dimensional bundle and there our condition is* always to be considered as verified. At the maximum, λ_p is the *standard representation of S^1 on* $\mathbf{R}^2$, whose Euler class e generates $H^2(BS^1) = \mathbf{Z}[e]$. Thus our principle applies and immediately implies that z was *equivariantly perfect.*

In some sense this simple phenomenon occurs over and over again for the Yang-Mills functional, for unitary bundles. Partly this is due to the fact that *the stability groups of a connection A for P, are always of the form*

$$H_A = U(n_1) \times \cdots \times U(n_k), \qquad \sum n_i = n.$$

Hence all $H^*(BH)$'s are polynomial rings over which the injectivity criterion is easily checked. Note by the way how much simpler these cohomology groups are compared to what one encounters in the closed geodesic problem. Nevertheless this principle is also applicable there and helps in the computation of $H_*^{O(2)}(\Lambda S^n)$.

Finally a word on the application of (4.22). As we are primarily concerned with $U(n)$-bundles we will discuss them in terms of the vector bundles of rank n they define.

In view of the work of Narasimhan and Seshadri, the following assertions are then easily verified.

(4.26) Let L be a line bundle with first Chern-class n. Then the only extremum of S is the minimum of S and corresponds one-to-one to the Jacobian torus of M: $J = (S^1)^{2g}$.

(4.27) Let E be a vector bundle of rank 2, and first Chern-class 1. Then the minimum of S corresponds to a smooth variety isomorphic to the *space of moduli* $\mathbf{M}_2$ *of stable bundles* in the sense of algebraic geometry.

(4.28) All other extrema correspond to products of extrema of the 1-dimensional case. Thus they correspond to varieties $J \times J$ one for each decomposition

$$E = L_1 \oplus L_2, \quad c_1(L_1) < c_1(L_2), \quad c_1(L_1) + c_1(L_2) = 1.$$

One next computes the index of each of these varieties, to be given by

$$\mathrm{index}(J \times J) = 2 \dim H^1(M; L_2^* \otimes L_1) = 2g + 4(c_1(L_2) - 1).$$

(Here the cohomology is taken in the holomorphic sense.) So that finally one computes

$$(4.29) \qquad \mathfrak{M}_t^g(S) = \frac{P_t(\mathbf{M}_2)}{1 - t^2} + \frac{t^{2g}(1 + t)^{4g}}{(1 - t^2)^2(1 - t^4)}.$$

Note that here $1/(1 - t^2)$ should be thought of as $P_t(BU_1)$, with U_1 the centralizer of points in $\mathbf{M}_2$, while $1/(1 - t^2)^2$ corresponds to the centralizer of the $J \times J$ varieties. Finally $1/(1 - t^4)$ occurs in the second factor instead of $1 + t^4 + t^8 + \cdots$ and thus accounts for the *sum over all* critical points of the (4.27) type.

Now, (4.21), (4.22) and (4.29) yield a formula for $P_t(\mathbf{M})$:

$$(4.30) \qquad P_t(\mathbf{M}_2) = \frac{(1 + t)^{2g}(1 + t^3)^{2g}}{(1 - t^2)(1 - t^4)} - \frac{t^{2g}(1 + t)^4}{(1 - t^2)(1 - t^4)}.$$

Similarly for higher n, because (4.22) leads to a relation which recursively determines $P_t(\mathbf{M}_n)$.

The remarkable fact is, as mentioned before, that this recursive procedure for $P_t(\mathbf{M}_n)$—at least over $\mathbf{Q}$—was already known when M. Atiyah and I discovered it in the context of Morse theory. The formula (4.30) is implicit in Newsteads work and comes from a direct computation [N]. Precisely the formula (4.30) then occurs in Harder's paper [H], but deduced from theorems of C. L. Siegel and the Weil conjecture, which at the time were still conjectural. The general case was then taken up by Narasimhan and Harder—again from this number theoretic point of view.

Actually (4.22) refines all these results in the sense that it implies that all the varieties $\mathbf{M}_n$ are *free of torsion*, and such statements are out of range of the Harder methods.

This has of necessity been a very brief glimpse into the Morse theory in this context, and I have not had time to go into details, or to even come to grips with the now essential infinite dimensionality of the problem.

A paper with M. Atiyah in preparation will hopefully be finished soon on the subject, where the $P_t(\mathbf{M}_n)$ are also computed by a direct equivariant stratification of a space equivalent to $\mathfrak{a}(P)$. In a sense one can use the basic structure theorems of Narasimhan and Seshadri to describe the negative bundle stratification of this space, without even mentioning the Morse theory.

Still, in the final analysis, these two points of view fit together beautifully, and are then seen to be analogous to the Morse theory and Algebraic Geometry approach to the cell structures on the compact homogeneous spaces G/P, of the complex semisimple Lie groups.

In any case, the interested reader can find some of the details in a preliminary version of our paper, [A-B] available as a preprint.

BIBLIOGRAPHY

[A-B] M. Atiyah and R. Bott, *On the Yang-Mills equations over Riemann surfaces*, Inst. Hautes Études Sci. Publ. Math., Preprint.

[B1] R. Bott, *Nondegenerate critical manifold*, Ann. of Math. (2) **60** (1954), 248–261.

[B2] ______, *The periodicity theorem for the classical groups*, Ann. of Math. (2) **70** 2 (1959), 179–203.

[B3] ______, *On the iteration of closed geodesics and the Sturm intersection theory*, Comm. Pure Appl. Math. **9** (1956), 171–206.

[G-M] D. Gromoll and W. Meyer, *On differentiable functions with isolated critical points*, Topology **8** (1969), 361–369.

[H] G. Harder, *Eine Bemerkung Zu einer Arbeit von P. E. Newstead*, J. für Math. **242** (1970), 16–25.

[H-N] G. Harder and M. S. Narasimhan, *On the cohomology groups of moduli spaces of vector bundles over curves*, Math. Ann. **212** (1975), 215–248.

[K] W. Klingenberg, *Lectures on closed geodesics*, Die Grundlehren der Mat. Wissenschaften, vol. 230, Springer-Verlag, Berlin and New York, 1978.

[L-S] L. Lyusternik and L. Schnirelmann, *Topological methods in the calculus of variations*, Gosndarstv. Izdat. Tehn-Teor. Lit., Moscow, 1930.

[N] P. E. Newstead, *Stable bundles of rank 2 and odd degree over a curve of genus 2*, Topology **7** (1968), 205–215.

[S] C. S. Seshadri, *Space of unitary vector bundles on a compact Riemann surface*, Ann. of Math. (2) **85** (1967), 303–336.

[S-B] H. Samelson and R. Bott, *Applications of the theory of Morse to symmetric spaces*, Amer. J. Math. **80** (1968), 965–1029.

[S-S] S. Smale, *Differentiable dynamical systems*, Bull. Amer. Math. Soc. **73** (1967), 747–817.

[T] R. Thom, *Sur une partition en cellules associée àune fonction sur une varité*, C. R. Acad. Sci. Paris Sér. A-B **228** (1949), 973–975.

DEPARTMENT OF MATHEMATICS, HARVARD UNIVERSITY, CAMBRIDGE, MASSACHUSETTS 02138

Proceedings of Symposia in Pure Mathematics
Volume 39 (1983), Part 2

Periodic Solutions of Nonlinear Vibrating Strings
and Duality Principles

HAÏM BREZIS

Introduction. Our lecture deals with the study of T-periodic solutions for the nonlinear vibrating string equation:

$$(1) \qquad \begin{aligned} u_{tt} - u_{xx} + g(u) &= f(x, t), & 0 < x < \pi, t \in \mathbf{R}, \\ u(x, t) &= 0, & x = 0, x = \pi, t \in \mathbf{R}, \\ u(x, t + T) &= u(x, t), & 0 < x < \pi, t \in \mathbf{R}. \end{aligned}$$

Here g denotes a continuous function on $\mathbf{R}$ such that $g(0) = 0$ and $f(x, t)$ is a given T-periodic function of t.

Problem (1) may be viewed as an *infinite-dimensional Hamiltonian system* (let us recall that H. Poincaré has abundantly investigated the question of periodic solutions for finite-dimensional Hamiltonian systems; see [**50**]). Indeed if we set $p = u$ and $q = u_t$, then (1) becomes

$$\frac{\partial}{\partial t} \begin{pmatrix} p \\ q \end{pmatrix} = \begin{pmatrix} H_q \\ -H_p \end{pmatrix} + \begin{pmatrix} 0 \\ f \end{pmatrix}$$

where the Hamiltonian H is defined on the space $H_0^1(0, \pi) \times L^2(0, \pi)$ by

$$H(p, q) = \frac{1}{2} \int_0^\pi (p_x)^2 \, dx + \int_0^\pi G(p) \, dx + \frac{1}{2} \int_0^\pi q^2 \, dx$$

and G denotes a primitive of g.

We shall be concerned with two distinct questions.

Question 1. Existence of *forced vibrations*; that is, given $f(x, t)$ find at least one solution of (1).

Question 2. Existence of *free vibrations* (or "breathers"); that is, assume $f \equiv 0$ and find at least one *nonzero* solution of (1).

In what follows we consider only the *fixed period problem*: T is prescribed. In fact, we shall even assume *systematically* that T/π is rational and sometimes we shall use for simplicity $T = 2\pi$. Such an assumption plays a technical[1] but essential role in the proofs.

Reprinted from Bulletin Amer. Math. Soc. (N. S.) 8 (1983), 409–426.

1980 *Mathematics Subject Classification.* Primary 35K60.

[1] When T/π is not rational we are led to unsolved difficulties related to small divisors.

We point out that nothing is known for the *fixed energy problem*: that is, $f \equiv 0$ and $H(u, u_t)$—which is constant along solutions of (1)—is prescribed; even though much progress has been achieved in recent years on the fixed energy problem for finite-dimensional Hamiltonian systems. See e.g. A. Weinstein [65–67], J. Moser [48], P. Rabinowitz [54, 55, 58] (the last reference is a survey article with many references), I. Ekeland-J. M. Lasry [36], A. Ambrosetti-G. Mancini [6], H. Hofer [40], and H. Berestycki-J. M. Lasry [14].

In §I we state the main results. In §II, we discuss a dual variational principle which turns out to be extremely useful in solving (1). It is motivated by the dual variational formulation for finite-dimensional Hamiltonian systems discovered by F. Clarke and I. Ekeland (see [24, 25, 34]). In §III we present the proof of Theorem 1, which deals with forced vibrations. In §IV we sketch the proofs of Theorems 2.3 and 4 dealing with free vibrations.

Problem (1) has been approached in the sixties by purely *analytical methods*; we mention the works of O. Vejvoda [63], L. Cesari [21], J. Hale [38], P. Rabinowitz [51], H. Lovicarova [44], L. De Simon-G. Torelli [33] and the recent book of O. Vejvoda [64] which contains an extensive bibliography. In the seventies, the *combination of analytical estimates* with *topological and variational tools* (initiated by P. Rabinowitz in [52]) has shed a new light on Problem (1). We shall emphasize here this viewpoint.

I. The main results. Throughout this paper we shall assume that g is a continuous nondecreasing function on **R**, such that $g(0) = 0$. Similar results hold when g is nonincreasing. However, very little is known if we drop the monotonicity assumptions (see, however, Remarks 5 and 10).

I.1. *Forced vibrations.* For simplicity we assume throughout subsection I.1 that $T = 2\pi$. We denote by A the differential operator

$$Au = u_{tt} - u_{xx}$$

acting on functions which satisfy the Dirichlet boundary conditions in x and which are 2π-periodic in t. We denote by $N(A)$ its null space so that $N(A)$ consists exactly of functions of the form

$$N(A) = \{\varphi(x, t); \varphi(x, t) = p(t + x) - p(t - x)$$
$$\text{where } p \text{ is any } 2\pi\text{-periodic function}\}.$$

THEOREM 1. *Assume*

$$(2) \qquad |g(u)| \leqslant \gamma|u| + C, \qquad \forall u \in \mathbf{R}, \text{ for some constants } \gamma < 3 \text{ and } C.$$

Assume $f(x, t) \in L^\infty$ *admits a decomposition of the form*

$$f(x, t) = f^*(x, t) + f^{**}(x, t)$$

with

(3) $$\int_0^\pi \int_0^{2\pi} f^*\varphi \, dxdt = 0 \quad \text{for all } \varphi \in N(A)$$

and

(4) $g(-\infty) + \delta \leqslant f^{**}(x, t) \leqslant g(+\infty) - \delta \quad \text{for some } \delta > 0, \text{ for all } x, t.$

Then there exists a weak solution $u \in L^\infty$ of (1).

By a *weak solution* of (1) we mean a function $u \in L^\infty([0, \pi]) \times \mathbf{R}$ which is T-periodic and such that

$$\int_0^T \int_0^\pi u(v_{tt} - v_{xx}) \, dxdt + \int_0^T \int_0^\pi g(u)v \, dxdt = \int_0^T \int_0^\pi fv \, dxdt$$

for all $v \in C^2([0, \pi] \times \mathbf{R})$ satisfying the boundary and periodicity condition.

Theorem 1 is due to A. Bahri and myself [12]. If, in addition, g satisfies $g(\pm\infty) = \pm\infty$, then any $f \in L^\infty$ admits a decomposition of the form (3), (4) (choose $f^* = 0$). Therefore we obtain

COROLLARY 1. *Assume that (2) holds and that $g(\pm\infty) = \pm\infty$. Then for every $f \in L^\infty$ there exists a weak solution $u \in L^\infty$ of (1).*

Corollary 1 was first proved by Nirenberg and myself in [17] (some of the essential estimates devised in [17] will be presented in §III).

REMARK 1. The decomposability assumption on f provides an "almost" necessary and sufficient condition for the solvability of (1). Indeed, if (1) has a solution then f may be written as $f = f^* + f^{**}$ where $f^* = Au$ and $f^{**} = g(u)$. Thus, we find (3) and

(4′) $$g(-\infty) \leqslant f^{**}(x, t) \leqslant g(+\infty).$$

In case g is *strictly increasing* we may even conclude that (4) holds and then (3)–(4) is *exactly* a necessary and sufficient condition for the solvability of (1). Theorem 1 is related to the general "principle" which holds in many instances: the range of the sum is almost equal to the sum of the ranges, i.e. a nonlinear equation of the form $Au + Bu = f$ has a solution provided f admits a decomposition of the form $f = f^* + f^{**}$ with $f^* \in R(A)$ and $f^{**} \in R(B)$ (see [18, 29]).

Suppose now that $-g(-\infty) = g(+\infty) = g_\infty$; using the Hahn-Banach theorem it is easy to see that (3)–(4) hold if and only if

$$\left| \int_0^\pi \int_0^{2\pi} f\varphi \, dxdt \right| \leqslant (g_\infty - \delta) \int_0^\pi \int_0^{2\pi} |\varphi| dxdt$$

for every $\varphi \in N(A)$ and for some $\delta > 0$. Such a condition can be viewed as a nonlinear Fredholm alternative (of the same nature as the one used by Landesman-Lazer [42] in nonlinear equations at resonance; the main difference with their work is that the infinite-dimensional nullspace introduces further difficulties in our problem).

REMARK 2. *Uniqueness* can be proved (see [18, Corollary 1.6]) under the additional assumptions that g is strictly increasing and that

$$|g(u) - g(v)| \leq \gamma |u - v| \quad \text{for all } u, v \in \mathbf{R} \text{ and for some } \gamma < 3.$$

Nonuniqueness may otherwise occur (see Theorem 2 below).

REMARK 3. *Smoothness.* If f, g are C^∞ and furthermore g is strictly increasing then any weak solution $u \in L^\infty$ of (1) belongs to C^∞ (see [17, Theorem 2]).

REMARK 4. The restriction $\gamma < 3$ in (2) is essential. Indeed, if $g(u) = 3u$ equation (1) does not have a solution for any f: in fact (1) has a solution if and only if f is orthogonal to $N(A + 3I) \neq \{0\}$ since -3 is an eigenvalue of A. Assumption (2) asserts that the nonlinear term g is allowed to interact with the spectrum of A only through 0. A major difficulty arises here because 0 is an eigenvalue with infinite multiplicity. When g does not interact with the spectrum of A (the "nonresonant" case) or when g "interacts" with one nonzero eigenvalue of A, Problem (1), becomes much easier (see J. Mawhin [46, 47], H. Brezis-L. Nirenberg [18], G. Mancini [45], H. Amann-G. Mancini [1]). *Nothing is known in case g grows at infinity faster than linear, for example,* $g(u) = u^3$. In view of recent results[2] of A. Bahri [8], A. Bahri-H. Berestycki [9–11], M. Struwe [60] and P. Rabinowitz [59] it seems reasonable to conjecture that when $g(u) = u^3$, Problem (1) possesses a solution—even infinitely many solutions—for every f (or at least for a dense set of f's).

REMARK 5. We have assumed that g is *monotone*. This assumption plays an essential role in the proofs which involve an *interplay of monotonicity and compactness devices.* Such a combination has been extensively used in the past since the pioneering works of F. Browder [19] and J. Leray-J. L. Lions [43] dealing with nonlinear elliptic (and parabolic) problems. In that case the top order differential operator is monotone (i.e., elliptic) and there is much freedom on the lower order terms—which are handled by compactness. For Problem (1) the situation is reversed: the top order differential operator A is not monotone, but A^{-1} (off its nullspace) is compact; the monotonicity of the lower order term $g(u)$ is used in a crucial manner (because $N(A)$ is infinite dimensional and no compactness is available on $N(A)$).

Let us examine a simple example of a *nonmonotone* g: consider Problem (1) with $g(u) = u + 36 \sin u$. It is not known whether (1) possesses a solution for every f (even though the solutions of (1) are *a priori* bounded in L^∞, but such an estimate is too weak for proving existence). Recently, H. Hofer [39] has proved that (1) possesses a solution for a *dense set* of f's (see also an earlier work by

---

[2] Dealing with elliptic equations and finite-dimensional Hamiltonian systems.

M. Willem [**68**]). On the other hand, J. M. Coron [**32**] has proved that (1) possesses a solution for every f satisfying the additional *symmetry* assumption:

$$f(\pi - x, \pi + t) = f(x, t)$$

for all $x \in [0, \pi]$, and $t \in \mathbf{R}$.

I.2. *Free vibrations.* We assume that $f \equiv 0$. By a nontrivial solution of (1) we mean a function $u \in L^\infty$ satisfying (1) in the weak sense and such that $g(u) \neq 0$ on a set (x, t) of positive measure (in particular $u \neq 0$ on that set).

The main results are the following:

THEOREM 2. *Assume* (2) *and*

$$(5) \qquad \liminf_{u \to 0} \frac{g(u)}{u} > 3.$$

Then there exists a nontrivial 2π-periodic solution of (1).

THEOREM 3. *Assume $g \not\equiv 0$ and*

$$(6) \qquad \lim_{|u| \to \infty} \frac{g(u)}{u} = 0.$$

Then there exists some T_0 such that for each $T > T_0$ which is a rational multiple of π, equation (1) *admits a nontrivial T-periodic solution.*

THEOREM 4. *Assume*

$$(7) \qquad \lim_{|u| \to \infty} \frac{g(u)}{u} = \infty,$$

$$(8) \qquad \begin{cases} \text{\textit{there are constants} } \alpha > 0 \text{ \textit{and} } C \text{ \textit{such that}} \\ \tfrac{1}{2} u g(u) - G(u) \geq \alpha \,|g(u)| - C \quad \text{\textit{for all} } u \in \mathbf{R}, \\ \text{\textit{where} } G(u) = \displaystyle\int_0^u g(t)\,dt. \end{cases}$$

Then for each T which is a rational multiple of π, equation (1) *admits a nontrivial T-periodic solution.*

Theorem 2 is due to J. M. Coron [**29**]; Theorem 3 is due to J. M. Coron and myself [**15**]. Theorem 4 is due to P. Rabinowitz [**53**] under slightly more restrictive assumptions on g. As stated here, Theorem 4 is due to J. M. Coron-L. Nirenberg and myself [**16**]; the proof in [**16**] is simpler than the original proof of P. Rabinowitz. I shall sketch it in §IV.

REMARK 6. It is tempting to "visualize" Theorems 2, 3, and 4 as follows. In Theorem 2 replace Au by $-3u$ (-3 plays a special role as the first negative eigenvalue of A). Under the assumptions of Theorem 2 the scalar equation $-3u + g(u) = 0$ admits a nontrivial solution. More generally, denote by $\lambda_{-1}(T)$ the first negative eigenvalue of A acting in T-periodic functions. As we shall see (in §IV), $|\lambda_{-1}(T)| \to 0$ as $T \to \infty$. Under the assumptions of Theorem 3 the scalar equation $\lambda_{-1}(T)u + g(u) = 0$ admits a nontrivial solution if T is large enough. Finally, in Theorem 4 we may as well look for a T/n-periodic solution of (1) (n integer) instead of a T-periodic solution. Since $\lambda_{-1}(T/n) \to -\infty$ as $n \to \infty$, it is clear, with assumption (7), that the scalar equation $\lambda_{-1}(T/n)u + g(u) = 0$ has a nontrivial solution for n large enough.

REMARK 7. Under the assumptions of Theorem 3, and in addition if $\lim_{u \to 0}(g(u)/u) = 0$, then J. M. Coron [30] proves that there exist at least *two* nontrivial solutions of (1) for T large enough. Again, the scalar analogy is suggestive.

REMARK 8. Other existence and *multiplicity* results have been obtained by H. Amann-E. Zehnder [2, 3], K. C. Chang [22], K. C. Chang-S. P. Wu-S. Li [23], in case $g(u)$ has a linear behavior as $|u| \to \infty$. Roughly speaking, their results assert that if the interval joining $g'(0)$ to $g'(\infty)$ crosses m eigenvalues of $-A$, then (under some extra technical assumptions) Problem (1) possesses at least m nontrivial solutions; the proofs rely on the use of Morse and Lusternik-Schnirelmann arguments.

REMARK 9. P. Rabinowitz has obtained further results related to Theorem 4. In [56] he investigates the existence of subharmonic solutions of (1) (that is distinct solutions of period nT). In [57] he studies the case where g is monotone decreasing (instead of increasing).

REMARK 10. J. M. Coron [32] has recently succeeded in removing the monotonicity assumption (on g) in most of the previous results (at the expense of some extra technical assumptions). When g is not monotone the major difficulty arises from the infinite-dimensional null space $N(A)$ (see Remark 5). In order to overcome this difficulty Coron looks for a nontrivial solution of (1) within a *restricted class* H_1 of functions satisfying some *symmetry properties* and such that

(i) $N(A) \cap H_1 = \{0\}$,

(ii) H_1 is stable under A and g.

Consider (for example) $g(u) = \sin u$; Coron proves that given any integer m, Problem (1) has at least m distinct solutions provided T is large enough. Unfortunately, this device does not apply when $g(x, t, u)$ depends also on (x, t) —except if g satisfies some special symmetry properties.

REMARK 11. It is of great interest (see [20]) to study the existence of nontrivial solutions for a problem similar to (1), when x *varies on* **R** instead of the bounded interval $[0, \pi]$; that is

$$(1') \quad \begin{cases} u_{tt} - u_{xx} + g(u) = 0, & -\infty < x < +\infty, t \in \mathbf{R}, \\ u(x, t) \to 0, & \text{as } |x| \to \infty, t \in \mathbf{R}, \\ u(x, t + T) = u(x, t), & -\infty < x < +\infty, t \in \mathbf{R}. \end{cases}$$

When $g(u) = \sin u$ an explicit solution is known to exist for every period $T > 2\pi$ (see e.g. G. Lamb [41]):

$$u(x, t) = 4 \arctan\left[\frac{\varepsilon \sin(2\pi t/T)}{\cosh(2\pi\varepsilon x/T)}\right] \quad \text{where } \varepsilon = \left[\left(\frac{T}{2\pi}\right)^2 - 1\right]^{1/2}.$$

It has been proved by J. M. Coron [**31**], for a general function g, that nontrivial T-periodic solutions of (1′) can exist only when

$$g'(0) > (2\pi/T)^2.$$

In particular, if $g(u) = \sin u$, there is no T-periodic solution when $T < 2\pi$ (this is consistent with the above formula). When $g(u) = -\sin u$ or $g(u) = \pm u^3$ then $u \equiv 0$ is the only solution of (1′) (for any period T). It seems that the class of nonlinearities $g(u)$ for which (1′) has a solution is very limited (perhaps only multiples of $\sin u$!).

II. A duality principle. Recently, F. Clarke and I. Ekeland have discovered an interesting dual variational formulation[3] for Hamiltonian systems associated with a convex Hamiltonian H (see [**24, 25, 35**]). Such a duality principle has turned out to be extremely useful for various purposes (see [**4–6**], [**26–28**], [**35–37**]). While analyzing this principle and trying to adapt it to Problem (1) we were led to the following abstract scheme which fits both Hamiltonian systems (with convex H) and Problem (1) (with monotone g):

Let H be a Hilbert space and let $A\colon D(A) \subset H \to H$ be an unbounded linear operator such that $A^* = A$ and $R(A)$ is closed. Thus H admits an orthogonal decomposition $H = R(A) \oplus N(A)$; for each $u \in H$ we write $u = u_1 + u_2$ with $u_1 \in R(A)$, $u_2 \in N(A)$. A^{-1} is a well-defined bounded operator from $R(A)$ into $R(A)$.

Let J be a C^1 convex function on H and set $B = \nabla J$. Assume B is one-to-one and onto. Consider the equation

$$(9) \qquad\qquad Au + Bu = f.$$

Equation (9) has a natural variational structure: the solutions of (9) correspond to the critical points of the functional

$$(10) \qquad\qquad \Phi(u) = \tfrac{1}{2}(Au, u) + J(u) - (f, u)$$

on $D(A)$.

We introduce the *new function*

$$(11) \qquad\qquad v = Bu - f.$$

Solving (9) is clearly equivalent to finding v such that

$$(12) \qquad\qquad \begin{cases} v \in R(A), \\ A^{-1}v + B^{-1}(v + f) \in N(A). \end{cases}$$

---

[3] Prior to their works, J. F. Toland [**61, 62**] had introduced a somewhat related idea in the study of variational problems involving a difference of two convex functions.

Problem (1) has also a variational structure. Namely the solutions of (2) correspond to the critical points of the functional

(13)
$$\begin{cases} \Psi(v) = \tfrac{1}{2}(A^{-1}v, v) + J^*(v + f), \\ \text{subject to the constraint } v \in R(A) \end{cases}$$

where J^* denotes the conjugate convex function (i.e., Legendre transform) of J.

We say that (13) is the *dual variational formulation of Problem* (9). It is *much easier* in practice to find critical points of Ψ (on $R(A)$) than critical points of Φ. The reasons are the following:

(a) In general, A has an infinite sequence of eigenvalues going from $-\infty$ to $+\infty$, so that Φ is unbounded from above and below. Critical points of Φ can never be obtained by a simple minimization (or maximization). On the other hand, in general, A^{-1} is compact, and so Ψ satisfies a condition of the Palais-Smale type. Therefore one can make use of classical techniques in order to find critical points of Ψ. It may even happen in some cases that $J^*(v + f)$ "dominates" the negative part of $\tfrac{1}{2}(A^{-1}v, v)$; then Ψ is coercive and a critical point may be obtained by simply minimizing Ψ on $R(A)$. Note that Rabinowitz [53] works with Φ rather than Ψ; this explains why his argument is more intricate.

(b) Another reason is that, in problem (9), the component of u in $N(A)$ is difficult to estimate. It disappears from problem (12)—or rather it appears in (12) as a "harmless" Lagrange multiplier (a comparable device is used in solving the Navier-Stokes equation when the pressure term is turned into a Lagrange multiplier by projecting the equation on divergence free vector fields).

When applying this scheme to problem (1) we choose $H = L^2((0, \pi) \times (0, T))$,

$$\begin{cases} Au = u_{tt} - u_{xx}, & \text{under appropriate boundary and periodicity conditions,} \\ Bu = g(u). \end{cases}$$

For the study of the finite-dimensional Hamiltonian system

$$\begin{cases} -\dfrac{dp}{dt} + H_q = f_1, \\ \dfrac{dq}{dt} + H_p = f_2 \end{cases}$$

we choose $H = L^2(0, T) \times L^2(0, T)$, $u = \binom{p}{q}$,

$$Au = \begin{pmatrix} \dfrac{dq}{dt} \\ -\dfrac{dp}{dt} \end{pmatrix} = \begin{pmatrix} 0 & I \\ -I & 0 \end{pmatrix} \begin{pmatrix} \dfrac{du}{dt} \end{pmatrix} \quad \text{under periodic conditions and}$$

$$Bu = \begin{pmatrix} H_p \\ H_q \end{pmatrix} = \nabla H(u).$$

III. Proof of Theorem 1. The duality principle of §II applies provided B is one-to-one and onto. This is not always the case under the assumptions of Theorem 1. Therefore we consider first the perturbed equation

$$(14) \qquad Au_\varepsilon + g(u_\varepsilon) + \varepsilon u_\varepsilon = f.$$

The proof of Theorem 1 is divided into 4 steps.

Step 1. Existence of a solution u_ε of (14).

Step 2. "Soft" estimates for u_ε:

$$\|u_\varepsilon\|_{L^1} \leq C, \qquad \|Au_\varepsilon\|_{L^2} \leq C, \qquad \|g(u_\varepsilon)\|_{L^2} \leq C.$$

Step 3. A further estimate, $\|u_\varepsilon\|_{L^\infty} \leq C$.

Step 4. Passage to the limit as $\varepsilon \to 0$.

Step 1. Set $\Omega = (0, \pi) \times (0, 2\pi)$. We recall some properties of A in $L^2(\Omega)$ (see [17] and the references quoted therein):

$$(15) \qquad A^* = A,$$

$$(16) \qquad R(A) \text{ is closed in } L^2.$$

Set $K = A^{-1}$ defined from $R(A)$ into $R(A)$. The eigenvalues of A are $\{j^2 - k^2;\ j = 1, 2, 3, \ldots, k = 0, 1, 2, \ldots\}$. The corresponding eigenfunctions are $\sin jx \sin kt$ and $\sin jx \cos kt$. In particular -3 is the first negative eigenvalue of A and so

$$(17) \qquad (Au, u) \geq -\frac{1}{3}\|Au\|_{L^2}^2 \quad \text{for all } u \in D(A)$$

or equivalently

$$(17') \qquad (Kf, f) \geq -\frac{1}{3}\|f\|_{L^2}^2 \quad \text{for all } f \in R(A).$$

K is a compact operator in L^2 and in addition

$$(18) \qquad \|Kf\|_{L^\infty} \leq C\|f\|_{L^1} \quad \text{for all } f \in R(A),$$

$$(19) \qquad \|Kf\|_{H^1} \leq C\|f\|_{L^2} \quad \text{for all } f \in R(A).$$

We shall prove that for $\varepsilon < 3 - \gamma$, equation (14) has a solution u_ε. Indeed set $g_\varepsilon(u) = g(u) + \varepsilon u$ and let h_ε be the inverse function of g_ε. Set $H_\varepsilon(t) = \int_0^t h_\varepsilon(s)ds$.

In view of the duality principle of §II applied to $Bu = g(u) + \varepsilon u$, equation (14) has a solution provided we can find a critical point of

$$\Psi_\varepsilon(v) = \tfrac{1}{2}\int_\Omega Kv \cdot v + \int_\Omega H_\varepsilon(v + f)$$

subject to the constraint $v \in R(A)$.

We need the following:

LEMMA 1. *There are constants $\alpha > 0$ and C such that*

$$(20) \qquad \Psi_\varepsilon(v) \geq \alpha\|v\|_{L^2}^2 - C \quad \textit{for all } v \in R(A).$$

PROOF. We have

$$|g_\varepsilon(u)| \le (\gamma + \varepsilon)|u| + C \quad \text{for all } u \in \mathbf{R}$$

i.e.,

$$|v| \le (\gamma + \varepsilon)|h_\varepsilon(v)| + C \quad \text{for all } v \in \mathbf{R}$$

and so

$$(21) \qquad H_\varepsilon(v) \ge \frac{1}{2(\gamma + \varepsilon)}|v|^2 - C|v| \quad \text{for all } v \in \mathbf{R}.$$

We easily derive (20) from (17′) and (21).

It follows from Lemma 1 that Ψ_ε achieves its minimum on $R(A)$ (note that Ψ_ε is lower semicontinuous for the weak topology).

Step 2. For any solution u_ε of (14) we have

$$(22) \qquad \|u_\varepsilon\|_{L^1} \le C, \qquad \|Au_\varepsilon\|_{L^2} \le C, \qquad \|g(u_\varepsilon)\|_{L^2} \le C.$$

In what follows we denote by C various constants independent of ε. We deduce from (2) and (4) that

$$(23) \qquad (g(u) - f^{**}(x, t))u \ge \frac{\delta}{2}|u| - C, \quad \text{for all } u \in \mathbf{R} \text{ and } (x, t) \in \Omega,$$

$$(24) \qquad (g(u) - f^{**}(x, t))u \ge \frac{1}{\gamma}|g(u)|^2 - C|g(u)| - C$$

$$\text{for all } u \in \mathbf{R}, (x, t) \in \Omega.$$

Choosing $u = u_\varepsilon$ in (24) and using (17) we find

$$(25) \quad \varepsilon\|u_\varepsilon\|_{L^2}^2 + \frac{1}{\gamma}\|g(u_\varepsilon)\|_{L^2}^2 \le \frac{1}{3}\|Au_\varepsilon\|_{L^2}^2 = C\|Au_\varepsilon\|_{L^2} + C\|g(u_\varepsilon)\|_{L^2} + C.$$

Replacing $g(u_\varepsilon)$ by $(f - Au_\varepsilon - \varepsilon u_\varepsilon)$ in (25) we conclude that

$$\|Au_\varepsilon\|_{L^2} \le C \quad \text{and} \quad \|g(u_\varepsilon)\|_{L^2} \le C.$$

Finally we choose $u = u_\varepsilon$ in (23) and we obtain $\|u_\varepsilon\|_{L^1} \le C$.

Step 3. We shall prove that

$$(26) \qquad \|u_\varepsilon\|_{L^\infty} \le C.$$

We write $u_\varepsilon = u_{1\varepsilon} + u_{2\varepsilon}$ with $u_{1\varepsilon} + u_{2\varepsilon}$ with $u_{1\varepsilon} \in R(A)$, $u_{2\varepsilon} \in N(A)$. By (22) and (18) we already know that

$$(27) \qquad \|u_{1\varepsilon}\|_{L^\infty} \le C.$$

Therefore it suffices to prove that $\|u_{2\varepsilon}\|_{L^\infty} \le C$.

Since $u_{2\varepsilon} \in N(A)$, $u_{2\varepsilon}$ has the form

$$u_{2\varepsilon}(x, t) = p_\varepsilon(t + x) - p_\varepsilon(t - x)$$

where p_ε is a 2π-periodic function with $\int_0^{2\pi} p_\varepsilon = 0$, given by

$$p_\varepsilon(t) = \frac{1}{2\pi}\int_0^\pi [u_{2\varepsilon}(x, t - x) - u_{2\varepsilon}(x, t + x)]\,dx$$

so that, by (22) we obtain

$$\|p_\varepsilon\|_{L^1} \leqslant C. \tag{28}$$

For simplicity we write $p = p_\varepsilon$.

On the other hand we observe that a function $\psi \in L^2$ belongs to $N(A)^\perp$ if and only if

$$\int_0^\pi \left[\psi(x, t - x) - \psi(x, t + x)\right] dx = 0 \quad \text{for a.e. } t.$$

Indeed $\psi \in N(A)^\perp$ if and only if $\int_\Omega \psi(x, t)[q(t + x) - q(t - x)] dx\, dt = 0$ for every 2π-periodic function q. Since $g(u_\varepsilon) + u_{2\varepsilon} - f^{**} \in N(A)^\perp$ we have

$$\begin{cases} \varepsilon p(t) + \dfrac{1}{2\pi} \displaystyle\int_0^\pi \left[g(u_\varepsilon(x, t - x)) - g(u_\varepsilon(x, t + x))\right] dx \\[2mm] = \dfrac{1}{2\pi} \displaystyle\int_0^\pi \left[f^{**}(x, t - x) - f^{**}(x, t + x)\right] dx \quad \text{for a.e. } t. \end{cases} \tag{29}$$

From (27) we deduce that

$$u_\varepsilon(x, t - x) \geqslant -C + p(t) - p(t - 2x), \qquad u_\varepsilon(x, t + x) \leqslant C + p(t + 2x) - p(t)$$

and by (4) we have

$$f^{**}(x, t - x) \leqslant g(+\infty) - \delta, \qquad f^{**}(x, t + x) \geqslant g(-\infty) + \delta.$$

We derive from (29) that

$$\varepsilon p(t) + \frac{1}{2\pi} \int_0^\pi \left[g(-C + p(t) - p(t - 2x)) - g(C + p(t + 2x) - p(t))\right] dx$$

$$\leqslant \tfrac{1}{2}\left[g(+\infty) - g(-\infty)\right] - \delta \quad \text{for a.e. } t.$$

Finally we set $\tilde{g}(u) = \tfrac{1}{2}(g(u) - g(-u))$ and we find

$$\varepsilon p(t) + \frac{1}{2\pi} \int_0^{2\pi} \tilde{g}(-C + p(t) - p(s)) dx \leqslant \tilde{g}(+\infty) - \delta \quad \text{for a.e. } t. \tag{30}$$

Let $M = \operatorname{ess\,sup}_{t \in (0, 2\pi)} p(t) \geqslant 0$; our objective is to obtain a bound for M independent of ε.

By (30) we have

$$\frac{1}{2\pi} \int_0^{2\pi} \tilde{g}(-C + M - p(s)) dx \leqslant \tilde{g}(+\infty) - \delta. \tag{31}$$

On the other hand let

$$\Sigma = \{s \in (0, 2\pi); p(s) \geqslant M/2\}.$$

By (28) we know that $Mm(\Sigma)/2 \leqslant C$. Also, we have

$$\int_0^{2\pi} \tilde{g}(-C + M - p(s)) ds = \int_\Sigma \int_{c\Sigma} \geqslant m(\Sigma)\tilde{g}(-C) + m(^c\Sigma)\tilde{g}\left(-C + \frac{M}{2}\right).$$

Suppose $M \geqslant 2C$; we deduce from (31) that

$$\frac{1}{2\pi}\left(2\pi - \frac{2C}{M}\right)\tilde{g}\left(\frac{M}{2} - C\right) \leqslant \tilde{g}(C)\frac{2C}{M} + \tilde{g}(+\infty) - \delta$$

—which prevents M from going to infinity as $\varepsilon \to 0$. Therefore $p \leqslant C$; similarly $p \geqslant -C$ and we conclude that $\| p \|_{L^\infty} \leqslant C$.

Step 4. Passage to the limit.

Using the estimates of Steps 2 and 3 we may extract a sequence $\varepsilon_n \to 0$ such that

$$u_{\varepsilon_n} \to u \qquad \text{weakly in } L^2,$$

$$Au_{\varepsilon_n} \to Au \qquad \text{weakly in } L^2,$$

$$u_{1\varepsilon_n} \to u_1 \qquad \text{strongly in } L^2.$$

We need Minty's device in order to pass to the limit in the nonlinear term $g(u_\varepsilon)$.

We have

$$\left(g(u_\varepsilon) - g(\varphi), u_\varepsilon - \varphi \right) \geqslant 0 \quad \text{for all } \varphi \in L^2,$$

i.e.,

$$(32) \qquad \left(f - Au_\varepsilon - \varepsilon u_\varepsilon - g(\varphi), u_\varepsilon - \varphi \right) \geqslant 0.$$

Note that $(Au_\varepsilon, u_\varepsilon) = (Au_\varepsilon, u_{1\varepsilon})$ and so $(Au_{\varepsilon_n}, u_{\varepsilon_n}) \to (Au, u)$. Passing to the limit in (32) we find

$$\left(f - Au - g(\varphi), u - \varphi \right) \geqslant 0 \quad \text{for all } \varphi \in L^2.$$

We conclude in the usual manner, choosing $\varphi = u + t\psi$, that u satisfies $Au + g(u) = f$.

IV. Proofs of Theorems 2.3 and 4.

PROOF OF THEOREM 2. We start with the perturbed problem

$$(33) \qquad Au_\varepsilon + g(u_\varepsilon) + \varepsilon u_\varepsilon = 0.$$

As we know, the solutions of (33) correspond to critical point of

$$\Psi_\varepsilon(v) = \frac{1}{2} \int_\Omega Kv \cdot v + \int_\Omega H_\varepsilon(v)$$

on $R(A)$—through the relation $v = g(u) + \varepsilon u$. Clearly 0 is a critical point of Ψ_ε; we shall obtain a nontrivial critical point and then pass to the limit.

We already know (see the proof of Theorem 1), that $\mathrm{Inf}_{R(A)} \psi_\varepsilon$ is achieved, say at $v_\varepsilon \in R(A)$.

We end the following

LEMMA 2. $\mathrm{Inf}_{R(A)} \psi_\varepsilon \leqslant -m$, *where* $m > 0$ *is a constant independent of ε.*

PROOF OF LEMMA 2. From (5) we deduce that $| g(u) | \geqslant \theta | u |$ for some constant $\theta > 3$ and for u near 0 and so $| g_\varepsilon(u) | \geqslant \theta | u |$ for u near 0. Hence $| v | \geqslant \theta | h_\varepsilon(v) |$ for v near 0 and $H_\varepsilon(v) \leqslant | v |^2/2\theta$ for v near 0. Let $v_0 = \sin x \sin 2t$ be an eigenfunction of K corresponding to the eigenvalue $-1/3$. We have

$$\psi_\varepsilon(tv_0) = -\frac{t^2}{6} \| v_0 \|_{L^2}^2 + \int_\Omega H_\varepsilon(tv_0)$$

and for t near 0

$$\psi_\varepsilon(tv_0) \leqslant \left(-\frac{1}{6} + \frac{1}{2\theta}\right) t^2 \|v_0\|_{L^2}^2.$$

The conclusion of Lemma 2 follows directly.

PROOF OF THEOREM 2 CONCLUDED. From Lemma 2 we obtain a nontrivial solution u_ε of (33) related to v_ε by the relation $g(u_\varepsilon) + \varepsilon u_\varepsilon = v_\varepsilon$. We know (see the proof of Theorem 1) that $\|u_\varepsilon\|_{L^\infty} \leqslant C$. Therefore we may extract a sequence $\varepsilon_n \to 0$ such that $u_{\varepsilon_n} \to u$, $Au_{\varepsilon_n} \to Au$, $g(u_{\varepsilon_n}) \to g(u)$, $v_{\varepsilon_n} \to g(u)$ weakly in L^2.

By Lemma 2 we have $\frac{1}{2}\int_\Omega Kv_\varepsilon \cdot v_\varepsilon \leqslant -m$. Hence, at the limit $\frac{1}{2}\int_\Omega Kg(u)\cdot g(u) \leqslant -m$, and u is a nontrivial solution of $Au + g(u) = 0$.

SKETCH OF THE PROOF OF THEOREM 3. We consider now the problem of finding T-periodic solutions. Since T is a rational multiple of π we may write $T = 2\pi b/a$ where a and b are coprime integers.

Set $\Omega = (0, \pi) \times (0, T)$; we consider in $L^2(\Omega)$ the operator $Au = u_{tt} - u_{xx}$ acting on functions satisfying the boundary and T-periodicity conditions. As in §III we have (15), (16), (18) and (19). There are two differences with the case $T = 2\pi$. Here,

$$N(A) = \{p(t+x) - p(t-x), \text{ where } p \text{ has period } 2\pi/a = T/b\};$$

the eigenvalues of A consist of

$$\left\{j^2 - (2\pi k/T)^2; j = 1, 2, 3, \ldots, k = 0, 1, 2, \ldots\right\}$$

and they correspond to eigenfunctions $\sin jx \sin(2\pi kt/T)$, $\sin x \cos(2\pi kt/T)$. Let $\lambda_{-1}(T)$ be the first negative eigenvalue of A. Instead of (17) we have

$$(Au, u) \geqslant \frac{1}{|\lambda_{-1}(T)|} \|Au\|_{L^2} \quad \text{for all } u \in D(A).$$

Note that $\lambda_{-1}(T) \to 0$ as $T \to \infty$. Indeed let $\mu = j^2 - (2\pi k/T)^2$ with $j = 1$ and $k = [T/2\pi] + 1$.

We have $1 - (1 + 2\pi/T)^2 \leqslant \mu < 0$ and so $|\lambda_{-1}(T)| \leqslant |\mu| \leqslant 4\pi(1 + \pi/T)/T$. We use the same approach as in the proof of Theorem 2, i.e., minimize Ψ_ε on $R(A)$. We shall need the following lemma.

LEMMA 3. *There exists T_0 such that if $T > T_0$ and T is a rational multiple of π, then*

$$\mathop{\text{Inf}}_{R(A)} \Psi_\varepsilon \leqslant -1 \qquad (\varepsilon > 0).$$

PROOF OF LEMMA 3. Since $g(u) \not\equiv 0$ we may assume for example that

$$g(u) \geqslant \delta > 0 \quad \text{for } u \geqslant R.$$

Therefore

$$h_\varepsilon(v) \leqslant R \quad \text{for } 0 \leqslant v \leqslant \delta$$

and

$$H_\varepsilon(v) \leqslant C \quad \text{for } 0 \leqslant v \leqslant \delta.$$

Let $v_0 = \sin jx \sin(2\pi kt/T)$ be an eigenfunction of A corresponding to the eigenvalue $\lambda_{-1}(T)$. Let $v = \delta(1 + v_0)/2$. We have $K(1) = -x(x - \pi)/2$ and $Kv_0 = v_0/\lambda_{-1}(T)$. Thus

$$(Kv, v) = \frac{\delta^2}{4} \iint \left[-\frac{x}{2}(x - \pi) + \frac{1}{\lambda_{-1}(T)} v_0^2 \right] dxdt = \frac{1}{16} \delta^2 \pi \left[\frac{\pi^2}{3} + \frac{1}{\lambda_{-1}(T)} \right]$$

and

$$\Psi_\varepsilon(v) \leqslant \frac{T}{32} \delta^2 \pi \left[\frac{\pi^2}{3} + \frac{1}{\lambda_{-1}(T)} \right] + CT \to -\infty \quad \text{as } T \to +\infty.$$

Hence $\operatorname{Inf}_{R(A)} \Psi_\varepsilon \leqslant -1$ provided $T \geqslant T_0$ for some large T_0.

In what follows we fix $T \geqslant T_0$.

PROOF OF THEOREM 3 CONCLUDED. By minimizing Ψ_ε on $R(A)$ we obtain—as in the proof of Theorem 2—a solution of the equation

$$Au_\varepsilon + g(u_\varepsilon) + \varepsilon u_\varepsilon = 0$$

such that $v_\varepsilon = g(u_\varepsilon) + \varepsilon u_\varepsilon$ satisfies

$$\frac{1}{2} \int Kv_\varepsilon \cdot v_\varepsilon \leqslant -1.$$

[Note that Lemma 1 holds since, by (6), we have $|g(u)| \leqslant \gamma |u| + C$, for all u, with $\gamma < |\lambda_{-1}(T)|$.]

Using the same technique as in the proof of Theorem 1 we show that $\|u_\varepsilon\|_{L^\infty} \leqslant C$ and we pass to the limit as $\varepsilon \to 0$. (The only difference is that, instead of (23) we have

$$(23') \qquad\qquad g(u) \cdot u \geqslant \delta |u^+| - C \quad \text{for all } u \in \mathbf{R}$$

which leads to $\int u_\varepsilon^+ \leqslant C$ and then $\int |u_\varepsilon| = 2\int u_\varepsilon^+ - \int u_\varepsilon = 2\int u_\varepsilon^+ - \int u_\varepsilon \leqslant C$ since $\int u_{2\varepsilon} = 0$.)

SKETCH OF THE PROOF OF THEOREM 4. For simplicity I shall describe the proof in the special case where $g(u) = |u|^{m-2} u$, $m > 2$ and $T = 2\pi$. In the general case the proof is technically more complicated, but along the same lines (see [**16**] for the details). Set $v = g(u) = |u|^{m-2} u$. Using the dual formulation we have to find a nontrivial critical point for the functional

$$\Psi(v) = \frac{1}{2} \int Kv \cdot v + \frac{1}{m'} \int |v|^{m'}$$

subject to the constraint $v \in E$, where $1/m + 1/m' = 1$ and $E = \{v \in L^{m'}(\Omega);$ $\int v\varphi = 0$ for all $\varphi \in N(A)\}$. Note that, by (18), K maps E into L^∞ and that K is compact from E into $L^m(\Omega)$. Clearly Ψ is C^1 on E; in fact

$$\langle \Psi'(v), \zeta \rangle_{E^*, E} = \int Kv \cdot \zeta + \int |v|^{m'-2} v\zeta \quad \text{for all } v, \zeta \in E.$$

Using the Hahn-Banach theorem we may write

$$(34) \qquad\qquad Kv + |v|^{m'-2} v = w + \varphi$$

with $w \in L^m$, $\|w\|_{L^m} = \|\Psi'(v)\|_{E^*}$ and $\varphi \in N(A) \cap L^m$. Here, $\mathrm{Inf}_E \Psi = -\infty$ and $\mathrm{Sup}_E \Psi = +\infty$; thus we can no longer find a critical point by minimization of maximization. We shall, instead, rely on the following geometrical result due to A. Ambrosetti-P. Rabinowitz [7]. (See also [13] and [49].)

LEMMA (MOUNTAIN PASS LEMMA). *Assume Ψ is a C^1 function on a Banach space E and satisfies the Palais-Smale condition, i.e.*

(PS)
$$\begin{cases} \text{whenever a sequence } \{v_j\} \text{ in } E \text{ satisfies } |\Psi(v_j)| \leqslant C \\ \text{and } \Psi'(v_j) \to 0 \text{ in } E^* \text{ there exists a subsequence of } v \\ \text{which converges in } E. \end{cases}$$

Assume also

(35)
$$\begin{cases} \text{there are constants } r > 0 \text{ and } \rho > 0 \text{ such that } \Psi(v) \geqslant \rho \\ \text{for every } v \in E \text{ with } \|v\| = r, \end{cases}$$

(36) $$\Psi(0) < \rho \quad \text{and} \quad \Psi(v_0) < \rho \quad \text{for some } v_0 \in E \text{ with } \|v_0\| \geqslant r.$$

Then there is a critical point v of Ψ such that $\Psi(v) \geqslant \rho$.

VERIFICATION OF (PS). We write

$$Kv_j + |v_j|^{m'-2} v_j = w_j + \varphi_j$$

with $w_j \in L^m$, $\|w_j\|_{L^m} \to 0$, $\varphi_j \in N(A) \cap L^m$. We have

$$\frac{1}{2} \int Kv_j \cdot v_j + \frac{1}{2} \int |v_j|^{m'} = \frac{1}{2} \int w_j v_j,$$

$$\left| \frac{1}{2} \int Kv_j \cdot v_j + \frac{1}{m'} \int |v_j|^{m'} \right| \leqslant C.$$

Therefore

$$\left(\frac{1}{m'} - \frac{1}{2} \right) \int |v_j|^{m'} \leqslant C + \frac{1}{2} \|w_j\|_{L^m} \|v_j\|_{L^{m'}}$$

and so v_j remains bounded in $L^{m'}$. We extract a subsequence—still denoted by v_j—which converges weakly to v in E. By the convexity of the function $|t|^{m'}$ we have

$$\frac{1}{m'} |v|^{m'} - \frac{1}{m'} |v_j|^{m'} \geqslant |v_j|^{m'-1} v_j (v - v_j) = (w_j + \varphi_j - Kv_j)(v - v_j).$$

Thus

(37) $$\frac{1}{m'} \int |v|^{m'} - \frac{1}{m'} \int |v_j|^{m'} \geqslant \int (w_j - Kv_j)(v - v_j).$$

It follows from the compactness of K that the right-hand side of (37) goes to zero and thus $\limsup \|v_j\|_{L^{m'}} \leqslant \|v\|_{L^{m'}}$. Hence $v_j \to v$ strongly in $L^{m'}$.

VERIFICATION OF (35). By (18) we have

$$\Psi(v) \geqslant -C\|v\|_{L^1}^2 + \frac{1}{m'}\|v\|_{L^{m'}}^{m'} \geqslant -C\|v\|_{L^{m'}}^2 + \frac{1}{m'}\|v\|_{L^{m'}}^{m'}$$

and the conclusion follows immediately since $m' < 2$.

VERIFICATION OF (3.6). Choose v_0 of the form $v_0 = av_1$ where v_1 is any element in E such that $\int Kv_1 \cdot v_1 < 0$ and a is large enough.

We now deduce from the Mountain Pass Lemma that there exists $v \in E$, $v \neq 0$, such that $\Psi'(v) = 0$. Then $u = |v|^{m'-2}v$ is a nontrivial solution of the equation $Au + g(u) = 0$ with $u \in L^m$. Finally we show that $u \in L^\infty$ using the same argument as in the proof of Theorem 1 (Step 3).

REMARK 12. Instead of using the Mountain Pass Lemma one can give a very elementary argument. Indeed, minimize $\int Kv \cdot v$ on the set $\{v \in E;\ \|v\|_{L^{m'}} \leqslant 1\}$. The minimum is achieved at some $v_0 \in E$ with $\|v_0\|_{L^{m'}} = 1$ and $\int Kv_0 \cdot v_0 < 0$. Clearly we have $Kv_0 + \lambda |v_0|^{m'-2}v_0 = \varphi$ for some constant $\lambda > 0$ and some $\varphi \in N(A) \cap L^m$. Then $v = \alpha v_0$ is a critical point of Ψ on E provided α is an appropriate constant. Unfortunately this argument relies heavily on the fact that $g(u)$ is *homogeneous* and does not extend to the general case while the previous argument does.

REFERENCES

1. H. Amann and G. Mancini, *Some applications of monotone operator theory to resonance problems* (to appear).

2. H. Amann and E. Zehnder, *Nontrivial solutions for a class of nonresonance problems and applications to nonlinear differential equations*, Ann. Scuola Norm. Sup. Pisa **8** (1980), 539–603.

3. ______, *Multiple periodic solutions for a class of nonlinear autonomous wave equations*, Houston J. Math. (to appear).

4. A. Ambrosetti, *Recent advances in the study of the existence of periodic orbits of Hamiltonian systems* (to appear).

5. A. Ambrosetti and G. Mancini, *Solutions of minimal period for a class of convex Hamiltonian systems*, Math. Ann. **255** (1981), 405–421.

6. ______, *On a theorem of Ekeland and Lasry concerning the number of periodic Hamiltonian trajectories*, J. Differential Equations (to appear).

7. A. Ambrosetti and P. Rabinowitz, *Dual variational methods in a critical point theory and applications*, J. Funct. Anal. **14** (1973), 345–381.

8. A. Bahri, *Topological results on a certain class of functionals and applications*, J. Funct. Anal. (to appear).

9. A. Bahri and H. Berestytcki, *A perturbation method in critical point theory and applications*, Trans. Amer. Math. Soc. **267** (1981), 1–32.

10. ______, *Forced vibrations of superquadratic Hamiltonian systems*, Acta Math. (to appear); see also C. R. Acad. Sci. Paris Sér. A **292** (1981), 315–318.

11. ______, *Existence of periodic solutions for some second order systems of nonlinear ordinary differential equations*, Comm. Pure Appl. Math. (to appear).

12. A. Bahri and H. Brezis, *Periodic solutions of a nonlinear wave equation*, Proc. Roy. Soc. Edinburgh **85** (1980), 313–320.

13. V. Benci and P. Rabinowitz, *Critical point theorems for indefinite functionals*, Invent. Math. **52** (1979), 241–273.

14. H. Berestycki and J. M. Lasry, *Multiple periodic solutions of Hamilton's equations on a given energy surface* (to appear).

15. H. Brezis and J. M. Coron, *Periodic solutions of nonlinear wave equations and Hamiltonian systems*, Amer. J. Math. **103** (1981), 559–570.

16. H. Brezis, J. M. Coron and L. Nirenberg, *Free vibrations for a nonlinear wave equation and a theorem of P. Rabinowitz*, Comm. Pure Appl. Math. **33** (1980), 667–689.

17. H. Brezis and L. Nirenberg, *Forced vibrations for a nonlinear wave equation*, Comm. Pure Appl. Math. **31** (1978), 1–30.

18. ______, *Characterizations of the ranges of some nonlinear operators and applications to boundary value problems*, Ann. Scuola Norm. Sup. Pisa **5** (1978), 225–326.

19. F. Browder, *Nonlinear elliptic boundary value problems*, Bull. Amer. Math. Soc. **69** (1963), 862–874.

20. D. Campbell, Personal communication and papers to appear.

21. L. Cesari, *Existence in the large of periodic solutions of hyperbolic partial differential equations*, Arch. Rational Mech. Anal. **20** (1965), 170–190.

22. K. C. Chang, *Solutions of asymptotically linear operator equations via Morse theory* (to appear).

23. K. C. Chang, S. P. Wu and Shujie Li, *Multiple periodic solutions for an asymptotically linear wave equation* (to appear).

24. F. Clarke, *Periodic solution to Hamiltonian inclusions*, J. Differential Equations (to appear).

25. ______, *A classical variational principle for periodic Hamiltonian trajectories*, Proc. Amer. Math. Soc. **76** (1979), 186–188.

26. ______, *The dual action, optimal control and generalized gradients* (to appear).

27. F. Clarke and I. Ekeland, *Hamiltonian trajectories having prescribed minimal period*, Comm. Pure Appl. Math. **33** (1980), 103–116.

28. ______, *Nonlinear oscillations and boundary-value problems for Hamiltonian systems*, Arch. Rational Mech. Anal. (to appear).

29. J. M. Coron, *Résolution de l'équatin $Au + Bu = f$ où A est linéaire et B dérive d'un potentiel convexe*, Ann. Fac. Sci. Toulouse **1** (1979), 215–234 and C. R. Acad. Sci. Paris Sér. A **288** (1979), 805–808.

30. ______, *Solutions périodiques non triviales d'une équation des ondes*, Comm. Partial Differential Equations **6** (1981), 829–848.

31. ______, *Période minimale pour une corde vibrante de longueur infinie*, C. R. Acad. Sci. Paris Sér. A **294** (1982), 127–129.

32. ______, *Periodic solutions of a nonlinear wave equation without assumption of monotonicity* (to appear).

33. L. De Simon and G. Torelli, *Soluzione periodiche di equazioni a derivate parziali di tipo iperbolico nonlineari*, Rend. Sem. Mat. Univ. Padova **40** (1968), 380–401.

34. I. Ekeland, *Periodic solutions of Hamiltonian equations and a theorem of P. Rabinowitz*, J. Differential Equations **34** (1979), 523–534.

35. ______, *A perturbation theory near convex Hamiltonian systems* (to appear).

36. I. Ekeland and J. M. Lasry, *On the number of periodic trajectories for a Hamiltonian flow on a convex energy surface*, Ann. of Math. (2) **112** (1980), 283–319.

37. ______, *Duality in nonconvex variational problem* (to appear).

38. J. Hale, *Periodic solutions of a class of hyperbolic equations containing a small parameter*, Arch. Rational Mech. Anal. **23** (1967), 380–398.

39. H. Hofer, *On the range of a wave operator with nonmonotone nonlinearity* (to appear).

40. ______, *A new proof for a result of Ekeland and Lasry concerning the number of periodic Hamiltonian trajectories on a prescribed energy surface* (to appear).

41. G. Lamb, *Elements of soliton theory*, Wiley, New York, 1980.

42. E. Landesman and A. Lazer, *Nonlinear perturbations of linear elliptic boundary value problems at resonance*, J. Math. Mech. **19** (1970), 609–623.

43. J. Leray and J. L. Lions, *Quelques résultat de Visik sur les problèmes élliptiques nonlinéaires par les méthodes de Minty-Browder*, Bull. Soc. Math. France **93** (1965), 97–107.

44. H. Lovicarova, *Periodic solutions of a weakly nonlinear wave equation in one dimension*, Czech. Math. J. **19** (1969), 324–342.

45. G. Mancini, *Periodic solutions of some semilinear autonomous wave equations*, Boll. Un. Mat. Ital. **5** (1978), 649–673.

46. J. Mawhin, *Solutions périodiques d'équations aux dérivées partielles hyperboliques nonlinéaires*, Mélanges Th. Vogel, Presse Univ. Bruxelles, Bruxelles, 1978, pp. 301–319.

47. ______, *Compacité, monotonie et convexité dans l'étude de problèmes aux limites semi-linéaires*, Lecture Notes Univ. de Sherbrooke, 1981.

48. J. Moser, *Periodic orbits near an equilibrium and a theorem by Alan Weinstein*, Comm. Pure Appl. Math. **29** (1976), 727–747.

49. L. Nirenberg, *Variational and topological methods in nonlinear problems*, Bull. Amer. Math. Soc. (N.S.) **4** (1981), 267–302.

50. H. Poincaré, *Les méthodes nouvelles de la mécanique céleste*, Gauthier-Villars, Paris, 1982.

51. P. Rabinowitz, *Periodic solutions of nonlinear hyperbolic partial differential equations*, Comm. Pure Appl. Math. **20** (1967), 145–205.

52. ______, *Time periodic solutions of nonlinear wave equations*, Manuscripta Math. **5** (1971), 165–194.

53. ______, *Free vibrations for a semilinear wave equation*, Comm. Pure Appl. Math. **31** (1978), 31–68.

54. ______, *Periodic solutions of Hamiltonian systems*, Comm. Pure Appl. Math. **31** (1978), 157–184.

55. ______, *Periodic solutions of a Hamiltonian system on a prescribed energy surface*, J. Differential Equations **33** (1979), 336–352.

56. ______, *Subharmonic solutions of a forced wave equation*, Amer. J. Math. (to appear).

57. ______, *On nontrivial solutions of a semilinear wave equation*, Ann. Scuola Norm. Sup. Pisa (to appear).

58. ______, *Periodic solutions of Hamiltonian systems: a survey*, SIAM J. Math. Anal. (to appear).

59. ______, *Multiple critical points of perturbed symmetric functionals*, Trans. Amer. Math. Soc. **272** (1982), 753–769.

60. M. Struwe, *Infinitely many critical points for functionals which are not even and applications to superlinear boundary value problems*, Manuscripta Math. **32** (1980), 335–364.

61. J. Toland, *A duality principle for nonconvex optimisation and the calculus of variations*, Arch. Rational Mech. Anal. **71** (1979), 41–61.

62. ______, *Duality in nonconvex optimization*, J. Math. Anal. Appl. (to appear).

63. O. Vejvoda, *Periodic solutions of a linear and weakly nonlinear wave equation in one dimension. I*, Czech. Math. J. **14** (1964), 341–382.

64. O. Vejvoda et al., *Partial differential equations: time-periodic solutions*, Noordhoff, Groningen, 1981.

65. A. Weinstein, *Normal modes for a nonlinear Hamiltonian system*, Invent. Math. **20** (1973) 47–57.

66. ______, *Periodic orbits for convex Hamiltonian systems*, Ann. of Math. (2) **108** (1978), 507–518.

67. ______, *On the hypotheses of Rabinowitz' periodic orbit theorems*, J. Differential Equations **33** (1979), 353–358.

68. M. Willem, *Density of the range of potential operators*, Proc. Amer. Math. Soc. **83** (1981), 341–344.

DEPARTMENT OF MATHEMATICS, UNIVERSITÉ DE PARIS VI, 4PL. JUSSIEU, 75230 PARIS CÉDEX 05, FRANCE

Proceedings of Symposia in Pure Mathematics
Volume 39 (1983), Part 2

Fixed Point Theory and Nonlinear Problems

FELIX E. BROWDER

Introduction. Among the most original and far-reaching of the contributions made by Henri Poincaré to mathematics was his introduction of the use of topological or "qualitative" methods in the study of nonlinear problems in analysis. His starting point was the study of the differential equations of celestial mechanics, and in particular of their periodic solutions. His work on this topic began with his thesis in 1879, and was developed in detail in his great three-volume work, *Méthodes nouvelles de la méchanique céleste*, which appeared in the early 1890s and summarized his many memoirs of the intervening period. It continued until his memoir shortly before his death in 1912 in which he put forward the unproved fixed point result usually referred to as "Poincaré's last geometric theorem".

The ideas introduced by Poincaré include the use of fixed point theorems, the continuation method, and the general concept of global analysis. The writer's acquaintance with Poincaré's influence came through contact with Solomon Lefschetz and Marston Morse, both of whom were very explicit as to the role of Poincaré as an initiator in this direction of mathematical development. In 1934, in the Foreword to his Colloquium volume on *The calculus of variations in the large*. Morse had put this forward very forcefully in the first paragraph:

> "For several years the research of the writer has been oriented by a conception of what might be termed macro-analysis. It seems probable to the author that many of the objectively important problems in mathematical physics, geometry, and analysis cannot be solved without radical additions to the methods of what is now strictly regarded as pure analysis. Any problem which is nonlinear in character, which involves more than one coordinate system or more than one variable, or whose structure is initially defined in the large, is likely to require considerations of topology and group theory in order to arrive at its meaning and its solution. In the solution of such problems classical analysis will frequently appear as an instrument in the small, integrated over the whole problem with the aid of group theory or topology. Such conceptions

---

1980 *Mathematics Subject Classification.* Primary 47H.

> are not due to the author. It will be sufficient to say that
> Henri Poincaré was among the first to have a conscious
> theory of macro-analysis, and of all mathematicians was
> doubtless the one who most effectively put such a theory into
> practice."

I know off-hand of no such written testimony by Lefschetz, though his advice in conversation included the recommendation that everyone strongly interested in nonlinear problems in analysis should read through the Collected Works of Poincaré. How seriously this advice was meant is relatively difficult to determine after all this time. It always seemed to me that this was advice meant to frighten, not to encourage effort in this area, somewhat like the recommendation of André Weil that every aspiring algebraic number theorist should read through Hilbert's *Zahl-Bericht* once a year. However, we can associate Lefschetz with the analysis offered by Liapounoff in the classical memoir on stability theory of 1892 which Lefschetz caused to be republished in the Annals of Mathematics Studies in 1947. In this work, which was translated and published in French in 1907 under the title *Problème général de la stabilité du mouvement*, Liapounoff wrote

> "L'essai unique, autant que je sache, de solution rigoreuse de
> la question appartient à M. Poincaré, qui dans le Mémoire
> remarquable sous bien des rapports *Sur les courbes définies
> par les equations différentielles* et, en particulier, dans les deux
> dernières Parties, considère des questions de stabilité relatives
> au cas d'équations différentielles du second ordre et s'arrête
> aussi à quelques questions voisines, se rapportant à des
> systèmes du troisième ordre.
>
> Bien que M. Poincaré se borne a des cas très particuliers,
> les méthodes dont il se sert permettent des applications
> beaucoup plus générales et peuvent encore conduire à
> beaucoup de nouveaux résultats. C'est ce qu'on verra par ce
> qui va suivre, car, dans une grande partie des mes recherches,
> je me suis guidé par les idées développés dans le Mémoire
> cité."

Let me conclude this kind of textual testimony by one more quotation, this time from Jurgen Moser's Annals of Mathematics Study, *Stable and random motions in dynamical systems*, published in 1973. In the Introduction (page 4), he writes

> "However, the mathematical difficulties connected with
> this field, inspired more and more the study of basic theoreti-
> cal problems leading to the development of new mathematical

tools. The main influence in this direction came from Poincaré who started a number of new lines of thought, like the qualitative theory of differential equations, the quest for the topology of the energy manifold, he formulated and proved fixed point theorems to establish the existence of periodic solutions. In this connection one may recall that the fixed point theorem by P. Bohl and Poincaré also had its origin in this field, to which Bohl devoted his entire mathematical life."

The topic to which the present paper is devoted, degree theory for nonlinear mappings, is one to which Poincaré made an early and important contribution. His first results appear in a note in the Comptes Rendus [**32**] and a more detailed development [**33**] published in the Bulletin Astronomique, both in 1883 only four years after his Thesis. In this pair of papers, Poincaré announces the following result (in my translation): "Let $\xi_1, \xi_2, \ldots, \xi_n$ be n continuous functions of n variables $x_1, x_2, \ldots, x_n$: the variable x_i is subjected to vary between the limits $+a_i$ and $-a_i$. Let us suppose that for $x_i = a_i$, ξ_i is constantly positive, and that for $x_i = -a_i$ constantly negative; I say that there will exist a system of values of x for which all the ξ's vanish."

For the proof, he refers to the celebrated paper by Kronecker in 1869 [**25**] which began the theory of the topological degree of a mapping. (For a detailed analysis of this paper and of the subsequent literature on the Kronecker index, see the recent historical paper by Siegberg [**35**]. The reference to the 1883 papers by Poincaré, I owe to Mahwin [**27**]. Most accounts of Poincaré's role begin with his paper [**34**] in 1886.) No account of the proof is given in the 1883 papers, but in 1886, Poincaré [**34**] published the argument on the continuation invariance of the index which is the basis for the proof. This is the memoir cited by Liapounoff, and like the 1883 announcements, applies what amounts to a fixed point argument to prove the existence of periodic solutions of a system of ordinary differential equations.

We can describe the basic idea in the following general terms: Suppose we are given a system of differential equations

$$\frac{du}{dt}(t) = T(u(t))$$

where the right-hand side is independent of time t. Suppose this system has the property that for each initial value u_0, the system has one and only one solution defined over an interval $[0, T]$. Then the problem of finding a periodic solution $u(t)$ of period T is equivalent to finding an initial value u_0 such that the solution $u(t)$ with that given initial value has $u(T) = u_0$. If we set

$$S(t)u_0 = u(t),$$

this means finding a fixed point of the mapping $S(T)$.

The continuation method, a favorite technique of Poincaré, consists of imbedding the problem in a one-parameter family of problems depending upon an auxiliary parameter s and considering the solvability of the problem as s varies. The link between these two devices is the index or topological degree, which assures under appropriate hypotheses that the solvability of a family of problems is invariant under continuous perturbations. This continuity or homotopy invariance is the decisive property of the topological degree of mapping. Thus, in the 1880s, Poincaré was effectively using the basic properties of the topological degree, usually associated with L. E. J. Brouwer [7] in 1912. The result stated by Poincaré has come to be known as the theorem of Miranda because it was proved by C. Miranda [28] in 1940, who showed that it was equivalent to the Brouwer fixed point theorem. (It had earlier been applied by Michal Golomb [21] in 1935, who observed that its proof was implicit in another paper [6] by Brouwer in 1911.)

In the present paper, we give a simple exposition of the classical theory of the degree of a mapping as given by Kronecker and Brouwer, and of its extension by Leray and Schauder [26] in 1934 to mappings in infinite-dimensional Banach spaces of the form $I - g$, with g compact. We discuss in detail the existence and uniqueness of these degrees as defined by the additional properties of additivity, homotopy invariance, and normalization. We also present a self-contained exposition of the recent extension given by the writer [11–14] of these existence and uniqueness results for the degree functions for nonlinear mappings of monotone type from a reflexive Banach space X to its conjugate space X^*. We show how such mappings arise from the combination of the ideas of fixed point theory and the somewhat different circle of ideas associated with the direct method of the calculus of variations. The concept of degree of mapping in all these forms is one of the most effective tools for studying the properties of existence and multiplicity of solutions of nonlinear equations.

1. Degree theory in the finite-dimensional case. In the present section, we wish to develop as intuitively as possible, the fundamental properties of the classical topological degree as they were explicitly formulated by Brouwer in 1912 (and implicitly used by Poincaré, Bohl, and others (e.g. Picard [31]) on the basis of Kronecker's earlier results about the index).

Let us begin with some conventions on notation. We shall consider mappings with domain in a topological space X and with values lying in a topological space Y. If G is an open subset of X, we denote by $\mathrm{cl}(G)$ its closure in X, and by ∂G, its boundary in X. Thus $f: \mathrm{cl}(G) \to Y$ will be the prototype of the maps for which a degree function is to be defined. If $f: \mathrm{cl}(G_0) \to Y$ is such a map and G is an open subset of G_0, then by f_G we mean the restriction of the given map f to $\mathrm{cl}(G)$.

THEOREM 1. *Let $X = R^n = Y$ for a given positive integer n. For bounded open subsets G of X, consider continuous mappings $f: \mathrm{cl}(G) \to Y$, and points y_0 in Y such that y_0 does not lie in $f(\partial G)$. Then to each such triple (f, G, y_0), there corresponds an integer $d(f, G, y_0)$ having the following properties:*

(a) *If $d(f, G, y_0) \neq 0$, then $y_0 \in f(G)$. If f_0 is the identity map of X onto Y, then for every bounded open set G and $y_0 \in G$, we have*

$$d(f_{0,G}, G, y_0) = +1.$$

(b) (*Additivity*). *If f: $\mathrm{cl}(G) \to Y$ is a continuous map with G a bounded open set in X, and G_1 and G_2 are a pair of disjoint open subsets of G such that*

$$y_0 \notin f(\mathrm{cl}(G)\backslash(G_1 \cup G_2))$$

then

$$d(f, G, y_0) = d(f_{G_1}, G_1, y_0) + d(f_{G_2}, G_2, y_0).$$

(c) (*Invariance under homotopy*). *Let G be a bounded open set in X, and consider a continuous homotopy $\{f_t: 0 \leq t \leq 1\}$ of maps of $\mathrm{cl}(G)$ into Y. Let $\{y_t: 0 \leq t \leq 1\}$ be a continuous curve in Y such that $y_t \notin f_t(\partial G)$ for any t in $[0, 1]$. Then $d(f_t, G, y_t)$ is constant in t on $[0, 1]$.*

THEOREM 2. *The degree function $d(f, G, y_0)$ is uniquely determined by the three conditions of Theorem 1.*

Theorem 1 is an appropriately formalized version of the properties of the classical Brouwer degree. Theorem 2 contains an observation made independently in 1972 and 1973 by Fuhrer [20] and Amann and Weiss [2], respectively.

What is this degree function which we describe in Theorems 1 and 2? Intuitively, it is intended to be an algebraic count of the number of solutions x in G for the equation $f(x) = y_0$. We speak of algebraic count because (as we specify more concretely in a moment) some solutions are counted positively, others negatively. The paradigmatic case is that of a mapping f of class C^1 with only regular points x as solutions of the equation $f(x) = y_0$, i.e. at each solution $f'(x)$ is a nonsingular linear transformation of R^n. The number of such solutions is then finite, and we describe such solutions as positive if $f'(x)$ preserves orientation and negative if it reverses orientation. Then $d(f, G, y_0)$ equals the number of positive solutions minus the number of negative solutions. Thus defined, the degree function for such mappings is certainly at least an integer. It is also clearly additive and satisfies the normalization condition. The main problem is to show that is extendable to all mappings and is homotopy invariant.

One way of establishing these facts is to identify $d(f, G, y_0)$ for such maps with an integral of the type studied by Kronecker. To do this, let us assume first that G has a C^1 boundary $S = \partial G$ which we orient in an appropriate way so as to satisfy the conditions for Stokes theorem:

$$\int_G d\alpha = \int_S \alpha$$

for any C^1 $(n - 1)$-form α on $\mathrm{cl}(G)$. We shall describe an $(n - 1)$-form ω on

$R^n \setminus \{0\}$ in terms of the fundamental solution $e_n(r)$ of radial type for the Laplacian on R^n. Thus for $n > 2$, $e_n(r) = c_n r^{2-n}$, while for $n = 2$, $e_2(r) = c_2 \ln(1/r)$. Thus, setting $\Delta = \sum_{j=1}^{n} (\partial/\partial x_j)^2$, we have

$$\Delta(e_n) = \delta,$$

where δ is the Dirac delta function at 0. We set

$$\omega = \sum_{j=1}^{n} (-1)^j \frac{\partial}{\partial x_j}(e_n(r)) \, d\hat{x}_j$$

with $r = (\sum_{j=1}^{n} x_\gamma^2)^{1/2}$ and $d\hat{x}_j = dx_1 \cdots dx_j \cdots dx_n$. Then

$$\int_{\partial G} \omega = \begin{cases} +1 & \text{if } 0 \in G, \\ 0 & \text{if } 0 \notin \mathrm{cl}(G). \end{cases}$$

Using the form $f_*(\omega)$ on S induced by the map f of S into R^n, we assert that for maps f having only regular points in $f^{-1}(0)$

$$d(f, G, 0) = \int_{\partial G} f_*(\omega).$$

We need only observe that on any domain G_0 for which f has no zeroes, $df_*(\omega) = f_*(d\omega) = 0$. Hence

$$\int_{\partial G_0} f_*(\omega) = 0$$

for such G_0. We can take G_0 to be the complement in G of the union of a family of small balls around the various points of the finite set $\{x_1, \ldots, x_r\} = f^{-1}(0)$. As the balls are taken smaller and smaller (if B_j is the ball around the jth point x_j)

$$\int_G f_*(\omega) = \sum_{j=1}^{r} \int_{\partial B_j} f_*(\omega)$$

while

$$\int_{\partial B_j} f_*(\omega) - \int_{\partial B_j} L_{j*}(\omega) \to 0$$

where L_j is the linear map $f'(x_j)$. Finally,

$$\int_{\partial B_j} L_{j*}(\omega) = \mathrm{sgn}(\det(L_j)).$$

If we combine these facts, we identify $d(f, G, 0)$ with the corresponding integral.

A similar representation holds for $d(f, G, y_0)$ if we note that $d(f, g, y_0) = d(f^{(y_0)}, G, 0)$, where $f^{(y_0)}(x) = f(x) - y_0$ for all x in $\mathrm{cl}(G)$.

The point of identifying $d(f, G, y_0)$ with an integral is that the integral makes sense for any C^1 mapping f, not just for the ones with regular points in $f^{-1}(y_0)$, provided only that $f(S)$ does not contain y_0. Moreover, it is continuous under C^1 deformations of the mapping f, under the sole restriction that during the deformation y_0 never appears in $f(\partial G)$.

We now apply the theorem of Sard-Morse in the case of maps of R^n into R^n to approximate f in the C^1 topology by a map g having only regular points in its zero set with $\| f - g \|_{C^1(\mathrm{cl}(G))} < \varepsilon$. It follows that for any f and the corresponding g, the integrals for f and g are very close. Since g is one of the paradigmatic maps we were treating earlier, $d(g, G, 0)$ is an integer. Hence, the integral for f defining $d(f, G, 0)$ being arbitrarily close to an integer, must itself be an integer. However, an integer-valued function of a parameter which is continuous in that parameter must be constant. Hence $d(f_s, G, y_0)$ is constant under any C^1 deformation during which it remains defined.

Thus, we have a degree function $d(f, G, y_0)$ defined for f of class C^1 and G having a C^1 boundary. We now proceed to remove these restrictions on the smoothness of f and G.

Let G be any bounded open subset of R^n and consider a continuous map f: $\mathrm{cl}(G) \to R^n$ which is of class C^2 in G and such that $0 \notin f(\partial G)$. Then we can find $\gamma > 0$ such that on the γ-neighborhood of ∂G,

$$|f(x)| > \gamma.$$

We consider a C^1 function $\varphi(r)$ which is 1 for $r \geq \gamma$, 0 for $r < \gamma/2$. Set

$$\tilde{\omega} = \varphi(r)\omega.$$

For any open subset G_1 of G which has a C^1 boundary and contains the complement in G of the γ-neighborhood of ∂G, the boundary G_1 is contained in the γ-neighborhood of ∂G. Hence

$$d(f_{G_1}, G_1, 0) = \int_{\partial G_1} f_*(\omega) = \int_{\partial G_1} f_*(\tilde{\omega}).$$

However, the form $f_*(\tilde{\omega})$ is of class C^1 in $\mathrm{cl}(G_1)$. Hence

$$\int_{\partial G_1} f_*(\tilde{\omega}) = \int_{G_1} df_*(\tilde{\omega}) = \int_{G_1} f_*(d\tilde{\omega}) = \int_{G_1} f_*(d\varphi \wedge \omega).$$

Since

$$f_*(d\varphi \wedge \omega) = 0$$

on the γ-neighborhood of ∂G, we see that the resulting integral is independent of the choice of G_1 and can be expressed as an integral over $G \setminus N_\gamma(\partial G)$, with $N_\gamma(\partial G)$ the γ-neighborhood.

Finally, suppose $f: \mathrm{cl}(G) \to R^n$ is merely continuous. Then we can approximate it arbitrarily closely in the C^0-norm by C^2 maps g. If h is another such approximation, then so is

$$g_t = (1 - t)g + th$$

for any t in $[0, 1]$, and for any t, $g_t(x) \neq 0$ for x in ∂G. Hence

$$d(g_t, G, 0)$$

is independent of t in $[0, 1]$ since $\{g_t\}$ is a C^1 homotopy of C^1 maps. We define

$$d(f, G, 0) = d(g, G, 0),$$

and more generally, $d(f, G, y_0) = d(g, G, y_0)$ for such approximations g.

The degree function thus defined is immediately seen to satisfy the three conditions of Theorem 1.

For the normalization conditions, this follows immediately for the maps with only regular solutions of $f(x) = y_0$ by the definition of the degree function for such maps. (Moreover, the identity map is immediately seen to have the appropriate value for its degree function.) For more general maps f, our process of definition yields a map g with only regular solutions such that g is close to f in the C^0-norm and

$$d(f, G, y_0) = d(g, G, y_0).$$

Hence, if $\| f - g \|_{C^0} < \varepsilon$, since $d(f, G, y_0) \neq 0$, we see that g must have y_0-points in G, so that there exist points in $\mathrm{cl}(G)$ such that $\| f(x) - y_0 \| < \varepsilon$ for any $\varepsilon > 0$. Since $f(\mathrm{cl}(G))$ is closed, it follows that $y_0 \in f(\mathrm{cl}(G))$. Since y_0 does not lie in $f(\partial G)$ by assumption, y_0 must belong to $f(G)$. Additivity on domain follows immediately by the corresponding additivity for maps with regular y_0-points. Finally, given a continuous homotopy $\{f_t\}$ in the C^0 topology and a continuous curve $\{y_t\}$ in R^n, we can consider $\{f_t - y_t\}$, and show that $d(f_t, G, 0)$ is constant after the modification of the f_t. We can find $\gamma > 0$ such that $|f_t(x)| > \gamma$ for x on ∂G. We now choose a sequence $0 = t_0 < t_1 < \cdots < t_r = 1$ such that $\| f_{t_j} - f_{t_{j+1}} \|_{C^0} < \gamma/2$. For each j, choose a C^1-map $g_j \colon \mathrm{cl}(G) \to R^n$ such that

$$\| f_{t_j} - g_j \|_{C^0(\mathrm{cl}(G))} < \gamma/2.$$

Then for each j, we have

$$d(f_{t_j}, G, 0) = d(g_j, G, 0),$$

and it suffices to show that for each j,

$$d(g_j, G, 0) = d(g_{j+1}, G, 0).$$

We consider the linear homotopy

$$h_t = (1 - t)g_j + t g_{j+1}$$

between g_j and g_{j+1}. This homotopy is of class C^1, so that to obtain the constancy of the degree under the homotopy, it suffices to show that for all x in ∂G and t in $[0, 1]$, $h_t(x) \neq 0$. Both points $g_j(x)$ and $g_{j+1}(x)$ are contained in the open ball of radius γ about the point $f_{t_j}(x)$, and hence so is their convex linear combination $h_t(x)$. Since this ball does not contain 0, $h_t(x) \neq 0$.

REMARK. Analytical proofs for the existence and properties of the degree function have been given a number of times in the literature in the last forty years, for the first time explicitly by Nagumo in 1941. A recent version with bibliographical references is that of Fenske [19].

We now observe that the argument given above, together with one further observation, yields the proof of Theorem 2. Suppose that we are given a degree function $d_1(f, G, y_0)$ which satisfies the three conditions of Theorem 1. We wish to identify it with the degree function d already constructed. We may approximate f as closely as we please in the C^0-norm by a map g of class C^1 which has only finitely many points $\{x_1,\ldots,x_r\}$ in G at which $f(x) = y_0$, and all of these points are regular points. Since f and g are close, the linear homotopy

$$h_t = (1 - t)f + tg$$

is permissible in that it produces no points x on ∂G such that $h_t(x) = y_0$ for any t in $[0, 1]$. By homotopy invariance of the degrees,

$$d_1(f, G, y_0) = d_1(g, G, y_0), \qquad d(f, G, y_0) = d(g, G, y_0).$$

Hence, we need only show the equality of the two degrees on such mappings g. Moreover, using the additivity on domain, we can replace G by the union of small balls B_j around the various points x_j. It thus suffices to take $G = B$ where B is the ball of radius ε about the point x_0 which is the unique solution of $g(x) = y_0$ in B. If $\varepsilon > 0$ is sufficiently small and if L is the derivative of g at x_0, the homotopy between g and g_0 with $g_0(x) = y_0 + L(x - x_0)$ given by

$$k_t = (1 - t)g + tg_0$$

has the property that it produces no y_0 points on ∂B. Hence, we need only verify that

$$d(g_0, B, y_0) = d_1(g_0, B, y_0).$$

Furthermore, if we deform L through the nonsingular linear mappings of R^n, both degrees remain constant. Since the nonsingular linear mappings of L have two path components, the maps with positive determinant and the maps with negative determinant, it suffices to consider a single map in each class. For the mappings with positive determinant (the orientation-preserving maps), we choose the identity map I, and the equality of the two degrees follows from the normalization condition. In the other case, we choose the linear mapping L given by

$$Lx_1 = -x_1, \qquad Lx_j = x_j \quad \text{for } j \geqslant 2.$$

To handle this last case, let us observe that it really amounts to considering the same problem in the one-dimensional case, $n = 1$. Indeed, let B_{n-1} be the unit ball in R^{n-1} and for any mapping $f_1: \mathrm{cl}(G_1) \to R^1$ with G_1 a bounded open subset in R^1, let us define

$$G_2 = x_0 + G_1 \times B_{n-1}, \qquad f_2(x) = y_0 + h(x), \qquad h(x) = (f_1(x_1), x_2,\ldots,x_n).$$

We define two degrees d and d_2 on the one-dimensional maps by setting

$$d(f_1, G_1, u) = d(f_2, G_2, y_0 + u), \qquad d_2(f_1, G_1, u) = d_1(f_2, G_2, y_0 + u)$$

where u is considered as a vector $(u, 0,\ldots,0)$. The degree function d is the conventional one on R^1; we must show it identical to the degree function d_2

defined in terms of d_1. For $n = 1$, the problem is best resolved by the following picture:

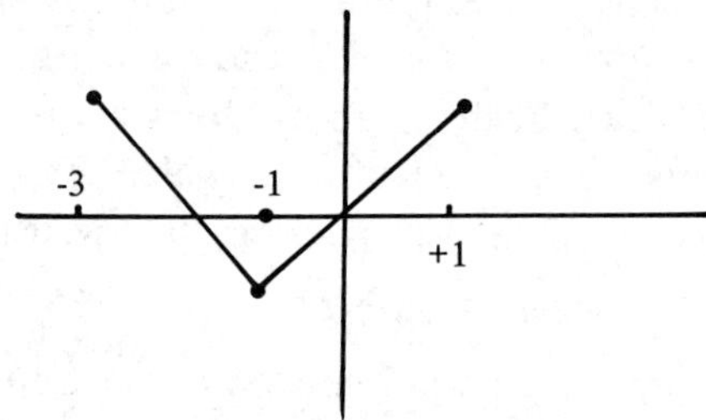

The function h whose graph is pictured on the interval $[-3, +1]$ has two zeros at 0 and at (-2). Its total degree over $G_0 = (-3, +1)$ with respect to 0 must be 0, since it is linearly homotopic to the constant function $h_0(x) \equiv 1$ without zeroes on the boundary. Since

$$d_2(h, G_0, 0) = d_2(h, G_1, 0) + d_2(h, G_2, 0)$$

where $G_1 = (-3, -1)$ and $G_2 = (-1, +1)$ and since $d_2(h, G_2, 0) = +1$, $d_2(h, G, 0) = 0$, we have

$$d_2(h, G_1, 0) = -1$$

which verifies our assertion for L.

This completes the proof of Theorem 2.

2. General degree theories and their elementary properties. Using the result of §1 as a model, we may formulate the problem of constructing more general degree theories in the following terms:

DEFINITION 1. *We are given a domain space X and a range space Y, both topological spaces.*

We are given a class O of open subsets G of X.

For each G in O, we consider a family of maps f: $\mathrm{cl}(G) \to Y$; the collection of all such maps for the various G of O, we call F, the family of maps over which the degree theory is to be defined.

For each G in O, we consider a family of homotopies $\{f_t: 0 \leq t \leq 1\}$ of maps in F, all having the common domain $\mathrm{cl}(G)$; the total collection of all such homotopies for the various G in O, we call H, the class of permissible homotopies for the degree theory.

Then by a degree theory over the class F which is invariant with respect to the homotopies of H and which is normalized by a given map f_0 of X into Y, we mean: For each y_0 in Y and for each f in F, f: $\mathrm{cl}(G) \to Y$ for which $y_0 \notin f(\partial G)$, an integer $d(f, G, y_0)$ is to be prescribed and the prescription is to satisfy the following three conditions:

(a) (*Normalization*) *If $d(f, G, y_0) \neq 0$, then $y_0 \in f(G)$. For each G in O, $f_{0,G}$ lies in F, and if $y_0 \in f_0(G)$, then $d(f_{0,G}, G, y_0) = +1$.*

(b) (*Additivity on domain*) *Suppose that* $f \in F$; f: $\mathrm{cl}(G) \to Y$, *and that* G_1 *and* G_2 *are a pair of disjoint subsets of* O *contained in* G. *Let* $y_0 \notin f(\mathrm{cl}(G) \backslash (G_1 \cup G_2))$. *Then*

$$d(f, G, y_0) = d(f_{G_1}, G_1, y_0) + d(f_{G_2}, G_2, y_0).$$

(c) (*Invariance under homotopy*) *If* $\{f_t: 0 \leqslant t \leqslant 1\}$ *is a homotopy in* H *with domain* $\mathrm{cl}(G)$ *for* G *in* O, *and if* $\{y_t: 0 \leqslant t \leqslant 1\}$ *is a continuous curve in* Y, *with* $y_t \notin f_t(\partial G)$ *for all* t *in* $[0, 1]$, *then* $d(f_t, G, y_t)$ *is constant in* t *on* $[0, 1]$.

The existence of a degree theory for a given class of mappings is not a trivial assertion. Let us verify this fact through several elementary observations.

PROPOSITION 1. *Let* X *and* Y *be equal to the same Hilbert space* H, *let* O *be a class of open subsets of* H *which includes open balls, and consider a class of mappings* F *and a class of homotopies* H *which includes the affine homotopies in* F (*i.e. homotopies of the form* $f_t = (1 - t)f + tf_1$ *for* f *and* f_1 *in* F). *Suppose a degree theory exists for* F *invariant with respect to* H, *and normalized by* $f_0 = I$. *Then*

For any f *in* F *with domain a closed ball* B *about the origin in* H *such that* $(f(u), u) \geqslant 0$ *for all* u *in* B, f *must have a zero in* B.

PROOF OF PROPOSITION 1. Let G be the interior of B. G lies in O and by hypothesis, f lies in the class F. If f has a zero on ∂B, the assertion follows. Otherwise, $d(f, G, 0)$ is well defined. Let f_0 be the restriction of the identity map I to B. Then f_0 lies in F and $d(f_0, G, 0) = +1$. Consider the affine homotopy,

$$f_t = (1 - t)f_0 + tf \qquad (0 \leqslant t \leqslant 1).$$

By assumption, this homotopy lies in the class H of permissible homotopies. Hence, if we can show that for all t in $(0, 1)$, $f_t(x) \neq 0$ on ∂B, then $d(f_t, G, 0)$ is constant in t and hence equals $+1$ for all t in $[0, 1]$. Suppose $f_t(x) = 0$ for $0 < t < 1$, with x in ∂B. Then

$$0 = (f_t(x), x) = (1 - t)(f_0(x), x) + t(f(x), x).$$

Since $(f(x), x) \geqslant 0$ for all x in ∂G, it follows that

$$0 \geqslant (1 - t)(f_0(x), x) = (1 - t)\|x\|^2 > 0,$$

which is impossible. Hence $d(f, G, 0) = +1$, and by the normalization property of the degree function, $0 \in f(G)$.   Q.E.D.

A corollary of Proposition 1 is the following:

PROPOSITION 2. *Suppose that a degree theory satisfying the hypotheses of Proposition 1 exists over a class of mappings* F, *and that for a ball* B *about the origin,* g: $B \to B$ *is a mapping such that* $I - g$ *lies in* F. *Then* g *has a fixed point in* B.

PROOF OF PROPOSITION 2. Since g maps B into B, it follows that for x on ∂B, $\|g(x)\| \leqslant \|x\|$. Hence for $f = I - g$,

$$(f(x), x) = \|x\|^2 - (g(x), x) \geqslant \|x\|^2 - \|g(x)\| \cdot \|x\| \geqslant 0.$$

Applying Proposition 1, f must have a zero in B, i.e. g must have a fixed point in B.   Q.E.D.

We draw two conclusions from Proposition 2. First, because we have a degree theory for all continuous mappings on the finite-dimensional space $H = R^n$, all continuous self-mappings of balls in that space have a fixed point, i.e. the Brouwer fixed point theorem. The second conclusion is that for an infinite-dimensional Hilbert space H, we cannot have a degree theory in the sense of Proposition 1 over all continuous mappings since the only case in which all continuous self-maps of the unit ball of a Banach space have a fixed point is that in which the space is finite dimensional.

A variant of Proposition 1 gives us the Poincaré-Miranda theorem stated in the Introduction.

PROPOSITION 3. *Let $X = Y$ be a Banach space, and suppose that for a class of mappings F and a class of homotopies H, we have a degree function defined on F, invariant under H, and normalized by the identity mapping, where H includes all affine homotopies in F. Let G be a set in O, with $0 \in G$, and suppose that f: $\mathrm{cl}(G) \to X$ lies in F. Suppose that for each x in ∂G, there exists a linear functional w_x in X^* such that $\langle w, f(x) \rangle \geq 0$, $\langle w, x \rangle > 0$.*

Then f has a zero in $\mathrm{cl}(G)$.

PROOF OF PROPOSITION 3. Again, we may assume that $0 \notin f(\partial G)$. We construct the homotopy $f_t = (1 - t)I + tf$: $\mathrm{cl}(G) \to X$. By our hypotheses, this homotopy lies in the class H. For $t = 0$, $d(f_0, G, 0) = +1$ and hence to show that $d(f, G, 0) = d(f_1, G, 0) = +1$, it suffices to show that $d(f_t, G, 0)$ is constant in t on $[0, 1]$. This will follow from the property of invariance under homotopy if we can verify that for each t in $(0, 1)$ and all x in ∂G, $f_t(x) \neq 0$.

Suppose, however, that for a given x in ∂G and some t in $(0, 1)$, we have $f_t(x) = 0$. Then

$$0 = \langle w_x, f_t(x) \rangle = (1 - t)\langle w_x, x \rangle + t\langle w_x, f(x) \rangle \geq (1 - t)\langle w_x, x \rangle > 0,$$

which is a contradiction.   Q.E.D.

COROLLARY TO PROPOSITION 3. *Let G be the cube $\{x\colon -a_j < x_j < +a_j, j = 1,\ldots,n\}$ in R^n, and let f be a mapping of $\mathrm{cl}(G)$ into R^n such that on the face $C_j^+ = \{x_j = a_j\}$, $f_j(x) \geq 0$, and on the face $C_j^- = \{x_j = -a_j\}$, $f_j(x) \leq 0$. Then f has a zero in $\mathrm{cl}(G)$.*

PROOF OF THE COROLLARY. The boundary of G consists of the union of the faces C_j^+ and C_j^- for various j. We consider R^n as a Hilbert space, and on the faces C_j^+ and C_j^-, we choose w_x to be the element which is 0 except in the jth place, and x_j in the jth place. Then $\langle w_x, f(x) \rangle = x_j f_j(x) \geq 0$, while $\langle w_x, x \rangle = x_j^2 = a_j^2 > 0$. (On the intersections we use whichever of the vectors w_x we please). Hence the hypotheses of Proposition 3 are satisfied, and f has a zero.   Q.E.D.

The classical example of an extension of the degree theory for the finite-dimensional case is the Leray-Schauder degree theory which is defined for the case in which $X = Y$ is an arbitrary Banach space, 0 is the class of bounded open subsets of X, F is the class of continuous maps f: $\mathrm{cl}(G) \to X$ with $(I - f)(\mathrm{cl}(G))$ relatively compact in X. Here the class of homotopies $\{f_t: 0 \leqslant t \leqslant 1\}$ is restricted by the assumption that there exists a fixed compact set K in X such that $(I - f_t)(\mathrm{cl}(G)) \subset K$ for all t in $[0, 1]$.

A possible natural extension of the finite-dimensional and the Leray-Schauder degree theories might be a degree theory for proper Fredholm mappings of index zero. Such a theory in the sense of Definition 1 cannot exist.

PROPOSITION 4. *There can be no degree theory in the sense of Definition 1 for proper Fredholm mappings of index zero of class C^∞ in any infinite-dimensional Hilbert space.*

PROOF OF PROPOSITION 4. The problem, of course, is that all the nonsingular linear mappings of an infinite-dimensional Hilbert space form a single path component (in fact are contractible to a point by a theorem of Kuiper). Suppose that we had a degree theory of the type described. Then given any finite-dimensional subspace R^n of the Hilbert space, we could *suspend* all the differentiable maps on R^n to get Fredholm maps of index zero. The resulting degree theory for maps on R^n obtained by taking the degree of their suspension must be the same as the original degree in R^n, since the uniqueness argument actually works for differentiable mappings of any class. If L_0 is a nonsingular linear mapping of R^n, we would then have

$$d(L, B_n, 0) = d(S(L_0), B, 0) = d(I, B, 0) = +1,$$

since there is a path of nonsingular linear maps leading from any given one to I. This contradicts the fact that for suitably chosen L_0, $d(L_0, B_n, 0) = -1$.   Q.E.D.

The proof indicates clearly why the appropriate degree theory over the class of all proper Fredholm operators of index zero must have values in Z_2. (For details of such theories, we refer to the article [18].)

3. The Leray-Schauder theory. The most celebrated form of degree theory for application to nonlinear problems in partial differential equations has been that introduced by Leray and Schauder [26] in 1934 for mappings f in infinite-dimensional Banach spaces such that $I - f$ is compact. We shall give a reasonably complete exposition of this theory based upon the results already derived for the finite-dimensional degree.

The basic general result is the following:

THEOREM 3. *Let X be a Banach space, and consider the family F of continuous mappings f: $\mathrm{cl}(G) \to X$, with G a bounded open subset of X, and with $(I - f)(\mathrm{cl}(G))$ relatively compact in X. Let H be the family of continuous homotopies of maps $\{f_t: 0 \leqslant t \leqslant 1\}$ in F with a common domain $\mathrm{cl}(G)$ such that there exists a compact subset K of X with $(I - f_t)(\mathrm{cl}(G)) \subset K$ for all t in $[0, 1]$. Then there exists one and only one degree function $d(f, G, y_0)$ in the sense of Definition 1 with the identity mapping I as normalizing mapping.*

The uniqueness holds even if the homotopies are restricted to affine homotopies and the mappings to differentiable maps of the form $I - f$ with f having a finite-dimensional range.

The proof of Theorem 3 uses several auxiliary devices which we introduce in the following proposition.

PROPOSITION 5. *Let X be a finite-dimensional Banach space, X_0 a subspace of X, such that X is the direct sum of X_0 and another subspace X_1. Let B_1 be the unit ball about zero in X_1, and for each bounded open subset G_0 of X_0, let $G = G_0 \times B_1$. Let f be a continuous map of $\mathrm{cl}(G_0)$ into X_0. Then we define the suspension $S(f)$ mapping $\mathrm{cl}(G)$ into X by setting*

$$S(f)(x_0 + x_1) = f(x_0) + x_1, \qquad (x_0 \in \mathrm{cl}(G_0),\ x_1 \in B_1).$$

Let y_0 be a point of X_0 such that $y_0 \notin f(\partial G_0)$. Then

(1) $y_0 \notin S(f)(\partial G)$.

(2) $d_0(f, G_0, y_0) = d(S(f), G, y_0)$, *where the first degree is calculated in X_0 and the second in X.*

PROOF OF PROPOSITION 5. We set $d_1(f_0, G_0, y_0)$ to be equal to $d(S(f), G, y_0)$. To do this, we first verify (1). Suppose that $x = x_0 + x_1$ has the property that $S(f)(x) = y_0$. This means that

$$f(x_0) + x_1 = y_0$$

or that $x_1 = y_0 - f(x_0)$ lies in X_0. Since x_1 also lies in X_1 and the two subspaces are complementary, $x_1 = 0$ and $f(x_0) = y_0$. Hence x lies in G_0 which does not intersect G. Thus $d(S(f), G, y_0)$ is well defined.

We now examine this new function f_1 on the class of mappings f and note that it satisfies the three conditions for an index function. By Theorem 2, it therefore coincides with the degree function $d(f, G_0, y_0)$.   Q.E.D.

PROPOSITION 6. *Let X be a finite-dimensional Banach space, X_0 a subspace of X, G a bounded open set in X such that $G_0 = G \cap X_0$ is nonempty. Let f be a continuous mapping of $\mathrm{cl}(G)$ into X, such that $f = I - g$ and $g(\mathrm{cl}(G)) \subset X_0$. Let y_0 be a point of X_0 such that $y_0 \notin f(\partial G)$, and let $f_0: \mathrm{cl}(G_0) \to X_0$ be the restriction of f to $\mathrm{cl}(G_0)$.*

Then $d(f, G, y_0) = d(f_0, G_0, y_0)$.

PROOF OF PROPOSITION 6. Since ∂G_0 is a subset of ∂G, both degrees are well defined since $y_0 \notin f(\partial G)$.

Let X_1 be a subspace of X complementary to X_0, B_1 the unit ball about 0 in X_1, $G_2 = G_0 \times B_1$. Let $S(f_0)$ be the suspension of f_0 in the sense defined in Proposition 5, with $S(f_0)$ mapping $\mathrm{cl}(G_2)$ into X by the prescription

$$S(f)(x_0 + x_1) = f_0(x_0) + x_1.$$

We shall now consider explicitly for what points x in their respective domains $f(x) = y_0$ and $S(f_0) = y_0$. In the first case, we have

$$x - g(x) = y_0$$

so that if $x = x_0 + x_1$, $x_1 = y_0 - g(x) - x_0 \in X_0$, so that $x_1 = 0$, x lies in G_0, and $f_0(x) = y_0$. Similarly, if $S(f)(x) = y_0$, then

$$x_1 = y_0 - f_0(x_0) \in X_0,$$

$x_1 = 0$, and x lies in G_0 and is a solution of $f_0(x) = 0$. Since $G_0 \subset G \cap G_2$, it follows that

$$d(S(f_0), G_2, y_0) = d(S(f_0), G \cap G_2, y_0); \qquad d(f, G, y_0) = d(f, G \cap G_2, y_0).$$

We now set up the affine homotopy

$$h_t = (1 - t)f + tS(f_0)$$

on $\mathrm{cl}(G \cap G_2)$. If $h_t(x) = 0$ for any t, then we have

$$(1 - t)(x - g(x)) + t(x_1 + f_0(x_0)) = y_0,$$

which may be rewritten as

$$x_1 = y_0 + (1 - t)\{g(x) - x_0\} - tf_0(x_0) \in X_0.$$

Hence $x_1 = 0$, $x = x_0$, $y_0 = f_0(x)$. Thus $x_0 \in G_0$, $x \notin \partial(G_1 \cap G_2)$. Using the invariance of the degree function under homotopy, we see that

$$d(f, G \cap G_2, y_0) = d(S(f_0), G \cap G_2, y_0).$$

Finally, we have

$$
\begin{aligned}
d(f_0, G_0, y_0) &= d(S(f_0), G_2, y_0) = d(S(f_0), G \cap G_2, y_0) \\
&= d(f, G \cap G_2, y_0) = d(f, G, y_0). \quad \text{Q.E.D.}
\end{aligned}
$$

LEMMA. *Let K be a compact subset of a Banach space. Then given any $\varepsilon > 0$, there exists a finite-dimensional subset K_ε of X and a continuous mapping p of K into K_ε such that for every x of K, $\|x - p(x)\| < \varepsilon$.*

PROOF OF THE LEMMA. We may form the covering of K by open balls of radius ε. We can find a finite subcovering by compactness and a partition of unity corresponding to this subcovering, i.e., a finite set of points $\{x_1, \ldots, x_r\}$ in K and a family of continuous functions $\alpha_j: K \to [0, 1]$ such that each α_j has its support in the ball of radius ε about x_j and on K, $\Sigma_{j=1}^r \alpha_j = 1$. We set $p(x) = \Sigma_{j=1}^r \alpha_j(x)x_j$. Then $p(x)$ lies in the simplex spanned by the x_j, and for each x, $p(x)$ is a convex linear combination of those x_j lying in the open ball of radius ε about x. Hence $p(x)$ also lies in that ball. Q.E.D.

PROOF OF THEOREM 3. Each f is a proper map of $\mathrm{cl}(G)$ into X. Indeed, suppose K is a compact subset of X, and consider x in $f^{-1}(K)$. Then $x \in g(\mathrm{cl}(G)) + K \subset K_1$, where K_1 is a compact subset of X. Hence $f^{-1}(K)$ is closed in X and a subset of a compact subset of X, and therefore is compact itself. Thus f is proper, and maps closed subsets of $\mathrm{cl}(G)$ into closed subsets of X_0. In particular, $f(\partial G)$ is closed in X and does not contain y_0. Therefore there exists $r > 0$ such that $f(\partial G)$ does not intersect the ball of radius r about y_0.

Using the preceding Lemma, for any $\varepsilon < r$, if we take the compact set K which is the closure of $g(\mathrm{cl}(G))$, we may find a continuous mapping p of K into a finite-dimensional subset of X such that $\| p(u) - u \| < \varepsilon$ for every u in K. We now define a new mapping

$$\tilde{f} = I - \tilde{g} : \mathrm{cl}(G) \to X$$

where $\tilde{g} = p \circ g$. For every x in $\mathrm{cl}(G)$, $\| f(x) - \tilde{f}(x) \| < \varepsilon$. In particular, $y_0 \notin \tilde{f}(\partial G)$.

Take any finite-dimensional subspace X_0 of X which contains g_0 and $\tilde{g}(\mathrm{cl}(G))$, and let $G_0 = G \cap X_0$. Then by Proposition 6, $d(f_{G_0}, G_0, y_0)$ is independent of the choice of the subspace. We propose to show that it is also independent of the choice of the approximation.

Take two such approximations $f_1 = I - g_1$ and $f_2 = I - g_2$ with $\| g_1(x) - g(x) \| < r$, $\| g_2(x) - g(x) \| < r$ for every x in $\mathrm{cl}(G)$ and with the values of g_1 and g_2 lying in finite-dimensional subspaces of X. If we set $g_t = (1 - t)g_1 + t g_2$, then each g_t is another such approximation with the values of all g_t lying in a fixed finite-dimensional subspace X_0. If we let X_0 also contain y_0, then $I - g_t$ restricted to $\mathrm{cl}(G_0)$ with $G_0 = G \cap X_0$, is a permissible homotopy, and $d(I - g_t, G_0, y_0)$ is independent of t in $[0, 1]$. It follows that $d(f_1, G_0, y_0) = d(f_2, G, y_0)$.

We define the degree function $d(f, G, y_0)$ as the common value of $d(f, G_0, y_0)$ for any of the approximating mappings and the corresponding finite-dimensional subspaces X_0 containing y_0 and $g(\mathrm{cl}(G))$. The normalization condition and the additivity on domain follow easily by choosing the approximating mappings sufficiently close to f. For a permissible homotopy $f_t = I - g_t$, with $g_t(\mathrm{cl}(G)) \subset K$ for a fixed compact set K and all t in $[0, 1]$, we choose $r > 0$ such that the ball of radius r about y_t does not meet $f_t(\partial G)$ for any t in $[0, 1]$ and then apply the approximation operator p for K with $\varepsilon < r$ to all the g_t. Then $\{\tilde{g}_t : 0 \leqslant t \leqslant 1\}$ is a continuous homotopy of mappings with values in a finite-dimensional subspace X_0 of X. If $G_0 = G \cap X_0$, and we have absorbed y_t into the g_t, we see that $d(\tilde{f}_t, G_0, 0)$ is independent of t in $[0, 1]$. Thus all the conditions for the degree function have been satisfied.

Suppose now that we have two degree functions d and d_1. By the finite-dimensional argument, d and d_1 must give the same values for suspensions of mappings of finite-dimensional subspace of X. Hence by the proof of Proposition 6, the two degree functions give the same result for mappings $f = I - g$, with $g(\mathrm{cl}(G))$ lying in a finite-dimensional subspace of X. Since every f of the form $I - g$ can be uniformly approximated by mappings of this last type and is affinely homotopic to such nearby mappings with homotopy paths on the boundary that avoid y_0, it follows that the two degree functions must coincide. Thus the proof of Theorem 3 is complete.   Q.E.D.

4. The direct method of the calculus of variations and mappings of monotone type. Let X be a Banach space, G an open subset of X. If φ is a real-valued differentiable function on G, then its derivative at any point x_0 of G is given by

$$\lim_{\varepsilon \to 0} \frac{\varphi(x_0 + \varepsilon v) - \varphi(x_0)}{\varepsilon} = \langle \varphi'(x_0), v \rangle$$

which depends linearly and continuously upon the direction v in X and hence is an element of the conjugate space X^* of X.

We use here some of the standard notation of this kind of functional analysis. The duality between the real Banach spaces X^* and X is given by $\langle w, x \rangle$ for w in X^* and x in X. We shall use the symbol $\to$ for strong convergence, $\rightharpoonup$ for weak convergence.

The derivative φ' thus gives us a mapping from G, a subset of X, to X^*, the conjugate space of X. Thus mappings from subsets of X to X^* are a rather natural framework for any approach from functional analysis which attempts to encompass the direct method of the calculus of variations. The latter term, which goes back to the Dirichlet principle of Riemann and its rigorization by Hilbert in 1900, refers to the process of trying to get solutions of the equation $\varphi'(x) = 0$ by finding local maxima or minima of the functional $\varphi(x)$. The basic assumption which is necessary for the case when the problem is imposed in an infinite-dimensional context goes back in its origins to Hilbert and involves using a convexity assumption (or some form of modified convexity) upon the functional φ. The argument appears almost truistic in its modern form because it has been essentially absorbed into the structure of our basic functional analysis.

DEFINITION 2. *Let X be a Banach space, G a subset of X, f a mapping of G into X^*. Then*

(a) *f is said to be monotone if for all u and v of G,*

$$\langle f(u) - f(v), u - v \rangle \geqslant 0.$$

(b) *f is said to be of class $(S)_+$ if for any sequence $\{x_j\}$ in G which converges weakly to x in X and for which $\varlimsup \langle f(x_j), x_j - x \rangle \leqslant 0$, we have $x_j \to x$.*

(c) *f is said to be pseudo-monotone, if for any sequence $\{x_j\}$ in G for which $x_j \rightharpoonup x$ for some x in X while $\varlimsup \langle f(x_j), x_j - x \rangle \leqslant 0$, we have $\lim \langle f(x_j), x_j - x \rangle = 0$, and if $x \in G$, then $f(x_j) \rightharpoonup f(x)$.*

PROPOSITION 7. (a) *If G is a convex open subset of X and φ is a C^1 real-valued function on G, then φ is convex on G if and only if $f = \varphi'$ is a monotone mapping of G into X.*

(b) *Suppose that φ is a C^1 functional on G, $f = \varphi'$. If f is bounded and pseudo-monotone, then φ is weakly sequentially lower semicontinuous on G, i.e. if $\{x_j\}$ is a sequence in G which converges weakly to x in G, then $\varphi(x) \leqslant \varliminf \varphi(x_j)$.*

PROOF OF PROPOSITION 7. *Proof of* (a). Suppose first that φ' is monotone. For u and v two points of G and s in $[0, 1]$, set

$$q(s) = \varphi(su + (1 - s)v) - s\varphi(u) - (1 - s)\varphi(v).$$

To show that φ is convex, we must show that for every u, v and s, $q(s) \leqslant 0$. We see that $q(0) = 0 = q(1)$. On the other hand, q is continuously differentiable and

$$q'(s) = \langle \varphi'(v + s(u - v)), u - v \rangle + \varphi(v) - \varphi(u).$$

Hence for $0 \le s < t \le 1$, we have

$$q'(t) - q'(s) = \langle \varphi'(v_t) - \varphi(v_s), u - v \rangle$$

where $v_t = v + t(u - v)$, $v_s = v + s(u - v)$. Since $v_t - v_s = (t - s)(u - v)$, it follows from monotonicity that

$$q'(t) - q'(s) = (t - s)^{-1} \langle \varphi'(v_t) - \varphi'(v_s), v_t - v_s \rangle \ge 0.$$

Suppose that max $q(s) > 0$. Then q has an interior maximum at s_0 with $q'(s_0) = 0$. Since $q(s_0) > 0$, and $q'(t) \ge 0$ for all $t > s_0$, it is impossible that $q(1)$ could be zero. Hence $q(s) \le 0$ for all s in $[0, 1]$ and q is a convex function.

Suppose conversely that φ is convex. Then for u and v in G and $0 < s < 1$, we would have

$$\frac{\varphi(v + s(u - v)) - \varphi(v)}{s} \le \varphi(u) - \varphi(v).$$

Letting $s \to 0+$, we obtain

$$\langle \varphi'(v), u - v \rangle \le \varphi(u) - \varphi(v).$$

Interchanging u and v, we also have

$$\langle \varphi'(u), v - u \rangle \le \varphi(v) - \varphi(u).$$

Adding we find that

$$\langle \varphi'(u) - \varphi(v), u - v \rangle \ge 0,$$

i.e. φ' is monotone.

PROOF OF (b). For each t in $[0, 1]$, set $v_{j,t} = x + t(x_j - x)$. Then $v_{j,t} \to x$ as $j \to +\infty$ for each fixed t, while $v_{j,t} - x = t(x_j - x)$. For $t > 0$, we have therefore by the pseudo-monotonicity of φ',

$$\underline{\lim} \, \langle \varphi'(v_{j,t}), x_j - x \rangle = \underline{\lim} \, \langle \varphi'(v_{j,t}), t^{-1}(v_{j,t} - x) \rangle \ge 0.$$

For each j, we have

$$\varphi(x_j) - \varphi(x) = \int_0^1 \langle \varphi'(v_{j,t}), x_j - x \rangle \, dt \ge \int_0^1 \{ \langle \varphi'(v_{j,t}), x_j - x \rangle \}^- \, dt,$$

where the integrand in the last integral is uniformly bounded and converges to 0 as $j \to +\infty$ for each $t > 0$. Hence $\lim \varphi(x_j) - \varphi(x) \ge 0$.   Q.E.D.

The primary interest of the definitions for the class $(S)_+$ and for pseudo-monotonicity arises from the fact that one can verify these properties under suitable concrete hypotheses for the maps of a Sobolev space $W_0^{m,p}(\Omega)$ into its conjugate space $W^{-m,p}(\Omega)$ obtained from an elliptic operator in generalized divergence form

$$A(u) = \sum_{|\alpha| \le m} (-1)^{|\alpha|} D^\alpha A_\alpha(x, u, \ldots, D^m u).$$

Here Ω is an open subset of a Euclidean space R^n, $n \ge 1$, and p is an exponent in the reflexive range $1 < p < +\infty$. The Sobolev space $W^{m,p}(\Omega)$ is the space of u in $L^p(\Omega)$ (with respect to Lebesgue n-measure) with all the derivatives $D^\alpha u$ also in

$L^p(\Omega)$ for all distribution derivatives of order $\leqslant m$. The norm of this space is obtained by injecting it into the product of L^p-spaces, one for each derivative, by the jet-mapping

$$u \to \{ D^\alpha u \colon |\alpha| \leqslant m \}.$$

The closed subspace $W_0^{m,p}(\Omega)$ is obtained by taking the closure in $W^{m,p}(\Omega)$ of the testing functions with compact support in Ω.

The operator A given by the differential expression written above makes sense as a mapping of $W_0^{m,p}(\Omega)$ into its conjugate space $W^{-m,p'}(\Omega)$, with $p' = p/(p-1)$ the conjugate exponent to p, provided that if we replace u and its derivatives by the algebraic variable

$$\xi = \{ \xi_\alpha \colon |\alpha| \leqslant m \},$$

then $A_\alpha(x, \xi)$ is measurable in x for fixed ξ and continuous in ξ for fixed x in Ω, and satisfies an inequality of the form

$$|A_\alpha(x, \xi)| \leqslant c \big(|\xi|^{p-1} + k_0(x) \big)$$

with k_0 a function in $L^{p'}(\Omega)$. Under these conditions, if we substitute for u and its derivatives in the formal expression, functions in $L^p(\Omega)$, $A_\alpha(x, u, \ldots, D^m u)$ yields an element of $L^{p'}(\Omega)$. Its distribution derivative $D^\alpha A_\alpha(x, \xi(u))$ then becomes an element of the space of distributions $W^{-m,p'}(\Omega)$.

Further hypotheses of some sort of ellipticity and of weak semiboundedness must be imposed before this operator A from $W = W_0^{m,p}(\Omega)$ to X is either pseudo-monotone or of class $(S)_+$. Such conditions for example are

$$\sum_{|\alpha|=m} A_\alpha(x, \eta, \zeta) - A_\alpha(x, \eta, \zeta^\#)\big(\zeta_\alpha - \zeta_\alpha^\#\big) > 0$$

for $\zeta \neq \zeta^\#$, where ξ is broken up into the mth order piece ζ and the lower order piece η.

$$\sum_{|\alpha| \leqslant m} A_\alpha(x, \xi)\xi_\alpha \geqslant c_0 |\xi|^p - k_1(x) \qquad \big(k_1 \in L^1(\Omega) \big).$$

Under these hypotheses, A is of class $(S)_+$. Under very much weaker conditions replacing the semiboundedness in the second hypothesis, A remains pseudo-monotone.

If $X = X^*$ is a Hilbert space H, the mappings of type $(S)_+$ contain as a special case the Leray-Schauder maps $I - g$, with g compact. This follows from the fact that each strongly monotone mapping f is of type $(S)_+$ and the fact that the class $(S)_+$ is always invariant under compact perturbations. We say that f is strongly monotone if there exists a continuous positive increasing function $p(r)$ for $r > 0$ such that

$$\langle f(u) - f(v), u - v \rangle \geqslant p(\|u - v\|).$$

For the detailed development of the discussion of later sections, we need to develop the properties of our *canonical* monotone mapping, the duality mapping J, which under suitable hypotheses is also a map of class $(S)_+$. In our further discussion, we shall always assume that the Banach space X with which we are dealing is reflexive. This is certainly true for the Sobolev spaces $X = W_0^{m,p}(\Omega)$ which in fact are uniformly convex with duals which are also uniformly convex. It is not always true that every reflexive Banach space has an equivalent norm which is uniformly convex, and a fortiori that X is also uniformly convex. By results due to Lindenstrauss, Asplund, and Trojanski however, it is true that X can be renormed so that X and X^* are both locally uniformly convex. We shall use only one consequence of this renorming: In the resulting norms on both X and X^*, if a sequence $\{x_j\}$ converges weakly to x in X and $\|x_j\| \to \|x\|$, then x_j converges strongly to x, with a similar property for sequences $\{w_j\}$ in X^*.

PROPOSITION 8. *Let X be a reflexive Banach space which is normed so that both X and X^* are locally uniformly convex (and in particular are strictly convex). Then there exists a unique bicontinuous mapping J of X onto X^* which is given by the conditions that for each x of X, $\|J(x)\| = \|x\|$ and $\langle J(x), x \rangle = \|x\|^2$. This mapping J, called the duality mapping corresponding to the given norm on X, is both monotone and a mapping of class $(S)_+$.*

PROOF OF PROPOSITION 8. For each x in X, it follows from the Hahn-Banach theorem that there exists w in X such that $\|w\| = \|x\|$ and $\langle w, x \rangle = \|w\| \cdot \|x\|$. Such elements w lie on the sphere $\|w\| = \|x\|$ and are also characterized by the inequalities

$$\langle w, x \rangle = \|x\|^2, \qquad \|w\| \leqslant \|x\|.$$

Hence they form a convex subset of a sphere in X^*. If X^* is strictly convex, such an element w is unique, and this unique element is $J(x)$ for the given x.

For each x and u in X, we may compute

$$\begin{aligned}
\langle J(x) - J(u), x - u \rangle &= \|x\|^2 + \|u\|^2 - \langle J(x), u \rangle - \langle J(u), x \rangle \\
&= (\|x\| - \|u\|)^2 + \{\|u\| \cdot \|J(x)\| - \langle J(x), u \rangle\} \\
&\quad + \{\|J(u)\| \cdot \|x\| - \langle J(u), x \rangle\}.
\end{aligned}$$

All the parentheses are nonnegative. Hence J is monotone. Moreover, if we replace u by x_j, we obtain

$$\begin{aligned}
\langle J(x_j) - J(x), x_j - x \rangle &\geqslant (\|x_j\| - \|x\|)^2 \\
&\geqslant \left| \langle J(x_j), x \rangle - \|J(x_j)\| \cdot \|x\| \right| \\
&\geqslant \left| \langle J(x), x_j \rangle - \|J(x)\| \cdot \|x_j\| \right|.
\end{aligned}$$

Suppose $x_j \rightharpoonup x$. Then $\langle J(x), x_j - x \rangle \to 0$. Suppose moreover that

$$\overline{\lim} \, \langle J(x_j), x_j - x \rangle \leqslant 0.$$

Then $\|x_j\| \to x$, and it follows that $x_j \to x$. Hence J lies in the class $(S)_+$.

Suppose that $x_j \to x$. Then $(J(x_j) - J(x), x_j - x) \to 0$. Hence $\langle J(x_j), x \rangle \to \|x\|^2$. Choose any weakly convergent subsequence $J(x_j) \rightharpoonup w$. Then $\langle w, x \rangle = \|x\|^2$

and $\|w\| \leqslant x$. Hence $w = J(x)$. Moreover $\|w\| = \lim\|J(x_j)\|$ and therefore $J(x_j) \to w = J(x)$. Thus J is continuous from X to X^*. If on the other hand, $J(x_j) \to J(x)$, it also follows that $\langle J(x_j) - J(x), x_j - x \rangle \to 0$, so that $\|x_j\| \to \|x\|$, while

$$\langle J(x), x_j \rangle \to \|x\|^2.$$

It follows by the same argument that $x_j \to x$. Hence J^{-1} is continuous from X^* to X. Q.E.D.

5. The degree of mapping for mappings of class $(S)_+$. In the present section, we devote our efforts to the proof of the following theorem.

THEOREM 4. *Let X be a reflexive Banach space, and consider the family F of maps f: $\mathrm{cl}(G) \to X^*$, where G is a bounded open subset of X and f is a mapping of class $(S)_+$ with f demicontinuous (i.e. continuous from the strong topology of X to the weak topology of X^*). Let H be the class of affine homotopies in F, and let J be the duality mapping from X to X^* corresponding to an equivalent norm on X in which both X and X^* are locally uniformly convex.*

Then there exists one and only one degree function on F which is invariant under H and normalized by the map J.

The homotopies described in Theorem 4 are too weak for some important applications of this result. We therefore introduce the following broader class.

DEFINITION 3. *Let G be a bounded open subset of X, $\{f_t: 0 \leqslant t \leqslant 1\}$ a one-parameter family of maps of $\mathrm{cl}(G)$ into X^*. Then $\{f_t\}$ is said to be a homotopy of class $(S)_+$ if it satisfies the following condition: For any sequence $\{x_j\}$ in $\mathrm{cl}(G)$ converging weakly to some x in X and for any sequence $\{t_j\}$ in $[0, 1]$ converging to t for which*

$$\overline{\lim} \left\langle f_{t_j}(x_j), x_j - x \right\rangle \leqslant 0,$$

we have x_j converging strongly to x and $f_{t_j}(x_j) \to f_t(x)$.

THEOREM 5. *The degree function described in Theorem 4 is invariant under homotopies of class $(S)_+$.*

The construction of the degree function for maps of class $(S)_+$ and the proofs of Theorems 4 and 5 rest upon the general technique of Galerkin approximation.

DEFINITION 4. *Let X be a Banach space, X_0 a closed subspace of X, G an open subset of X such that $G \cap X_0 = G_0$ is nonempty. Let f: $\mathrm{cl}(G) \to X$ be a given mapping. Then the Galerkin approximant for the map f is the mapping f_0: $\mathrm{cl}(G_0) \to X_0$, where φ is the injection mapping of X_0 into X, φ^* the corresponding projection of X^* onto X_0^*, and f_0 is given by*

$$f_0(x) = \varphi^*(f(x)) \qquad (x \in \mathrm{cl}(G_0)).$$

In a sort of inverse relationship to the concept of the Galerkin approximant is the following notion of a generalized *suspension* operator.

DEFINITION 5. *Let X be a reflexive Banach space, J a duality mapping of X into X^* of class $(S)_+$. Suppose that X_0 is a closed subspace of X having a closed complement X_1 in X, P the projection of X on X_0 which corresponds to this splitting so that $(I - P)$ is the projection on X_1. We consider P as an element of the space of bounded linear maps of X, with P_1 the same projection considered as a bounded linear map of X on X_0. Then P^* is a bounded linear map of X^* into itself, P_1^* the injection of X_0^* into X^* whose image is the same as the range of the projection P^*. Let B_1 be the unit ball about 0 in X_1.*

Then

(1) For each bounded open subset G_0 of X_0, we define its suspension $S(G_0)$ as the bounded open subset of X given by

$$S(G_0) = P^{-1}(G_0) \cap (I - P)^{-1}(B_1).$$

Thus each element x of $S(G_0)$ can be written uniquely in the form

$$x = x_0 + x_1 \qquad (x_0 \in G_0, x_1 \in B_1)$$

where $x_0 = Px$, $x_1 = (I - P)(x)$.

(2) If $g: \mathrm{cl}(G_0) \to X_0^*$ is given, we define its suspension S_g as the map of $\mathrm{cl}(S(G_0)) \to X^*$ given by

$$S_g(x) = P_1^* g P(x) + (I - P)^* J(I - P)(x).$$

PROPOSITION 9. *Under the circumstances described in Definition 5:*

(a) If g is demicontinuous and of class $(S)_+$ from $\mathrm{cl}(G_0)$ to X_0^*, then S_g is demicontinuous and of class $(S)_+$ from $\mathrm{cl}(S(G_0))$ to X^*.

(b) If $\{g_t\}$ is an affine homotopy on $\mathrm{cl}(G_0)$, then $\{S_{g_t}\}$ is an affine homotopy on $\mathrm{cl}(S(G_0))$.

(c) If the point y_0 in X_0^* does not lie in $g(\partial G_0)$, then $P_1^*(y_0)$ does not lie in $S_g(\partial S(G_0))$.

PROOF OF PROPOSITION 9. *Proof of* (a). Let $\{u_j\}$ be a sequence in $\mathrm{cl}(S(G_0))$ such that $u_j \rightharpoonup u$ while $\overline{\lim}\langle S_g(u_j), u_j - u \rangle \leq 0$. Let $v_j = Pu_j$, $x_j = (I - P)u_j$. Then $v_j \rightharpoonup v = Pu$, $x_j \rightharpoonup x = (I - P)u$, with v in X_0, x in X_1. Using the definition of S_g, we see that

$$\langle S_g(u_j), u_j - u \rangle = \langle g(v_j), v_j - v \rangle + \langle J(x_j), x_j - x \rangle.$$

Since $\lim\langle J(x_j), x_j - x \rangle \geq 0$, it follows that $\overline{\lim}\langle g(v_j), v_j - v \rangle \leq 0$. Hence $v_j \to v$ and $g(\overline{v_j}) \rightharpoonup g(v)$, implying that $\langle g(v_j), v_j - v \rangle \to 0$. Hence

$$\overline{\lim} \langle J(x_j), x_j - x \rangle \leq 0.$$

Since J is of class $(S)_+$, $x_j \to x$. Therefore $u_j = x_j + v_j \to x + v = u$.   Q.E.D.

Proof of (b). This follows immediately from the linearity of P and P and the definition of the suspension of a map.

Proof of (c). Suppose for x in $\mathrm{cl}(S(G_0))$ that $S_g(x) = P_1^*(y_0)$. We remark that $P^*(X^*) = P_1^*(X_0^*)$ is a closed complement to $(I - P)^*(X^*)$. Thus if

$$P_1^*(g(Px) - y_0) + (I - P)^* J(I - P)(x) = 0,$$

it follows that

$$P_1^*(g(Px) - y_0) = 0, \qquad (I - P)^*J(I - P)(x) = 0.$$

Hence, from the second equation, we find that

$$0 = \langle (I - P)^*J(I - P)(x), x \rangle = \langle J(I - P)x, (I - P)x \rangle = \|(I - P)x\|^2,$$

so that $x = Px$. From the other equation, we see that since P_1^* is injective, $g(x) - y_0 = g(Px) - y_0 = 0$. By the assumption that $y_0 \notin g(\partial G_0)$, and since $x = Px$ lies in $\mathrm{cl}(G_0)$, we see that x lies in G_0. In particular, $P_1^*(y_0)$ does not lie in $S_g(\partial S(G_0))$ since $G_0 \subset S_g(G_0)$.   Q.E.D.

We can use Proposition 9 in the following way to obtain an important methodological conclusion about degree functions for maps of class $(S)_+$.

PROPOSITION 10. *Under the circumstances of Definition 5 and Proposition 9, suppose that for both of the spaces X and X_0, we have degree functions d and d_0 defined on the demicontinuous mappings of class $(S)_+$ and invariant with respect to affine homotopies in each case, with the degree functions normalized by J and J_0, respectively, where J and J_0 are the appropriate duality maps. Suppose moreover that the degree function d_0 is uniquely characterized on the maps of class $(S)_+$ by these properties. Then for each map g: $\mathrm{cl}(G_0) \to X_0^*$ demicontinuous and of class $(S)_+$ and for each y_0 in X_0^* such that $y_0 \notin g(\partial G_0)$, we have*

$$d_0(g, G_0, y_0) = d\big(S_g, S(G_0), P_1(y_0)\big).$$

PROOF OF PROPOSITION 10. For each g and y_0 as above, we define a new degree function using Proposition 9 by setting

$$d_1(g, G_0, y_0) = d\big(S_g, S(G_0), P_1^*(y_0)\big).$$

This gives a degree function which is invariant under affine homotopies on the demicontinuous maps g of class $(S)_+$. We remark that the normalization condition is always equivalent to

$$d_1(J_0, B_0, 0) = +1$$

for the unit ball B_0 with center 0 in X_0. The map $f = S_{J_0}$ satisfies the condition $\langle f(u), u \rangle > 0$ for $u \neq 0$. If $G = S(B_0)$, $0 \in G$ and the affine homotopy $\{f_t = (1 - t)f + tJ\}$ has no zeroes except at $u = 0$. Hence

$$d\big(S_{J_0}, S(B_0), 0\big) = d\big(J, S(B_0), 0\big) = +1.$$

Since d_0 is the unique degree function for maps of class $(S)_+$ on X_0, it follows that $d_0 = d_1$.   Q.E.D.

We apply this result to the case in which X_0 is a finite-dimensional Banach space. Then X_0 is equivalent to a Hilbert space and X_0^* can be identified with X_0. By the remark we just made, normalization by the identity mapping on a finite-dimensional space is equivalent to normalization by any mapping f which is bicontinuous and such that $\langle f(u), u \rangle > 0$ for $u \neq 0$, and in particular by the duality map corresponding to the original norm on X_0. Hence

COROLLARY TO PROPOSITION 10. *Let X be a reflexive Banach space, X_0 a finite-dimensional subspace of X. Suppose that there exists a degree function d on the demicontinuous maps of class $(S)_+$ in the space X. Then for each continuous map g:* $\mathrm{cl}(G_0) \to X_0^*$ *where G_0 is a bounded open subset of X_0 and for each y_0 outside* $g(\partial G_0)$, *we have S_g demicontinuous and of class $(S)_+$ mapping* $\mathrm{cl}(S(G_0))$ *into X^* and*

$$d_0(g, G_0, y_0) = d\big(S_g, S(G_0), P_1^*(y_0)\big).$$

PROOF OF THE COROLLARY. Since X_0 is of finite dimension, it has a closed complement X_1 in X, so that the suspension operator of Definition 5 is well defined for some projection map P of X on X_0. Moreover, for finite-dimensional spaces X_0, each continuous map g is demicontinuous and of class $(S)_+$ (indeed the two classes coincide). Moreover, as we already observed, we know the uniqueness of the degree function by the results of §1. Hence the desired conclusion follows from Proposition 10.   Q.E.D.

We use the Corollary to Proposition 10 as an essential tool in the proof of our basic auxiliary result, Proposition 11.

PROPOSITION 11. *Let X be a reflexive Banach space, X_0 a finite-dimensional subspace of X, and suppose that a degree function d is given on the demicontinuous maps of class $(S)_+$ defined on the closures of bounded open sets G in X with values in X^*. Suppose that this degree is invariant under affine homotopies and is normalized by a duality map J of class $(S)_+$.*

Let f: $\mathrm{cl}(G) \to X^*$ be a map in the given class, with $0 \in G$ and let f_0 be the Galerkin approximant of f with respect to X_0, so that f_0: $\mathrm{cl}(G_0) \to X_0^*$, with $G_0 = G \cap X_0$. *Suppose either that $d(f, G, 0)$ is not defined, or $d_0(f_0, G_0, 0)$ is not defined, or (if both are well defined), then*

$$d(f, G, 0) \neq d_0(f_0, G_0, 0)$$

where we use d_0 to denote the finite-dimensional degree in X_0.

Then: There exists u in ∂G such that

$$\langle f(u), u \rangle \leqslant 0,$$

and for all v in X_0,

$$\langle f(u), v \rangle = 0.$$

PROOF OF PROPOSITION 11. Suppose first that for some u_0 in ∂G, $f(u_0) = 0$. Then u_0 satisfies our conclusion. Suppose next that for some u_0 in ∂G_0, $f_0(u_0) = 0$. Then again u_0 satisfies the conclusion. Thus we may assume without loss of generality that both $d(f, G, 0)$ and $d_0(f_0, G_0, 0)$ are well defined (i.e. $0 \notin f(\partial G)$, $0 \notin f_0(\partial G_0)$), and

$$d(f, G, 0) \neq d_0(f_0, G_0, 0).$$

If we apply the Corollary of Proposition 10, we see that

$$d_0(f_0, G_0, 0) = d\big(S_{f_0}, S(G_0), 0\big).$$

We define an auxiliary mapping $f_1 \colon \mathrm{cl}(G) \to X^*$ by setting

$$f_1(x) = P^*f(x) + (I - P)^*J(I - P)(x).$$

We have already noted in the proof of Proposition 9 that since 0 lies in X_0, all the solutions x of the equation $S_{f_0}(x) = 0$ must lie in G_0 and hence in $G \cap S(G_0)$. Using the additivity property of the degree, we see that

$$d\big(S_{f_0}, S(G_0), 0\big) = d\big(S_{f_0}, S(G_0) \cap G, 0\big).$$

Similarly, suppose for x in $\mathrm{cl}(G)$, we have $f_1(x) = 0$. Then we have

$$(I - P)^*J(I - P)(x) = 0,$$

from which it follows that $(I - P)x = 0$, x lies in $\mathrm{cl}(G_0)$, and $P^*f(x) = 0$. For each v in X_0, it follows that

$$\langle f(x), v \rangle = \langle P^*f(x), v \rangle = 0,$$

so that $f_0(x) = 0$. Thus x must lie in G_0, since by our assumption $0 \notin f_0(\partial G_0)$.

We note next that f_1 is a mapping of class $(S)_+$ from $\mathrm{cl}(G)$ to X. The first summand is compact since P is of finite dimension, so that it suffices to show that $(I - P)^*J(I - P)$ is of class $(S)_+$. Suppose that $u_j \rightharpoonup u$, and that

$$\overline{\lim} \, \langle (I - P)J(I - P)u_j, u_j - u \rangle \leqslant 0.$$

If $x_j = (I - P)u_j$, then $x_j \rightharpoonup x = (I - P)u$ by the fact that $(I - P)$ is a continuous linear map. Moreover

$$\langle (I - P)^*J(I - P)u_j, u_j - u \rangle = \langle Jx_j, x_j - x \rangle.$$

Thus

$$\overline{\lim} \, \langle Jx_j, x_j - x \rangle \leqslant 0.$$

Since J is of class $(S)_+$, it follows that $x_j \to x$. Since P is compact, we know also that $Pu_j \to Pu$. Hence $u_j = x_j + Pu_j \to x + Pu = u$, and we have shown that f_1 is of class $(S)_+$. Since f_1 is obviously also continuous, the degree function d applies to the mapping f_1. Hence

$$d(f_1, G, y_0) = d(f_1, G \cap G_2, y_0).$$

We now consider the affine homotopy $h_t = (1 - t)f_1 + tS_{f_0}$ on $G \cap G_2$. For any t in $(0, 1)$, suppose that for some x, $h_t(x) = 0$. This means that

$$(1 - t)P^*f(x) + tP_1^*f_0(Px) + (I - P)^*J(I - P)x = 0.$$

Again this means that both summands must equal 0, i.e. in particular that $(I - P)^*J(I - P)x = 0$. This means once more that $x = Px$. Thus for every v in X_0,

$$0 = \langle (1 - t)P^*f(x) + tP^*f_0(x), v \rangle = \langle f_0(x), v \rangle.$$

Thus x lies in $\mathrm{cl}(G_0)$ and $f_0(x) = 0$, so that by our initial assumption x lies in G_0. In particular, h_t has no zeroes for any t on the boundary of $G \cap G_2$. Hence

$$d(f_1, G \cap G_2, 0) = d\big(S_{f_0}, G \cap G_2, 0\big).$$

Combining the various equalities we have proved, we obtain

$$d(f_0, G_0, 0) = d(S_{f_0}, G_2, 0) = d(S_{f_0}, G \cap G_2, 0)$$
$$= d(f_1, G \cap G_2, 0) = d(f_1, G, 0).$$

If we combine this with our initial assumption that $d(f_0, G_0, 0) \neq d(f, G, 0)$, we find that

$$d(f, G, 0) \neq d(f_1, G, 0).$$

Now form the affine homotopy $k_t = (1 - t)f + tf_1$. By the inequality of the degrees on G for f and f_1 over 0, we see that there must exist u_0 in ∂G and t in $(0, 1)$ such that $k_t(u_0) = 0$.

For any v in X_0, we see that

$$0 = \langle k_t(u_0), v \rangle = (1 - t)\langle f(u_0), v \rangle + t\langle f(u_0), v \rangle = \langle f(u_0), v \rangle.$$

Hence, we obtain the desired equality

$$\langle f(u_0), v \rangle = 0 \qquad (v \in X_0).$$

Now consider the equality

$$0 = \langle k_t(u_0), (I - P)u_0 \rangle$$
$$= (1 - t)\langle f(u_0), (I - P)u_0 \rangle + t\langle J(I - P)u_0, (I - P)u_0 \rangle$$
$$= (1 - t)\langle f(u_0), u_0 \rangle + t\|(I - P)u_0\|^2.$$

Hence

$$\langle f(u_0), u_0 \rangle = -t(1 - t)^{-1}\|(I - P)u_0\|^2 \leqslant 0. \quad \text{Q.E.D.}$$

PROOF OF THEOREM 4. We carry through the proof both for the existence and for the uniqueness of the degree function for maps of class $(S)_+$ by applying the Galerkin approximation method over the partially ordered set of finite-dimensional subspaces of X.

Let Λ be this partially ordered set of finite-dimensional subspaces X_λ ordered by inclusion. For each λ, let φ_λ be the injection map of X_λ into X and φ_λ^* the corresponding projection map of X^* onto X_λ^*. We consider only finite-dimensional spaces such that $G_\lambda = G \cap X_\lambda$ is nonempty. We assume that $y_0 = 0$, which does not affect the generality of the argument since we can replace each map f by $f - y_0$. We now assert

There exists λ_0 in Λ such that for all $\lambda > \lambda_0$, $0 \notin f_\lambda(\partial G_\lambda)$ and $d(f_\lambda, G_\lambda, 0)$ is independent of λ.

Suppose otherwise. Then for each X_λ, there would exist $X_\mu \supset X_\lambda$ such that

$$d(f_\mu, G_\mu, 0) \neq d(f_\lambda, G_\lambda, 0)$$

(or one of the two degrees is not defined). In all these cases, we may apply Proposition 11 with X replaced by X_μ and X_λ considered as a subspace of X_μ. Then f_λ is obviously the Galerkin approximant of f_μ with respect to X_λ. Since both X_μ and X_λ have degree functions defined, and the uniqueness of degree holds for

X_λ, we apply Proposition 11 and obtain a point u in ∂G_μ such that

$$\langle f(u), u \rangle = \langle f_\mu(u), u \rangle \leq 0,$$

while for all v in X_λ,

$$\langle f(u), v \rangle = \langle f_\mu(u), v \rangle = 0.$$

Let us now define a subset V_λ of ∂G by

$$V = \{ u \mid u \in \partial G, \langle f(u), u \rangle \leq 0; \langle f(u), v \rangle = 0 \text{ for all } v \text{ in } X_\lambda \}.$$

By our preceding paragraph, each V_λ is nonempty. The family $\{V_\lambda\}$ is contained in a fixed bounded set ∂G and it has the finite intersection property. If we denote by w-cl(V_λ) the closure of V in the weak topology of X, then w-cl(V_λ) is weakly compact, and

$$\bigcap_\lambda \text{w-cl}(V_\lambda) \neq \varnothing.$$

Let x_0 be an arbitrary point in this intersection.

Let v be an arbitrary point in X, and choose a subspace X_λ in the ordered set Λ such that X_λ contains both x_0 and v. Since x_0 lies in w-cl(V_λ) and the space X is reflexive, there exists a sequence $\{u_j\}$ in V_λ which converges weakly to x_0 in X. Such u_j must lie in ∂G by the definition of V_λ. Moreover, for each j,

$$\langle f(u_j), u_j \rangle \leq 0, \qquad \langle f(u_j), x_0 \rangle = 0, \qquad \langle f(u_j), v \rangle = 0.$$

In particular,

$$\langle f(u_j), u_j - x_0 \rangle \leq 0,$$

and a fortiori, $\overline{\lim}\langle f(u_j), u_j - x_0 \rangle \leq 0$. Using the assumption that f lies in the class $(S)_+$, we see that u_j converges strongly to x_0 and hence x_0 lies in ∂G. By the demicontinuity of f, $f(u_j) \to f(x_0)$. Hence

$$\langle f(x_0), v \rangle = \lim \langle f(u_j), v \rangle = 0.$$

Since v was an arbitrary element of X, it follows that $f(x_0) = 0$. Thus we reach a contradiction with the assumption that $0 \notin f(\partial G)$.

We define $d(f, G, 0)$ as the common value of $d(f_\lambda, G_\lambda, 0)$ for X_λ sufficiently large.

The normalization properties and the additivity property of the degree function thus defined follow immediately from the definition. The invariance under homotopies which are affine as well as homotopies of class $(S)_+$ asserted in Theorem 5 follows from the following two assertions:

PROPOSITION 12. *Each affine homotopy between two demicontinuous maps f and f_1 of class $(S)_+$ is a homotopy of class $(S)_+$.*

PROPOSITION 13. *If $\{f_t\}$ is a homotopy of class $(S)_+$ on cl(G) and suppose that $0 \notin f_t(\partial G)$ for any t in $[0, 1]$. Then there exists λ_0 in Λ such that for any t in $[0, 1]$ and all λ in Λ, $\lambda > \lambda_0$, we have $d(f_{t,\lambda}, G_\lambda, 0)$ well defined and independent of both t and λ.*

PROOF OF PROPOSITION 12. Suppose that $\{u_j\}$ is a sequence in $\mathrm{cl}(G)$ with $u_j \to u$, $\{t_j\}$ a sequence in $[0, 1]$ with $t_j \to t$, and such that

$$\overline{\lim}\left\{(1 - t_j)\langle f(u_j), u_j - u\rangle + t_j\langle f_1(u_j), u_j - u\rangle\right\} \leq 0.$$

Since f and f_1 are demicontinuous maps of class $(S)_+$, we know that

$$\underline{\lim}\langle f(u_j), u_j - u\rangle \geq 0, \qquad \underline{\lim}\langle f_1(u_j), u_j - u\rangle \geq 0.$$

Suppose without loss of generality that $t > 0$. Since

$$\underline{\lim}(1 - t_j)\langle f(u_j), u_j - u\rangle \geq 0$$

we see that

$$t\,\overline{\lim}\langle f_1(u_j), u_j - u\rangle = \overline{\lim}\left\{t_j\langle f_1(u_j), u_j - u\rangle\right\} \leq 0.$$

Since f is of class $(S)_+$, it follows that u_j converges strongly to u. Therefore

$$(1 - t_j)f(u_j) + t_j f_1(u_j) \to (1 - t)f(u) + tf_1(u),$$

and the proof of Proposition 12 is complete.   Q.E.D.

PROOF OF PROPOSITION 13. Suppose that the assertion of Proposition 13 were false. Then by Proposition 11, for each X_λ in Λ, there would exist $X_\mu \supset X_\lambda$, a point u in ∂G_μ and a parameter value t in $[0, 1]$ such that

$$\langle f_t(u), u\rangle \leq 0, \qquad \langle f_t(u), v\rangle = 0 \quad (v \in X_\lambda).$$

For each λ in Λ we form a subset W_λ of the Cartesian product of $\partial G \times [0, 1]$, where

$$W = \left\{[u, t] \mid u \in \partial G, t \in [0, 1]; \langle f_t(u), u\rangle \leq 0; \langle f_t(u), v\rangle = 0\,(v \in X_\lambda)\right\}.$$

By our preceding consideration, each W_λ would be a nonempty subset of the bounded set $\partial G \times [0, 1]$, and the family $\{W_\lambda\}$ has the finite intersection property. If we take the closure of W_λ in the product of the weak topology on ∂G and the ordinary topology on $[0, 1]$ and denote this closure by w-cl(W_λ), then this set is compact in its appropriate topology. Hence

$$\bigcap_\lambda \text{ w-cl}(W_\lambda) \neq \varnothing.$$

Let $[u_0, t_0]$ be a point of this intersection.

For an arbitrary point v of X, choose X_λ in Λ such that X_λ contains both u_0 and v. Since X is reflexive, we may choose a sequence $[u_j, t_j]$ in the corresponding W_λ such that $u_j \to u_0$, $t_j \to t$. For each index j, we have

$$\langle f_{t_j}(u_j), u_j\rangle \leq 0, \qquad \langle f_{t_j}(u_j), u_0\rangle = 0, \qquad \langle f_{t_j}(u_j), v\rangle = 0.$$

In particular it follows that

$$\overline{\lim}\langle f_{t_j}(u_j), u_j - u_0\rangle \leq 0.$$

If we apply the assumption that $\{f_t\}$ is a homotopy of class $(S)_+$, it follows that $u_j \to u_0$ and $f_{t_j}(u_j) \to f_{t_0}(u_0)$. Hence $u_0 \in \partial G$, and $\langle f_t(u_j), v\rangle \to \langle f_{t_0}(u_0), v\rangle = 0$. Since v is arbitrary, $f_{t_0}(u_0) = 0$ which contradicts our assumption that $0 \notin f_t(\partial G)$ for all t.   Q.E.D.

We have now completed the proof of the existence of the degree in Theorem 5. It remains to prove the uniqueness of the degree in Theorem 4. We shall prove this fact in the following proposition.

PROPOSITION 14. *Let X be a reflexive Banach space and let d_1 be a degree function on the class F of demicontinuous mappings of class $(S)_+$ of the closures of bounded open sets G in X into X^*. Suppose that d_1 is invariant under the class of affine homotopies in F. Then d_1 coincides with the degree d defined above in terms of Galerkin approximants.*

PROOF OF PROPOSITION 14. Suppose the assertion were false. Then there would exist a bounded open set G in X, a demicontinuous map f of $\text{cl}(G)$ into X^* of class $(S)_+$ and a point y_0 in X^*, $y_0 \notin f(\partial G)$ such that

$$d(f, G, y_0) \neq d_1(f, G, y_0).$$

We may assume without loss of generality that $y_0 = 0$.

By the definition of d in terms of Galerkin approximants, it follows from this assumption that for each λ in Λ, there would exist $X_\mu \supset X_\lambda$ such that $d_0(f_\mu, G_\mu, 0) \neq d_1(f, G, 0)$. We now apply the conclusion of Proposition 11 to the degree function d_1 on X and derive the existence of u in ∂G such that

$$\langle f(u), u \rangle \leqslant 0; \qquad \langle f(u), v \rangle = 0 \quad \text{for all } v \text{ in } X_\mu.$$

We define the subset V_λ of ∂G by

$$V = \left\{ u \mid u \in \partial G; \langle f(u), u \rangle \leqslant 0; \langle f(u), v \rangle = 0 \text{ for all } v \text{ in } X_\lambda \right\}.$$

By the last paragraph, each V_λ is nonempty and the family $\{V_\lambda\}$ has the finite intersection property. By the weak compactness of $\text{w-cl}(V_\lambda)$, it follows that $\bigcap_{\lambda \in \Lambda} \text{w-cl}(V_\lambda) \neq \varnothing$. Let u_0 be a point of this intersection.

Let X_λ be an element of Λ which contains both u_0 and a given arbitrary point v of X. By the reflexivity of X, there exists a sequence $\{u_j\}$ from the corresponding V_λ with $u_j \rightharpoonup u_0$. For each index j,

$$\langle f(u_j), u_j \rangle \leqslant 0; \qquad \langle f(u_j), u_0 \rangle = 0; \qquad \langle f(u_j), v \rangle = 0$$

and $u_j \in \partial G$. A fortiori $\overline{\lim} \langle f(u_j), u_j - u_0 \rangle \leqslant 0$, so that $u_j \to u_0$ and $f(u_j) \rightharpoonup f(u_0)$. Thus $u_0 \in \partial G$, and $\langle f(u_0), v \rangle = \lim \langle f(u_j), v \rangle = 0$. Since v was an arbitrary element of X, $f(u_0) = 0$, with $u_0 \in \partial G$. This is a contradiction, and the proof of Proposition 14 is complete.

Thus the proofs of Theorems 4 and 5 are complete.

If we seek to extend the degree function we have constructed to the broader class of pseudo-monotone mappings, we encounter the significant difficulty that the pseudo-monotone maps are not proper in the general case, and for an arbitrary bounded open subset G, it is not true for a pseudo-monotone map f of $\text{cl}(G)$ into X^*, that $f(\partial G)$ is closed nor that $f(\text{cl}(G))$ is closed. The latter will indeed be the case if G is convex, but is false in general otherwise. This necessitates a serious modification of the definition of a degree function if we are to obtain one for such mappings. We will not restate the formal definition of a degree function in its complete form to show the salient features of the new definition, but merely the modifications that are necessary.

DEFINITION 6. *We shall speak of a degree function in the extended (or weak) sense if the following modifications are made in the characterizing properties of the degree in Definition 1.*

(1) *The degree function $d(f, G, y_0)$ is to be defined only if $y_0 \notin \mathrm{cl}(f(\partial G))$.*

(2) *If $d(f, G, y_0) \neq 0$, then $y_0 \in \mathrm{cl}(f(G))$.*

(3) *The normalizing map f_0 is assumed to be proper on bounded sets.*

(4) *In the definition of the additivity property on domains, we must assume that*

$$y_0 \notin \mathrm{cl}\big(f(\mathrm{cl}(G) \backslash (G_1 \cup G_2))\big).$$

(5) *In the property of homotopy invariance, we must assume that if Y is a metric space, then a fixed ball $B_r(y_t)$ does not intersect $f_t(\partial G)$ for all t in $[0, 1]$.*

THEOREM 6. *There exists one and only one degree function in the extended sense on the class F of maps $f: \mathrm{cl}(G) \to X^*$, where X is a reflexive Banach space and the maps f are pseudo-monotone, with the degree invariant under affine homotopies and normalized by the duality mapping J.*

THEOREM 7. *The degree function of Theorem 6 is actually invariant under pseudo-monotone homotopies, where a homotopy $\{f_t\}$ is said to be pseudo-monotone if for a sequence $\{u_j\}$ converging weakly to u in X and a sequence $\{t_j\}$ converging to t in $[0, 1]$ for which $\overline{\lim} \langle f_{t_j}(u_j), u_j - u \rangle \leq 0$, we have $\lim \langle f_{t_j}(u_j), u_j - u \rangle = 0$, and if u lies in $\mathrm{cl}(G)$, then $f_{t_j}(u_j) \to f_t(u)$.*

We shall not give the detailed proofs of Theorems 6 and 7 here except to observe that in this case, the proof by Galerkin approximants does not suffice. We must instead approximate each pseudo-monotone mapping f by approximating mappings f_ε, $\varepsilon > 0$, in the class $(S)_+$ where $f_\varepsilon = f + \varepsilon J$ and J is the duality mapping of class $(S)_+$ which we have considered above. For each f which is pseudo-monotone and demicontinuous, the approximating map f_ε is demicontinuous and of class $(S)_+$. If $y_0 \notin \mathrm{cl}(f(\partial G))$, then for $\varepsilon > 0$ and sufficiently small, $y_0 \notin f_\varepsilon(\partial G)$. On the other hand, on any compact subinterval of $(0, \varepsilon_0)$, $\{f_\varepsilon\}$ is an affine homotopy of maps of class $(S)_+$. Therefore, by the invariance of the degree for maps of class $(S)_+$ under affine homotopies, $d(f_\varepsilon, G, y_0)$ will be independent of ε for $0 < \varepsilon < \varepsilon_0$. This common value, we designate as $d(f, G, y_0)$, and it will be our extended degree function for the class of pseudo-monotone mappings. The invariance under pseudo-monotone homotopies follows from the fact that for each $\varepsilon > 0$, $f_t + \varepsilon J$ yields a homotopy of class $(S)_+$. It does not follow in this case that each affine homotopy is pseudo-monotone (unless f and f_1 are bounded), but it is still the case that for each $\varepsilon > 0$, $\{f_t + \varepsilon J\}$ is an affine homotopy of maps of class $(S)_+$ and hence a homotopy of class $(S)_+$.

Restricted to the class of maps of class $(S)_+$, any degree in the extended sense just becomes a degree in the original sense of Definition 1 since all the maps and homotopies involved are proper. Thus an extended degree is unique by Theorem 4 on the dense subclass of maps of class $(S)_+$. By continuity of the degree under the homotopy $f + \varepsilon J$ as ε ranges over $[0, \varepsilon_0)$, we find the uniqueness of the extended degree on the class of pseudo-monotone mappings.

6. The degree for more general mappings of monotone type. There are a number of classes of mappings of monotone type with domain in a reflexive Banach space X and with values in the conjugate space X^* for which suitable extensions of the arguments which we developed above can be used to develop an existence and uniqueness theory for a suitably defined degree of mapping. To complete our presentation here, we consider the simplest and most generally useful of such classes, that of maps of the form $T + f$ with T maximal monotone and f bounded and of class $(S)_+$.

Let X be a reflexive Banach space. We consider maps T from X with values subsets of X^*. With each such map, we associate its graph $G(T)$ in $X \times X^*$, i.e.

$$G(T) = \big\{ [u, w] \mid u \in X, w \in T(x) \big\}.$$

Then

The mapping T is said to be monotone if for any pair of elements $[u, w]$ and $[x, y]$ in $G(T)$, we have the inequality

$$\langle w - y, u - x \rangle \geq 0.$$

T is said to be maximal monotone if it is monotone and maximal in the sense of graph inclusion among monotone maps (multivalued) from X to X^*. An equivalent version of the last clause is that for any $[u_0, w_0]$ in $X \times X^*$ for which

$$\langle w_0 - y, u_0 - x \rangle \geq 0$$

for all $[x, y]$ in $G(T)$, we have $[u_0, w_0]$ in $G(T)$. If the Banach space X is reflexive, an equivalent statement is that the mapping $T + J$ has all of X^* as its range. (We assume as before that X has been given an equivalent norm in which the duality mapping J is bicontinuous and lies in the class $(S)_+$.)

Maximal monotone mappings occur in a number of useful and important contexts. First of all, all demicontinuous monotone maps of X into X^* are maximal monotone. Second, for any maximal monotone map T of X into the subsets of X^*, its inverse T^{-1} is maximal monotone from X^* to the subsets of X (this makes maximal monotonicity useful in the study of Hammerstein integral equations). Finally if $\varphi \colon X \to R \cup \{+\infty\}$ is a proper lower-semi-continuous convex function, its subgradient $\partial \varphi$ is maximal monotone, where

$$(\partial \varphi)(u) = \big\{ w \mid w \in X^*; \text{ For all } v \text{ in } X, \varphi(v) - \varphi(u) \geq \langle w, v - u \rangle \big\}.$$

This fact makes maximal monotone mappings a useful tool in the study of variational inequalities.

THEOREM 8. *Let X be a reflexive Banach space, T a maximal monotone mapping from X to 2^{X^*} with $0 \in T(0)$. Let G be a bounded open subset of X, and let f be a bounded mapping of $\mathrm{cl}(G)$ into X^* of class $(S)_+$. For each $\varepsilon > 0$, consider the* *generalized Yosida transformation T_ε corresponding to T [3] given by*

$$T_\varepsilon = \left(T^1 + \varepsilon J^{-1}\right)^{-1}.$$

(A) *Suppose that for a given y_0 in X^*, y_0 does not lie in $(T + f)(\partial G)$. Then there exists $\varepsilon_0 > 0$ such that for $0 < \varepsilon < \varepsilon_0$, y_0 does not lie in $(T_\varepsilon + f)(\partial G)$.*

(B) *For each $\varepsilon > 0$, the mapping $T_\varepsilon + f$ of $\mathrm{cl}(G)$ into X^* is of class $(S)_+$. Hence for $0 < \varepsilon < \varepsilon_0$, the degree function $d(T_\varepsilon + f, G, y_0)$ is defined by the results of the preceding section. Moreover, for $0 < \varepsilon < \varepsilon_1$, the values of all these degrees coincide.*

DEFINITION 7. *We set the degree function $d(T + f, G, y_0)$ to be the common value of $d(T_\varepsilon + f, G, y_0)$ for ε sufficiently small.*

PROPOSITION 15. *Let X be a reflexive Banach space, $\{T_t : 0 \leqslant t \leqslant 1\}$ a family of maximal monotone maps from X to 2^{X^*} with $0 \in T_t(0)$ for all t. Let J be a duality mapping from X to X^* which corresponds to a norm on X in which both X and X^* are locally uniformly convex. Consider the following four conditions on the family $\{T_t\}$:*

(1) (*generalized pseudo-monotonicity*) *Suppose that for a sequence $\{t_j\}$ converging to t in $[0, 1]$, we have a sequence $[u_u, w_j]$ in $G(T_{t_j})$ with u_j converging weakly to u in X, w_j converging weakly to w in X^*. Suppose further $\overline{\lim}\langle w_j, u_j \rangle \leqslant \langle w, u \rangle$. Then $w \in T_t(u)$, and $\langle w_j, u_j \rangle \to \langle w, u \rangle$.*

(2) *Consider the mapping φ of $X^* \times [0, 1]$ into X given by*

$$\varphi(w, t) = (T_t + J)^{-1}(w).$$

Then φ is continuous from $X^* \times [0, 1]$ to X (with both X and X^* given their strong topologies).

(3) *For each w in X^*, the mapping φ_w of $[0, 1]$ into X given by*

$$\varphi_w(t) = (T_t + J)^{-1}(w)$$

is continuous from $[0, 1]$ to the strong topology on X.

(4) (*Strong lower-semi-continuity of $G(T_t)$ in t*) *Given $[x, y]$ in $G(T_t)$ and a sequence $\{t_j\}$ converging to t in $[0, 1]$, there exists a sequence $[x_j, w_j]$ with each $[x_j, w_j]$ in $G(T_{t_j})$ such that x_j converges strongly to x in X, y_j converges strongly to y in X^*.*

Then, conditions (1), (2), (3), *and* (4) *are mutually equivalent.*

DEFINITION 8. *A family $\{T_t\}$ of maximal monotone mappings which satisfies the mutually equivalent conditions* (1)–(4) *is called a pseudo-monotone homotopy of maximal monotone maps.*

PROOF OF PROPOSITION 15. We shall prove that $(4) \Rightarrow (1) \Rightarrow (2)$, $(3) \Rightarrow (4)$. Obviously $(2) \Rightarrow (3)$.

PROOF THAT (4) IMPLIES (1). Let $[x, y]$ be an element of $G(T_t)$. For the sequence $\{t_j\}$ given in (1), choose a family $[x_j, y_j]$ as described in condition (4) with $[x_j, y_j]$ in $G(T_{t_j})$ and $[x_j, y_j]$ converging strongly to $[x, y]$. For each j, the monotonicity of T_{t_j} implies that

$$\langle w_j - y_j, u_j - x_j \rangle \geqslant 0.$$

Hence

$$\langle w, u \rangle \geq \overline{\lim} \langle w_j, u_j \rangle \geq \underline{\lim} \langle w_j, u_j \rangle \geq \lim\{\langle w_j, x_j \rangle + \langle y_j, u_j - x_j \rangle\}$$
$$= \langle w, x \rangle + \langle y, u - x \rangle.$$

Thus, $\langle w - y, u - x \rangle \geq 0$. Since this inequality holds for all $[x, y]$ in $G(T_t)$, the maximal monotonicity of T_t implies that $[u, w]$ lies in $G(T_t)$. On the other hand, setting $[x, y] = [u, w]$ in the inequality above, we see that

$$\langle w, u \rangle \geq \overline{\lim} \langle w_j, u_j \rangle \geq \underline{\lim} \langle w_j, u_j \rangle \geq \langle w, u \rangle.$$

Hence $\langle w_j, u_j \rangle \to \langle w, u \rangle$.   Q.E.D.

PROOF THAT (1) IMPLIES (2). Let $u_j = (T_{t_j} + J)^{-1}(w_j)$ with w_j converging strongly to w in X^* and t_j converging to t in $[0, 1]$. Then

$$w_j = y_j + J(u_j)$$

for elements y_j in $G(T_{t_j})$. Since $[0, 0] \in G(T_{t_j})$ for each j, we have

$$\langle w_j, u_j \rangle = \langle y_j, u_j \rangle + \|u_j\|^2 \geq \|u_j\|^2.$$

Hence the sequence $\{u_j\}$ is bounded, as is the sequence $J(u_j)$. We wish to show that u_j converges strongly to $u = (T_t + J)^{-1}w$. To do so, it suffices to assume that u_j converges weakly to u, and show that u must be this given element and that the convergence is strong. We may also assume that $J(u_j)$ converges weakly to z in X^*, while $y_j = w_j - J(u_j)$ converges weakly to $w - z = y$.

For each j, we have

$$\langle w_j, u_j - u \rangle = \langle y_j, u_j - u \rangle + \langle J(u_j), u_j - u \rangle.$$

Since w_j converges strongly to w while $u_j - u$ converges weakly to 0, it follows that $\langle w_j, u_j - u \rangle \to 0$. On the other hand, since J is pseudo-monotone,

$$\underline{\lim} \langle J(u_j), u_j - u \rangle \geq 0.$$

Hence

$$\overline{\lim} \langle y_j, u_j - u \rangle \leq 0$$

i.e.

$$\overline{\lim} \langle y_j, u_j \rangle \leq \langle y, u \rangle.$$

If we apply condition (1), we find that $y \in T_t(u)$, that $\langle y_j, u_j \rangle \to \langle y, u \rangle$, i.e. $\langle y_j, u_j - u \rangle \to 0$. Hence

$$\langle J(u_j), u_j - u \rangle \to 0.$$

Since J is a map of class $(S)_+$ by hypothesis, u_j must converge strongly to u and $J(u_j)$ must converge strongly to $J(u)$. Hence $y = w - J(u)$, i.e. $w \in (T_t + J)(u)$, or $u = (T_t + J)^{-1}(w)$.   Q.E.D.

PROOF THAT (3) IMPLIES (4). Let $[x, y] \in G(T_t)$. Then $y + J(x) \in (T_t + J)(x)$, i.e. $x = (T_t + J)^{-1}(y + J(x))$. If $\{t_j\}$ is a sequence converging to t in $[0, 1]$, let

$$x_j = \left(T_{t_j} + J\right)^{-1}(y + J(x)).$$

By condition (3), x_j converges strongly to x in X, so that $J(x_j)$ converges strongly to $J(x)$ in X^*. On the other hand, there exists y_j in $T(x_j)$ such that $y + J(x) = y_j + J(x_j)$. However, $y_j = y + J(x) - J(x_j)$ converges strongly to y in X^*.   Q.E.D.

THEOREM 9. *Let X be a reflexive Banach space, G a bounded open subset of X, $\{ f_t\colon 0 \leqslant t \leqslant 1 \}$ a homotopy of class $(S)_+$ of maps of $\mathrm{cl}(G)$ into a bounded subset of X^*. Let $\{ T_t\colon 0 \leqslant t \leqslant 1 \}$ be a family of maximal monotone mappings of X into 2^{X^*} with $0 \in T_t(0)$ for all t. Suppose $\{ T_t \}$ is a pseudo-monotone homotopy in the sense of Definition 1. Let $\{ y(t)\colon t \in [0,1] \}$ be a continuous path in X^* such that $y(t) \notin (T_t + f_t)(\partial G)$ for all t in $[0,1]$. Then*

(a) *For each $\varepsilon > 0$, $T_{t,\varepsilon} = (T_t^{-1} + \varepsilon J^{-1})^{-1}$ is a bounded pseudo-monotone mapping of X into X^*, while the family $\{ T_{t,\varepsilon} + f_t\colon 0 \leqslant t \leqslant 1 \}$ is a homotopy of class $(S)_+$ from $\mathrm{cl}(G)$ to X^*.*

(b) *There exists $\varepsilon_0 > 0$, such that for $0 < \varepsilon < \varepsilon_0$,*

$$y(t) \notin (T_{t,\varepsilon} + f_t)(\partial G),$$

$d(T_{t,\varepsilon} + f_t, G, y(t))$ *is well defined and independent of t and ε.*

THEOREM 10. *The degree function defined by Definition 1 on the class of maps of the form $T + f$ with T maximal monotone and f bounded and of class $(S)_+$ is a classical degree function with respect to the class of homotopies described in Theorem 2 with the canonical map $f_0 = J$.*

PROOF OF THEOREMS 8 AND 9. As stated, Theorem 9 includes Theorem 8 as a special case since the constant homotopy $T_t = T$ for all t is indeed pseudo-monotone in the sense of Definition 8. Hence it suffices to prove Theorem 9. The proof rests on the following auxiliary result.

PROPOSITION 16. *Let X be a reflexive Banach space, $\{ T_t\colon t \in [0,1] \}$ a pseudo-monotone homotopy of maximal monotone maps from X to 2^{X^*} with $0 \in T_t(0)$ for all t. Let $\{ u_j \}$ be a sequence in X converging weakly to u, and for sequences $\{ t_j \}$ in $[0,1]$ converging to t, $0 < \varepsilon_j$, $0 < \delta_j$, $\varepsilon_j \to 0$, $\delta_j \to 0$, let*

$$v_j = T_{t_j,\varepsilon_j}(u_j), \qquad z_j = T_{t_j,\delta_j}(u_j).$$

Suppose further that for another sequence $\{ s_j \}$ in $[0,1]$, and for

$$w_j = (1 - s_j)v_j + s_j z_j,$$

we have w_j converging weakly to w in X^*, while $\overline{\lim} \langle w_j, u_j \rangle \leqslant \langle w, y \rangle$.
Then $w \in T_t(u)$, and $\langle w_j, u_j \rangle \to \langle w, u \rangle$.

PROOF OF PROPOSITION 16. Since $v_j = T_{t_j,\varepsilon_j}(u_j)$, it follows that

$$v_j \in T_{t_j}\!\left(u_j - \varepsilon_j J^{-1}(v_j) \right).$$

Similarly, $z_j \in T_{t_j}(u_j - \delta_j J^{-1}(z_j))$. Since $0 \in T_{t_j}(0)$, and T_{t_j} is monotone,

$$\left(v_j, u_j - \varepsilon_j J^{-1}(v_j) \right) \geqslant 0, \qquad \left(z_j, u_j - \delta_j J^{-1}(z_j) \right) \geqslant 0.$$

Hence

$$\varepsilon_j \| v_j \|^2 \leqslant \langle v_j, u_j \rangle, \qquad \delta_j \| z_j \|^2 \leqslant (z_j, u_j).$$

If we multiply each of these inequalities by $(1 - s_j)$ and s_j, respectively and add, we obtain

$$(1 - s_j)\varepsilon_j \| v_j \|^2 + s_j \delta_j \| z_j \|^2 \leqslant \langle w_j, u_j \rangle \leqslant M.$$

Hence $(1 - s_j)\varepsilon_j \| v_j \|^2$ is bounded, as is $s_j \delta_j \| z_j \|^2$, so that

$$(1 - s_j)\varepsilon_j \| v_j \| \to 0, \qquad s_j \delta_j \| z_j \| \to 0.$$

Let $[x, y]$ be any element of $G(T_t)$. By the condition (4) for the pseudo-monotone homotopy $\{T_t\}$, for each j, there exist elements $[x_j, y_j]$ of $G(T_{t_j})$ with x_j converging strongly to x, y_j converging strongly to y. By the monotonicity of T_{t_j},

$$\left\langle v_j - y_j, u_j - \varepsilon_j J^{-1}(v_j) - x_j \right\rangle \geqslant 0, \qquad \left\langle z_j - y_j, u_j - \delta_j J^{-1}(z_j) - x_j \right\rangle \geqslant 0.$$

Hence,

$$\left\langle v_j - y_j, u_j - x_j \right\rangle \geqslant \varepsilon_j \left\langle v_j - y_j, J^{-1}(v_j) \right\rangle \geqslant - \| y_j \| \varepsilon_j \| v_j \|,$$

and

$$\left\langle z_j - y_j, u_j - x_j \right\rangle \geqslant \delta_j \left\langle z_j - y_j, J^{-1}(z_j) \right\rangle \geqslant - \| y_j \| \delta_j \| z_j \|.$$

If we multiply these inequalities by $(1 - s_j)$ and s_j, respectively, and add, we get

$$\left\langle w_j - y_j, u_j - x_j \right\rangle \geqslant - M\{(1 - s_j)\varepsilon_j \| v_j \| + s_j \delta_j \| z_j \|\} \to 0.$$

Thus

$$\langle w, u \rangle \geqslant \overline{\lim} \left\langle w_j, u_j \right\rangle \geqslant \underline{\lim} \left\langle w_j, u_j \right\rangle \geqslant \underline{\lim} \left\{ \left\langle w_j - y_j, +x_j \right\rangle + \left\langle y_j, +u_j \right\rangle \right\}$$
$$= \left\langle w - y, x \right\rangle + \left\langle y, u \right\rangle.$$

Hence

$$\left\langle w - y, u - x \right\rangle \geqslant 0 \qquad ([x, y] \text{ in } G(T_t)),$$

and by the maximal monotonicity of T_t, $w \in T_t(u)$. Substituting $[u, w]$ for $[x, y]$ in the preceding chain of inequalities, we see that

$$\langle w, u \rangle \geqslant \overline{\lim} \left\langle w_j, u_j \right\rangle \geqslant \underline{\lim} \left\langle w_j, u_j \right\rangle \geqslant \langle w, u \rangle.$$

Thus, finally, $\left\langle w_j, u_j \right\rangle \to \langle w, u \rangle$.   Q.E.D.

PROOF OF THEOREM 9 COMPLETED. By Proposition 12, it follows that if $\{T_t\}$ is a pseudo-monotone homotopy, then T_t^{-1} is a pseudo-monotone homotopy. Similarly, if $\{T_t^{-1}\}$ is a pseudo-monotone homotopy and if $\{g_t\}$ is a pseudo-monotone homotopy of bounded monotone maps, then $\{T_t^{-1} + g_t\}$ is a pseudo-monotone homotopy. In particular, $\{T_t^{-1} + \varepsilon J^{-1}\}$ is a pseudo-monotone homotopy, and $T_{\varepsilon,t} = (T_t^{-1} + \varepsilon J^{-1})^{-1}$ is also a pseudo-monotone homotopy. Hence $T_{\varepsilon,t} + f_t$ is a pseudo-monotone homotopy of class $(S)_+$ for each $\varepsilon > 0$. Thus part (a) holds.

Suppose the conclusion of part (b) were false. Then there would exist a sequence $\{t_j\}$ in $[0, 1]$ and two sequences of positive numbers $\varepsilon_j, \delta_j \to 0$ and points

u_j on ∂G, such that for suitable s_j in $[0, 1]$

$$y_{t_j} = (1 - s_j)T_{t_j, \varepsilon_j}(u_j) + s_j T_{t_j, \delta_j}(u_j) + f_{t_j}(u_j).$$

We may assume that t_j converges to t, and that y_{t_j} converges strongly to y_t. If we set $v_j = T_{t_j, \varepsilon_j}(u_j)$, $z_j = T_{t_j, \delta_j}(u_j)$, and $w_j = y_{t_j} - f_{t_j}(u_j)$, we may assume that u_j converges weakly to u and that w_j converges weakly to w. Moreover,

$$\overline{\lim} \left\langle w_j, u_j - u \right\rangle = \lim \left\langle y_{t_j}, u_j - u \right\rangle - \lim \left\langle f_{t_j}(u_j), u_j - u \right\rangle \leq 0$$

since

$$\underline{\lim} \left\langle f_{t_j}(u_j), u_j - u \right\rangle \geq 0$$

by the pseudo-monotonicity of the homotopy $\{f_t\}$. If we apply Proposition 2, it follows that $w \in T_t(u)$, and that $\langle w_j, u_j - u \rangle \to 0$. Hence

$$\left\langle f_{t_j}(u_j), u_j - u \right\rangle \to 0,$$

and since the homotopy $\{f_t\}$ is of class $(S)_+$, u_j converges strongly to u, and $f_{t_j}(u_j)$ converges weakly to $f_t(u)$. Thus u lies in ∂G, $y_t - f_t(u) = w \in T_t(u)$, i.e. $y_t \in (T_t + f_t)(\partial G)$, which contradicts the hypothesis.   Q.E.D.

THEOREM 11. *The degree function described by Theorem 3 can be extended to the class of mappings of the form $T + f$, with f pseudo-monotone and bounded by letting $d(T_t + f, G, y_0) = \lim_{\delta \to 0} d(T + f + \delta J, G, y_0)$, where the mapping $f + \delta J$ is of class $(S)_+$. This extension has properties similar to the corresponding extension of the degree function to pseudo-monotone mappings in §5.*

We turn now to the uniqueness result for the degree function over the class $T + f$, with T maximal monotone, f of class $(S)_+$ and bounded.

THEOREM 12. *Let X be a reflexive Banach space. Then there exists exactly one degree function on the class of maps $T + f$, with T maximal monotone and f bounded and of class $(S)_+$, which is normalized by J and such that the degree is invariant under all affine homotopies of the form*

$$(1 - t)(T + f) + tf_1$$

with T maximal monotone, f and f_1 of class $(S)_+$.

PROOF OF THEOREM 12. Let d_1 be such a degree function. When restricted to the maps of class $(S)_+$, it coincides with the unique degree function on that class. Hence, we must show that this unique identification persists when we pass to the broader class $(T + f)$.

Suppose $y_0 \notin (T + f)(\partial G)$. Consider the affine homotopy between $(T + f)$ and the approximating map $(T_\varepsilon + f)$ for small $\varepsilon > 0$ which we employed above in proving the existence of the degree. If $d_1 \neq d$, then for arbitrary small $\varepsilon > 0$, we can find u_ε on ∂G and t_ε in $[0, 1]$ such that

$$(1 - t_\varepsilon)T(u_\varepsilon) + t_\varepsilon T_\varepsilon(u_\varepsilon) + f(u_\varepsilon) \ni y_0.$$

We may choose a sequence u_j in ∂G, $u_j \rightharpoonup u$, $t_j \to t$, $\varepsilon_j \to 0$, $f(u_j) \to v_0$. For each j, we have $w_j \in T(u_j)$ such that

$$(1 - t_j)w_j + t_j T_{\varepsilon_j}(u_j) + f(u_j) = y_0.$$

We now apply a variant of the argument used in the proof of Proposition 16. If $v_j = T_{\varepsilon_j}(u_j)$, then

$$v_j \in T\big(u_j - \varepsilon_j J^{-1}(v_j)\big).$$

Similarly, $w_j \in T(u_j)$. Hence

$$\varepsilon_j \|v_j\|^2 \leq \langle v_j, u_j \rangle, \qquad 0 \leq \langle w_j, u_j \rangle.$$

If we multiply these inequalities by $(I - t_j)$ and t_j, respectively, and add, we obtain

$$t_j \varepsilon_j \|v_j\|^2 \leq \langle y_0 - f(u_j), u_j \rangle \leq M.$$

Hence $t_j \varepsilon_j \|v_j\|^2$ is bounded, so that $t_j \varepsilon_j \|v_j\| \to 0$.

Let $[x, y]$ be any element of $G(T)$. Then

$$\langle v_j - y, u_j - \varepsilon_j J^{-1}(v_j) - x \rangle \geq 0, \qquad \langle w_j - y, u_j - x \rangle \geq 0.$$

Thus,

$$\langle v_j - y, u_j - x \rangle \geq \langle v_j - y, \varepsilon_j J^{-1}(v_j) \rangle \geq \langle -y, \varepsilon_j J^{-1}v_j \rangle \geq -\varepsilon_j \|v_j\| \|y\|.$$

If we multiply the inequalities by t_j and $(1 - t_j)$, respectively, and add, we obtain

$$\langle y_0 - f(u_j) - y, u_j - x \rangle \geq -\varepsilon_j t_j \|v_j\| \|y\|,$$

where the term on the right approaches zero as $j \to +\infty$. Hence, if $z_j = y_0 - f(u_j)$, then $z_j \rightharpoonup z = y_0 - v$, and

$$\varliminf \langle z_j - y, u_j - x \rangle \geq 0.$$

On the other hand, since $z_j + f(u_j) = y_0$, we have

$$\langle z_j, u_j - u \rangle + \langle f(u_j), u_j - u \rangle = \langle y_0, u_j - u \rangle,$$

i.e.

$$\varlimsup \langle z_j, u_j - u \rangle + \varliminf \langle f(u_j), u_j - u \rangle \leq 0.$$

Since f is of class $(S)_+$, $\varliminf \langle f(u_j), u_j - u \rangle \geq 0$. Hence

$$\varlimsup \langle z_j, u_j - u \rangle \leq 0, \qquad \varlimsup \langle z_j, u_j \rangle \leq \langle z, u \rangle.$$

Thus

$$\langle z, u \rangle \geq \varlimsup \langle z_j, u_j \rangle \geq \varliminf \langle z_j, u_j \rangle \geq \langle y, u - x \rangle + \langle z, x \rangle.$$

It follows that $\langle z - y, u - x \rangle \geq 0$ for all $[x, y]$ in $G(T)$. By the maximal monotonicity of T, $z \in T(u)$. Replacing $[x, y]$ by $[u, z]$, we find that $\langle z_j, u_j \rangle \to \langle z, u \rangle$. Hence

$$\varlimsup \langle f(u_j), u_j - u \rangle \leq \varlimsup \langle z_j, u - u_j \rangle = 0.$$

Since f is of class $(S)_+$, $u_j \to u$ so that u lies on ∂G, and $f(u_j) \to f(u)$. Thus $f(u) = y_0 - z$, or in other words, $y_0 \in (T + f)(u)$. This contradicts the assumption that $y_0 \notin (T + f)(\partial G)$. The contradiction is to the assumption that the two degrees are different for $(T + f)$. Hence the two degrees coincide.   Q.E.D.

BIBLIOGRAPHY

1. P. Alexandroff and H. Hopf, *Topologie*, Springer-Verlag, Berlin, 1935.

2. H. Amann and S. Weiss, *On the uniqueness of the topological degree*, Math. Z. **130** (1973), 39–54.

3. H. Brezis, M. G. Crandall, and A. Pazy, *Perturbations of nonlinear maximal monotone sets in Banach spaces*, Comm. Pure Appl. Math. **23** (1970), 123–144.

4. P. Bohl, *Uber die Bewegung eines mechanischen Systems in der Nähe einer Gleichgewichtslage*, J. Reine Angew. Math. **127** (1904), 179–276.

5. Ju. G. Borisovich and Yu. I. Sapronov, *A contribution to the topological theory of condensing operators*, Soviet Math. Dokl. **9** (1968), 1304–1307.

6. L. E. J. Brouwer, *Beweis der Invarianz der Dimensionzahl*, Math. Ann. **70** (1911).

7. ______, *Uber Abbildung der Mannigfaltigkeiten*, Math. Ann. **71** (1912), 97–115.

8. F. E. Browder, *Topological methods for nonlinear elliptic equations of arbitrary order*, Pacific J. Math. **17** (1966), 17–31.

9. ______, *Topology and nonlinear functional equations*, Studia Math. **31** (1968), 189–204.

10. ______, *Nonlinear equations of evolution and nonlinear operators in Banach spaces*, Nonlinear Functional Analysis, Proc. Sympos. Pure Math., vol. 18, Part II, Amer. Math. Soc., Providence, R. I., 1975.

11. ______, *Degree of mapping for nonlinear mappings of monotone type*. I, II, III, Proc. Nat. Acad. Sci. USA, March, April 1983 (to appear).

12. ______, *The degree of mapping and its generalizations*, Proc. Special Session on Fixed Point Theory (Summer Meeting, 1982), Amer. Math. Soc., Providence, R. I., (to appear).

13. ______, *The theory of degree of mapping for nonlinear mappings of monotone type*, Nonlinear Partial Differential Equations and Applications, IV, Pitman (to appear).

14. ______, *L'unicité du degré topologique pour des applications de type monotone*, C. R. Acad. Sci. Paris, January 10, 1983 (to appear).

15. F. E. Browder and R. D. Nussbaum, *The topological degree for non-compact nonlinear mappings in Banach spaces*, Bull. Amer. Math. Soc. **74** (1968), 671–676.

16. F. E. Browder and W. V. Petryshyn, *Approximation methods and the generalized topological degree for nonlinear mappings in Banach spaces*, J. Funct. Anal. **3** (1969), 217–245.

17. J. Cronin, *Fixed points and topological degree in nonlinear analysis*, Math. Surveys, No. 11, Amer. Math. Soc., Providence, R. I., 1964.

18. K. D. Elworthy and A. J. Tromba, *Differential structures and Fredholm maps on Banach manifolds*, Global Analysis, Proc. Sympos. Pure Math., vol. 15, Amer. Math. Soc., Providence, R. I., 1969.

19. C. Fenske, *Analytische Theorie des Abbildungsgrades für Abbildungen in Banachräumen*, Math. Nachr. **48** (1971), 279–290.

20. L. Fuhrer, *Ein elementarer analytischer Beweis zur Eindeutigkeit des Abbildungsgrades im R^n*, Math. Nachr. **54** (1972), 259–267.

21. M. Golomb, *Zur Theorie der Nichtlinearen Integralgleichungen, Integralgleichungsysteme und allgemeine Funktionalgleichungen*, Math. Z. **39** (1934–35), 45–75.

22. A. Granas, *The Leray-Schauder index and the fixed point theory for arbitrary ANR's*, Bull. Soc. Mat. France **100** (1972), 209–228.

23. J. Hadamard, *Sur quelques applications de l'indice de Kronecker*, Introduction to: J. Tannery, La Théorie des Fonctions d'une Variable, Paris, 1910.

24. M. A. Krasnoselski, *Topological methods in the theory of nonlinear integral equations*, Pergamon, Oxford, 1963.

25. L. Kronecker, *Uber Systeme von Funktionen mehrer Variablen*, Monatsb. Berlin Akad. (1869), 159–193, 688–698.

26. J. Leray and J. Schauder, *Topologie et equations fonctionnelles*, Ann. Sci. École. Norm. Sup. **51** (1934), 45–78.

27. J. Mahwin, *Topology and nonlinear boundary value problems*, Dynamical Systems, Vol. 1, Academic Press, 1976, pp. 51–82.
28. C. Miranda, *Un'osservazione su una teorema di Brouwer*, Boll. Unione Mat. Ital. (1940), 527.
29. R. D. Nussbaum, *Degree theory for local condensing maps*, J. Math. Anal. Appl. **37** (1972), 741–766.
30. ______, *On the uniqueness of the topological degree for k-set contractions*, Math. Z. **137** (1974), 1–6.
31. E. Picard, *Sur les nombres des racines communes q plusieurs équations simultanes*, J. Math. Pures Appl. **8** (1892), 5–24.
32. H. Poincaré, *Sur certaines solutions particulieres du problème des trois corps*, C. R. Acad. Sci. Paris **97** (1883), 251–252.
33. ______, *Sur certaines solutions particulieres du problème des trois corps*, Bull. Astronomique **1** (1884), 63–74.
34. ______, *Sur les courbes définies par une équation différentielle. IV*, J. Math. Pures Appl. **85** (1886), 151–217.
35. H. W. Siegberg, *Some historical remarks concerning degree theory*, Amer. Math. Monthly (1981), 125–139.
36. J. T. Schwartz, *Nonlinear functional analysis*, Gordon and Breach, New York, 1969.
37. I. V. Skrypnik, *Nonlinear elliptic equations of higher order*, Kiev, 1973.

DEPARTMENT OF MATHEMATICS, UNIVERSITY OF CHICAGO, CHICAGO, ILLINOIS 60637

Proceedings of Symposia in Pure Mathematics
Volume 39 (1983), Part 2

Variational and Topological Methods in Nonlinear Problems[1]

L. NIRENBERG

These lectures are meant as an informal introduction to some of the techniques used in proving existence of solutions of nonlinear problems of the form

$$F(u) = y. \tag{1}$$

Here F is a continuous (and usually smooth) mapping from one topological space X to another Y, and the spaces are usually infinite dimensional. The model to keep in mind is one in which these spaces are function spaces defined in domains on finite-dimensional manifolds, and F is a system of nonlinear partial differential operators–or integral operators.

A number of special topics will be presented–in three parts:

I. Global methods: homotopy, in particular topological degree theory, and extensions. Applications to nonlinear boundary value problems.

II. Variational methods, in which a solution is a stationary point of some functional. Applications.

III. Local study, perturbation about a solution.

An important analytic aspect of all these problems is that of finding a priori estimates for the solutions. How one does that varies from problem to problem and I will barely touch on these technical aspects. I will try, rather, to avoid technicalities and stress the topological and variational ideas.

The lectures are not addressed to the experts in these fields–for them there will be little new. They are given with the hope of attracting others to the subject. Up to now, the topological and abstract ideas used are rather primitive, and I am confident that there will be enormous further development–involving more and more sophisticated topology.

A condensed version of some of this material was presented in [**48**].

Here is a more specific list of the topics treated.

I.1 Some classical things. Continuity method. Degree theory.

I.2 Some recent extensions of Leray-Schauder degree theory.

I.3 Extension of degree theory to Fredholm maps by Elworthy and Tromba.

Reprinted from Bulletin Amer. Math. Soc. (N.S.) 4 (1981), 267–302.

1980 *Mathematics Subject Classification.* Primary 35-02, 35-A15, 35A05, 35B10, 4902, 4602, 58F05, 58-G16; Secondary 35B32, 35J65, 35L20, 58E15.

[1]Based on four lectures given in March 1980 at the Institute for Advanced Study in the Hermann Weyl lecture series. The work was partially supported by National Science Foundation Grants MCS-7900813 and INT-77-20878 and by U. S. Army Research Office Grant DAA-29-78-G-0127.

I.4 Fredholm maps with positive index.

I.5 Modifications of degree associated to invariant orbits of ordinary differential equations.

II.1 The Palais-Smale condition (PS) and an elliptic boundary Value problem.

II.2 The mountain pass lemma. Solution of the boundary value problem.

II.3 Generalizations and variants of the mountain pass lemma.

II.4 A theorem of P. Rabinowitz on periodic solutions for a Hamiltonian system.

II.5 Periodic solutions of a nonlinear string equation.

III.1 Remarks on bifurcation theory for Fredholm operators.

III.2 The Nash-Moser Implicit Function technique.

III.3 Klainerman's result on the existence of smooth solutions for all time of a nonlinear hyperbolic initial value problem.

I. Global topological methods

For convenience we will assume that the spaces X, Y are Banach spaces and that F is a smooth mapping. We begin with some very classical things.

I.1 A standard, but still very useful, method, is to try to show that

(i) Range of F is open,

(ii) Range of F is closed.

For (i), a natural tool is the implicit function theorem. If $y = F(u_0)$, in order to show that a neighbourhood of y_0 is contained in the image of one of u_0, one looks at the continuous linear operator $A = F'(u_0)$: $X \to Y$. If A is bijective the desired result follows from the implicit function theorem (IFT). It suffices, in fact, that A be surjective.

We shall make use of the following

DEFINITION. If $A = F'(u_0)$ is bijective, the point u_0 is called a *regular point* of F.

To establish (ii) one usually tries to show that the map F is proper, i.e., the preimage of every compact set is compact. It is here that a priori estimates for solutions of the equation enter in a crucial way. This is usually where the hard work comes in.

A variant of this approach is the

CONTINUITY METHOD. The operator F is continuously connected by a one-parameter family of operators F_t, $0 \leqslant t \leqslant 1$, $F_1 = F$, to an operator F_0 for which it is known that a solution of (1) exists. Then one tries to prove

(i') the set of t for which a solution exists is open.

(ii') the set of t for which a solution exists is closed.

Again one leans on the implicit function theory and on a priori estimates.

A word of caution: For a nonlinear partial differential operator F these methods usually fail–unless the operator is elliptic (to be explained later for a special case). The reason is that the implicit function theorem may not be applicable. There are of course different ways in which it may fail: (a) The linear operator $A = F'(u_0)$ may be injective, and so invertible on its range, but the range may not be the whole space Y. For example the range may be dense but not closed in Y. This may happen in the following way. It may be that one can solve (in some sense) $Au = y$ for every $y \in Y$, but the "solution" x doesn't lie in the space X, but in some larger space. For example, X and Y may be spaces of functions with some given degrees of smoothness–defined on some bounded domain in R^n. It may be that we can solve $Au = y$, $\forall y \in Y$, but u does not have the desired smoothness, i.e. we "lose derivatives". This is typical for nonelliptic differential operators. In such cases one tries to use the generalized implicit function theorems, such as the Nash-Moser method. This will be taken up in part III.

(b) The IFT may fail in a simpler way. One may have

$$\ker F'(u_0) \neq 0$$

but

$$\text{Range } F'(u_0) \text{ is a closed subspace of } Y.$$

In such a situation the local problem is usually called a bifurcation problem (splitting of solutions may occur). The cases that have received the most intensive study are those in which

$$\dim \ker F'(u_0) = d < \infty$$

and

$$\text{codim Range } F'(u_0) = d' < \infty$$

i.e. Y may be decomposed as Range $F'(u_0) \oplus Y_2$, dim $Y_2 = d'$. If F has this property at u_0 we say that F is a *Fredholm operator* there, and its index is $d - d'$.

Using well-known properties of linear Fredholm operators (see for example [56]) one finds that F is Fredholm everywhere in a neighbourhood of u_0 and its index is constant there.

Turning to topological methods, recall that homotopy theory began with attempts to find conditions on maps F to guarantee that (1) has a solution. Consider a continuous map F from the closed unit ball B in R^n into R^k for which one wishes to solve the equation

$$F(u) = 0.$$

Let ϕ be the restriction of F to ∂B and assume $\phi : \partial B \to R^k \setminus \{0\}$. Homotopy theory yields a sufficient condition on ϕ so that *every* continuous extension F of ϕ to the inside of B has, necessarily, a solution of $F(u) = 0$. Namely, the homotopy class of $\phi : S^{n-1} \to R^k \setminus \{0\}$ should be nontrivial. An equivalent formulation, for the normalized map

$$\psi(u) = \frac{\phi(u)}{|\phi(u)|}$$

is that $\psi : S^{n-1} \to S^{k-1}$ should be homotopically nontrivial. In case $k = n$ this is equivalent to the assertion that the degree of the map $\psi =$ the degree of the map F at the origin, is nonzero. For generic y with $F : \partial B \to R^n \setminus \{y\}$, the degree of the map F at y,

$$\deg(F, B, y) = \nu, \quad \text{an integer.}$$

ν represents the number of times y is covered, i.e. the number of preimages of y with each preimage counted with $+$ or -1 according as, locally, F preserves or reverses orientation there.

The most frequently used topological tool in attacking global nonlinear problems is the Leray, Schauder [38] extension of degree theory to infinite dimensions. In Banach space X it applies to operators $F: X \to X$ of a special form

$$F = I - K \tag{2}$$

where I = Identity operator and K is a compact operator (i.e. the closure of the image under K of any bounded set is compact). If K is smooth then F is a Fredholm map everywhere with index = 0.

Let us consider equation (1) for such F defined in the closure of a bounded domain Ω in X, mapping into X. Assume

$$y \notin F(\partial\Omega). \tag{3}$$

Then the degree of the map F in $\overline{\Omega}$ at the point y is defined:

$$\deg(F, \Omega, y) = \nu = \text{integer.}$$

It is obtained from finite-dimensional degree by approximating K (within ε on $\overline{\Omega}$) by a map K_ε into a finite-dimensional subspace X_ε containing y. One then defines $\deg(F, \Omega, y)$ as the finite-dimensional degree

$$\deg((I - K_\varepsilon), \Omega \cap X_\varepsilon, y).$$

To verify that this is independent of ε, for small ε, one uses the fact that degree does not change under suspension of a map F, i.e. by extension of the map to a product space with another space by taking the product of F with the identity map in the other space.

We list a few important properties:

(a) If $\nu = \deg(F, \Omega, y) \neq 0$ then $F(u) = y$ has a solution in Ω. For $F = I$, $\nu = 1$.

(b) If $\overline{\Omega} = \bigcup_{i=1}^{\infty} \overline{\Omega}_i$, Ω_i open, nonoverlapping and $y \notin F(\partial\Omega_i)$, $\forall i$, then

$$\deg(F, \Omega, y) = \sum_i \deg(F, \Omega_i, y).$$

(c) The degree ν is invariant under homotopy of the map $F_t = I - K_t$, $0 \leqslant t \leqslant 1$, in our class, i.e., provided $\{K_t(x) | x \in \overline{\Omega}, 0 \leqslant t \leqslant 1\}$ has compact closure and

$$y \notin F_t(\partial\Omega), \qquad \forall t.$$

(d) Suppose that a solution u of $F(u) = u - K(u) = y$ is a regular point of F. (By IFT it is an isolated solution.) Then the local degree (index) of F at u is defined as $\operatorname{ind}(u) = \operatorname{loc} \deg F$ at $u = \deg(F, \varepsilon\text{-ball about } u; y)$. It is independent of ε for ε small, and equals $+1$ or -1 according as the sum of the algebraic multiplicities of the negative eigenvalues of $F'(u)$ is even or odd.

Presentations of degree theory and derivations on these properties may be found in [37], [57], [39], [47], [51] and many other places.

Degree theory is often used in the following way: To show that (1) has a solution with F as in (2), try to show that all solutions lie inside some bounded domain Ω–that is, obtain a priori estimates for all solutions. Then, to prove $\nu = \deg(F, \Omega, y) \neq 0$, construct a deformation of F, $F_t = I - K_t$ belonging to our class, and such that $y \notin F_t(\partial\Omega)$, deforming $F = F_1$ to F_0, for which one knows $\deg(F_0, \Omega, y) \neq 0$. For $F_0 = I$, we find $\nu = 1$.

Property (d) is sometimes useful in trying to determine the number of solutions of (1). Sometimes one can show that $F^{-1}(y)$ consists of a finite number of regular points in Ω, all having the same local degrees, say ± 1. If $\deg(F, \Omega, y) = k$ it follows that there are exactly $|k|$ solutions in Ω.

This completes the classical things and we now take up some recent extensions.

I.2 Considering, still, mappings of the form (2) and satisfying (3) we will describe two results related to (d).

The set of regular points Ω^r in Ω, of F, is open (by IFT). Let $\mathcal{C}$ be a connected component of Ω^r. For any $u \in \mathcal{C}$, define for small ε

$$\mathrm{ind}(u) = (\deg F, \varepsilon\text{-ball about } u, F(u)) = \pm 1.$$

It is easy to see that this is a constant in $\mathcal{C}$ and so may be considered as $\mathrm{ind}(\mathcal{C})$. In [7] Ambrosetti and Mancini showed (in a Hilbert space, though it works as well in a Banach space), that this index necessarily changes as we cross from one component of Ω^r to another–in a particular generic circumstance. They then applied this to prove that certain equations in Hilbert space have exactly three solutions. Their generic condition is easy to describe:

Let $u_0 \in \Omega$ be a nonregular point of F and assume that

(i) ker $F'(u_0)$ is spanned by a vector v. Consequently Range $F'(u_0)$ is a closed linear subspace X_1 of X, of codimension one. Consider next the second derivative (Hessian) $F''(u_0)$ of F. This is a symmetric bilinear map of $X \times X \to X$. Assume

(ii) there is a vector w such that

$$F''(u_0)(v, w) \text{ is not in } X_1.$$

In this case it is easy to see (Theorem 3.7.2 in [47]) that in a neighbourhood of u_0, the set Ω^r has two components which are separated by a smooth hypersurface. Their result is that under conditions (i), (ii), the indices of these two components are different.

We turn now to a recent generalization of (d). In bifurcation theory one sometimes encounters a finite-dimensional manifold of solutions of (1). Let us suppose that a connected component of $F^{-1}(y)$ consists of a compact finite-dimensional manifold M without boundary. It is natural to try to determine $\mathrm{ind}(M) = $ the local degree at y of F in a neighbourhood of M–assuming F is "regular" on M, i.e., assuming:

REGULARITY.

$$\ker F'(u) = T_u M, \qquad \forall u \in M.$$

Here $T_u M$ is the tangent space to M at u.

Assuming M is orientable, and F is regular on M, Sylvester [58] proved the following

THEOREM. *If* dim *M is odd then* ind *M* = 0. *If* dim *M is even then*

$$\mathrm{ind}(M) = \pm\chi(E)$$
$$= \pm \text{ Euler characteristic of a vector bundle } E \text{ over } M,$$

where for $u \in M$, *the fibre*

$$E_u = X/(\mathrm{Range}\ F'(u)).$$

In particular, if $T_u M \cap \mathrm{Range}\ F'(u) = 0, \forall u \in M$, *then*

$$\mathrm{ind}(M) = \pm\chi(M).$$

Furthermore one has + *or* − *according as the sum of algebraic multiplicities of negative eigenvalues of* $F'(u)$ *is even or odd, for* $u \in M$.

So far no application has been made of the result.

Degree theory extends to set-valued maps of the form (2): For $u \in \bar{\Omega}$, $K(u)$ is assumed to be a compact convex set which is upper semicontinuous in u. Furthermore $\bigcup_{u \in \bar{\Omega}} K(u)$ has compact closure. One wishes to solve

$$y \in u - K(u). \tag{1'}$$

If $\forall u$ on $\partial\Omega$, $u - y \notin K(u)$, then $\deg(F, \Omega, y)$ can be defined as before, with similar properties. A treatment of this may be found in the book of Lloyd [39].

Recently this has been used effectively by K. C. Chang [18] to solve several problems: (a) plasma problems, (b) an obstacle problem, in which one wishes to find solutions of an elliptic partial differential equation satisfying a side condition: it is to lie above a given obstacle.

I.3 Attempts have been made to extend degree theory to mappings F which do not have the form (2). Elworthy and Tromba [27], [28] have developed degree theory for smooth Fredholm mappings between Banach manifolds. (A Banach manifold modeled on a Banach space X is an infinite-dimensional manifold for which each point has a distinguished neighbourhood U and a chart map $\phi\colon U$ onto X.) For an overlapping neighbourhood V, with chart map $\psi\colon V \to X$, it is required that on $\psi(U \cap V)$, $\phi \circ \psi^{-1}$ have the form (2). For convenience, we will consider only Fredholm maps between Banach spaces, $F\colon X \to Y$ with F defined on the closure $\bar{\Omega}$ of a bounded domain Ω in X. F is assumed to be proper.

Using suitable orientations on X and Y, in terms of the admissible chart mappings, they defined for Fredholm maps F *of index zero*, an oriented degree

$$\deg(F, \Omega, y) = \nu, \qquad \nu \text{ an integer,}$$

in case $y \notin F(\partial\Omega)$.

They proved that this has the properties (a)–(d) above except that (c) holds in a weaker form. Namely, under a suitably restricted class of deformations, the degree ν is not necessarily invariant, but $|\nu|$ is.

Their theory makes use of the Sard-Smale lemma. This has been redone in a somewhat simpler way, via more systematic use of transversality, by Borisovich, Zvyagin and Sapronov. See their useful expository article [12]. Furthermore they extend the Elworthy-Tromba theory to maps of the form $F = F_1 + K$ where F_1 is a smooth Fredholm map of index zero and K is a continuous compact map.

This generalized degree has not received much application but I believe it will prove useful. As an indication of its use in a situation where the Leray-Schauder degree does not apply let us consider a second order quasilinear elliptic equation for a real function $u(x)$ defined in a bounded domain G in R^n with smooth boundary—under the boundary condition $u = 0$ on ∂G.

$$\sum a^{ij}(x, u, \operatorname{grad} u)u_{x_i x_j} + f(x, u, \operatorname{grad} u) = 0 \quad \text{on } G.$$

Ellipticity means that the quadratic form

$$a^{ij}(x, u, \operatorname{grad} u)\xi_i \xi_j$$

is always positive definite. We try to find a solution u as a fixed point of the map $K: u \mapsto v$, where v is the solution of

$$\sum a^{ij}(x, u, \operatorname{grad} u)v_{x_i x_j} + f(x, u, \operatorname{grad} u) = 0 \text{ in } G, \qquad v = 0 \text{ on } \partial G.$$

Thus we assume we know how to solve linear elliptic boundary value problems. In appropriate function spaces, the map K is a smoothing operator and therefore compact. For example if u belongs to the Sobolev space H^s of functions having square integrable derivatives in G up to order s, for s large then v belongs to H^{s+1}.

So we wish to solve

$$F(u) = u - K(u) = 0.$$

Suppose we can establish a priori estimates for solutions together with derivatives—guaranteeing that the solutions all lie inside some big ball B in our function space. Then $\deg(F, B, 0)$ is defined, and we may try to prove it is not zero.

In place of the quasilinear equation above let us consider a general nonlinear elliptic equation

$$A[u] \equiv A(x, u, \operatorname{grad} u, u_{x_i x_j}) = y(x) \text{ in } G, \qquad u = 0 \text{ on } \partial G.$$

Ellipticity means the quadratic forms

$$\sum \frac{\partial A}{\partial u_{x_i x_j}}(x, \ldots)\xi_i \xi_j$$

are all positive definite. It appears that the Leray-Schauder theory is not applicable since it does not seem possible to recast this in the form (1), (2). However in the space of functions in $H^s(G)$, s large, vanishing on the boundary, the Elworthy-Tromba theory of Fredholm maps may be applicable provided we have a priori bounds in H^s for all solutions. Their degree

$$\deg(A, B, y) = \nu$$

is then defined, and $\nu \neq 0$ ensures the existence of a solution. It seems surprising that up to now this method has not been employed.

I.4 Elworthy and Tromba have also treated Fredholm maps F with positive index. The corresponding degree is no longer an integer but a Pontrjagin framed cobordism class (see [27], [28], [46], [12], [11] which also contain further references). This is however difficult to work with. In some situations one may use stable homotopy or cohomotopy; in [46] some application to elliptic boundary value problems has been made. But so far no natural example has arisen in which these methods have been used. I am confident that some will occur.

To conclude the discussion of topological methods for (1), it should be mentioned that many mathematicians have made important contributions in this field. To mention a few: M. A. Krasnoselskii, F. Browder, R. Nussbaum, W. Petryshyn, and for more general use of homotopy theory K. Geba and A. Granas. See [11], [31], [47] for some useful references to recent developments.

I.5 In the study of periodic solutions of systems of ordinary differential equations much work has been done to find invariants associated with closed orbits, or isolated families of closed orbits, i.e., which are invariant under perturbation of the system. This is to guarantee that the perturbed equations also possess closed orbits near the original ones.

Let us recall the simplest situation. Consider a system of ordinary differential equations for a vector x in R^n.

$$\dot{x} = \frac{dx}{dt} = f(x). \tag{4}$$

Suppose we have a nonconstant closed (i.e. periodic) orbit $x_0(t)$. Through a point y_0 on the orbit consider a hyperplane P perpendicular to the orbit at y_0. On it, near y_0 consider the Poincaré map F of $P \to P$ defined as follows. From any point $y \in P$ near y_0 follow the solution curve of the system until it strikes P again for the first time near y_0–it is easily seen that it must. Let z be the point where it first strikes P. The map $y \mapsto z$ is called the Poincaré map. Clearly y_0 is a fixed point of this map.

Suppose y_0 is an isolated fixed point, set $F(y) = y - z(y)$. Then for a small ball B in P centered at y_0 the degree

$$\deg(F, B, 0) = \nu$$

is well defined. If $\nu \neq 0$ we may then infer that any slight perturbation of the system (4) has a periodic solution lying close to $x_0(t)$ (and in fact its period will be close to that of $x_0(t)$).

In [30] Fuller introduced a more general notion, now called the Fuller index, attached to a closed orbit of (4). This has proved very useful; Mallet-Paret and Yorke [40] modified this to study global connected sets of orbits in case f depends on a parameter. The reader may also find many useful references here.

Conley [21] has introduced a very interesting generalization of the Morse index for any isolated set which is invariant under the flow of (4), i.e. the maximal invariant set in some neighbourhood of itself. In Morse theory, one studies Morse functions: real functions f on n-dimensional manifolds having nondegenerate stationary points, i.e. at a stationary point the Hessian matrix of second derivatives

$$f_{xx} \text{ is nonsingular.}$$

The number k of negative eigenvalues of the Hessian is called the index of the stationary point. The global structure of the (compact) manifold is tied to the number of critical points of f with various indices.

Conley's generalized Morse index assigned to a compact isolated invariant set S is the homotopy type of a pointed topological space, i.e. a space with a distinguished point in it. Very roughly, it is defined in the following way. Let N be a neighbourhood containing S compactly in its interior. An index pair $\langle N_1, N_2 \rangle$ is a pair N_1, N_2 of compact subsets of N satisfying several conditions:

(i) relative to N they are invariant under forward flow i.e. for $i = 1, 2$, $\forall x_0 \in N_i$ and $\forall t > 0$, the point $x(t, x_0)$ (i.e. the point on the orbit starting at x_0, after time t has elapsed) is either in N_i or is not in N,

(ii) $S \subset \text{int}[N_1 \setminus (N_1 \cap N_2)]$,

(iii) N_2 is roughly the set of points through which orbits eventually leave N under the flow, i.e. if some point x in N_1 eventually (its orbit) flows out of N then it first passes through N_2.

The generalized Morse index of S is the homotopy class

$$h(S) = [N_1 / (N_1 \cap N_2)].$$

(Recall that two spaces A and B are homotopy equivalent if there exist two continuous mappings $\phi: A \to B$ and $\psi: B \to A$ such that $\phi \circ \psi$ and $\psi \circ \phi$ are homotopic to the identity in B and A respectively.)

One of the main properties is

THEOREM. $h(S)$ *is independent of the choice of* N_1 *and* N_2 *and also of the neighbourhood* S. *Furthermore suppose* $N \subset N'$, *and for* N' *we have an index pair* $\langle N_1', N_2' \rangle$, *and suppose*

$$[N_1 / (N_1 \cap N_2)] \neq [N_1' / (N_1' \cap N_2')].$$

Then N' *must contain some other invariant set besides* S.

Recently Amann and Zehnder [4] have applied this theorem to prove the existence of solutions of some nonlinear partial differential equations. They use the result in a rather special situation–which we now describe in the simplest case. Consider a real smooth function f defined on all of R^n, which, near the origin, is equal to a nondegenerate (i.e. having no zero eigenvalues) homogeneous quadratic having k negative eigenvalues. Near infinity f is also assumed to be a nondegenerate homogeneous quadratic having k' negative eigenvalues.

LEMMA. *If* $k' \neq k$ *then* f *must have a nontrivial critical point.*

REMARK. In case $k' - k$ is odd the result is easily proved using degree theory. Set $F = \text{grad } f$. The local degree of F near the origin,

$$\deg(F, |x| < \varepsilon, 0) = (-1)^k$$

as is easily seen. On the other hand for R large

$$\deg(F, |x| < R, 0) = (-1)^{k'}.$$

Thus if these numbers are different, F has to vanish at some point on $\varepsilon < |x| < R$. This proof does not require that F be a gradient map in $\varepsilon < |x| < R$. On the other hand if F is not a gradient map and $k' - k$ is even, the result need not hold. This argument also shows that if we just assume the condition on f near infinity, then it follows that f must have a stationary point.

Let us see how the Conley index is used to prove the lemma. We may suppose that near the origin,

$$f = \sum_1^{n-k} x_j^2 - \sum_{\alpha > n-k} x_\alpha^2.$$

Consider the gradient flow $\dot{x} = \operatorname{grad} f$. The origin is an isolated invariant set S. Let us take $N = \{|x| < R\}$, k large, $N_1 = |x| < \varepsilon$ small and the set generated by its flow in N, and $N_2 = N_1 \cap \{R - 1 \leqslant |x| < R\}$. In this case one finds

$$[N_1/(N_1 \cap N_2)] = [S^{n-k}, x_0]$$

where x_0 is a fixed point on S^{n-k}.

On the other hand if we take N_1' to be a large ball $|x| < R/2$ and the set generated by its flow in N and take N_2' correspondingly, using the behaviour near infinity we find

$$[N_1'/(N_1' \cap N_2')] = [S^{n-k'}, x_0].$$

Since the homotopy classes are different, for $k \neq k'$, we conclude from Conley's theorem that the region

$$\varepsilon < |x| < R$$

must contain another invariant set under the gradient flow. However, since it is a *gradient flow*, we see that f increases on a flow line. It follows easily that this other invariant set must consist of invariant points, i.e. stationary points of f.

K. C. Chang [19] has found a proof of the lemma which uses only Morse theory. It makes use of the Morse inequalities for a Morse function defined on a domain with boundary, and having suitable behaviour on the boundary.

We are coming closer to our second topic, variational methods, concerned with finding stationary points of real functions. To conclude this section, we mention one more useful result which is due to Castro and Lazer [17]. Though the result seems intuitively clear, I do not know an elementary proof of it.

THEOREM. *Let f be a smooth function defined on R^n having only a finite number of critical points and such that $f(x) \to +\infty$ as $|x| \to \infty$. Then for R large*

$$\deg(\operatorname{grad} f, B_R, 0) = 1.$$

They use this to prove the existence of more than one critical point:

LEMMA. *Let f be a smooth function in R^n with $f(x) \to +\infty$ as $|x| \to \infty$. Assume the origin is a nondegenerate critical point and that f achieves its minimum at a point $x_0 \neq 0$. Then f has at least three stationary points.*

This follows easily from the preceding theorem, using degree theory.

II. Variational methods

In the 1960's classical Morse theory was extended to real functions defined on infinite-dimensional spaces, such as Banach manifolds, by Palais, Smale and others–including the Morse inequalities. This material can be found in several sources, see for example [57], [49] and [50], and will not be taken up in these lectures.

We will discuss arguments for proving the existence of stationary points of real functions f defined on a Banach space X. In case the function f is bounded from above or below it is reasonable to try to show that it attains a maximum or minimum. For convex functions, a classical result in this direction is

THEOREM. *Let f be a lower-semi-continuous convex function, bounded from below, defined on a reflexive Banach space, such that $f(x) \to \infty$ as $x \to \infty$. Then f attains its minimum.*

Results on existence of stationary points make use of some kind of compactness. In the theorem just quoted the compactness is hidden. The proof uses the fact that a closed bounded convex set in a reflexive Banach space is compact in the weak topology. Here is the proof of the theorem. Let x_n be a minimizing sequence, i.e. $f(x_n) \to \inf f$. Clearly the $|x_n|$ are bounded, and so a subsequence which we still denote by x_n converges weakly to some x. By Mazur's lemma there is a sequence of convex combinations y_i of $(x_1, \ldots, x_i)$, $i = 1, 2, \ldots$, converging to x strongly. From the convexity of f we see that $f(y_i) \to \inf f$, and from the lower-semi-continuity we conclude that $f(x) = \inf f$. See [25] for variational problems for convex functions.

If f is not convex it need not achieve its infimum. However the following result of Ekeland [23] shows the existence of points which are almost points of minimum.

THEOREM. *Let X be a complete metric space and f a real lower semicontinuous function defined on X and bounded from below. Then there exists some x_0 in X such that*

$$f(x) > f(x_0) - \text{dist}(x, x_0) \quad \text{for all } x \in X, x \neq x_0.$$

II.1 A compactness-kind of condition that is usually employed in proving the existence of stationary points is the condition C of Palais and Smale, now called the (PS) condition, for a function $f \in C^1$.

(PS) *Any sequence $\{x_j\} \in X$ such that $|f(x_j)| \leq M$ and $f'(x_j) \to 0$ in norm in X^* (the dual space) has a strongly convergent subsequence.*

Here $f'(x)$ represents the derivative of f at x, and is an element of the dual space X^* of continuous linear functionals on X. Such a function always takes on its infimum:

LEMMA. *Let f be a real C^1 function on a Banach space satisfing* (PS) *and bounded from below. Then f achieves a minimum at some point.*

REMARKS. 1. In practice, given a function f for which we seek solutions of

$$F(x) \equiv f'(x) = 0,$$

the exact topological space on which one should work is not given to us. It should be chosen sufficiently small so that f is smooth, say in C^1. On the other hand it cannot be taken *too* small for then the condition (PS) will not hold. One has to strike just the right balance.

2. It is natural to ask how verification of (PS) differs from the usual task of obtaining a priori estimates, i.e., showing that the map $F(x) = f'(x)$ is proper. There is one additional useful item here; the fact that the $f(x_j)$ are all bounded.

Let us illustrate these remarks by discussing a simple semilinear boundary value problem for a real function $u(x)$ defined in a bounded domain G in R^n with smooth boundary ∂G.

$$\Delta u + g(x, u) = 0 \text{ in } G, \qquad u = 0 \text{ on } \partial G; \tag{5}$$

(here $\Delta = \Sigma \, \partial^2/\partial x_i^2$ is the Laplace operator). We seek a positive solution u assuming

$$\begin{cases} g \text{ is smooth and } g(x, u) > 0 \quad \text{for } u > 0, x \in \Omega, \\ g(x, u) = o(u) \quad \text{as } u \to 0 \text{ uniformly in } x, \\ g(x, u) = a(x)u^\sigma, a(x) > 0 \text{ in } \overline{G}, \sigma > 1, \text{ for } u \text{ large.} \end{cases} \tag{6}$$

If $\sigma < (n + 2)/(n - 2)$, (5) has a positive solution $u(x)$–as we will see. The solution will be obtained as a stationary point of the functional

$$f(u) = \int_G \left[\frac{1}{2} |\text{grad } u|^2 - G(x, u(x)) \right] dx \tag{7}$$

where $G(x, u) = \int_0^u g(x, s) \, ds$. Since we are looking for a positive solution there is no loss of generality in changing g for $u < 0$ and assuming that $g(x, u) > 0$ for $u = 0$ and

$$g(u) = a(x)|u|^\sigma \quad \text{for } |u| \text{ large.}$$

For such a g, any nontrivial solution of (5) will automatically be positive, by the maximum principle, and so a solution of our original equation.

What space of functions should we work in? and will condition (PS) hold?

We wish f to be of class C^1 in the space. From the form of f it seems reasonable to work in a space contained in H_0^1, i.e. the space of functions in the Sobolev space H^1 which vanish on the boundary ∂G. According to the

Sobolev embedding theorem,

$$\begin{cases} \text{the injection map } H_0^1 \to L^p(G), p > 1, \text{ is} \\ \text{continuous} \Leftrightarrow p \leqslant \dfrac{2n}{n-2}, \text{ and compact} \Leftrightarrow p < \dfrac{2n}{n-2}. \end{cases} \tag{8}$$

Now

$$G(x, u) \sim \frac{a(x)u^{\sigma+1}}{\sigma+1} \quad \text{for } u \text{ large,}$$

so for f to be well defined on H_0^1 we should require $\sigma + 1 \leqslant 2n/(n-2)$, i.e. $\sigma \leqslant (n+2)/(n-2) = \sigma_0$. In case $\sigma > \sigma_0$, one might be tempted to work in, say, $H^s \cap H_0^1$ for some $s > 1$. But after a little reflection, one concludes that control on f cannot yield control on the H^s norm, so that (PS) cannot hold. Thus it seems sensible to work in the space $X = H_0^1$ and to require $\sigma \leqslant \sigma_0$. We may then take $\|u\|_{H_0^1}^2 = \int_G |\text{grad } u|^2$. But then does (PS) hold?

LEMMA. *Condition* (PS) *holds* $\Leftrightarrow \sigma < \sigma_0$.

SKETCH OF PROOF OF $\Leftarrow$. We leave the other part as an exercise. Let $\{u_j\}$ be a sequence in H_0^1 such that

$$|f(u_j)| \leqslant M, \qquad f'(u_j) \to 0 \quad \text{in norm.}$$

The last condition means that

$$\int_\Omega \text{grad } u_j \text{ grad } \zeta - g(x, u_j)\zeta = o(1) \cdot \|\zeta\|_{H_0^1}, \forall \zeta \in H_0^1. \tag{9$_j$}$$

Inserting $\zeta = u_j$ we find, setting $\|u_j\|_{H_0^1} = c_j$,

$$\frac{1}{c_j} \int_G |\text{grad } u_j|^2 - u_j g(x, u_j) \to 0. \tag{10}$$

Since $\int_G \frac{1}{2} |\text{grad } u_j|^2 - G(x, u_j)$ is bounded and

$$\int G(x, u_j) - \frac{1}{\sigma+1} u_j g(x, u_j) \text{ is bounded}$$

we find

$$\int \frac{\sigma+1}{2} |\text{grad } u_j|^2 - u_j g(x, u_j) \text{ is bounded.} \tag{11}$$

CLAIM. The norms $c_j = \|u_j\|_{H_0^1}$ are bounded. If not, for a suitable subsequence, we may suppose $c_j \to \infty$. Combining (10) and (11) we find

$$\left(\frac{\sigma+1}{2} - 1 \right) c_j \text{ is bounded}$$

–a contradiction since $\sigma > 1$; the claim is proved.

We may select a subsequence, still denoted by u_j, which converges weakly to u in X, and converges strongly to u in L^β, for some β in $\sigma + 1 < \beta < \sigma_0 + 1$. This is possible in view of the Sobolev embedding theorem quoted above. One then finds easily that $g(x, u_j(x))$ converges to $g(x, u(x))$ in L^{σ_0+1}. It follows easily, considering $(9)_j$–$(9)_k$, with $\zeta = u_j - u_k$, that

$$\|u_j - u_k\|^2_{H_0^1} \to 0 \quad \text{as } j, k \to \infty$$

and the lemma is proved.

So for $\sigma < (n + 2)/(n - 2)$, our functional f given in (7) satisfies (PS). Next we will see how one finds nontrivial stationary points.

II.2 We shall present several arguments for finding stationary points of a functional f which is unbounded from above and below–leading us to look for saddle points. These are found by some min max argument. One considers a certain class A of sets Σ and forms

$$\max_{u \in \Sigma} f(u)$$

for any Σ in A and then

$$c = \inf_{\Sigma \in A} \max_{u \in \Sigma} f(u),$$

and tries to show that the number c is a stationary value of f, i.e., is a value of f at some stationary point.

In case f is an even function, $f(u) = f(-u)$, there is the well-developed category theory of Liusternik-Schnirelman and the related notion of genus to find solutions (often infinitely many). In this approach one takes for A the class of all sets Σ of category $\geqslant$ some number. An excellent presentation of this theory may be found in [50], and we will not discuss this method. Analogous theories in case other group actions are present have been developed by various authors (see Benci [8], [9]).

The methods we will take up here are all variations on a basic result known to everyone who has done any walking in the hills: the mountain pass lemma.

Let us first describe this for a real function $f(x)$, defined on the plane–representing the height of the land above sea level over the point $x \in R^2$. Suppose the origin lies in a valley surrounded by a ring of mountains, i.e. there is an open neighbourhood Ω of the origin such that

$$f(x) \geqslant c_0 > f(0) \quad \text{for } x \in \partial\Omega.$$

Suppose that there is some point x_0 outside, i.e., $x_0 \notin \overline{\Omega}$, such that $f(x_0) < c_0$. We wish to walk from x_0 to 0 climbing as little as possible, i.e. keeping f as low as possible. Naturally the way to do this is to take a path crossing the mountain over the lowest mountain pass. The top of the mountain pass corresponds to a stationary point of f and there the value of f, i.e. the stationary value, equals

$$c = \inf_P \max_{x \in P} f(x) \geqslant c_0.$$

Here P represents any continuous path from x_0 to 0 and inf is taken with respect to all such paths. Clearly every such path has to intersect $\partial\Omega$, and so $\max_P f \geqslant c_0$. This is the assertion of the mountain pass lemma–that the number c so defined is a stationary value of f. Note that c will, in general, be less than sup f.

This formulation is not quite correct since the plane is not compact, and one should add a compactness condition, for example, (PS). The lemma is then true and holds even in Banach space. In this form it is due to Ambrosetti and Rabinowitz [6], though similar arguments had been employed many years ago in proving the existence, in special situations, of several minimal surfaces spanning some curve.

MOUNTAIN PASS LEMMA [6]. *Let f be a C^1 real function defined on a Banach space X and satisfying (PS). Assume there is an open neighbourhood Ω of 0 and a point $x_0 \notin \bar{\Omega}$ such that*

$$f(0), f(x_0) < c_0 \leqslant \inf_{\partial\Omega} f.$$

Then the following number is a critical value of f,

$$c = \inf_P \max_P f \geqslant c_0$$

where P represents any continuous path from x_0 to 0.

In recent years extensions and variations of this lemma have proved very useful and we will present several in the next section. But first we show, with its aid, how to obtain a positive solution of (5) for $\sigma < \sigma_0 = (n + 2)/(n - 2)$.

THEOREM. *For g satisfying (6) with $\sigma < \sigma_0$ there exists a positive solution of* (5).

PROOF. Consider the functional (7) in $X = H_0^1$. We have shown that it satisfies (PS) and, for g smooth, one verifies that it is of class C^1. Note that $f(0) = 0$. To find a nontrivial solution via the mountain pass lemma we will show that for some $r > 0$, $c_0 > 0$ we have

$$f(u) \geqslant c_0 \quad \text{if } \|u\|_{H_0^1} = r. \tag{18}$$

Furthermore for some u_0, $\|u_0\|_{H_0^1} > r$, $f(u_0) \leqslant 0$. From the mountain pass lemma it will follow that there is a stationary point $u \in H_0^1$ of f, with $f(u) \geqslant c_0 > 0$. Thus $u \neq 0$.

The function u is a generalized solution of (5) but by a standard boot strap argument one shows that it is smooth, and then, by the maximum principle, is positive.

To prove (11) we now make use of (6) from which we may infer that for any $\varepsilon > 0$

$$G(x, u) \leqslant \varepsilon |u|^2 + c_\varepsilon |u|^{\sigma + 1}.$$

Thus for $\|u\|_{H_0^1} = r$,

$$\int G(x, u(x)) \leqslant \int \varepsilon u^2 + C_\varepsilon |u|^{\sigma + 1}$$

$$\leqslant C(\varepsilon r^2 + C_\varepsilon r^{\sigma + 1}), \qquad C \text{ independent of } \varepsilon,$$

by (8). Consequently for $\|u\|_{H_0^1} = r$,

$$f(u) = \frac{1}{2} r^2 - \int G(x, u) \geqslant \frac{1}{2} r^2 - C(\varepsilon r^2 + C_\varepsilon r^{\sigma + 1})$$

and (12) follows for ε and then r, chosen small.

Finally we have to choose u_0 so that

$$\|u_0\|_{H_0^1} > r \quad \text{and} \quad f(u_0) \leqslant 0.$$

Simply take $u_0 = \lambda v$, for v a fixed smooth positive function vanishing on the boundary, and λ sufficiently large. The existence theorem is proved.

REMARKS. Suppose we consider in place of (5) the boundary value problem

$$Lu + g(x, u) = 0 \text{ in } G, \qquad u = 0 \text{ on } G, \tag{5$'$}$$

with g satisfying (6) and L a linear elliptic operator with smooth coefficients

$$Lu = a^{ij}(x)u_{x_i x_j} + a^i(x)u_{x_i} + a(x)u,$$

with $a^{ij}(x)\xi_i\xi_j$ positive definite, and $a(x) \leqslant 0$. If L is not formally selfadjoint the problem cannot be formulated as a variational one. In [13] Brézis and Turner solved this (permitting g to depend also on grad u) in case $\sigma < n/(n-2)$. In this case they were able to obtain the appropriate a priori estimates for *all* positive solutions of (5$'$). In case $L = \Delta$ the corresponding a priori estimates for the general case $\sigma < (n+2)/(n-2)$ have recently been established by De Figueiredo, Lions and Nussbaum [22], and for general L by Gidas and Spruck [32].

For $\sigma = \sigma_0 = (n+2)/(n-2)$ a nonzero solution of (5) need not exist, as was shown by Pokhozhaev for $g = u^{\sigma_0}$ in a star-shaped domain (see [32]). On the other hand in a ring shaped domain $r_1 < |x| < r_2$, and $g = u^{\sigma_0}$, there *is* a positive solution.

Several problems of current interest are just on the borderline where (PS) fails. Among these are the Yang-Mills equations of Physics, and the Yamabe problem in Differential Geometry. In the latter, one is given a compact Riemannian manifold without boundary and wishes to find a new metric on it which is conformally equivalent to the given one, and having constant scalar curvature.

II.3 We now take up some extensions of the mountain pass lemma. The following is due to Rabinowitz [52]; we shall present a more general form due to Ni [45]. Let us first describe it for a function f defined in R^3, of class C^1 and satisfying (PS).

GENERALIZATION. Assume there are two disjoint simple closed curves Γ_0, Γ_1 in R^3 which link each other and such that the values of f on one are bounded away from its values on the other, i.e.,

$$f(x) \leqslant c_0 < c_1 \leqslant f(y), \qquad \forall x \in \Gamma_0, \forall y \in \Gamma_1.$$

Then the following number is a critical value of f.

$$c = \inf_{\Sigma} \ \max_{x \in \Sigma} f(x) \geqslant c_1$$

where Σ represents any surface spanning the curve Γ_0 and inf is taken over all such Σ.

The linking of the two curves is simply to ensure that any surface Σ spanning Γ_0 necessarily intersects Γ_1, so that $\max_\Sigma f \geqslant c_1$. This result seems much less intuitive than the preceding. Here is the general form.

GENERALIZED MPL. *Let f be a C^1 real function defined on R^N satisfying* (PS). *Let ϕ and ψ be continuous mappings of spheres S^k and S^{N-k-1} into R^N whose images are disjoint and have nontrivial linking. Assume that*

$$f(\phi(x)) \leqslant c_0 < c_1 \leqslant f(\psi(y)), \qquad \forall x \in S^k, \forall y \in S^{N-k-1}.$$

Then

$$c = \inf_{h} \ \max_{x \in B^{k+1}} \ f(h(x)) \geqslant c_1$$

is a critical value of f. Here h represents any continuous extension of ϕ inside the unit ball B^{k+1} in R^{k+1}, and the inf *is taken over all such h.*

REMARKS. If we reverse the roles of ϕ and ψ we see that there is also a critical value $\leqslant c_0$. The condition on linking is simply to ensure that for every extension h, $h(B^{k+1})$ necessarily intersects the image $\psi(S^{N-k-1})$. In fact the theorem holds if the set $\psi(S^{N-k-1})$, on which f is assumed to be $\geqslant c_1$, is any set with the property that it intersects every image $h(B^{k+1})$. Furthermore, in place of R^N we may have an N-dimensional manifold, and in place of S^k, the boundary of a compact $(k+1)$-dimensional manifold. The original mountain pass lemma corresponds essentially to the case $k = 0$. See [45] for an infinite-dimensional version and [54], [10] and [8] for related results.

Here is still another variant of these ideas due to A. Castro [16].

THEOREM. *Let f be a C^1 real function defined on a Banach space X satisfying* (PS). *Assume X has a direct sum decomposition*

$$X = X_1 \oplus X_2, \qquad \dim X_1 = k < \infty.$$

Let $S_1(B_1)$ be the unit sphere (ball) in X_1, and S_2 a sphere $\|x\| = R$ in X_2. Assume for some constants $\beta < \alpha$; $c_0 < c_1$,

$$f(x) \leqslant c_0 \quad \text{for } x \in X_1, \|x\| < 1,$$
$$f(x) \leqslant \beta \quad \text{for } x \in X_1, \|x\| = 1 \text{ i.e. } x \in S_1,$$
$$f(x) \geqslant \alpha \quad \text{for } x \in X_2, \|x\| \leqslant R,$$
$$f(x) \geqslant c_1 \quad \text{for } x \in X_2, \|x\| = R \text{ i.e. } x \in S_2.$$

Then the following is a critical value of f:

$$c = \inf_{\Sigma} \ \max_{x \in \Sigma} f(x) \geqslant \alpha.$$

Here Σ is any k-dimensional surface in X spanning S_1 and homotopic to B_1 in $X \setminus S_2$, and the inf *is taken with respect to all such Σ.*

The mountain pass lemma and all the extensions given here, as well as the lemma near the beginning of §II.2, are proved in the same way by a well-known procedure which is presented in a systematic way in Palais [49]. We will indicate it by sketching the proof of the Generalized MPL. To avoid some technicalities let us consider the case that X is a Hilbert space and $f \in C^2$.

PROOF. Assume that the number c defined there is not a critical value.

Then for $0 < \varepsilon$ small, it follows from (PS) that for some $\alpha > 0$

$$\|f'(x)\| \geq \alpha > 0 \quad \text{if } c - \varepsilon < f(x) < c + \varepsilon.$$

We may take $\varepsilon < c_1 - c_0$. Choose an extension h of ϕ inside B^{k+1} with

$$f(h(x)) \leq c + \frac{\varepsilon}{2}, \qquad \forall x \in B^{k+1}.$$

We shall make use of a modification of the gradient flow associated to $-f$. Let $\eta(u)$ be a locally Lipschitz function satisfying

$$\eta \equiv 1 \quad \text{on } f^{-1}\left(c - \frac{\varepsilon}{2}, c + \frac{\varepsilon}{2}\right),$$

$$\eta \equiv 0 \quad \text{outside } f^{-1}(c - \varepsilon, c + \varepsilon).$$

Consider the flow associated with the differential equation in X:

$$\frac{dx}{dt} = -\eta(x)\frac{f'(x)}{\|f'(x)\|^2}, \qquad x(0) = y,$$

and denote the solution, which exists locally, by $x(t) = x(t, y)$. Note that on a solution curve $x(t)$ we have

$$\frac{d}{dt} f(x(t)) = -\eta(x(t)) \leq 0.$$

Thus within a time interval ε, every point x in the region

$$f^{-1}(c - \varepsilon/2, c + \varepsilon/2),$$

where $\eta \equiv 1$, flows into the region where $f \leq c - \varepsilon/2$. In particular, since f decreases under the flow, it follows that the image

$$\tilde{h}(y) = x(\varepsilon, h(y)), \qquad y \in B^{k+1},$$

lies in the region where $f \leq c - \varepsilon/2$. However, for $y \in S^k = \partial B^{k+1}$, we have $f(\phi(y)) \leq c_0 < c_1 - \varepsilon \leq c - \varepsilon$. Hence $\eta(y) = 0$ and the point y does not move under the flow. Consequently

$$\tilde{h}(y) = h(y) = \phi(y) \quad \text{for } y \in S^k,$$

and $\tilde{h}$ is an admissible map in our competition. But then $\tilde{h}(B^{k+1})$ cannot lie in the region where $f \leq c - \varepsilon/2$. Contradiction!

II.4 This section will be devoted to two results of Rabinowitz [53] on periodic solutions for a Hamiltonian system–as illustration of the use of some of the results in the preceding section.

A Hamiltonian system is a system of ordinary differential equations for a pair of vectors p, q in R^n, depending on time

$$\dot{p} = \frac{dp}{dt} = -H_q, \quad \text{i.e.,} \quad \frac{dp^i}{dt} = -\frac{\partial H}{\partial q^i},$$

$$\dot{q} = \frac{dq}{dt} = H_p, \quad \text{i.e.,} \quad \frac{dq^i}{dt} = \frac{\partial H}{\partial p^i}.$$

Here the Hamiltonian $H(p, q)$ is a given real function defined on R^{2n}. The problem is to prove the existence of nontrivial, i.e. nonconstant, time periodic solutions.

It is trivial to verify that on any solution curve $p(t)$, $q(t)$, the Hamiltonian $H(p(t), q(t))$ is constant. Thus it makes sense to look for a periodic solution on a given level surface of H. The first result of Rabinowitz is the following (for convenience we take H to be smooth).

THEOREM A. *Assume that* grad $H \neq 0$ *on the level set* Σ *where* $H = 1$ (*so that* Σ *is a regular hypersurface*). *Assume that* Σ *is compact and strictly star shaped about the origin, i.e., any ray from the origin hits* Σ *at just one point and nontangentially. Then there is a nontrivial periodic solution of* (13) *on* Σ.

This is a beautiful result and we will give, essentially, a complete proof of it. There are earlier perturbation results about the existence of nontrivial periodic solutions near a trivial one, and also a global result by Weinstein in case H is convex–by quite different methods. (See [**53**] for references.) It is natural to ask whether the result is true if Σ is merely diffeomorphic to a sphere. This problem is open.

First a useful

REMARK. If grad $H \neq 0$ on a level set Σ of H, then the set of integral curves of (13) on Σ depend *only* on the hypersurface Σ–they are otherwise independent of H.

This is easily proved. Suppose H' is another Hamiltonian having Σ as level set and with grad $H' \neq 0$ on Σ. Then

$$\text{grad } H' = f \text{ grad } H \quad \text{on } \Sigma,$$

with f a positive function on Σ. Suppose $p(t)$, $q(t)$ is an integral curve for H' on Σ, i.e.

$$\dot{p} = -H'_q = -fH_q, \qquad \dot{q} = H'_p = fH_p.$$

If we reparametrize this curve by a new variable s with $ds = f \, dt$, we see that

$$dp/ds = -H_q, \qquad dq/ds = H_p$$

on the curve, and the remark is proved.

Theorem A is a simple consequence of a very general result in [**53**]:

THEOREM B. *Consider a* C^1 *Hamiltonian* H *satisfying*
(i) $H \geqslant 0$ *and* $H(p, q) = o(|p|^2 + |q|^2)$ *near the origin*,
(ii) $0 < H < \theta(p \cdot H_p + q \cdot H_q)$, $0 < \theta < \frac{1}{2}$ *for* $|p|^2 + |q|^2$ *large*.
Then for this H, *and any given* $T > 0$, *there exists a nontrivial periodic solution of* (13) *with period* T.

REMARK. By integrating condition (ii) we find

$$H \geqslant C(|p|^2 + |q|^2)^{1/2\theta} \text{ near infinity}, C > 0, \tag{ii$'$}$$

i.e. H grows faster than quadratically near infinity.

DERIVATION OF THEOREM A. By the remark, we may replace the given Hamiltonian by any other provided the set where it equals 1 is Σ. Choose such a new H to be positive homogeneous of degree 4. The existence of such an H follows immediately from the strict star-shaped property, and in fact it is only here that the property is used.

Applying Theorem B with $T = 1$ to this new H we infer that there is a nontrivial solution $p(t)$, $q(t)$ with period 1. It lies on *some* level surface $H = c$, but not necessarily the desired one Σ: $c = 1$. To obtain such a solution we simply perform a suitable scaling (stretching) using the homogeneity of H, and the theorem is proved.

We are going to sketch some of the main points in the proof of Theorem B. By a stretching of the time variable, and of H, we may suppose that $T = 2\pi$. In the space of 2π periodic functions $p(t)$, $q(t)$, consider the functional

$$F(p, q) = \int_0^{2\pi} \left[p \cdot \frac{dq}{dt} - H(p(t), q(t)) \right] dt.$$

It is readily verified that a stationary point (a pair of 2π periodic functions $p(t)$, $q(t)$) of F is a solution of (13). That is, (13) is the corresponding Euler equation. The pair $p(t) \equiv q(t) \equiv 0$ is such a trivial solution but we seek a nontrivial one. Since H grows faster than quadratically at infinity we see, on taking $p = q = $ large constant vectors, that F is unbounded from below. Taking bounded and highly oscillatory functions p, q, we see that F is also unbounded from above. We wish, therefore, to find a saddle point of F–other than the trivial one.

The procedure in [53] is the following: (a) approximate the space by a suitable finite-dimensional one; (b) in the finite-dimensional space, prove the existence of a nontrivial stationary point; (c) go to the limit in the approximation. The last part is rather technical. In particular, in going to the limit, one has to make sure that the stationary point obtained in step (b) does not run off to infinity nor to the origin.

The finite-dimensional approximation is the following: Let us consider the space of 2π-periodic functions $(p(t), q(t))$ in L^2. Then the action integral

$$Q(p, q) = \int^{2\pi} p \cdot \frac{dq}{dt} \, dt$$

is a quadratic form and associated with it are its eigenvalues λ, which are integers, and eigenvectors spanned by finite Fourier series. Take as approximating space, the space R^N spanned by the eigenfunctions associated with the eigenvalues in $|\lambda| \leqslant M$ for large M. Restricting F to this space we will show how to obtain a nontrivial stationary point–using just some simple properties of F. First decompose $R^N = E_1 \oplus E_2$ where E_1 is the subspace spanned by all eigenvectors of Q having nonpositive eigenvalues, E_2 is the subspace spanned by the eigenvectors with positive eigenvalues. These spaces are mutually orthogonal in L^2. From the fast growth property (ii$'$) we see that outside a large ball of radius R (we are using the L^2 norm), the term H in the integrand of F dominates the other term, which has quadratic growth. Thus $F < 0$ outside a ball of radius R. (This is true only in finite dimensions; the reasoning would be false in infinite-dimensional space.) In fact, we see that $F \to -\infty$ as we go to infinity, so F satisfies (PS) in R^N.

Since $H \geqslant 0$ we see that $F \leqslant 0$ on E_1. Consider the behaviour of F on E_2, on which $Q \geqslant \lambda_1(|p|^2 + |q|^2)$ for some $\lambda_1 > 0$. From property (i) of H we see that for small $\varepsilon > 0$,

$$F(p, q) \geqslant c_1 > 0 \quad \text{on } \|(p, q)\|_{L^2} = \varepsilon.$$

For convenience we will denote the pair (p, q) by z.

From these simple properties we will now derive the existence of a nontrivial stationary point of F on R^N. The point where F achieves its maximum is of course such a point. But it is useless. For as we go to the limit on the approximation, the maximum of F on R^N will go to $+\infty$.

Instead we may apply the Generalized MPL of the last section. For the sphere S^{N-k-1} and the map ψ we take the ε sphere about the origin in E^2 and the injection map. For the image of the sphere S^k, $\phi(S^k)$, we take the plane E_1. One will naturally object that this is not topologically a sphere. Well, just make it one by wrapping it around, i.e., closing it off, near infinity outside the ball of radius R. It is intuitively clear (and easy to prove, using degree theory) that these two spheres really link. Hence by the Generalized MPL the following number is a stationary value of F.

$$c = \inf_h \max_{x \in B^{k+1}} F(h(x)) \geqslant c_1 > 0.$$

This may be much lower than the maximum of F, in fact we will see that it is bounded from above by a constant independent of M.

When closing off E_1 near infinity, let us do this in such a way that the resulting "sphere" lies in the space E', spanned by E_1 and a fixed unit eigenvector $\phi = (p_0, q_0)$ corresponding to the lowest positive eigenvalue of Q. We may take for B^{k+1} the "ball" in E_1' bounded by the "sphere".

First we will obtain an upper bound for c independent of M. If we take for h, in the definition of c, the identity (i.e., injection) map, then

$$0 < c \leqslant \max_{B^{k+1}} \left(Q - \int H\right). \tag{14}$$

This maximum is taken on at some point $z = (p(t), q(t))$ of the form

$$z = a\phi + z_1, \qquad z_1 \in E_1$$

and $\|z\|^2 = a^2 + \|z_1\|^2$. Since $H \geqslant 0$ we see, by our choice of E_1, that

$$c \leqslant Q(z) = a^2\lambda_1 + Q(z_1) \leqslant a^2\lambda_1 \leqslant \lambda_1\|z\|^2.$$

Furthermore we see from (14) that

$$\int H(z) < Q(z) \leqslant \lambda_1\|z\|^2.$$

Using (ii'), recall that $\theta \leqslant \frac{1}{2}$, we obtain easily a bounded on $\|z\|$, and hence on c, independent of M.

The last part of the argument, the limit process, is technical and we will confine the rest of our discussion to the simple case of a function H which is positive homogeneous of degree 4. In this way, at any rate, we include a complete proof of Theorem A.

Observe first that a stationary point $z = (p(t), q(t))$ in R^N where F has the stationary value c, satisfies the Euler equation

$$\int_0^{2\pi} \left[p \cdot \dot{\psi} + \phi \cdot \dot{q} - H_p(p, q) \cdot \phi - H_q(\ ) \cdot \psi \right] dt = 0 \qquad (15)$$

for every function $(\phi(t), \psi(t))$ lying in our space R^N. Here $\dot{\psi} = d\psi/dt$ etc. If we set $\phi = p, \psi = q$ we find

$$\int 2p \cdot \dot{q} = \int p \cdot H_p + q \cdot H_q = 4 \int H \qquad (16)$$

by the homogeneity hypothesis. On the other hand,

$$c = \int p \cdot \dot{q} - H,$$

and hence

$$\int H = c \leqslant C.$$

(The letter C will be used to denote various constants independent of M.) By the homogeneity, it follows that

$$\|z\|_{L^4} \leqslant C. \qquad (17)$$

Next we wish to establish the estimate

$$\|\dot{z}\|_{L^2} \leqslant C \qquad (18)$$

and to this end we set $(\phi, \psi) = (\dot{q}, -\dot{p})$ (which is indeed in E_1) in (15). After a partial integration we find

$$\|\dot{z}\|_{L^2}^2 = \int H_p \cdot \dot{q} - H_q \cdot \dot{p} \leqslant \|\dot{z}\|_{L^2} \|\mathrm{grad}\ H\|_{L^2}$$

so that

$$\|\dot{z}\|_{L^2} \leqslant \|\mathrm{grad}\ H\|_{L^2} \leqslant C\|z\|_{L^6}^3. \qquad (19)$$

Inserting (19) and (17) in the standard interpolation inequality

$$\|z\|_{L^\infty} \leqslant C\left[\int |z|^2 + |\dot{z}|^2 \right]^{1/6} \|z\|_{L^4}^{2/3},$$

we find

$$\|z\|_{L^\infty} \leqslant C\left(1 + \int |z|^6 \right)^{1/6} \leqslant C(1 + C\|z\|_{L^\infty}^2)^{1/6} \quad \text{by (17).}$$

From this it follows that $\|z\|_{L^\infty} \leqslant C$, and (18) then follows from (19).

From these inequalities we may draw a further conclusion. Set $\|z\|_{L^\infty} = \rho$. According to (16) we have for our solution z

$$2 \int H = \int p \cdot \dot{q} = \int (p - \bar{p}) \cdot \dot{q}$$

where $\bar{p}$ is the average of p on $(0, 2\pi)$,

$$\leqslant \|\dot{q}\|_{L^2} \|p - \bar{p}\|_{L^2} \leqslant C\|\dot{q}\|_{L^2} \|\dot{p}\|_{L^2};$$

thus

$$\int H \leqslant C\|\dot{z}\|_{L^2}^2. \qquad (20)$$

On the other hand, by (19),

$$\|\dot{z}\|_{L^2}^2 \leqslant C\int |z|^6 \leqslant C\rho^2 \int |z|^4 \leqslant C\rho^2 \int H.$$

Inserting this in (20) we find, since $\int H > 0$,

$$1 \leqslant C\rho^2$$

i.e.

$$\rho = \|z\|_{L^\infty} \geqslant \delta \qquad\qquad (21)$$

where δ is a fixed positive constant independent of M.

We are now in a position to carry out the limit process. Let $M \to \infty$ through the integers. By (18) we may assert that a subsequence of the $z = z_M$ converge weakly in $H^1(S^1)$ and also (by the Sobolev embedding theorem) *uniformly* to a function $z = (p(t), q(t))$ which is a solution of (13). By (21) we have $\|z\|_{L^\infty} \geqslant \delta > 0$ and so z is a nontrivial solution. A standard regularity argument then shows that the solution is smooth. Theorem B for the special case, and hence Theorem A, are proved.

REMARKS. 1. By Theorem 2, for every positive integer k, (13) has a nontrivial solution with period T/k. Hence (13) has infinitely many solutions with period T. It is an open problem whether there are solutions of period T for which T is the smallest period.

2. Remarkably, Rabinowitz [54] has shown that Theorem B holds even if assumption (i) is dropped. The proof, again via a minimax argument, is based on a cohomological index theory developed by Fadell and Rabinowitz [29]. Furthermore, there exist solutions of (13) with given period T and having arbitrarily large L^∞ norms.

3. Clarke and Ekeland [20] have proved an analogue of Theorem B but under somewhat opposite conditions on H. They assume H to be convex and to satisfy

$$\frac{H}{|p|^2 + |q|^2} \to \begin{pmatrix} \infty \\ 0 \end{pmatrix} \quad \text{at} \begin{pmatrix} 0 \\ \text{infinity} \end{pmatrix},$$

and prove that there exists a nontrivial solution having any given $T > 0$ as *minimal* period. Their proof is based on a different variational problem.

They introduce a different Lagrangian $F(p, q)$ for which the Euler equation is still (13). It is, essentially, a dual to the F treated above. In [24] Ekeland treated a Hamiltonian behaving like that of Rabinowitz using a dual variational problem.

4. In connection with Theorem A, Ekeland and Lasry [26] consider a Hamiltonian H for which the level set Σ: $H = 1$ is a closed convex set lying in an annular region

$$0 < \frac{r}{2} < |p|^2 + |q|^2 < r.$$

They prove that there are at least n different periodic solutions of (13) on Σ. (A closed solution curve may of course be traversed several times, but all these solutions are not considered to be different.)

II.5 In Remark 3 at the end of the preceding section we mentioned the use of a kind of dual Lagrangian in finding periodic solutions. In this section we will describe an adaptation of it applied to a different problem: that of finding time-periodic solutions $u(x, t)$ of a nonlinear vibrating string equation on an interval $(0, \pi)$:

$$u_{tt} - u_{xx} + g(u) = 0, \quad 0 < x < \pi; \qquad u(0, t) = u(\pi, t) = 0. \qquad (22)$$

Here we assume $g(0) = 0$, so that $u = 0$ is a solution, but we seek a nontrivial solution having a given time period T.

This problem bears some resemblance to the one treated in the preceding sections and similar methods are used. Until a few years ago most results for this problem were local: g was assumed to be appropriately small (perhaps with a small factor ε attached) and one sought solutions close to zero. In the last few years new methods have been introduced enabling one to treat large g. However, up to now the results are mainly confined to the case that g is monotone (either increasing or decreasing). Thus we cannot yet treat the sine-Gordon equation: $g = \sin u$–for which explicit time periodic solutions are known in case the string goes to infinity in both directions.

There is also a restriction on the given time period T. For technical reasons we can, essentially, till now, only treat the case that $T = 2\pi/\lambda$ is a rational multiple of the length of the string, i.e., λ is rational. The reason for this will be explained presently.

Much work has been done in the case that g grows at most linearly as $|u| \to \infty$ (see [14] and [5] for some recent results). The case that g grows superlinearly is, perhaps, more interesting. The main result in this case is due to Rabinowitz [52]. Assuming g as above and satisfying

$$g(u) = o(u) \quad \text{as } u \to 0,$$

$$G(u) \leqslant \theta u g(u), \qquad |u| \text{ large, some } \theta \text{ in } \left(0, \tfrac{1}{2}\right)$$

he proved there is a nontrivial periodic solution with any given time period $T = 2\pi/\lambda$, λ rational. A particular, and interesting, example is $g = u^3$. The solution is obtained as a nontrivial stationary point of the Lagrangian, defined for functions $u(x, t)$ satisfying the boundary condition and with time period $2\pi/\lambda$,

$$F(u) = \int_0^{2\pi} \int_0^{\pi} \left[\frac{1}{2}(u_x^2 - u_t^2) + G(u(x, t)) \right] dx \, dt \qquad (23)$$

where

$$G(u) = \int_0^u g(s) \, ds.$$

His proof follows the scheme of the one in the preceding section and makes use of the Generalized MPL–which indeed he introduced in [52].

We will describe a different variational problem using a sort of dual Lagrangian, which leads to a simpler proof of the result. This is due to Brézis, Coron, Nirenberg [15]. The case $g = u^3$ admits of a particularly simple treatment. Ω will denote the period rectangle $0 < x < \pi, 0 < t < T$.

First some remarks about the condition that λ is rational. This comes from an analysis of the linear problem for u with time-period $2\pi/\lambda$:

$$Au = u_{tt} - u_{xx} = f, \qquad u(0, t) = y(\pi, t) = 0.$$

Fourier expansion provides a complete analysis of this–a sine series in x being appropriate because of the boundary condition. Arguing formally, let us set

$$u = \sum_{\substack{j>0 \\ k}} a_{jk} \sin jx \, e^{i\lambda kt}, \qquad f = \sum f_{jk} \sin jx \, e^{i\lambda kt}.$$

Then

$$Au = \sum (j^2 - \lambda^2 k^2) a_{jk} \sin jx \, e^{i\lambda kt}$$

and, for a solution,

$$a_{jk} = \frac{f_{jk}}{j^2 - \lambda^2 k^2}$$

provided the denominator is not zero; if it is then we should have $f_{jk} = 0$. In general, but not always, if λ is irrational the denominators $j^2 - \lambda^2 k^2$ may be arbitrarily small for infinitely many values of j and k. In that case if we take u in $L^2(\Omega)$ or $H^s(\Omega)$ or $C^\infty(\overline{\Omega})$, it is very difficult to characterize the range of A. In general it will not be closed (in some reasonable topology) and for this reason the case of general T is still intractable. We have here a small divisor problem. If λ is rational however, for convenience let us confine ourselves to the case $\lambda = 1$, the denominator, if $\neq 0$, is an integer and hence $\geqslant 1$ in absolute value. Furthermore $|j^2 - k^2| \to \infty$ as $|j| + |k| \to \infty$, $j \neq k$. In this case, $\lambda = 1$, however, the null space $N = N(A)$ is infinite dimensional, spanned by $\sin jx \cos jt$, $\sin jx \sin jt$, $j = 1, 2 \ldots$. (We will use N, sloppily, to denote the set of C^∞ functions in ker A, or its closure in some topology.) In fact any function in N has the form

$$q(t + x) - q(t - x)$$

where q is periodic of period 2π.

In $L^2(\Omega)$ the range $R(A)$ is then $N^\perp$ and if we factor out $N(A)$ we may consider $A^{-1}: R(A) \to R(A)$. There is a simple explicit integral formula for $A^{-1}: R(A) \to R(A)$ and it is not hard to verify the following properties of A^{-1}.

(i) In $L^2(\Omega)$, A^{-1} is compact.
(ii) $\|A^{-1}f\|_{L^\infty} \leqslant C\|f\|_{L^1}, f \in R(A)$.
(iii) Set

$$E_p = \left\{ v \perp_{L^2} N \,|\, v \in L^p \right\} \quad \text{with the } L^p \text{ norm.}$$

Then $A^{-1}: E_p \to L^{p'}$ is compact for $1 < p < \infty$. Here $1/p + 1/p' = 1$.
The result in [15] is slightly more general than that in [52].

THEOREM [15]. *Let g be as above and monotone increasing and let λ be rational. Assume*

$$\lim_{|t|\to\infty} \frac{g(t)}{t} = \infty, \qquad \tfrac{1}{2} g(t) - G(t) \geqslant \alpha|g(t)| - C$$

$\forall t$ *and some constants C, $\alpha > 0$. Then there is a nontrivial solution $u \in L^\infty$ of* (22) *with time period 2π.*

For g strictly increasing and C^∞, it was shown in [14] and [52] that $u \in C^\infty(\overline{\Omega})$.

The variational approach of [15] was motivated by the work in [20], [24]. We will assume g is strictly increasing; denote g^{-1} by h and set

$$H(v) = \int_0^v h(s)\, ds.$$

First we rewrite equation (22)–arguing formally. Suppose u is a solution; decompose it

$$u = u_1 + u_2,$$

$u_1 \in R(A)$, $u_2 \in N$, and set

$$v = g(u_1 + u_2).$$

The equation then has the form

$$Au_1 + v = 0.$$

So $v \in R(A)$,

$$u_1 + u_2 = h(v),$$

and the equation is equivalent to

$$u_1 + A^{-1}v = 0$$

or, after adding u_2 to both sides

$$A^{-1}v + h(v) = u_2 \in N.$$

To this form we now associate the following Lagrangian *defined on $R(A)$ alone*, i.e. we throw away the infinite-dimensional nullspace N,

$$F(v) = \int\!\!\int_\Omega \left[\frac{1}{2}\, v \cdot A^{-1}v + H(v) \right], \qquad v \in R(A). \tag{24}$$

Note, indeed, that if $v \in R(A)$ is a stationary point of F then

$$0 = F'(v)(\zeta) = \int\!\!\int_\Omega (A^{-1}v + h(v))\zeta, \qquad \forall \zeta \in R(A).$$

It follows that $A^{-1}v + h(v) = u_2$ belongs to $R^\perp = N$. Reversing the steps above we obtain a solution of (22). Thus the u_2 component of u appears as a kind of Lagrange multiplier.

The Lagrangian (24) looks much more tractable than (23) since A^{-1} is compact. The case $g(u) = |u|^q u$, $q \neq 1$, is particularly simple. Consider $g = u^3$ (this proof works as well for other values of q). In this case $H = \text{const } |v|^{4/3}$.

We may obtain a solution via a still different variational problem. Namely

$$\min_{\substack{v \in E_{4/3} \\ \|v\|_{L^{4/3}} \leqslant 1}} \int\int_{\Omega} v \cdot Av.$$

Because of the compactness property (iii), with $p = 4/3$, and the fact that $v \cdot A^{-1}v$ may be negative one finds that the minimum occurs at some point v, which is necessarily on the boundary of the ball. Consequently for some Lagrange multiplier $\lambda > 0$ we have

$$A^{-1}v + \lambda|v|^{-2/3}v \in N.$$

We want $\lambda = 1$, and this is achieved simply by stretching v. Unfortunately this argument works only in the case that g is homogeneous.

As an exercise for the reader we suggest reproving the result for $g = u^3$ using the mountain pass lemma in the space $E_{4/3}$.

The general case is treated in [15] with the aid of the mountain pass lemma in the space E_1. In order to work in this space the function $h(v)$ is first changed for $|v|$ large so as to be constant on the intervals $v > k$; $v \leqslant -k$. The space L^1 is not reflexive, making it difficult to verify (PS). In fact we were not able to, and we were led to a modified form of the condition. The modification is employed in just the same way as condition (PS) and we believe it will prove useful in other problems. In this form, c is a fixed real number, and we consider a real function F defined on a Banach space X which is Gâteaux differentiable (in our particular case in [15] the function is not Fréchet differentiable).

CONDITION $(PS)_c$. *If there is a sequence $\{u_j\}$ in X such that*

$$F(u_j) \to c \quad and \quad F'(u_j) \to 0 \text{ strongly,}$$

then c is a critical value of F.

Clearly $(PS) \Rightarrow (PS)_c$ for every c.

The mountain pass lemma then holds in the following form, [15],

LEMMA. *Let F be Gâteaux differentiable and $F': X \to X^*$ continuous from the strong topology of E into the weak $*$ topology of E^*. Assume that there is an open neighbourhood U of 0, a point $u_0 \notin \overline{U}$, and a constant ρ such that $F(u_0)$, $F(0) < \rho \leqslant F(u)$ for all $u \in \partial U$. Assume F satisfies $(PS)_c$ for*

$$c = \inf_{P} \max_{u \in P} F(u) \geqslant \rho.$$

Here P is a continuous path from 0 to u_0, and the inf is taken over all such paths. Then c is a critical value of F.

III. Local theory

III.1 This part will be concerned with the local study of equation (1). We will suppose as before that the mapping F is from one Banach space X to another, Y, and that

$$F(u_0) = y_0.$$

One wishes to study the set of solutions near u_0 of a perturbed equation

$$F(u) = y \quad \text{near } y_0 \tag{25}$$

–or F may depend on one or more parameters λ

$$F(u, \lambda) = y, \qquad F(u_0, \lambda_0) = y_0. \tag{26}$$

Our remarks will refer to the case that the implicit function theorem (IFT) is not applicable.

In case F is a Fredholm operator the well-known Lyapounoff-Schmidt procedure (see for example Chapter 3 in [47]) reduces the local study to one in finite-dimensional spaces. In case dim ker $F'(u_0) = d$, codim Range $F'(u_0) = d'$, the procedure leads to a mapping from R^d to $R^{d'}$. But the problem remains a difficult one; many different kinds of phenomena can occur.

One procedure in the finite-dimensional case, is to try to find a suitable local change of variables after which the (finite-dimensional) map F reduces to some simple canonical form admitting rather easy analysis. For example, if F is a real smooth function defined on R^d near the origin $F(0) = 0$, the simplest case is the case: $dF(0) \neq 0$. Then by the IFT we can find new coordinates so that F has the form $F(x) = x_1$. If $dF(0) = 0$, the next simple generic case, is to suppose that 0 is a nondegenerate critical point of F. In that case the well-known Morse lemma says that after a suitable change of coordinates, F is equal to a homogeneous quadratic function. More generally, the theory of singularities of differentiable maps, stability of maps and, in particular, catastrophe theory of Thom and Mather is very useful in the study of qualitative behaviour of solutions of (26). In this theory one makes changes of variables (x, λ) and y to reduce the mappings to simple forms. Sometimes one wishes to distinguish the parameter variable λ, and not permit general change of variables which mix those with x. A suitable corresponding form of the theory has been developed by Golubitsky and Schaeffer [33], and applied to problems in elasticity as well as others.

Local problems in which F is a Fredholm operator, but Range $F'(u_0) \neq Y$, are often called bifurcation problems. Even in such problems, global topological techniques are very useful. We describe one example:

A classical bifurcation problem is one concerning periodic solutions for ordinary differential equations; here x is a vector in R^n.

$$\frac{dx}{dt} = F(x, \lambda), \tag{27}$$

where F depends smoothly on x and a real parameter λ, near $\lambda = 0$, and

$$F(0, \lambda) \equiv 0, \qquad \forall \lambda.$$

We are interested in finding nontrivial periodic solutions near the origin for $|\lambda|$ small, which bifurcate from the origin. Let

$$F_x(0, \lambda) = L(\lambda),$$

and assume the matrix $L(0)$ is nonsingular. Then for $|\lambda|$ small, the only stationary point of (27) is $x = 0$. It is easy to see that no periodic solutions will bifurcate from the origin unless $L(0)$ has a purely imaginary eigenvalue.

Suppose $i\beta$, $\beta > 0$, is an eigenvalue of $L(0)$. Then the linearized equation

$$\frac{dx}{dt} = L(0)x$$

has a nonconstant periodic solution of period $2\pi\beta^{-1}$. (For $|\lambda|$ small, $L(\lambda)$ has a unique eigenvalue $\alpha(\lambda) + i\beta(\lambda)$ which is close to $i\beta$, and it depends continuously on λ.) We are interested in finding nonconstant periodic solutions with period close to $2\pi\beta^{-1}$ in case $|\lambda|$ is small.

In this direction there is the classical result of Hopf. Here is a well-known formulation (see [41] for a general survey).

HOPF BIFURCATION THEOREM. *Let $i\beta$ be as above and assume that no other integral multiple of $i\beta$ is an eigenvalue of $L(0)$. Assume that $\alpha(\lambda) \neq 0$ for $\lambda \neq 0$ and changes sign as λ crosses 0. Then for $|\lambda|$ small, there exists a nontrivial small periodic solution of (27) with period close to $2\pi\beta^{-1}$.*

Alexander and Yorke [2] discovered a generalization. Assume as before that $i\beta$, $\beta > 0$, is an eigenvalue of $L(0)$ and that $\mathrm{Mult}(i\beta) = \{ik_1\beta, ik_2\beta, \ldots, ik_r\beta\}$, $1 \leqslant k_1 \leqslant \cdots \leqslant k_r$, positive integers, are all the eigenvalues of $L(0)$ which are positive integral multiples of $i\beta$, counted with multiplicity. For $|\lambda|$ small there is then a unique set of r eigenvalues $\alpha_j(\lambda) + i\beta_j(\lambda)$, $j = 1, \ldots, r$ close to the set, and varying continuously with λ.

EXTENSION OF THE HOPF BIFURCATION THEOREM [2]. *Assume that for $\lambda \neq 0$, $\alpha_j(\lambda) \neq 0$, $j = 1, \ldots, r$. Let r^+ (r^-) be the number of these which are positive for $\lambda > 0$ $(\lambda < 0)$. Assume that $r = r^+ - r^-$, the number of changes of signs of the α_j as λ passes through 0, is odd. Then for $|\lambda|$ small there is a nontrivial, small periodic solution of (27) with period close to $2\pi\beta^{-1}$.*

The proof in [2] relies on some topological machinery. Subsequently Ize [35] presented a simpler proof which uses less topology. It makes use of the Hopf map of S^3 to S^2. [2] and [35] also contain more global results, concerning a connected family of periodic solutions.

The paper [35] has many results on bifurcation in particular when a real or complex parameter is involved. In papers [1], [3] further topological tools are used in analyzing bifurcation. Applications of bifurcation theory may be found in [55].

III.2 As we mentioned in §I.1 the implicit function theorem may fail because of the "loss of derivatives" on applying $F(u_0)^{-1}$–a situation which is common in nonelliptic problems. For example, consider the wave equation for a function $u(x, t)$, $x \in R^n$, $t \in R^1$,

$$u_{tt} - \Delta_x u = f, \qquad u = u_t = 0 \quad \text{at } t = 0.$$

If, say, f has compact support then for f in the Sobolev space H^1 the solution u belongs to H^{s+1}–not to H^{s+2}. If we consider then a perturbation problem of the form

$$F(u) = u_{tt} - f(u_{x_k}, u_t, u_{x_i x_j}) = y; \qquad u = u_0 = 0 \quad \text{at } t = 0,$$

which is hyperbolic, and $u = 0$ is a solution for $y = 0$, and wish to solve this for y small, the implicit function theorem is not applicable.

In [44], Nash introduced a procedure for attacking local problems in which such a loss of derivatives occurs. Since then several variations have been developed. In particular Moser [42] introduced a scheme which is based on the Newton iteration scheme for solving a nonlinear problem. This method is now commonly called the Nash-Moser technique. See also Hörmander [34] for another modification of the Nash method. This technique is to my mind a truly fundamental development in the study of nonlinear problems. For the first time it enabled one to solve problems involving "loss of derivatives", which up to then had appeared completely intractable.

Before describing the Nash-Moser procedure let us first recall the Newton iteration scheme for solving a problem $F(u) = g$ where F maps $R^N \to R^N$. Suppose $F(u_0) = y_0$ and y is close to y_0. Assuming $F'(u_0)$ is invertible, the Picard iteration scheme is the following,

$$u_{p+1} = u_p + F'(u_0)^{-1}(y - F(u_p)), \qquad p = 0, 1, \ldots,$$

and it converges like a geometric series. Newton's scheme is

$$u_{p+1} = u_p + F'(u_p)^{-1}(y - F(u_p)), \qquad p = 0, 1, \ldots,$$

assuming that $F'(u)^{-1}$ exists for u near u_0. This scheme converges much more rapidly than Picard's.

In case the operators $F'(u)^{-1}$ "lose derivatives", u_{p+1} is less regular than u_p, and so there is hope for convergence in a function space with some prescribed regularity.

The Nash-Moser scheme makes use of smoothing operators depending on a parameter. To be specific, say we are working in Sobolev spaces H^s of some manifold and denote the corresponding norms by $\| \ \|_s$. For $\theta > 1$, S_θ is supposed to be a linear operator mapping H^s into C^∞, $\forall s$, and satisfying

$$\|S_\theta u\|_k \leqslant C\theta^{k-j}\|u\|_j \quad \text{for } j < k,$$

$$\|(I - S_\theta)u\|_k \leqslant C\theta^{k-j}\|u\|_j \quad \text{for } j > k.$$

The scheme is to choose a sequence $\theta_j = 2^{\alpha^j}$, $1 < \alpha < 2$, and work with a modified Newton scheme

$$u_{p+1} = u_p + S_{\theta_p}F'(u_p)^{-1}(y - F(u_p)).$$

Because of the smoothing, all the iterates are C^∞-functions. Under suitable conditions on F–in particular one needs $F'(u)$ to be invertible not only at u_0 but nearby–one shows that the scheme converges. The error introduced by the smoothing operators is compensated for by the rapid convergence of the Newton scheme. We refer the reader to [43], Chapter 2 in [57] and the last chapter in [47].

III.3 Recently Klainerman [36] has adapted this scheme to study the following problem posed by F. John. Consider the initial value problem (we will limit ourselves to a special case) for $u(x, t)$

$$u_{tt} - a^{ij}(\nabla u)u_{x_i x_j} = 0,$$
$$u = f_0, \qquad u_t = g_0 \quad \text{at } t = 0. \tag{28}$$

The coefficients a^{ij} are assumed to be smooth and to satisfy

$$a^{ij} - \delta^{ij} = b^{ij} \quad \text{and} \quad b^{ij}(0) = 0.$$

The functions $f_0, g_0 \in C^\infty$ and (with many derivatives) are assumed to be small. For convenience let us suppose they have compact support. The problem is whether there exist global (i.e. for all time) smooth solutions. If we do not assume f_0, g_0 to be small then simple examples show that the answer can be no. On the other hand, in one space dimension, even if f_0, g_0 are small but not zero, if the a^{ij} are not constant, one knows that shocks will occur. Namely, the solution will exist and be smooth for small t, but eventually singularities must develop. In higher dimensions there is more space for the characteristics (which carry singularities) to spread and so one might hope that smooth solutions may exist for all time.

Klainerman proved this to be the case if $n \leqslant 6$.

Naturally one is interested in knowing what happens for smaller n, in particular $n = 3$. Very recently F. John has shown that for the equation

$$u_{tt} - \frac{\Delta u}{1 - u_t} = 0,$$

no matter how small are $(f_0, g_0) \not\equiv 0$, there is no global smooth solution for all time.

We will conclude these lectures by a brief and very informal description of Klainerman's procedure. The method we describe is the first one he devised. In the published paper [36] he used a modification. The proof is highly technical so we can only sketch some of the ideas.

A classical result is that one can solve the initial value problem, and obtain a C^∞ solution in a small time interval, $0 \leqslant t \leqslant T$, where T depends only on the following (all integral norms are over the space variable in R^n):

$$\sum_{|\alpha| \leqslant [n/2]+2} \|\partial^\alpha f_0\|_{L^2} + \sum_{|\alpha| \leqslant [n/2]+1} \|\partial^\alpha g_0\|_{L^2}.$$

Here α is a multi-index of nonnegative integers $\alpha = (\alpha_1, \ldots, \alpha_n)$, $\partial^\alpha = \partial_1^{\alpha_1} \ldots \partial_n^{\alpha_n}$, $\partial_j = \partial/\partial x_j$, these involve only space derivatives. Thus to find a global smooth solution it suffices to find a solution for which the following norms (in R^n) are bounded in $0 \leqslant t \leqslant T$:

$$\sum_{|\alpha| \leqslant [n/2]+2} \|\partial^\alpha u\|_{L^2} + \sum_{|\alpha| \leqslant [n/2]+1} \|\partial^\alpha u_t\|_{L^2} \leqslant C(T)$$

for every T; $C(T)$ may however grow as $T \to \infty$ in any way.

(a) In order to attack the problem we expect to make use of results for corresponding linear equations. Consider a linear problem

$$u_{tt} - a^{ij}(x, t)u_{x_i x_j} + b^i u_{x_i} = g$$

with the matrix $a = \{a^{ij}\}$ positive definite. Bounds on the L^2 norms (in R^n) of derivatives of the solution are usually obtained with the aid of energy estimates. If one defines the energy at time t as

$$\int_{R^n} u_t^2 + a^{ij} u_{x_i} u_{x_j}$$

then one finds, using Green's theorem, that

$$\frac{d}{dt} E(t) = \int a_t^{ij} u_i u_j - a_{x_j}^{ij} u_i u_t + 2gu_t - 2b^i u_i u_t$$

$$\leqslant \left(\max_{R^n} |\nabla a| + \max_{R^n} |b| \right) E + 2\| g \|_{L^2} E^{1/2}$$

–in rather obvious notation. Integrating from 0 to time t we find

$$E^{1/2}(t) \leqslant c + \int_0^t \exp \left[\int_0^s \left(\max_x |\nabla a(x, \sigma)| + \max_x |b| \right) d\sigma \right] \| g \|_{L^2}(s) \, ds.$$

If we want the energy to be finite for all time it seems reasonable to require that for all t,

$$\max_x (|\nabla a| + \max |b|)(t) \leqslant \frac{K}{(1 + t)^{1+\varepsilon}} \quad \text{for some } \varepsilon > 0.$$

This linear problem is the kind that will arise in the Nash-Moser scheme when we linearize about some iterate v. In view of the way the coefficients in the linear operator depend on v it seems reasonable to work with functions whose derivatives decay as $t \to \infty$, uniformly in x.

(b) The next crucial fact is that for solutions of the classical wave equation such decay estimates exist. Namely, suppose u satisfies

$$u_{tt} - \Delta u = 0, \quad u = 0, \qquad u_t = g \quad \text{at } t = 0.$$

Then

$$|u(x, t)| \leqslant Ct^{(1-n)/2} \sum_{|\alpha| < (n-1)/2} |\partial^\alpha g|_{L^1},$$

$$|\nabla u(x, t)| \leqslant Ct^{(1-n)/2} \sum_{|\alpha| < (n+1)/2} |\partial^\alpha g|_{L^1}. \tag{29}$$

In addition, by the standard energy estimates (applied to spatial derivatives of u), and by the Sobolev embedding theorem, we have

$$|\nabla u(x, t)| \leqslant C \sum_{|\alpha| < n/2 + 1} \| \partial^\alpha g \|_{L^2}. \tag{30}$$

This leads to a corresponding decay for a solution of the inhomogeneous equation

$$u_{tt} - \Delta u = g, \qquad u = u_t = 0 \quad \text{at } t = 0,$$

in case g decays in time. In terms of the appropriate solution operator $V(t)$, u is given by (here we suppress the space variables)

$$u(t) = \int_0^t V(t - s)g(s) \, ds$$

so that

$$|\nabla u(x, t)| \leq \left| \int_0^{t/2} \nabla V(t - s)g(s)\, ds \right| + \left| \int_{1/2}^t \nabla V(t - s)g(s)\, ds \right|.$$

We have divided the t integral into two parts. In the first we use the estimate (29) which is useful for large, not small, t, while in the second we use (30). We thus obtain the estimate (where we have thrown something away)

$$|\nabla u(x, t)| \leq Ct^{(1-n)/2} \int_0^{t/2} \sum_{|\alpha| < n/2 + 1} |\partial^\alpha g|_{L^1}(s)\, ds$$

$$+ C \int_{t/2}^t \sum_{|\alpha| < n/2 + 1} \|\partial^\alpha g\|_{L^2}\, ds.$$

Since we want solutions to decay, this suggests introducing decay norms, for $k \geq 1$,

$$|u|_{k,r,L^1} = \sup_{t>0} (1 + t)^k \sum_{|\alpha| < r} |\partial^\alpha u|_{L^1},$$

$$\|u\|_{k,r} = \sup_{t>0} (1 + t)^k \sum_{|\alpha| < r} \|\partial^\alpha u\|_{L^2}.$$

Then from the preceding inequality we find for $k > 1$,

$$|\nabla u(x, t)| \leq Ct^{(3-n)/2}|g|_{0,n/2+1,L^1} + C(1 + t)^{1-k}\|g\|_{k,n/2+1}. \tag{31}$$

Now in our application of this to be the nonlinear problem, via the Nash-Moser scheme, the function g will involve products of derivatives of iterates. In view of energy estimates, we cannot expect the L^2 norms of these derivatives to decay. They may only, at best, remain bounded. So if the max norm in R^n of the derivatives decay like $O(t^{-k})$ then, at best, $\|g\|_{k,n/2+1}$ will be bounded. But then the last term on the right of (31) decays only like $O(t^{1-k})$. Thus there is a *loss in decay*—as well as the usual loss in smoothness. Furthermore, the first term on the right of (31) shows that one cannot expect to establish faster decay than $O(t^{(3-n)/2})$. In particular, for $n = 3$, this doesn't give any decay at all.

(c) Now we turn to the iteration scheme. To overcome the loss in decay (as well as regularity), Klainerman introduces a modified "smoothing operator" in the Nash-Moser scheme: His operator involves smoothing and, *at the same time, cut off after large time*: If $\zeta(t) \geq 0$ is a C^∞ function on $t \geq 0$, $\zeta \equiv 1$ in $0 \leq t \leq 1$, $\zeta(t) = 0$ for $t > 2$, define the operator

$$C_\theta = \text{multiplication by } \zeta\!\left(\frac{t}{\theta}\right), \qquad \theta > 1.$$

The new "smoothing operator" will be a product

$$C_\theta \cdot S_{\theta'}, \qquad \theta, \theta' > 1.$$

Next, the iteration scheme used is a modification of the Nash-Moser scheme. For a suitable choice of $\theta_p \to \infty$ and $\theta_p' \to \infty$, set

$$\bar{S}_p = C_{\theta_p} S_{\theta_p'}.$$

In fact $\theta_{p+1} = \theta_p^\tau$ for some τ in $1 < \tau < 2$, and $\theta_p' = \theta_p^\varepsilon$ for some small $\varepsilon > 0$. The $(p + 1)$st iterate is defined as the solution of

$$L_p(u_{p+1} - u_p) = y - \bar{S}_p F(u_p)$$

with zero Cauchy data on $t = 0$. Here L_p is the linear operator (a modification of $F'(u_p)$)

$$L_p = \partial_{tt} - \Delta - \bar{S}_p\left[b^{ij}(\nabla u_p)\right]\partial_i\partial_j + \bar{S}_p\left(u_{p_{x_i x_j}} b^{ij}(\nabla u_p)\right) \cdot \nabla$$

is rather obvious notation.

Making use of the energy estimates and the decay estimates described above (applied to spatial derivatives of the iterates) one establishes after much work a series of recursive estimates. This involves the use of still more norms and at this point we have to refer to [36] for the details.

Our description of Klainerman's treatment is perhaps too sketchy to be useful. It has been included as an indication of the flexibility of the Nash-Moser method. In particular, I want to call attention to the use of the "smoothing operators"–they need not merely smooth, they might operate in some other way–in this case by cut off for large time.

References

1. J. C. Alexander, *Bifurcation of zeros of parametrized functions* (to appear).

2. J. C. Alexander and J. A. Yorke, *Global bifurcation of periodic orbits*, Amer. J. Math. **100** (1978), 263–292.

3. ______, *Calculating bifurcation invariants as elements in the homotopy of the general linear group*, J. Pure Appl. Algebra **13** (1978), 1–8.

4. H. Amann and E. Zehnder, *Nontrivial solutions for a class of nonresonance problems and applications to nonlinear differential equations* (to appear).

5. ______, *Multiple periodic solutions for a class of nonlinear autonomous wave equations* (to appear).

6. A. Ambrosetti and P. H. Rabinowitz, *Dual variational methods in critical point theory and applications*, J. Funct. Anal. **14** (1973), 349–381.

7. A. Ambrosetti and G. Mancini, *Sharp nonuniqueness results for some nonlinear problems*, Nonlinear Anal. Theory Math. Appl. **3** (1979), 635–645.

8. V. Benci, *Some critical point theorems and applications*, Comm. Pure Appl. Math. **33** (1980), 147–172.

9. ______, *On the critical point theory for indefinite functionals in the presence of symmetries*, Ist. Mat. Appl. U. Dini, Univ. di Pisa, March 1980.

10. V. Benci and P. H. Rabinowitz, *Critical point theorems for indefinite functionals*, Invent. Math. **52** (1979), 241–273.

11. M. S. Berger, *Nonlinearity and functional analysis*, Academic Press, New York, 1978.

12. Yu. G. Borisovich, V. G. Zvyagin and Yu. I. Sapronov, *Nonlinear Fredholm maps and the Leray-Schauder theorem*, Uspehi Mat. Nauk **32** (1977), 3–54; English transl. in Russian Math. Surveys **32** (1977), 1–54.

13. H. Brézis and R. E. L. Turner, *On a class of superlinear elliptic problems*, Comm. Partial Differential Equations **2** (1977), 601–614.

14. H. Brézis and L. Nirenberg, *Forced vibrations for a nonlinear wave equation*, Comm. Pure Appl. Math. **31** (1978), 1–30.

15. H. Brézis, J. M. Coron and L. Nirenberg, *Free vibrations for a nonlinear wave equation and a theorem of P. Rabinowitz*, Comm. Pure Appl. Math. **33** (1980), 667–684.

16. A. Castro, *A two point boundary value problem with jumping nonlinearities*, Proc. Amer. Math. Soc. (to appear).

17. A. Castro and A. C. Lazer, *Critical point theory and the number of solutions of a nonlinear Dirichlet problem*, Ann. Mat. Pura Appl. (IV) **70** (1979), 113–137.

18. K. C. Chang, *The obstacle problem and partial differential equations with discontinuous nonlinearities*, Comm. Pure Appl. Math. 33 (1980), 117–146.

19. ______, *Solutions of asymptotically linear operator equations via Morse theory*, Comm. Pure Appl. Math. (to appear).

20. F. H. Clarke and I. Ekeland, *Hamiltonian trajectories having prescribed minimal period*, Comm. Pure Appl. Math. 33 (1980), 103–116.

21. C. C. Conley, *Isolated invariant sets and the Morse index*, CBMS Regional Conf. Ser. in Math., no. 38, Amer. Math. Soc., Providence, R. I., 1978.

22. D. DeFigueiredo, P. L. Lions and R. D. Nussbaum, *Estimations à priori pour les solutions positives de problèmes elliptiques semilinéaires*, C. R. Acad. Sci. Paris Ser. A. **290** (1980), 217–220.

23. I. Ekeland, *On the variational principle*, J. Math. Anal. Appl. **47** (1974), 324–353.

24. ______, *Periodic solutions of Hamiltonian equations and a theorem of P. Rabinowitz*, J. Differential Equations **34** (1979), 523–534.

25. I. Ekeland and R. Témam, *Convex analysis and variational problems*, Studies in Math. and Appl., Vol. 1, North-Holland, Amsterdam, American Elsevier, New York, 1976.

26. I. Ekeland and J. M. Lasry, *On the number of periodic trajectories for a Hamiltonian flow on a convex energy surface*, MRC Technical Report 2050, Math. Research Center, Univ. of Wisconsin, 1980.

27. K. D. Elworthy and A. J. Tromba, *Differential structures and Fredholm maps*, Proc. Sympos. Pure Math. (Berkeley, California, 1968), vol. 15, Amer. Math. Soc., Providence, R. I., 1970, pp. 45–94.

28. ______, *Degree theory on Banach manifolds*, Proc. Sympos. Pure Math., vol. 18, part 1, *Nonlinear functional analysis*, Amer. Math. Soc., Providence, R. I., 1970, pp. 86–94.

29. F. R. Fadell and P. H. Rabinowitz, *Generalized cohomological index theories for Lie group actions with an application to bifurcation questions for Hamiltonian systems*, Invent. Math. **45** (1978), 134–174.

30. F. B. Fuller, *An index of fixed point type for periodic orbits*, Amer. J. Math. **89** (1967), 133–148.

31. K. Geba and A. Granas, *Infinite dimensional cohomology theories*, J. Math. Pure Appl. **52** (1973), 145–270.

32. B. Gidas and J. Spruck, *A priori bounds for positive solutions of nonlinear elliptic equations*, Comm. Partial Differential Equations (submitted).

33. M. Golubitsky and D. Schaeffer, *A theory for imperfect bifurcation via singularity theory*, Comm. Pure Appl. Math. **32** (1979), 21–98.

34. L. Hörmander, *The boundary value problems of physical geodesy*, Arch. Rational Mech. Anal. **62** (1976), 1–52.

35. J. Ize, *Bifurcation theory for Fredholm operators*, Mem. Amer. Math. Soc. No. 174 (1976).

36. S. Klainerman, *Global existence for nonlinear wave equations*, Comm. Pure Appl. Math. **33** (1980), 43–101.

37. M. A. Krasnoselskii, *Topological methods in the theory of nonlinear integral equations*, Macmillan, New York, 1964.

38. J. Leray and J. Schauder, *Topologie et équations fonctionelles*, Ann. Sci. École Norm. Sup. **51** (1934), 45–78.

39. N. G. Lloyd, *Degree theory*, Cambridge Tracts in Math., No. 73, Cambridge Univ. Press, London, 1978.

40. J. Mallet-Paret and J. A. Yorke, *Snakes: Oriented families of periodic orbits, their sources, sinks and continuation* (to appear).

41. J. E. Marsden and M. McCracken, *The Hopf bifurcation and its applications*, Appl. Math. Sciences, Vol. 19, Springer-Verlag, New York, 1976.

42. J. Moser, *A new technique for the construction of solutions of nonlinear differential equations*, Proc. Nat. Acad. Sci. U.S.A. **47** (1961), 1824–1831.

43. ______, *A rapidly convergent iteration method and nonlinear partial differential equations. I*,

II, Ann. Scuola Norm. Sup. Pisa **20** (1966), 265–315; 499–535.

44. J. Nash, *The imbedding problem for Riemannian manifolds*, Ann. of Math. (2) **63** (1956), 20–63.

45. W. M. Ni, *Some minimax principles and their applications in nonlinear elliptic equations*, J. d'Analyse Math. **37** (1980), 248–275.

46. L. Nirenberg, *An application of generalized degree to a class of nonlinear problems*, 3rd Colloq. Anal. Fonct. Liège Centre Belge de Rech. Math., 1971, pp. 57–73.

47. ______, *Topics in nonlinear functional analysis*, Lecture Notes, Courant Inst., 1974.

48. ______, *Remarks on nonlinear problems*, The Chern Symposium, 1979 (W. Y. Hsiang et al., eds.) Springer-Verlag, Berlin and New York, 1980, pp. 189–197.

49. R. S. Palais, *Critical point theory and the minimax principle*, Proc. Sympos. Pure Math., vol. 15, Amer. Math. Soc., Providence, R. I., 1970, pp. 185–212.

50. P. H. Rabinowitz, *Variational methods for nonlinear eigenvalue problems* (CIME, Verona, 1974), Ediz. Cremonese Rome, 1974, pp. 141–195.

51. ______, *Théorie du degré topologique et applications à des problèmes aux limites nonlinéaires*, Lecture Notes, Analyse Numerique Fonctionelle, Univ. Paris VI, 1975.

52. ______, *Free vibrations for a semilinear wave equation*, Comm. Pure Appl. Math. **31** (1978), 31–68.

53. ______, *Periodic solutions of Hamiltonian systems*, Comm. Pure Appl. Math. **31** (1978), 157–184.

54. ______, *A variational method for finding periodic solutions of differential equations*. Nonlinear Evolution Equations (M. G. Crandall, ed.), Academic Press, New York, 1978, pp. 225–251.

55. D. H. Sattinger, *Topics in stability and bifurcation theory*, Lecture Notes in Math. no. 309, Springer-Verlag, Berlin and New York, 1973.

56. M. Schechter, *Principles of functional analysis*, Academic Press, New York, 1971.

57. J. Schwartz, *Nonlinear functional analysis*, Gordon and Breach, New York, 1969.

58. J. Sylvester, Ph.D. Thesis, Courant Inst. Math. Sci., New York Univ., 1980.

COURANT INSTITUTE OF MATHEMATICAL SCIENCES, NEW YORK UNIVERSITY, NEW YORK, NEW YORK 10012

Section 6

MECHANICS AND DYNAMICAL SYSTEMS

Proceedings of Symposia in Pure Mathematics
Volume **39** (1983), Part 2

The Meaning of Maslov's Asymptotic Method: The Need of Planck's Constant in Mathematics

JEAN LERAY

ABSTRACT. H. Poincaré defined asymptotic expansions. Their use by the W. K. B. method introduced a new kind of solution of linear differential equations. Maslov showed their singularities to be merely apparent. The clarification of those results leads to the introduction of "Lagrangian functions", of their scalar product and of "Lagrangian operators", which constitutes a new structure: the "Lagrangian analysis". The last step of its definition requires the choice of a constant. That constant has to be Planck's constant, when the equation is the Schrödinger or the Dirac equation describing the hydrogen atom–the study of atoms with several electrons is very incomplete.

1. Henri Poincaré's main field, more precisely the one where the number of his publications is the highest, happens to be celestial mechanics. For instance, he tried to establish the convergence of the series by means of which the motion of the solar system is computed; it was a failure. He proved indeed the opposite: the divergence of those series, whose numerical values furnished the most impressive, precise and famous predictions in science during the last century! Henri Poincaré explained that paradox: those series give a very good approximation of the wanted result, provided only their first terms, namely, a reasonable number of them, are taken into account. Of course, demanding mathematicians to be reasonable is dubious but Henri Poincaré [4] made it clear by defining *the asymptotic expansion* $\sum_{n=0}^{\infty} a_n x^n$ of a function of x at the origin: it is a *formal series* such that for each natural number N there exists a positive number c_N such that

$$\left| f(x) - \sum_{n=0}^{N} a_n x^n \right| \leqslant c_N |x|^{N+1} \quad \text{for } x \text{ near } 0. \tag{1.1}$$

Thus an asymptotic expansion of f is a formal series able to give a very good approximation of $f(x)$, when x is small, but unable to supply the exact value of $f(x)$.

2. The W.K.B. method constructs *asymptotic solutions* of a linear differential equation

$$H\left(x, \frac{1}{\nu} \frac{\partial}{\partial x}\right) u(\nu, x) = 0 \qquad \left(x \in X = \mathbf{R}^l; \nu \in i\,[0, \infty[\right), \tag{2.1}$$

whose unknown is the function u and whose parameter ν *tends to* $i\infty$.

Reprinted from Bulletin Amer. Math. Soc. (N.S.) 5 (1981), 15–27.

1980 *Mathematics Subject Classification*. Primary 47B99, 81C99; Secondary 35S99, 42B99.

Key words and phrases. Fourier transform, Metaplectic group, Asymptotic solution, Lagrangian, Operator, Schrödinger, Dirac equation.

Assume that (2.1) describes the evolution of a mechanical system; if that system were a finite set of particles, i.e., if (2.1) were an ordinary differential system, then Cauchy's problem should be studied. But let us assume it is a continuous mechanical system; therefore the physicists cannot impose its initial position and velocity: they have no more interest in Cauchy's problem.

What they are interested in are the *waves*

$$U(\nu, x) = \alpha(\nu, x)e^{\nu\phi(x)}, \tag{2.2}$$

solutions of the equation (2.1); in (2.2), the *phase* ϕ is a real-valued function, ν is near $i\infty$; therefore $e^{\nu\phi}$ rapidly oscillates, whereas the *amplitude* α is a complex-valued function which slowly varies.

Therefore ϕ has to satisfy the first order nonlinear differential equation

$$H(x, \phi_x) = 0 \qquad (\phi_x = \partial\phi/\partial x). \tag{2.3}$$

Assume H to be a real-valued function of

$$(x, p) \in X \oplus X^*, X^* \text{ being the dual of } X;$$

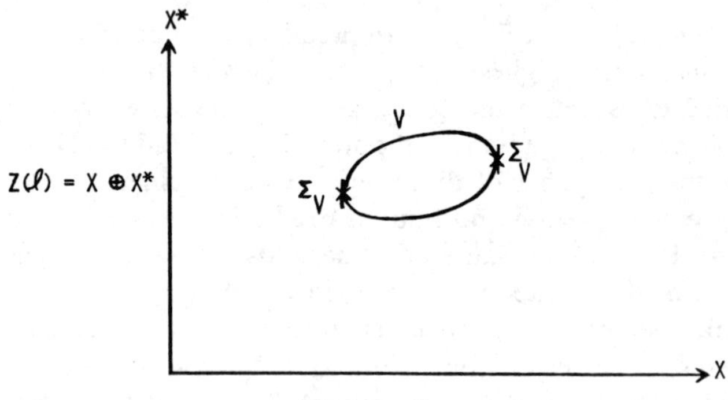

FIGURE 1

Denote by W the hypersurface of $X \oplus X^*$ where H vanishes:

$$W: H(x, p) = 0; \text{ assume } (H_x, H_p) \neq 0 \text{ on } W; \tag{2.4}$$

denote the graph of ϕ_x by

$$V = \{(x, \phi_x)\} \subset X \oplus X^*. \tag{2.5}$$

Then V is any l-dimensional subvariety of W on whose universal covering space $\check{V}$ there exists a function ϕ such that

$$d\phi = \langle p, dx \rangle \qquad (\langle p, x \rangle: \text{value of } p \in X^* \text{ at } x \in X); \tag{2.6}$$

i.e., V is any l-dimensional subvariety of W on which

$$\sum_{j=1}^{l} dp_j \wedge dx_j = 0 \tag{2.7}$$

$$[(x_j), (p_j): \text{dual coordinates of } x \in X \text{ and } p \in X^*].$$

Let $Z(l)$ be the $2l$-dimensional vector space $X \oplus X^*$ provided with the *symplectic structure* $[\cdot, \cdot]$ defined as follows

$$\text{if } z = (x, p) \text{ and } z' = (x', p'), \text{ then } [z, z'] = \langle p, x' \rangle - \langle p', x \rangle. \tag{2.8}$$

By definition, the *Lagrangian* subspaces of $Z(l)$ are its l-dimensional subspaces on which $[\cdot, \cdot]$ identically vanishes; the Lagrangian varieties of $Z(l)$ are its l-dimensional varieties whose tangent planes are Lagrangian. The definition (2.7) of V means

$$V \text{ is any Lagrangian variety contained in the hypersurface } W. \qquad (2.9)$$

The solution of the first order nonlinear differential equation (2.3) rests on the following fundamental property of V: the variety V is generated by *characteristic curves* of W, i.e., by curves of W satisfying the Hamiltonian system

$$\frac{dx_j}{H_{p_j}(x, p)} = -\frac{dp_k}{H_{x_k}(x, p)} \qquad (2.10)$$

$[j, k \in \{1, \ldots, l\}; H$ is a first integral of that system].

Without discussing more completely the construction of V, assume V and hence ϕ chosen: then (2.2) satisfies (2.1) mod $1/\nu$.

Now $U(\nu, x)$ satisfies

$$H\left(x, \frac{1}{\nu}\frac{\partial}{\partial x}\right)U(\nu, x) = 0 \quad \text{mod } 1/\nu^N \qquad (2.11)$$

[for any natural number N]

if and only if $\alpha(\nu, x)$ is a *formal series* in $1/\nu$

$$\alpha(\nu, x) = \sum_{r=0}^{\infty} \frac{1}{\nu^r}\alpha_r(x), \qquad (2.12)$$

the derivatives of $\alpha_0, \ldots, \alpha_r, \ldots$ along the characteristics generating V having a known value, $\ldots$, a value depending on $(\alpha_0, \ldots, \alpha_{r-1}), \ldots$. Then $U(\nu, x)$, defined by (2.2), is a formal (with respect to $1/\nu$) function of x satisfying (2.11): it is said to be an *asymptotic solution* of (2.1); its above construction is called: *W.K.B. method*.

By that method *wave mechanics* is related to *particle mechanics*: indeed, the Hamiltonian system (2.10) could describe the motion of particles and the computation of α_0 defines a density of those particles such that the conservation of their mass holds. Now the physicists hesitated between a corpuscular (Descartes and Newton, 17th century) and a wave (Huygens and Fresnel, 17th–19th centuries) structure of the light, before they concluded that light (Einstein, 1905: photons and quanta) and matter (de Broglie, 1924) have both structures. Henri Poincaré died[1] in 1912 some years before it became clear that the *quanta* require a new physics, which has not yet been attained even today. So let us continue to study the asymptotic solutions of differential equations and thus apply to problems which arose before Henri Poincaré's death some of the concepts owing to him: algebraic topology, first homotopy group, covering spaces and groups.

---

[1] He was only 58 years old; it was 3 years before "General Relativity" finally made "Relativity" acceptable by the astronomers.

An asymptotic solution U happens to be defined not on X but on V, where it has *singularities* on the *apparent contour* Σ_V of V, which is the set of the points of V where $dx_1 \wedge \cdots \wedge dx_l = 0$, i.e., where x cannot be used as local coordinate on V; the α_r become infinite on Σ_V; the order of their singularity increases with r.

For instance, let us use geometrical optics, i.e., use asymptotic solutions of the wave equation of light, for describing the propagation of the light issued from a monochromatic source through a steady optical system: the projections on X of the characteristic curves generating V are the rays of light, whose envelope, the "caustic", is the projection of Σ_V on X; that caustic plays the role of image of the source; on it the asymptotic solution is no more defined; therefore geometrical optics should have no meaning beyond the caustic; however, it still holds. That paradox has been solved by *V. P. Maslov* [3] as follows.

The singularities of an asymptotic solution U on the apparent contour Σ_V are merely *apparent singularities*: indeed they are suppressed by a convenient Fourier transform; Maslov's proof needs to be clarified by *V. I. Arnold's* topological results [1] and by *V. C. Buslaev's* use [2] of *I. E. Segal's* metaplectic group [5], which contains the Fourier transforms used by Maslov. Those improvements finally do not give some deeper knowledge of the asymptotic behaviour of the functions; but they lead to a new structure, defining, for instance, quite a new type of solutions of differential equations; §3 describes that *structure*, called *Lagrangian analysis*; §4 discusses its applications.

3. Lagrangian analysis.

The symplectic geometry. Denote by Z the symplectic space of dimension $2l$: i.e., $\mathbf{R}^{2l}$ provided by a bilinear antisymmetric real-valued form $[\cdot, \cdot]$ of maximal rank. A *frame*[2] is any isomorphism

$$R: Z \to Z(l) = X \oplus X^*$$

consistent with the symplectic structures of Z and $Z(l)$ [see §2].

If R' is another frame, then the *change of frame* RR'^{-1} is any automorphism of $Z(l)$, that is, any element s of the symplectic group $\mathrm{Sp}(l)$:

$$RR'^{-1} \in \mathrm{Sp}(l).$$

Each real-valued quadratic form A of $(x, x') \in X \oplus X'$, such that $\det(A_{x_j x'_k}) \neq 0$, defines an element s_A of $\mathrm{Sp}(l)$ as follows:

$$(x, p) = s_A(x', p') \text{ if and only if}$$
$$p = A_x(x, x'), \qquad p' = -A_{x'}(x, x'); \tag{3.1}$$

---

[2] R is the initial of the French translation Repère of Frame. Let us recall that an orthonormed frame of the Euclidean space E^3 is nothing else but an isomorphism $E^3 \to \mathbf{R} \oplus \mathbf{R} \oplus \mathbf{R}$, that direct sum being provided by the Euclidean structure implied by the structure of $\mathbf{R}$. That isomorphism has to be consistent with the Euclidean structures.

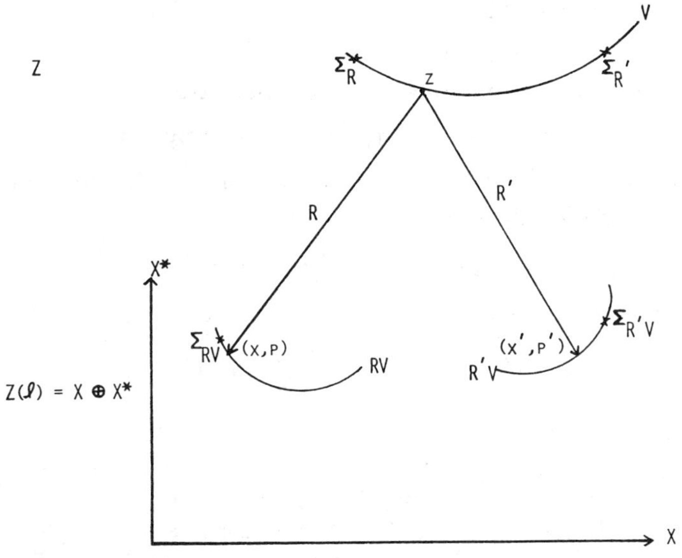

FIGURE 2

whence by Euler's formula

$$A(x, x') = \langle p, x \rangle - \langle p', x' \rangle; \tag{3.2}$$

s_A is the generic element s of $\mathrm{Sp}(l)$–but not any $s \in \mathrm{Sp}(l)$.

Lagrangian varieties V are defined in Z as they were in $Z(l)$ [see §2]; that definition can be expressed as follows:

$$d[z, dz] = 0 \quad \text{on } V;$$

i.e., on the universal covering space $\check{V}$ of V there exists a real-valued function ψ, the *Lagrangian phase*, such that

$$d\psi = \tfrac{1}{2}[z, dz]. \tag{3.3}$$

The image RV of V by R is a Lagrangian variety of $Z(l)$; denote its phase [§2] by $\phi_R \circ R^{-1}$; then, by a convenient choice of constants of integration

$$\phi_R(\check{z}) = \psi(\check{z}) + \tfrac{1}{2}\langle p, x \rangle, \tag{3.4}$$

where $z \in V$ is the projection of $\check{z} \in \check{V}$ and $(x, p) = Rz$.

Hence, from (3.2), if $RR'^{-1} = s_A$:

$$\phi_R(\check{z}) = \phi_{R'}(\check{z}) + A(x, x'), \tag{3.5}$$

where

$$(x, p) = Rz, \qquad (x', p') = R'z, \tag{3.6}$$

$z \in V$ being the projection of $\check{z} \in \check{V}$.

The *apparent contour $\check{\Sigma}_R$ of $\check{V}$ relative to R* is the subset of $\check{V}$ projecting on $\Sigma_R = R^{-1}\Sigma_{RV}$, which is the apparent contour of V, relative to R. In a sufficiently small neighborhood of each point of $\check{V} \setminus \check{\Sigma}_R \cup \check{\Sigma}_R$, the maps $\check{z} \mapsto z, z \mapsto x, z \mapsto x'$ are diffeomorphisms. Now, since RV is the graph of the gradient of its phase [see §2]

$$p = \partial\phi_R/\partial x, \qquad p' = \partial\phi_{R'}/\partial x';$$

thus, by $(3.1)_2$, the diffeomorphism $x \mapsto x'$ can be defined by

$$\left[A(x, x') + \phi_{R'} \right]_{x'} = 0, \quad \text{assuming (3.6)}. \tag{3.7}$$

The group $\mathrm{Sp}_2(l)$. Define a *formal function* on $\check{V} \setminus \check{\Sigma}_{R'}$ by the choice of a frame R' and by an expression

$$U_{R'}(\nu, \check{z}) = \alpha_{R'}(\nu, \check{z})e^{\nu\phi_{R'}(\check{z})}, \tag{3.8}$$

where $\alpha_{R'}$ is a formal series

$$\alpha_{R'}(\nu, \check{z}) = \sum_{r=0}^{\infty} \frac{1}{\nu^r} \alpha_r'(\check{z}),$$

the α_r' being infinitely derivable functions: $\check{V} \setminus \check{\Sigma}_{R'} \to \mathbf{C}$.

From now on, *let A be the datum of*
(i) a real-valued quadratic form A of $(x, x') \in X \oplus X$;
(ii) a choice of $\Delta(A) = \pm \left[\det(A_{x_j x_k'}) \right]^{1/2}$; we assume: $\Delta(A) \neq 0$.
Let $R = s_A R'$; i.e., $RR'^{-1} = s_A \in \mathrm{Sp}(l)$; define a linear map $S_A: U_{R'} \mapsto U_R$ by

$$(S_A U_{R'})(\nu, \check{z}) = U_R(\nu, \check{z})$$

$$= \left(\frac{|\nu|}{2\pi i} \right)^{1/2} \Delta(A) \int_{\check{V} \setminus \check{\Sigma}_{R'}} e^{\nu A(x, x')} U_{R'}(\nu, \check{z}') \, d'x', \tag{3.9}$$

where

$$z \in V \text{ is the projection of } \check{z} \in \check{V}; \ (x, p) = Rz;$$

$$z' \in V \text{ is the projection of } \check{z}' \in \check{V}; \ (x', p') = R'z'.$$

Obviously that map S_A can be decomposed into:
(i) the product by $e^{\nu A(0, \cdot)}$;
(ii) a map of type (3.9), where A is a bilinear, real-valued function of $(x, x') \in X \oplus X$, so that it is essentially a *Fourier transform*, which is unitary thanks to the numerical factor in (3.9);
(iii) the product by $e^{\nu A(\cdot, 0)}$.

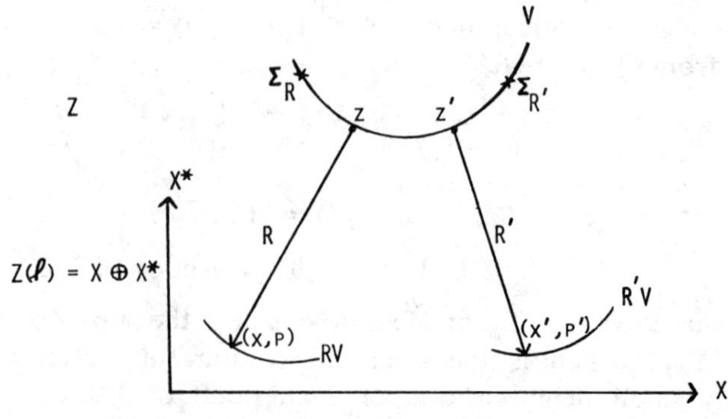

FIGURE 3

According to Maslov, the integral in (3.9) has to be computed by the method of the stationary phase (neglecting the terms more rapidly decreasing than any power of $1/\nu$, when ν tends to $i\infty$). The phase of the integrand is $A(x, x') + \phi_{R'}(\check{z}')$; assuming that $\mathrm{Supp}(U_{R'})$ belongs to $\check{V} \setminus \check{\Sigma}_R \cup \check{\Sigma}_{R'}$ and is small, the assumption that this phase is stationary means that (3.7) holds; then $z = z'$; hence, from (3.5): U_R *is a formal function on* $\check{V}$, *related to* R; the linear transform S_A is *local* on $\check{V}$: $\mathrm{Supp}(U_R) = \mathrm{Supp}(U_{R'})$.

It can be shown that the map S_A of formal functions is the generic–but not any–element of *the covering group of order* 2, $\mathrm{Sp}_2(l)$, *of* $\mathrm{Sp}(l)$; the projection $\mathrm{Sp}_2(l) \to \mathrm{Sp}(l)$ maps the two S_A corresponding to opposite choices of $\Delta(A)$ onto the same s_A. Hence

LEMMA. *Any* $S \in \mathrm{Sp}_2(l)$, *with projection* $s \in \mathrm{Sp}(l)$, *maps each formal function* $U_{R'}$ *defined on* $\check{V} \setminus \check{\Sigma}_R \cup \check{\Sigma}_{R'}$ *into such a formal function* U_R, *where* $R = sR'$. *That mapping is local.*

COMMENT. Of course the preceding account is not a proof of that lemma, which follows from the following results. Let $\mathcal{S}(X)$ be the Schwartz space of functions: $X \to \mathbf{C}$, all of whose derivatives decrease rapidly at infinity; its dual is the Schwartz space $\mathcal{S}'(X)$ of tempered distributions; let $\mathcal{K}(X)$ be the Hilbert space of square-integrable functions: $X \to \mathbf{C}$;

$$\mathcal{S}(X) \subset \mathcal{K}(X) \subset \mathcal{S}'(X).$$

Each $S \in \mathrm{Sp}_2(l)$ defines a continuous automorphism of $\mathcal{S}'(X)$; its restriction to $\mathcal{K}(X)$ is a unitary automorphism of $\mathcal{K}(X)$; its restriction to $\mathcal{S}(X)$ is a continuous automorphism of $\mathcal{S}(X)$, which is defined by (3.9) if $S = S_A$. Then each S defines also an automorphism of the vector space whose elements are the classes of functions of (x, ν) having the same *asymptotic behaviour* when ν tends to $i\infty$; any formal function U, defined on X by

$$U(\nu, x) = \alpha(\nu, x)e^{\phi\nu(x)}, \tag{3.10}$$

where $\alpha(\nu, x) = \sum_{r=0}^{\infty} \frac{1}{\nu^r} \alpha_r(x), \qquad \alpha_r: X \to \mathbf{C}, \phi: X \to \mathbf{R},$

specifies one of those asymptotic classes; now $\mathrm{Sp}_2(l)$ does *locally* operate [see: lemma], if lifted from X to $\check{V}$, by means of the use of formal functions U_R on $\check{V}$ and of their projections U defined by

$$U(\nu, x) = \sum_{\check{z}} U_R(\nu, \check{z}), \quad \text{where } x \text{ and } \check{z} \text{ have to satisfy (3.6)} \tag{3.11}$$

(that lifting assumes the support of U_R to be small enough).

COMMENT. The two $S \in \mathrm{Sp}_2(l)$ having the same projection $s \in \mathrm{Sp}(l)$ may be denoted by $\pm S$. Denote by $c \in S^1$ the product of a function by the complex number c of norm 1. The elements of *I. E. Segal's metaplectic* group [5] are the cS, assuming $cS = c'S'$, if and only if $c' = \pm c$, $S' = \pm S$ ($\pm$ being the same).

The q-symplectic geometry enables an improvement of the preceding lemma. Denote by Λ the *Lagrangian Grassmannian* of Z, i.e., the set of its Lagrangian vector subspaces; *V. I. Arnold* [1] established the isomorphism of its Poincaré group (i.e., first homotopy group) with the additive group of the integers

$$\pi_1[\Lambda] \simeq \mathbf{Z}. \tag{3.12}$$

Thus, for any $q \in \{1, 2, \ldots, \infty\}$, Λ has a unique *covering space Λ_q of order q*. The choice of a $\lambda_q \in \Lambda_q$ with a given projection λ on Λ is said to be the choice of a *q-orientation* of that λ; the 2-orientation is the usual Euclidean orientation. Similarly define $\Lambda(l)$ and $\Lambda_q(l)$ for $Z(l)$.

Moreover,

$$\pi_1[\mathrm{Sp}(l)] \simeq \mathbf{Z}; \tag{3.13}$$

thus, for any q, $\mathrm{Sp}(l)$ has a unique *covering group $\mathrm{Sp}_q(l)$ of order q*. Now $S \in \mathrm{Sp}_q(l)$ induces a homeomorphism of $\Lambda_{2q}(l)$, which has a projection on $\Lambda(l)$, namely, the homeomorphism of $\Lambda(l)$ induced by the projection s of S onto $\mathrm{Sp}(l)$.

Let us define a *q-frame R* as being

(i) an isomorphism $Z \to Z(l)$, consistent with the symplectic structures;

(ii) a homeomorphism $\Lambda_{2q} \to \Lambda_{2q}(l)$, having a projection which is the homeomorphism $\Lambda \to \Lambda(l)$ induced by the preceding isomorphism.

Then a change of *q-frame RR'^{-1}* is

(i) an element s of $\mathrm{Sp}(l)$;

(ii) a homeomorphism of $\Lambda_{2q}(l)$ induced by one of the $S \in \mathrm{Sp}_q(l)$, whose projection is $s \in \mathrm{Sp}(l)$.

Hence, that change of q-frame RR'^{-1} may be identified with that $S \in \mathrm{Sp}_q(l)$:

$$RR'^{-1} = S \in \mathrm{Sp}_q(l); \quad \text{i.e., } R = SR'. \tag{3.14}$$

Let us choose $q = 2$ and make more precise the definition of a *formal function $U_{R'}$ on $\check{V}$*: from now on, *R' has to be a 2-frame*; then the definition of $U_R = SU_{R'}$ has to be supplemented as follows:

$$R = SR'.$$

Thus the statement of the preceding lemma is simplified as follows:

THEOREM. *Any $S \in \mathrm{Sp}_2(l)$ maps any formal function $U_{R'}$ defined on $\check{V} \setminus \check{\Sigma}_R \cup \check{\Sigma}_{R'}$ into such a formal function U_R, where $R = SR'$. That mapping is local.*

The preceding theorem makes the following definition possible.

Lagrangian functions on $\check{V}$. Let V be a Lagrangian variety in Z. If for each 2-frame R a formal function U_R is given on $\check{V} \setminus \check{\Sigma}_R$ and if

$$(\forall\, R, R') \qquad U_R = SU_{R'} \quad \text{on } \check{V} \setminus \check{\Sigma}_R \cup \check{\Sigma}_{R'}, \text{ where } S = RR'^{-1},$$

then the set of those U_R constitutes a Lagrangian function $\check{U}$ on $\check{V}$:

$$\check{U} = \{U_R\}, \quad \text{by definition;}$$

U_R is said to be the expression of $\check{U}$ in the 2-frame R. Obviously $\check{U}$ has a support and a restriction to each open subset of $\check{V}$:

$$(\check{V} \setminus \check{\Sigma}_R) \cap \mathrm{Supp}(\check{U}) = \mathrm{Supp}(U_R).$$

COMMENT. U_R has on $\check{\Sigma}_R$ singularities, which can be described, using [3] *Maslov's index* m_R: it is a locally constant function $\check{V} \setminus \check{\Sigma}_R \to \mathbf{Z}$, whose topological definition (by means of Kronecker's index; i.e., intersection of chains) is due to *V. I. Arnold* [1] and which has also a geometric definition.

The scalar product $(\cdot \,\check{|}\, \cdot)$ of two Lagrangian functions with compact supports can be defined, using the facts that $\mathrm{Sp}_2(l)$ is unitary on $\mathcal{H}(X)$ and that there are partitions of the unity into sums of Lagrangian operators with supports in elements of any given open covering of Z.

Lagrangian operators. Let Ω be an open subset of Z; define on Ω a formal function, without phase

$$a^0(\nu, z) = \sum_{r=0}^{\infty} \frac{1}{\nu^r} a_r^0(z),$$

a_r^0 being an infinitely derivable function $\Omega \to \mathbf{C}$. For each frame R, define

$$a_R^0(\nu, x, p) = a^0(\nu, z), \quad \text{where } (x, p) = Rz,$$

and then

$$a_R^{\pm}(\nu, x, p) = \exp\left(\pm \frac{1}{2\nu} \langle \partial/\partial x, \partial/\partial p \rangle \right) a_R^0(\nu, x, p).$$

First, assume a^0, i.e., the a_r, to be polynomial; define the formal differential operator with polynomial coefficients

$$a_R = a_R^+\left[\nu, \overset{(2)}{x}, \frac{1}{\nu} \overset{(1)}{\frac{\partial}{\partial x}} \right] = a_R^-\left[\nu, \overset{(1)}{x}, \frac{1}{\nu} \overset{(2)}{\frac{\partial}{\partial x}} \right],$$

(1) specifying the operators to be applied at first; a_R locally maps a formal function U defined on X [see (3.10)] onto another one of same type and, by means of the projection (3.11) of U_R onto U, a formal function U_R defined on $\check{V} \setminus \check{\Sigma}_R$ onto another one. The set of the a_R constitutes a *Lagrangian operator*: $a = \{a_R\}$, by definition; if $\check{U} = \{U_R\}$ is a Lagrangian function defined on $\check{V}$, then $\{a_R U_R\}$ is also such a function, denoted by

$$a\check{U} = \{a_R U_R\}.$$

Now if formal polynomials tend on Ω to a formal function a^0, then their associated Lagrangian operators have a limit: the *Lagrangian operator* a *associated to the formal function* a^0 *defined on* Ω.

If $\check{U}$ is a Lagrangian function defined on $\check{V}$ and if $V \subset \Omega$, *then* a *maps locally* $\check{U}$ *onto* $a\check{U}$ *defined on* $\check{V}$.

Assume

$$(\forall z \in \Omega) \qquad a_0^0(z) \neq 0;$$

then a *has an inverse* a^{-1}, *which is still a Lagrangian operator.*

4. Lagrangian functions on V: The need of Planck's constant. *The definition of Lagrangian functions on* V makes use of Lagrangian functions on $\check{V}$ and of Poincaré's group $\pi_1[V]$ of V.

Each $\gamma \in \pi_1[V]$ defines a homeomorphism of $\check{V}$: if $\check{z} \in \check{V}$, then $\gamma\check{z} \in \check{V}$, $\check{z}$ and $\gamma\check{z}$ having the same projection on V. $\phi_R \circ \gamma - \phi_R$ is a constant, namely, the period $c_\gamma = \frac{1}{2}\int_\gamma [z, dz]$ of ψ. Therefore, if $U_R = \alpha_R e^{\nu\phi_R}$ is a formal function defined on $\check{V}$, then

$$U_R \circ \gamma^{-1} = \alpha_R \circ \gamma^{-1} e^{\nu\phi_R - \nu c_\gamma}$$

is not a formal function defined on $\check{V}$; indeed, its phase is not ϕ_R, but $\phi_R - c_\gamma$. Obviously the simplest definition of the image, γU_R, which has to be a formal function on $\check{V}$, of U_R by γ is

$$\gamma U_R = e^{(\nu - \nu_0)c_\gamma} U_R \circ \gamma^{-1}, \tag{4.1}$$

ν_0 *being a constant we have to choose*; since ν is a purely imaginary variable tending to $i\infty$, let us choose

$$\nu_0 \in i[0, \infty[\tag{4.2}$$

and note that there is no reason for making some more specific choice, for instance $\nu_0 = 0$.

In the quantum mechanics, the choice agreeing with the experimental results will be [see §5]

$$\nu_0 = \frac{i}{\hbar} = \frac{2\pi i}{h}, \qquad h \text{ being } Planck's \text{ } constant. \tag{4.3}$$

Once ν_0 is chosen, (4.1) shows how Poincaré's group $\pi_1[V]$ maps formal functions U_R, defined on $\check{V}$ and related to R, onto formal functions of the same kind; $\pi_1[V]$ commutes with $\mathrm{Sp}_2(l)$; therefore $\pi_1[V]$ maps Lagrangian functions $\check{U}$ on $\check{V}$ onto Lagrangian functions on $\check{V}$

$$\text{if } \check{U} = \{U_R\}, \quad \text{then } \gamma\check{U} = \{\gamma U_R\}. \tag{4.4}$$

The group $\pi_1[V]$ commutes with the Lagrangian operators: $a(\gamma\check{U}) = \gamma(a\check{U})$.

Define the *Lagrangian functions* on V as those Lagrangian functions $U = \{U_R\}$ on $\check{V}$ which are *invariant by* $\pi_1[V]$, i.e., such that for some R and hence for any R:

$U_R^0 = \alpha_R e^{\nu_0\phi_R}$ is a function defined on V, with formal numerical values.

A Lagrangian operator a transforms locally a Lagrangian function U on V; aU is a Lagrangian function on V. A Lagrangian function U on V has a support

$$\mathrm{Supp}(U) \subset V. \tag{4.5}$$

Define

$$\mathrm{Supp}(a) = \mathrm{Supp}(a^0) \subset Z, \tag{4.6}$$

where the Lagrangian operator a is associated with the formal function a^0. Then

$$\mathrm{Supp}(aU) \subset \mathrm{Supp}(a) \cap \mathrm{Supp}(U). \tag{4.7}$$

The scalar product $(\cdot \check{|} \cdot)$ of two Lagrangian functions on V and V' can be defined, when the intersection of their supports is compact; that definition makes use of $(\cdot \check{|} \cdot)$ which essentially differs from $(\cdot \check{|} \cdot)$.

$(\cdot\,|\,\cdot\,)$ defines only a quasi-norm: if $\check{U}$ is defined on $\check{V}$, then $(\check{U}\,|\,\check{U}) = 0$ does not imply $\check{U} = 0$.

$(\cdot\,|\,\cdot\,)$ defines a *norm* $\|U\|$, which is a *positive formal number without phase*

$$\|U\| = \sum_{r=0}^{\infty} \frac{1}{\nu^r} \alpha_r, \quad \text{where } \alpha_r \in \mathbf{C};$$

"positive" means $i^{-r}\alpha_r$ real, $i^{-s}\alpha_s > 0$, s being the smallest r such that $\alpha_r \neq 0$.

Schwarz's inequality and triangle inequality hold for those scalar products, quasi-norm and norm, strictly for $(\cdot\,|\,\cdot\,)$.

Now *the definition* of the *new structure* called *Lagrangian analysis* is *achieved*.

Nonhomogeneous Lagrangian equation. Let

$$aU = U' \tag{4.8}$$

be that equation, where a is the Lagrangian operator associated with the given formal function

$$a^0 = \sum_{r=0}^{\infty} \frac{1}{\nu^r} a_r^0 \qquad (a_r^0 \colon \Omega \to \mathbf{C}),$$

where U' is a given Lagrangian function on $V \subset \Omega$, the unknown Lagrangian function U having, obviously, to be defined on V. *Assume*

$$(\forall z \in \Omega) \qquad a_0^0(z) \neq 0;$$

then a has an inverse a^{-1}, which is a Lagrangian operator defined on Ω; therefore (4.8) has the unique solution

$$U = a^{-1}U'. \tag{4.9}$$

The homogeneous Lagrangian equation

$$aU = 0 \tag{4.10}$$

is much less trivial: U is defined on a Lagrangian variety V belonging to the variety W where a_0^0 vanishes; assume a_0^0 real-valued; denote $a_0^0 = H$; then W is the hypersurface of Z defined by the equation

$$W\colon H(z) = 0; \tag{4.11}$$

on each Lagrangian variety $V \subset W$, the solutions of (4.10) can be locally constructed by *the W.K.B. method*; the complete discussion of (4.10) is never easy, except for $l = 1$.

The homogeneous Lagrangian system

$$(\forall j = 1, 2, \ldots, l) \qquad a^{(j)}U = 0 \tag{4.12}$$

is especially interesting under the following assumptions: the l Lagrangian operators $a^{(j)}$ *commute* and are associated with *real-valued functions* $H^{(j)}$ defined on a domain Ω of Z. Then U has to be defined on the variety V of Z whose equations are

$$V\colon H^{(1)}(z) = \cdots = H^{(l)}(z) = 0; \tag{4.13}$$

that variety is Lagrangian; indeed the $H^{(J)}$ are in involution (i.e., their Poisson brackets vanish); the amplitude α_R of each expression U_R of U is obtained by the integration on V of closed Pfaffian forms.

5. Application to the quantum mechanics of the hydrogen atom in a constant magnetic field.

Schrödinger's equation. Let H be the Hamiltonian of the electron belonging to that atom; i.e., the Hamiltonian system (2.10) describes the nonquantified trajectories of that nonrelativistic or relativistic electron; then the Lagrangian operator a associated to H is the *Schrödinger* operator or the relativistic Schrödinger operator, i.e., the *Klein-Gordon* operator. H and, therefore, a depend on a parameter E, which is the *energy level* of the electron.

The quantum mechanics asserts that E has a value such that the equation

$$aU = 0 \tag{5.1}$$

has *a nontrivial solution.*

In wave quantum mechanics, the numerical value $\nu_0 = i/\hbar$ is given to ν, U is a function $X \to \mathbf{C}$, U and its gradient U_x have to be square-integrable. U represents a *not-observable* quantity.

That fact allows us to assume that U is a *Lagrangian function* defined on a Lagrangian variety V. Replace the assumption of square-integrability by the following one: *V is compact*. Note that the equation (5.1) has always to be supplemented by two other equations connected to the length of the momentum of the impulse and its component in the direction of the magnetic field. Therefore U has to satisfy a system of three Lagrangian equations. That system belongs to the type (4.12).

That Lagrangian system and the classical system studied in wave quantum mechanics are not equivalent, but the constructions of their solutions have some likeness and finally both of them define *the same energy levels E.*

Their computation becomes extremely *direct* and *simple* if the Lagrangian system is used mod $1/\nu^2$ and therefore U defined mod $1/\nu$.

Dirac's equation can be likewise solved in Lagrangian analysis; that gives anew the standard energy levels, even when the magnetic field is so strong that the Zeeman effect is replaced by the much more complicated Paschen-Back effect.

6. The study of atoms with several electrons by the wave quantum mechanics requires the computing of eigenvalues, which cannot be carried out, except in a few cases, or by using crude approximations suggested by the spirit of the first quantum mechanics. Until now the Lagrangian analysis did not offer better possibilities; see Maslov and his collaborators about the use of crude approximations.

BIBLIOGRAPHY

1. V. I. Arnold, *On a characteristic class intervening in quantum conditions*, Functional Anal. Appl. **1** (1967), 1–14. (Russian with English transl.)

2. V. C. Buslaev, *Quantization and the W. K. B. method*, Trudy Mat. Inst. Steklov **110** (1970), 5–28. (Russian)

3. V. P. Maslov, *Perturbation theory and asymptotic methods*, M. G. U., Moscow, 1965. (Russian)

4. H. Poincaré, *Sur les intégrales irrégulières des équations linéaires*, Acta Math. **8** (1886), 295–344.

5. I. E. Segal, *Foundations of the theory of dynamical systems of infinitely many degrees of freedom*. I, Mat.-Fys. Medd. Danske Vid. Selsk. **31** (1959), 1–39.

6. J. Leray, *Analyse lagrangienne et mécanique quantique*, Séminaire du Collège de France 1976–1977; R. C. P. **25**, Strasbourg, 1978.

The present paper is a comment of the preceding preprint, a revised version of which will be published in English by M. I. T. Press, Cambridge, Massachusetts and in Russian by M. I. R., Moscow.

COLLÈGE DE FRANCE, (MATH.) F-75231 PARIS CEDEX 05, FRANCE

Proceedings of Symposia in Pure Mathematics
Volume **39** (1983), Part 2

Differentiable Dynamical Systems
and the Problem of Turbulence

DAVID RUELLE

1. Conservative and dissipative dynamical systems. The mathematical study of differentiable dynamical systems has its origin in the desire to understand the time evolutions which occur in nature. This relation to physical reality has exercised a strong influence on the progress of the subject. Other strong influences resulted from the internal development of mathematics and physics.

The time evolutions which one observes in nature yield two large classes of dynamical systems. On one hand we have frictionless mechanical systems, or *conservative systems*. On the other hand we have all kinds of natural systems where some "noble" form of energy is converted to heat; these are the *dissipative systems*.

Classical mechanics–the study of conservative systems–was initiated by Isaac Newton and has played a central role in the development of Natural Philosophy. In particular it led to the *Méthodes nouvelles de la mécanique céleste* of Henri Poincaré, and later to the solution of small denominator problems by Kolmogorov, Arnold and Moser. The importance of classical mechanics is due to the existence of nontrivial conservative systems–those of celestal mechanics–which can be investigated with extreme precision both observationally and theoretically.

The dissipative systems have a priori no less interest than the conservative ones. Mathematically they are more general; physically they have widespread occurrence, and display a great variety of phenomena. Unfortunately, dissipative systems do not obey, with great precision, laws as simple as those of celestial mechanics. They are usually continuous systems, requiring an infinite number of parameters for their description. Because of these and other difficulties, a detailed dynamical study of dissipative systems is taking place only now.

2. Viscous flows. Diffusion of heat, chemical reactions, heating of an electrical resistor, are examples of dissipative phenomena. So is the flow of a viscous fluid, where internal friction dissipates mechanical energy into heat.

Viscous flows are among the most remarkable natural phenomena. Whirlpools of the river, waves of the sea, dancing flames of the fire, twisting shapes of the clouds have enchanted poets and fascinated philosophers. To the scientist also viscous flows have something to offer, since understanding

---

Reprinted from Bulletin Amer. Math. Soc. (N.S.) 5 (1981), 29–42.

1980 *Mathematics Subject Classification.* Primary 76A, 58F.

hydrodynamic turbulence is a major unsolved problem of theoretical physics. For historical reasons, however, hydrodynamics has fared rather differently from celestial mechanics.

The flow of viscous fluids is described with reasonable precision by the Navier-Stokes equation. This equation has been known for some time, but the mathematical tools and the experimental techniques needed for a detailed study of viscous flows are relatively recent. The mathematical theory of the Navier-Stokes equation begins only in the 1930s with the work of Jean Leray. The experimental study of viscous flow (in particular turbulent flow) is very difficult, and adequate techniques like laser anemometry are only a few years old.

Since the time of Newton it has become progressively more difficult for the same person to be a good mathematician and a good physicist. One man who showed extended competence was G. I. Taylor, who made a remarkable theoretical and experimental study of the flow between rotating concentric cylinders [69]. In general, however, the divorce between physics and mathematics has been unfavorable to the progress of hydrodynamics. At the same time, mathematicians were attracted by the more fundamental aspects of their science, and all the available energies of physicists were drawn by the development of quantum mechanics. Of course, major steps forward have been made by such high class scientists as J. Leray, A. N. Kolmogorov, or E. Hopf, but hydrodynamics has otherwise remained somewhat in the backwaters of the scientific storm of this century.

3. The Navier-Stokes equation. The components of the *Navier-Stokes equation* are

$$\frac{\partial v_i}{\partial t} + \sum_{j=1}^{d} v_j \frac{\partial v_i}{\partial x_j} = -\frac{\partial p}{\partial x_i} + \nu \Delta v_i + g_i \tag{1}$$

for $i = 1, \ldots, d$. Here d is the number of space dimensions ($d = 3$ usually, but $d = 2$ has been much studied). The v_i are the components of the velocity field, depending on position $x \in \mathbf{R}^d$ and time t. The left-hand side of the equation is the acceleration of a particle of fluid. The right-hand side contains the gradient of the pressure p, a "dissipative term" with the Laplace operator multiplied by a constant ν (the viscosity) and an external force g.[1] We supplement (1) by the *incompressibility condition*

$$\sum_{i=1}^{d} \frac{\partial v_i}{\partial x_i} = 0. \tag{2}$$

The fluid is enclosed in a region $\Omega \subset \mathbf{R}^d$, and its velocity is imposed on $\partial\Omega$; this is the *boundary condition*

$$v_i = a_i \quad \text{for } i = 1, \ldots, d \text{ on } \partial\Omega. \tag{3}$$

If Ω is bounded, we assume of course that the total flux of a through $\partial\Omega$ vanishes. Usually there is no flow through the boundaries and, since a viscous

---

[1] To be more precise, p and g are the pressure and force *divided by the density* of the fluid (which will be constant in view of (2)); ν is the *kinematic* viscosity.

fluid "sticks" to the wall, (2) will just express that the velocity of the fluid at the boundary is equal to the velocity of the wall (e.g. zero if the wall is fixed). Finally there is an *initial condition*

$$v_i = v_{i0} \quad \text{for } t = 0. \tag{4}$$

The Navier-Stokes equation contains partial derivatives and products which, a priori, make sense only if v is sufficiently smooth. If one multiplies (1) by a smooth function $\phi_i(x)$, sums over i and integrates over x, one obtains an equation $(1)_\phi$. The various terms of $(1)_\phi$ are linear or bilinear in v, and can be extended by continuity to functional spaces of not necessarily very differentiable functions. If an element of such a functional space verifies all equations $(1)_\phi$, in the sense of distributions in t, it is called a *weak solution* of (1). It is advantageous to look first for weak solutions of the Navier-Stokes equations, and later to try to prove their regularity.

For definiteness we assume from now on that Ω is bounded. To handle the boundary condition (3), one extends a to a divergence-free vector field in Ω, and inserts $v = a + u$ in (1), thus obtaining a new equation for u, with $u = 0$ on $\partial\Omega$. Let $\mathbf{H}$, resp. $\mathbf{V}$, be the completion of the space of smooth divergence-free vector fields with compact support in Ω, with respect to the L^2-norm

$$|u| = \left[\int_\Omega dx \sum_i u_i^2 \right]^{1/2},$$

resp. the Dirichlet norm

$$\|u\| = \left[\int_\Omega dx \sum_{ij} \left(\frac{\partial u_i}{\partial x_j} \right)^2 \right]^{1/2}.$$

The condition $u = 0$ on $\partial\Omega$ is expressed by $u \in \mathbf{V}$.

One shows that $\mathbf{H}$ is the orthogonal complement of the subspace of gradients in $L^2(\Omega)^3$. If one takes the projection of (1) orthogonal to gradients, the pressure term disappears and one is left with a functional equation of the form

$$du/dt = F_\mu(u) \tag{5}$$

for $u \in \mathbf{V}$. We have left a parameter μ to describe the intensity of the action exerted on the fluid via the force g and the boundary conditions. A parameter of this sort is usually present in experimental situations (Reynolds number). We have assumed that the action on the fluid is time independent, so that the right-hand side of (5) does not depend on t.

We indicate now the main results known on the existence and uniqueness of solutions of the Navier-Stokes equation[2] for $d = 2$ or 3. We assume that $\partial\Omega$ and a are sufficiently smooth, and that g is square-integrable.

3.1. THEOREM (EXISTENCE). *Given* $u_0 \in \mathbf{H}$ *there corresponds to it a weak solution* $u \in L^2_{\text{loc}}((0, \infty), \mathbf{V})$. *Furthermore* u *is weakly continuous* $[0, \infty) \mapsto \mathbf{H}$, *so that the initial condition* $u(0) = u_0$ *makes sense.*

---

[2] For details, see the monographs by Ladyzhenskaya [30], Lions [36], Temam [70], Girault and Raviart [16], and a paper by Foiaş and Temam [13].

3.2. THEOREM (UNIQUENESS). *If $d = 2$ the weak solution is unique; furthermore it is continuous $[0, \infty) \mapsto \mathbf{H}$, $(0, \infty) \mapsto \mathbf{V}$.*

If $d = 2$ and $u_0 \in \mathbf{V}$ the solution is continuous $[0, \infty) \mapsto \mathbf{V}$.

If $d = 3$ and $u_0 \in \mathbf{V}$ there is $t(u_0) > 0$ such that the weak solution is unique on $[0, t(u_0))$ and continuous $[0, t(u_0)) \mapsto \mathbf{V}$.

In general, we say that a solution is *regular* on an interval if it is continuous from that interval to $\mathbf{V}$. There can be only one regular solution u on an interval $[0, T]$ with $u(0) = u_0$. We write the solution as $u(t) = f^t u_0$. The maps (f^t), as far as they are defined, satisfy the semigroup property $f^{t+t'} = f^t \circ f^{t'}$. In order to state some analyticity properties of the family (f^t), we introduce the space $\mathbf{D}$ of elements of $\mathbf{V}$ with square-integrable second derivatives. We consider $\mathbf{D}$ as Hilbert space with the norm

$$\||u\|| = \left[\int_\Omega dx \sum_i \left(\sum_j \frac{\partial^2 u_i}{\partial x_j^2} \right)^2 \right]^{1/2}.$$

Notice that the inclusion $\mathbf{D} \hookrightarrow \mathbf{V}$ is compact.

3.3. THEOREM (ANALYTICITY).[3] *For each $R > 0$ there is $T(R) > 0$ such that the map $(u, t) \mapsto f^t u$ is defined and real-analytic*

$$\{ u \in \mathbf{V} : \|u\| < R \} \times (0, T(R)) \mapsto \mathbf{D}.$$

For $t \in (0, T(R))$, f^t maps $\{ u \in \mathbf{V} : \|u\| < R \}$ to a bounded subset of $\mathbf{D}$, and the derivative Df^t is an injective linear map $\mathbf{V} \mapsto \mathbf{D}$.

A dynamical system can be obtained by restricting the maps f^t to a suitable subset of $\mathbf{V}$. We shall say that an open set $M \subset \mathbf{V}$ is an *admissible set* (of initial conditions) if every $u \in M$ determines a regular solution on $[0, \infty)$ and there is $T_0 \geqslant 0$ such that the set $f^t M$ is relatively compact in M for $t > T_0$.[4] In particular, if $d = 2$, one can show that the ball $\{ u \in \mathbf{V} : \|u\| < R \}$ is admissible for sufficiently large R.[5]

3.4. COROLLARY. *Let M be an admissible set of initial conditions. For $t \geqslant T_0$, the maps $f^t : M \to M$ are defined, injective, and real-analytic. The semigroup property $f^t \circ f^{t'} = f^{t+t'}$ holds. The maps $(u, t) \mapsto f^t u$, $Df^t(u)$ are continuous $M \times (T_0, +\infty) \mapsto M, \mathfrak{B}(V)$.*

The intersection $\Lambda = \bigcap_{t > T_0} f^t M$ is compact. If $u \in \Lambda$, $t > T_0$, the operator $Df^t(u)$ is compact and injective.

4. The problem of turbulence. While turbulence is often, in practical terms, a well-defined phenomenon, its fundamental nature is still controversial. Call

---

[3] These analyticity properties are explicit or implicit in Fujita and Kato [15], Iooss [25], Foiaş and Temam [13].

[4] These conditions are satisfied if there is $T_1 > 0$ such that every $u \in M$ determines a regular solution on $[0, T_1]$ and $f^{T_1} M$ is relatively compact in M.

[5] There is $S > 0$ such that $\lim \neq \sup_{t \to \infty} \|f^t u\| \leqslant S$ for all $u \in V$ (see Foiaş and Temam [13, §2]). Take $R > R_1 > S$ and write $M = \{ u \in V : \|u\| < R \}$. If $t_0 = \frac{1}{2} T(R)$, the set $f^{t_0} M$ is bounded in $\mathbf{D}$, hence has compact closure in V. There is thus $t_1 > 0$ such that $f^{t_1}(f^{t_0} M) \subset \{ u \in V : \|u\| < R_1 \}$. Taking $T_1 = t_0 + t_1$, we see that $f^{T_1} M$ is relatively compact in M.

μ the *Reynolds number* describing the level of "excitation" of a hydrodynamical system (see equation (5)). When $\mu = 0$, the system is at rest (asymptotically). When μ is small, steady states and time periodic motions are observed. When μ is larger we have the *onset of turbulence*: the fluid motion becomes nonperiodic in time.

When μ is sufficiently large, one says that *fully developed turbulence* is present. The motion is very chaotic and appears to conform to statistical laws of some universality. Such laws have been proposed by Kolmogorov [29]: his "five-thirds law" on the spatial frequency distribution of velocities is well supported by observations. Kolmogorov's theory can be derived from dimensional arguments, and is thus of a somewhat superficial nature. It neglects in particular the phenomenon of *intermittency*: the fact that much of the velocity gradient $\partial v_i / \partial x_j$ is concentrated on a small part of physical space. Intermittency has received, at this time, only sketchy theoretical treatment (see Frisch, Sulem and Nelkin [14]). Fully developed turbulence is a three-dimensional phenomenon. For $d = 2$, chaotic flow is also observed (by numerical simulation) but the details are rather different.

In his early studies, Leray [34] proposed that turbulence was related to the possible nonuniqueness of weak solutions of the Navier-Stokes equation. A weak solution $t \to u(t)$ corresponding to a smooth initial condition may at some time t^* develop singularities and nonuniqueness. The blowing up of the velocity gradient at singularities may be interpreted as turbulent intermittency. It remains to be seen if nonuniqueness and singularities are really present. Singularities, if they exist, will be on a "small" set[6] as shown by Scheffer [64], [65] and therefore difficult to detect experimentally.

Leaving the difficult domain of fully developed turbulence, we shall now turn to the onset of turbulence: how does nonperiodicity develop? Here we take the point of view, corroborated by experiment, that the Navier-Stokes equation defines a dynamical system (no breakdown of uniqueness occurs). For $\mu = 0$ the dynamical system has an attracting fixed point corresponding to the state of rest. For small μ the existence of an attracting fixed point corresponding to a steady motion is ensured by the implicit function theorem. When μ is increased the Hopf bifurcation may give rise to an attracting periodic orbit corresponding to a time-periodic motion.[7] One may think that further bifurcations will produce attracting tori of increasing dimension k, corresponding to quasi-periodic motions with an increasing number k of basic frequencies. When k is large enough, the quasi-periodic motion is interpreted as turbulent. This point of view was proposed by Landau [32] in 1944, and independently by Hopf [23] in 1948. Looking at things more carefully, one finds that the bifurcation from a periodic solution (1-torus) to a 2-torus takes place according to expectations.[8] But the further bifurcations do not usually take place in a simple manner. If a quasi-periodic motion on a

---

[6] The Hausdorff dimension of this set is small.

[7] See Hopf [22], Iooss [24], and for a general discussion Marsden and McCracken [41] which contains an English translation of Hopf's article.

[8] See Naimark [43], Sacker [63], Ruelle and Takens [62], and again Marsden and McCracken for a general discussion.

k-torus does occur, it is unstable to perturbations, yielding periodic or (for $k \geqslant 3$) nonperiodic motions of all kinds.

In 1971, Takens and myself proposed that nonperiodic motions which are not quasi-periodic describe turbulence. We shall later discuss what is implied. Careful experiments have been made in the past few years. They show that real fluids may be in a steady state, or perform periodic motions, or quasi-periodic motions with commonly two (rarely three) basic frequencies. Furthermore non-quasi-periodic motions are seen as soon as the Reynolds number is large enough. All kinds of bifurcations are observed. *The general lesson seems to be that hydrodynamical systems at the onset of turbulence behave very much as generic differentiable dynamical systems in finite dimension.* Simple systems of differential equations with arbitrarily chosen coefficients, when studied by digital or analog computer, yield data so analogous to those of hydrodynamical experiments that it is not possible to tell them apart.

5. Sensitive dependence on initial condition and strange attractors. We have stated that a fluid at the onset of turbulence behaves very much like a generic differentiable dynamical system. This assertion is based in part on our mathematical understanding of dynamical systems, in part on the computer study of a number of simple examples. Our mathematical understanding includes the theory of simple bifurcations, and the theory of Smale's Axiom A diffeomorphisms. Computer studies have shown the existence of several new "mathematical phenomena" which would have been very difficult to guess otherwise, and which are in part not understood.

Of particular interest to us is the widespread phenomenon of *sensitive dependence on initial condition.* This means that a small change in initial condition produces an orbit which deviates exponentially (at least for a while) from the unperturbed orbit. If (f^t) defines a dynamical system on the manifold M (with discrete or continuous time t), we call *attractor* a compact set $\Lambda \subset M$ such that $f^{tx} \to \Lambda$ for x in some neighborhood of Λ. We assume that Λ satisfies some indecomposability condition (like topological transitivity). We shall say that Λ is a *strange* attractor if there is sensitive dependence on initial condition for orbits starting near Λ. This is *not* meant as a precise mathematical definition, but rather as a description (subject to revision) of various objects known either mathematically or from computer work.[9]

Reasonably well understood mathematically are the hyperbolic attractors:[10] the tangent bundle $T_\Lambda M$ has a continuous splitting into a *contracting* subbundle which contracts exponentially under Tf^t and an *expanding* subbundle which contracts under Tf^{-t} (in the continuous time case there is also a one-dimensional neutral subbundle in the flow direction). If the expanding subbundle is nontrivial, we have sensitive dependence on initial condition, and thus a *strange* hyperbolic attractor.

---

[9] Notice that Hamiltonian systems may exhibit sensitive dependence on initial conditions, but cannot have strange attractors because of Liouville's theorem.

[10] On hyperbolic attractors (and more generally Axiom A dynamical systems) see Smale [67] and Bowen [2].

A system of differential equations first considered by Lorenz [37] has suggested a mathematical model which is "almost hyperbolic" and fairly well understood.[11] Nothing has actually been proved, however, about the equations which suggested the model.

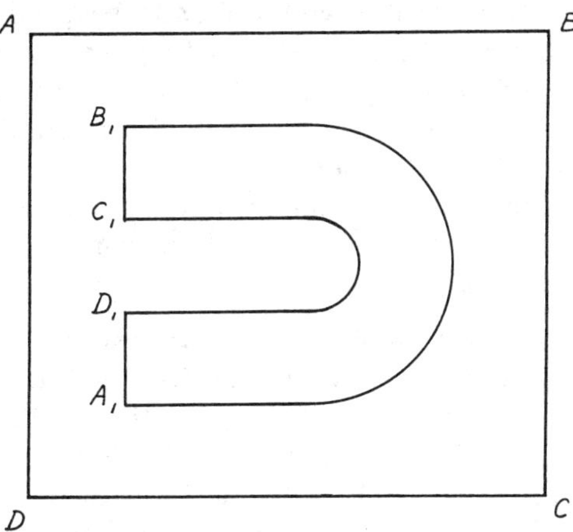

FIGURE 1.

There are strange hyperbolic attractors for diffeomorphisms of $\mathbf{R}^2$, as discovered by Plykin [50], but they are complicated and unobvious. On the other hand very simple folding maps (sending a rectangle $ABCD$ to a horseshoe $A_1B_1C_1D_1 \subset ABCD$; see Figure 1[12]) are found to produce sensitive dependence on initial condition. For instance the map f: $(x, y) \to (y + 1 - ax^2, bx)$ has been much studied numerically, and yields for $a = 1.4$, $b = 3$ the *Hénon attractor*.[13] The computer results suggest that sensitive dependence on initial condition is present for a set of positive Lebesgue measure of values of (a, b), but may be absent for a dense open set (where only periodic orbits would be attracting; see Feit [11]). Mathematically, little is known, but Newhouse[14] has shown that there is a residual subset of an open (nonempty) set of values of (a, b), for which an infinite number of attracting periodic orbits are present. If a strange attractor Λ is thus accompanied by infinitely many periodic attractors, there may be no neighborhood of Λ consisting of points x such that $f'x \to \Lambda$.

We shall not try to improve on our previous definition of a strange attractor to accommodate possible pathologies. It seems preferable to wait for a better mathematical understanding of the Hénon attractor (and other strange attractors known from computer work). What is important is that sensitive dependence on initial condition happens frequently in simple dynamical systems. These systems have a continuous frequency spectrum[15] by

---

[11] See Guckenheimer [18], Guckenheimer and Williams [19], Williams [72].

[12] This is not Smale's horseshoe (which is not an attractor).

[13] See Hénon [20], Feit [11], Curry [7].

[14] Newhouse [44], [45], and private communication.

[15] The frequency spectrum is obtained by frequency analysis of $t \mapsto f'x$. This is much more accessible experimentally than dependence on initial condition.

contrast with the discrete frequency spectrum observed in quasi-periodic systems. The quasi-periodic theory of Landau and Hopf had relied on a torus of large dimension to produce a continuous spectrum by accumulation of many different frequencies. It came somewhat as a surprise to hydrodynamicists that a continuous spectrum would already be produced by a strange attractor in three dimensions (see Monin [42] for an expression of this surprise). Recent experiments by Ahlers [1], Gollub and Swinney [17], and others are in favor of the "strange attractor" theory (first proposed in [62]) rather than that of Landau and Hopf.[16]

The above discussion suggests to call turbulent a hydrodynamical flow with sensitive dependence on initial condition. A quasi-periodic flow would then be nonturbulent (laminar).

6. Ergodic theory of differentiable dynamical systems. The idea of strange attractors has proved physically relevant and fruitful, but their direct geometrical study is discouragingly difficult. Ergodic theory has provided a more rewarding approach, which yields however only "almost everywhere" rather than "everywhere" statements. The first and fundamental result is the *multiplicative ergodic theorem* of Oseledec.[17] Consider a differentiable dynamical system (f^t), i.e., a semiflow or semigroup of differentiable maps $M \mapsto M$ where M is a compact manifold. The multiplicative ergodic theorem (or rather a corollary of it) asserts that there is an (f^t)-*invariant Borel set* Γ, *such that* $\rho(\Gamma) = 1$ *for every* (f^t)-*invariant probability measure* ρ *on* Γ, *and*

$$\lim_{t \to +\infty} \|Df^t(x)u\| = \chi(x, u)$$

exists for every $x \in \Gamma$. Given x, $\chi(x, u)$ takes only a finite number of values for $u \neq 0$. These values $\lambda_x^{(1)} < \cdots < \lambda_x^{(s)}$ are called *characteristic exponents*. More precisely, there is a filtration of $T_x M$ by linear subspaces:

$$0 = V_x^{(0)} \subset V_x^{(i)} \subset \cdots \subset V_x^{(s)} = T_x M$$

such that

$$\lim_{t \to +\infty} \|Df^t(x)u\| = \lambda_x^{(r)} \quad \text{if } u \in V_x^{(r)} \setminus V_x^{(r-1)}.$$

We call $m_x^{(r)} = \dim V_x^{(r)} - \dim V_x^{(r-1)}$ the *multiplicity* of $\lambda_x^{(r)}$. With respect to an ergodic measure ρ, the characteristic exponents and their multiplicities are almost everywhere constant.

The second fundamental result in the ergodic theory of differentiable dynamical systems is a *stable manifold theorem*. Taking for simplicity the case of a diffeomorphism, it is asserted that one can choose Γ as above and such that, if $x \in \Gamma$ and $\lambda_x^{(r)} < \lambda < \min(0, \lambda_x^{(r+1)})$, then

$$\mathcal{W}_x^\lambda = \left\{ y \in M: \limsup_{t \to \infty} \frac{1}{t} \log d(f^t x, f^t y) < \lambda \right\}$$

is contained in Γ, and is the image of $V_x^{(r)}$ by an injective immersion tangent

---

[16] See Swinney and Gollub [68] for a review.

[17] See Oseledec [46] and Raghunathan [53].

to the identity at x, and as smooth as the diffeomorphism f.[18] In other words, there are manifolds $\mathcal{W}_x^\lambda$ defined for almost all x, as smooth as f, and tangent at x to $V_x^{(r)}$. For $\lambda = 0$, we shall call

$$\mathcal{W}_x^- = \left\{ y \in M : \limsup_{t \to \infty} \frac{1}{t} \log d(f^t x, f^t y) < 0 \right\}$$

the *stable manifold* of x. The stable manifold for f^{-1} is called *unstable manifold*.

Stable and unstable manifolds have been much studied in the hyperbolic case; see Hirsch, Pugh, and Shub [21] and references quoted there. Pesin [47]–[49] extended the theory to general diffeomorphisms preserving a smooth measure. Neither the smooth measure nor the invertibility are actually necessary (see Ruelle [58], [59], Ruelle and Shub [61]). However, in the absence of invertibility, one obtains only *local* stable manifolds (or one has to make some transversality assumption), and the unstable manifold theorem is more complicated to formulate (or one has to assume injectivity).

Let us now leave the technicalities and ask what invariant probability measure. is relevant. Clearly the ergodic average $T^{-1}\Sigma_{t=0}^{T-1}\delta_{f^t x}$ (or $T^{-1}\int_0^T dt\, \delta_{f^t x}$) reproduces every ergodic measure ρ for suitable initial x. However, in many cases, the computer evaluation of the above ergodic average gives a single answer (and the probabilistic behavior of a turbulent fluid also often seems uniquely defined). The solution of this paradox is that the computer makes roundoff errors (and the time evolution of a fluid is affected by small random perturbations).

In the case of hyperbolic attractors, the occurrence of a single measure in the presence of small stochastic perturbations is well understood.[19] What happens is that a measure which is stationary under small stochastic perturbations yields, in the limit of zero perturbation, a measure "continuous along the unstable direction". This means that the conditional measures on unstable manifolds are absolutely continuous with respect to the Lebesgue measure on the unstable manifolds. There can be only one (f^t) invariant probability measure ρ with this property on a hyperbolic attractor; it is also characterized by the fact that the measure-theoretic entropy $h(\rho)$ is equal to the sum of the positive characteristic exponents $\Sigma_{\lambda^{(r)}>0} m^{(r)}\lambda^{(r)}$.

I believe that also for nonhyperbolic strange attractors, the stability under small stochastic perturbations drastically reduces the number of relevant ergodic measures.[20] Counterexamples show that the situation is not as simple as in the hyperbolic case, but some positive results are known. The inequality

$$h(\rho) \leqslant \int \rho(dx) \sum_{\lambda_x^{(r)} > 0} m_x^{(r)}\lambda_x^{(r)}$$

---

[18] The diffeomorphism f must be at least $C^{1+\epsilon}$; the theorem also holds in the C^∞ or C^ω category.

[19] The same measure is also obtained as ergodic average for almost all initial x near the attractor with respect to Lebesgue measure. See Sinai [66], Ruelle [54], Bowen and Ruelle [3], Kifer [27], [28].

[20] See Ruelle [56], [57].

holds in general.[21] If f is a diffeomorphism and ρ is continuous along unstable manifolds, Walters [71] and Katok [26] have shown that

$$h(\rho) = \int \rho(dx) \sum_{\lambda_x^{(r)} > 0} m_x^{(r)} \lambda_x^{(r)}.$$

Furthermore, Pugh and Shub [52] have announced that there is a set of positive Lebesgue measure on the manifold M, consisting of points x for which the ergodic average is equal to ρ (one assumes ρ ergodic, with no characteristic exponent equal to zero). It is tempting to believe that such a measure lives on the Hénon attractor. This would for instance explain why computer estimates of the upper characteristic exponent yields reproducibly a single value $\lambda \sim 0.42$. A priori one would expect a scatter of many different values, because the Hénon attractor undoubtedly carries many different ergodic measures with different characteristic exponents.

7. Ergodic theory of the Navier-Stokes equation. We revert to the notation of §3, and assume that there exists a nonempty admissible set $M \subset V$ of initial conditions (M always exists if $d = 2$). Then $\Lambda = \bigcap_{t > T_0} f^t M$ is compact. It would be most useful if one could prove (even under restrictive conditions) that Λ lies in a finite dimensional (f^t)-invariant manifold.[22] Weaker results have been obtained by Foiaş and Prodi [12], Ladyzhenskaya [31], and Mallet-Paret [38].

7.1. THEOREM (MALLET-PARET).[23] *Let V be a separable Hilbert space, and suppose that $\Lambda \subset M \subset V$ where Λ is compact and M open. Let $f: M \to V$ be C^1 and satisfy $f\Lambda \supset \Lambda$. Suppose finally that $Df(x)$ is a compact operator for all $x \in \Lambda$. Then Λ has finite Hausdorff dimension H.*

7.2. COROLLARY (MAÑÉ).[24] *If F is a linear subspace of V with dimension $> 2H + 1$, then there is a continuous projection $\pi: V \to F$ such that its restriction to Λ is injective.*

In view of Corollary 3.4, Mallet-Paret's theorem applies to the Navier-Stokes case (where we actually have $f\Lambda = \Lambda$). Mañé's corollary shows how to obtain a homeomorphism of Λ to a subset of a finite-dimensional Euclidean space.

Using the compactness of the set Λ and of the maps $Df(x)$, it is possible to extend the ergodic theory of §6 to the present infinite dimensional situation.[25]

---

[21] This was first proved by Margulis (quoted by Pesin [47]) for a diffeomorphism with smooth invariant measure. For the case of a general C^1 map, see Ruelle [55].

[22] Apart from the use of the center manifold theorem near a bifurcation, little is known on this problem.

[23] Foiaş and Temam [13] have derived a direct proof of this result for the Navier-Stokes case, and Mañé [39] for general Banach spaces.

[24] See Mañé [39].

[25] I have a Hilbert space extension of the multiplicative ergodic theorem and stable-unstable manifold theorems which weakens the compactness condition on the maps $Df(x)$ and does not assume that they are injective [59]. Independently, Mañé [40] has obtained an extension to compact injective maps in Banach spaces. The application to Navier-Stokes is discussed in Ruelle [60].

7.3. THEOREM. *For the Navier-Stokes semiflow, under the assumptions of §3, the characteristic exponents are defined, and form a sequence tending to $-\infty$ (therefore only a finite number of them can be positive). There are globally defined unstable manifolds; they are finite-dimensional and contained in Λ. There are locally defined stable manifolds (near Λ); they are finite-codimensional. Stable and unstable manifolds are real-analytic. If some ergodic measure ρ has no characteristic exponent equal to zero the stable and unstable manifolds are ρ-almost everywhere transverse to each other.*

For a more precise formulation see [59], [40], [60]. We have here assumed that the Navier-Stokes equations defined a smooth dynamical system. It may be, on the contrary, that turbulence is linked to the lack of smoothness of solutions of the Navier-Stokes equations, as proposed by Leray. This does however not seem likely in the weakly turbulent regime. The results quoted above are thus in agreement with the notion that a weakly turbulent viscous fluid behaves like a typical finite-dimensional smooth dynamical system.

8. Further remarks. Many dynamical phenomena have now been observed both in computer simulations and in hydrodynamical experiments. Of special interest is the "Feigenbaum bifurcation" first observed in computer studies of (noninvertible) maps of an interval. Apparently this is a new generic bifurcation which consists in the accumulation of "period doubling" bifurcations. An attracting fixed point can, by a well-understood bifurcation, lose its attracting character and be replaced by an attracting periodic orbit of order 2. Repeating a similar process n times produces an attracting orbit of order 2^n. What is surprising is that all these bifurcations are often really seen to follow each other, and to converge asymptotically as a geometric sequence. In other words in the space of maps of the interval there seems to exist a "Feigenbaum manifold" of codimension 1, which is the geometric limit of bifurcation manifolds corresponding to period doubling.

M. Feigenbaum, who discovered this phenomenon, has given it a heuristic but highly nontrivial explanation [8], [9] based on the study of an associated dynamical system in a Banach space. A rigorized version of Feigenbaum's arguments is now being worked out (see [6], [33], [4]). It appears that the Feigenbaum bifurcation also occurs in diffeomorphisms and flows (see [5], [10]). In any case, something very much like it is observed in some hydrodynamical experiments (see [35]: this is another example of the occurrence of the same phenomena in hydrodynamical experiments and computer studies, with the same doubts as to their detailed explanation).

To conclude our review it is appropriate to ask what the relation of all this is to the mathematical heritage of Henri Poincaré. His contribution to differentiable dynamical systems is sufficiently impressive, but what about turbulence? I think that it is to the credit of Poincaré that he saw that there was a problem, and kept away from it. That he saw a problem is visible from a section entitled "Explication d'un fait expérimental", at the end of his course on the *Théorie des tourbillons* [51]. He notices that in a liquid flow the vorticity is usually not diffuse, but rather tends to be concentrated in individual whorls. He says that this fact has not been properly explained mathematically, and attempts at justifying it by a stability calculation. (He mentions more calculations, which are not reproduced in the book.) The phenomenon considered by Poincaré is two dimensional, and therefore is not really turbulence, but it is clearly related to intermittency. The stability calculations just referred to seem to be as close as Poincaré came to a discussion of turbulence. It is now clear that the knowledge of his time would not allow him to do better. Mathematical physics tries to understand a world of unknown complexity with tools of known limitations. This requires boldness, and modesty. Obviously Henri Poincaré lacked neither of these two qualities.

REFERENCES

1. F. Ahlers, *Low-temperature studies of the Rayleigh-Bénard instability and turbulence*, Phys. Rev. Lett. **33** (1974), 1185–1188.

2. R. Bowen, *On Axiom A diffeomorphisms*, CBMS Regional Conf. Series, No. 35, Amer. Math. Soc., Providence, R. I., 1978.

3. R. Bowen and D. Ruelle, *The ergodic theory of Axiom A flows*, Invent. Math. **29** (1975), 181–202.

4. M. Campanino and H. Epstein, *On the existence of Feigenbaum's fixed point*, Comm. Math. Phys. (to appear).

5. P. Collet, J.-P. Eckmann and H. Koch, *Period doubling bifurcations for families of maps on* $\mathbf{R}^n$ (to appear).

6. P. Collet, J.-P. Eckmann and O. E. Lanford, *Universal properties of maps of an interval*, Comm. Math. Phys. **76** (1980), 211–254.

7. J. H. Curry, *On the Hénon transformation*, Comm. Math. Phys. **68** (1979), 129–140.

8. M. J. Feigenbaum, *Quantitative universality for a class of nonlinear transformations*, J. Statist. Phys. **19** (1978), 25–52.

9. ______, *The universal metric properties of nonlinear transformations*, J. Statist. Phys. **21** (1979), 669–706.

10. ______, *The transition to aperiodic behavior in turbulent systems*, Comm. Math. Phys. (to appear).

11. S. D. Feit, *Characteristic exponents and strange attractors*, Comm. Math. Phys. **61** (1978), 249–260.

12. C. Foiaş and G. Prodi, *Sur le comportement global des solutions nonstationnaires des équations de Navier-Stokes en dimension 2*, Rend. Sem. Mat. Univ. Padova **39** (1967), 1–34.

13. C. Foiaş and R. Temam, *Some analytic and geometric properties of the solutions of the evolution Navier-Stokes equations*, J. Math. Pures Appl. **58** (1979), 339–368.

14. U. Frisch, P.-L. Sulem and M. Nelkin, *A simple dynamical model of intermittent fully developed turbulence*, J. Fluid Mech. **87** (1978), 719–736.

15. H. Fugita and T. Kato, *On the Navier-Stokes initial value problem. I*, Arch. Rational Mech. Anal. **16** (1964), 269–315.

16. V. Girault and P.-A. Raviart, *Finite element approximation of the Navier-Stokes equations*, Lecture Notes in Math., vol. 749, Springer-Verlag, Berlin and New York, 1979.

17. J. P. Gollub and H. L. Swinney, *Onset of turbulence in a rotating fluid*, Phys. Rev. Lett. **35** (1975), 927–930.

18. J. Guckenheimer, *A strange, strange attractor*, The Hopf Bifurcation Theorem and Its Applications (J. E. Marsden and M. Mc Cracken, eds.) Springer-Verlag, Berlin and New York, 1976, pp. 368–381.

19. J. Guckenheimer and R. F. Williams, *Structural stability of Lorenz attractors*, Inst. Hautes Études Sci. Publ. Math. **50** (1979), 59–72.

20. M. Hénon, *A two-dimensional mapping with a strange attractor*, Comm. Math. Phys. **50** (1976), 69–77.

21. M. Hirsch, C. Pugh and M. Shub, *Invariant manifolds*, Lecture Notes in Math., vol. 583, Springer-Verlag, Berlin and New York, 1977.

22. E. Hopf, *Abzweigung einer periodischen Lösung von einer stationären Lösung eines Differentialsystems*, Ber. Math.-Phys. K. Sächs. Akad. Wiss. Leipzig **94** (1942), 1–22.

23. ______, *A mathematical example displaying the features of turbulence*, Comm. Pure Appl. Math. **1** (1948), 303–322.

24. G. Iooss, *Existence et stabilité de la solution périodique secondaire intervenant dans les problèmes d'évolution de type Navier-Stokes*, Arch. Rational Mech. Anal. **47** (1972), 301–329.

25. ______, *Sur la deuxième bifurcation d'une solution stationnaire de systèmes du type Navier-Stokes*, Arch. Rational Mech. Anal. **64** (1977), 339–369.

26. A. Katok, unpublished.

27. Ju. I. Kifer, *On the limiting behavior of invariant measures of small random perturbations for some smooth dynamical systems*, Dokl. Akad. Nauk SSSR **216** (5) (1974) = Soviet Math. Dokl. **15** (1941), 918–921.

28. ______, *On small random perturbations of some smooth dynamical systems*, Izv. Akad. Nauk SSSR Ser. Mat. **38** (5) (1974), 1091–1115 = Math. USSR-Izv. **8** (1974), 1083–1107.

29. A. N. Kolmogorov, *Local structure of turbulence in an incompressible fluid at very high Reynolds numbers*, Dokl. Akad. Nauk SSSR **30** (1941), 299–303.

30. O. A. Ladyzhenskaya, *The mathematical theory of viscous incompressible flow*, 2nd ed., Nauka, Moscow, 1970; 2nd English ed., Gordon and Breach, New York, 1969.

31. ______, *A dynamical system generated by the Navier-Stokes equations*, Zap. Naučn. Sem. Leningrad. Otdel. Mat. Inst. Steklov **27** (1972), 91–115 = J. Soviet Math. **3** (1976), 458–479.

32. L. D. Landau, *Turbulence*, Dokl. Akad. Nauk SSSR **44** (1944), 339–342.

33. O. E. Lanford, *Remarks on the accumulation of period-doubling bifurcations*, Mathematical Problems in Theoretical Physics, Lecture Notes in Physics, vol. 116, Springer-Verlag, Berlin and New York, 1980, pp. 340–342.

34. J. Leray, *Sur le mouvement d'un liquide visqueux emplissant l'espace*, Acta Math. **63** (1934), 193–248.

35. A. Libchaber et J. Maurer, *Une expérience de Rayleigh-Bénard de géométrie réduite; multiplication, accrochage et démultiplication de fréquences*, J. Physique (to appear).

36. J. L. Lions, *Quelques méthodes de résolution des problèmes aux limites non-linéaires*, Dunod, Paris, 1969.

37. E. N. Lorenz, *Deterministic non-periodic flow*, J. Atmosphere Sci. **20** (1963), 130–141.

38. J. Mallet-Paret, *Negatively invariant sets of compact maps and an extension of a theorem of Cartwright*, J. Differential Equations **22** (1976), 331–348.

39. R. Mañé, *On the dimension of the compact invariant sets of certain nonlinear maps*, preprint.

40. ______, *Lyapunov exponents and stable manifolds for compact transformations*, preprint.

41. J. E. Marsden and M. McCracken, *The Hopf bifurcation and its applications*, Applied Math. Sci., vol. 19, Springer-Verlag, Berlin and New York, 1976.

42. A. S. Monin, *On the nature of turbulence*, Uspehi Fiz. Nauk **125** (1) (1978), 97–122; English transl., Soviet Physiks-Uspekhi **21** (1978), 429–442.

43. Ju. I. Naimark, *On some cases of dependence of periodic solutions on parameters*, Dokl. Akad. Nauk SSSR **129** (4) (1959), 736–739.

44. S. Newhouse, *Diffeomorphisms with infinitely many sinks*, Topology **13** (1974), 9–18.

45. ______, *The abundance of wild hyperbolic sets and non-smooth stable sets for diffeomorphisms*, Inst. Hautes Études Sci. Publ. Math. **50** (1979), 102–151.

46. V. I. Oseledec, *A multiplicative ergodic theorem. Lyapunov characteristic numbers for dynamical systems*, Trudy Moskov. Mat. Obšč. **19** (1968), 197–221.

47. Ya. B. Pesin, *Lyapunov characteristic exponents and ergodic properties of smooth dynamical systems with an inveriant measure*, Dokl. Akad. Nauk SSSR **226** (4) (1976), 774–777 = Soviet Math. Dokl. **17** (1) (1976), 196–199.

48. ______, *Invariant manifold families which correspond to nonvanishing characteristic exponents*, Izv. Akad. Nauk SSSR Ser. Mat. **40** (6) (1976), 1332–1379 = Math. USSR-Izv. **10** (6) (1976), 1261–1305.

49. ______, *Lyapunov characteristic exponents and smooth ergodic theory*, Uspehi Mat. Nauk **32** (4) (196) (1977), 55–112 = Russian Math. Surveys **32** (4) (1977), 55–114.

50. R. V. Plykin, *Sources and currents of A-diffeomorphisms of surfaces*, Mat. Sb. **94 2(6)** (1974), 243–264.

51. H. Poincaré, *Théorie des tourbillons*, Georges Carré, Paris, 1893.

52. C. C. Pugh and M. Shub, *Differentiability and continuity of invariant manifolds*, preprint.

53. M. S. Raghunathan, *A proof of Oseledec' multiplicative ergodic theorem*, Israel J. Math. **32** (1979), 356–362.

54. D. Ruelle, *A measure associated with Axiom A attractors*, Amer. J. Math. **98** (1976), 619–654.

55. ______, *An inequality for the entropy of differentiable maps*, Bol. Soc. Brasil. Mat. **9** (1978), 83–87.

56. ______, *Dynamical systems with turbulent behavior*, Mathematical Problems in Theoretical Physics, Lecture Notes in Physics, vol. 80, Springer-Verlag, Berlin and New York, 1978, pp. 341–360.

57. ______, *Sensitive dependence on initial condition and turbulent behavior of dynamical systems*, Ann. N. Y. Acad. Sci. **316** (1978), 408–416.

58. ______, *Ergodic theory of differentiable dynamical systems*, Inst. Hautes Études Sci. Publ. Math. **50** (1979), 275–306.

59. ______, *Characteristic exponents and invariant manifolds in Hilbert space* (to appear).

60. ______, *Measures describing a turbulent flow*, Ann. N. Y. Acad. Sci. (to appear).

61. D. Ruelle and M. Shub, *Stable manifolds for maps*, Lecture Notes in Math., vol. 819, Springer, Berlin, 1980, pp. 389–392.

62. D. Ruelle and F. Takens, *On the nature of turbulence*, Comm. Math. Phys. **20** (1971), 167–192; note **23** (1971), 343–344.

63. R. J. Sacker, *On invariant surfaces and bifurcation of periodic solutions of ordinary differential equations*, Thesis, New York University, 1964.

64. V. Scheffer, *Turbulence and Hausdorff dimension*, Turbulence and Navier-Stokes Equation, Lecture Notes in Math., vol. 565, Springer-Verlag, Berlin and New York, 1979, pp. 94–112.

65. ______, *The Navier-Stokes equations on a bounded domain*, Comm. Math. Phys. **73** (1980), 1–42.

66. Ia. G. Sinai, *Gibbs measure in ergodic theory*, Uspehi Mat. Nauk **27 (4)** (1972), 21–64 = Russian Math. Surveys **27 (4)** (1972), 21–69.

67. S. Smale, *Differentiable dynamical systems*, Bull. Amer. Math. Soc. **73** (1967), 747–817.

68. H. L. Swinney and J. P. Gollub, *The transition to turbulence*, Physics Today **31** (1978), 41–49.

69. G. I. Taylor, *Stability of a viscous liquid contained between two rotating cylinders*, Philos. Trans. Roy. Soc. London Ser. A **223** (1923), 289–343.

70. R. Temam, *Navier-Stokes equations*, rev. ed., North-Holland, Amsterdam, 1979.

71. P. Walters, *Characteristic exponents and the entropy of diffeomorphisms*, preprint.

72. R. F. Williams, *The structure of Lorenz attractors*, Inst. Hautes Études Sci. Publ. Math. **50** (1979), 73–99.

INSTITUT DES HAUTES ÉTUDES SCIENTIFIQUES, BURES-SUR-YVETTE, FRANCE

Proceedings of Symposia in Pure Mathematics
Volume 39 (1983), Part 2

The Fundamental Theorem of Algebra
and Complexity Theory

STEVE SMALE[1]

PART I

1. The main goal of this account is to show that a classical algorithm, Newton's method, with a standard modification, is a tractable method for finding a zero of a complex polynomial. Here, by "tractable" I mean that the cost of finding a zero doesn't grow exponentially with the degree, in a certain statistical sense. This result, our main theorem, gives some theoretical explanation of why certain "fast" methods of equation solving are indeed fast. Also this work has the effect of helping bring the discrete mathematics of complexity theory of computer science closer to classical calculus and geometry.

A second goal is to give the background of the various areas of mathematics, pure and applied, which motivate and give the environment for our problem. These areas are parts of (a) Algebra, the "Fundamental theorem of algebra," (b) Numerical analysis, (c) Economic equilibrium theory and (d) Complexity theory of computer science.

An interesting feature of this tractability theorem is the apparent need for use of the mathematics connected to the Bieberbach conjecture, elimination theory of algebraic geometry, and the use of integral geometry.

Before stating the main result, we note that the practice of numerical analysis for solving nonlinear equations, or systems of such, is intimately connected to variants of Newton's method; these are iterative methods and are called fast methods and generally speaking, they are fast in practice. The theory of these methods has a couple of components; one, proof of convergence and two, asymptotically, the speed of convergence. But, not usually included is the total cost of convergence.

On the other hand, there is an extensive theory of search methods of solution finding. This means that a region where a solution is known to exist is broken up into small subsets and these are tested in turn by evaluation; the process is repeated. Here it is simpler to count the number of required steps and one has a good knowledge of the global speed of convergence. But, generally speaking, these are slower methods which are not used by the practicing numerical analyst.

The contrast between the theory and practice of these methods, in my

Reprinted from Bulletin Amer. Math. Soc. (N.S.) 4 (1981), 1–36.

1980 *Mathematics Subject Classification.* Primary 00-02, 12D10, 68C25, 65H05, 58-02; Secondary 01A05, 30D10.

[1]Partially supported by NSF Grant MCS-7717907 and the Miller institute at the University of California, Berkeley.

mind, has to do with the fact that search methods work inexorably and the analysis of cost goes by studying the worst case; but in contrast the Newton type methods fail in principle for certain degenerate cases. And near the degenerate cases, these methods are very slow. This motivates a statistical theory of cost, i.e., one which applies to most problems in the sense of a probabilistic measure on the set of problems (or data). There seems to be a trade off between speed and certainty, and a question is how to make that precise.

One clue can be taken from the problem of complexity in the discrete mathematics of theoretical computer science. The complexity of an algorithm is a measure of its cost of implementation. In these terms, problems (or algorithms) which depend on a "size" are said to be tractable provided the cost of solution does not increase exponentially as their size increases. The famous $P = NP$ problem of Cook and Karp lies in this framework.

In the case of a single polynomial the obvious "size" is the degree d. So these considerations pose the problem. Given $\mu > 0$, an allowable probability of failure, does the cost of computation via the modified Newton's method for polynomials in some set of probability measure $1-\mu$, grow at most as a polynomial in d? Moreover, one can ask that as μ varies, this cost be bounded by a polynomial in $1/\mu$. I was able to provide an affirmative answer to these questions.

Let me be more precise. The problem is to solve $f(z^*) = 0$ where $f(z) = \Sigma_{i=0}^{d} a_i z^i$, $a_i \in \mathbf{C}$ and $a_d = 1$. The algorithm is the modified Newton's method given by: let $z_0 \in \mathbf{C}$ and define inductively $z_n = T_h(z_{n-1})$ where $T_h(z) = z - hf(z)/f'(z)$ for some h, $0 < h \leqslant 1$. If $h = 1$, this is exactly Newton's method.

We will say that z_0 is an *approximate* zero provided if taking $h = 1$, the sequence z_n is well defined for all n, z_n converges to z^* as $n \to \infty$, with $f(z^*) = 0$ and $|f(z_n)/f(z_{n-1})| < \frac{1}{2}$ for all $n = 1, 2, \ldots$.

Practically and theoretically this is a reasonable definition. One could say that in this case, z_0 is in a strong Newton Sink.

Let $\mathcal{P}_d$ be the space of polynomials f, $f(z) = \Sigma_{i=0}^{d} a_i z^i$, $a_d = 1$. Thus $\mathcal{P}_d$ can be identified with $\mathbf{C}^d$, with coordinates $(a_0, \ldots, a_{d-1}) = a \in \mathbf{C}^d$.

Define

$$P_1 = \{ f \in \mathcal{P}_d | \; |a_i| < 1, i = 0, \ldots, d - 1 \}$$

and use normalized Lebesgue measure on P_1, for a probability measure.

MAIN THEOREM. *There is a universal polynomial $S(d, 1/\mu)$, and a function $h = h(d, \mu)$ such that for degree d and μ, $0 < \mu < 1$, the following is true with probability $1-\mu$. Let $x_0 = 0$. Then $x_n = T_h(x_{n-1})$ is well defined for all $n > 0$ and x_s is an approximate zero for f where $s = S(d, 1/\mu)$.*

More specifically we can say, if $s \geqslant [100(d + 2)]^9/\mu^7$, then with probability $1 - \mu$, x_s is well defined by the algorithm for suitable h and x_s is an approximate zero of f.

Note especially that h and s do not depend on the coefficients.

The use of probability is made more precise in the following very brief idea of the proof. There is a certain subset $W_* \subset \mathcal{P}_d$ such that $f \in W_*$, $z_0 = 0$ is a "worst case" for the algorithm "in the limit" $h \to 0$. We don't expect the algorithm to work in this case, no matter how small h is taken. But if $f \notin W_*$ the algorithm will converge for sufficiently small h. It will be shown that W_* is a real algebraic variety in $\mathcal{P}_d$ using elimination theory. Then a certain family of open neighborhoods Y_σ of W_*, $0 < \sigma < 1$, are described, decreasing to W_* as $\sigma \to 0$.

Using a theorem of Weyl on the volume of tubes, and formulae of integral geometry (Santalo) we are able to estimate the volume of Y_σ.

The idea then is that if Vol Y_σ/Vol $P_1 < \mu$ and $f \notin Y_\sigma$, then the algorithm with a suitable choice of h will arrive at an approximate zero after s steps.

The algorithm is "tracked" in the target space of f: $\mathbf{C} \to \mathbf{C}$. Thus we want to estimate $|f(z_n)/f(z_{n-1})|$ in terms of values $f(\theta)$ of critical points θ where $f'(\theta) = 0$.

What is needed can be seen more precisely in terms of the Taylor series expansion

$$\frac{f(z')}{f(z)} = 1 - h + \sum_{d=2}^{d} (-h)^k \frac{f^k(z)f}{n!} \frac{f(t)^{k-1}}{f'(z)^k}$$

where $z' = z - hf(z)/f'(z)$.

This motivates

THEOREM (1A OF §2, PART II). *If f is a polynomial with $f'(z) \neq 0$, then there is a critical point θ (i.e. $f'(\theta) = 0$) such that*

$$\frac{|f^k(z)|}{k!} \frac{|f(\theta) - f(z)|^{k-1}}{|f'(z)|^k} \leqslant 4^{k-1}, \qquad k = 2, 3, \ldots .$$

Thus if $z = 0$, $f(0) = 0$, $f(z) = \Sigma\, a_i z^i$, then

$$|f(\theta)|\, |a_k/a_1^k|^{1/(k-1)} \leqslant 4.$$

The proof uses mathematics related to the Bieberbach conjecture. The particular result used is due to Loewner.

In general, a number of related questions remain unsolved and new ones are suggested. Part III is devoted to these. The proof of the main result is in Part II, while the rest of Part I is devoted to background material.

Names which are italicized are listed in the bibliography at the end.

There is a final comment on the spirit of the paper. I feel one problem of mathematics today is the division into the two separate disciplines, pure and applied mathematics. Oftentimes it is taken for granted that mathematical work should fall into one category or the other. This paper was not written to do so.

I would like to acknowledge useful conversations, with a number of mathematicians including L. Blum, S. S. Chern, G. Debreu, D. Fowler, W. Kahan, R. Osserman, R. Palais, G. Schober and H. Wu.

Special thanks are due Moe Hirsch and Mike Shub.

2. There is a sense in which an important result in Mathematics is never finished. In particular one might ask "Has the fundamental theorem of algebra been proved satisfactorily?".

What do historians of mathematics say about this? They most frequently assert that the first proof of the Fundamental Theorem of Algebra was in Gauss' thesis. Here are some examples.

D. Struik (p. 115):

"GAUSS: THE FUNDAMENTAL THEOREM OF ALGEBRA. The first satisfactory proof of this theorem was presented by Carl Freidrich Gauss (1777–1855) in his Helmstädt doctoral dissertation of 1799 . . . ."

D. E. Smith (pp. 473–474):

"Fundamental Theorem . . . ." After these early steps the statement was repeated in one form or another by various later writers, including Newton (c. 1685) and Maclaurin (posthumous publication, 1748). D'Alembert attempted a proof of the theorem in 1746, and on this account the proposition is often called d'Alembert's theorem. Other attempts were made to prove the statement, notably by Euler (1749) and Lagrange but the first rigorous demonstration is due to Gauss (1799), with a simple treatment in (1849)."

H. Eves (p. 372):

"In his doctoral dissertation, at the University of Helmstädt and written at the age of twenty, Gauss gave the first wholly satisfactory proof of the *fundamental theorem of algebra* . . . ."

And finally Gauss himself, fifty years later, as related by D. E. Smith 1929, in *Source book in mathematics*, McGraw-Hill, New York, pp. 292–293:

"The significance of his first proof in the development of mathematics is made clear by his own words in the introduction to the fourth proof: 'the first proof] · · · had a double purpose, first to show that all the proofs previously attempted of this most important theorem of the theory of algebraic equations are unsatisfactory and illusory, and secondly to give a newly constructed rigorous proof'."

On the other hand, compare this with the following passages of Gauss' thesis translated into English from the Latin in *Struik*:

"Now it is known from higher geometry that every algebraic curve (or the single parts of an algebraic curve when it happens to consist of several parts) either runs into itself or runs out to infinity in both directions, and therefore if a branch of an algebraic curve enters into a limited space, it necessarily has to leave it again."[7] . . . and [7][footnote by Gauss], "It seems to be sufficiently well demonstrated that an algebraic curve can neither be suddenly interrupted (as e.g. occurs with the transcendental curve with equation $y = 1/\log x$), not lose itself after an infinite number of terms (like the logarithmic spiral), and nobody, to my knowledge, has ever doubted it. But if anybody desires it, then on another occasion I intend to give a demonstration which will leave no doubt . . . ."

These passages from Gauss are interesting for several reasons and we will return to them. But for the moment, I wish to point out what an immense gap Gauss' proof contained. It is a subtle point even today that a real algebraic plane curve cannot enter a disk without leaving. In fact even though Gauss redid this proof 50 years later, the gap remained. It was not until 1920 that

Gauss' proof was completed. In the reference *Gauss*, A. Ostrowski has a paper which does this and gives an excellent discussion of the problem as well (I am grateful to Horst Simon for providing me with an English translation of Ostrowski's paper).

One can understand the historical situation better perhaps from the point of view of *Imre Lakatos*. Lakatos in the tradition of Hegel, on one hand, and Popper, on the other, sees mathematics as a development which proceeds as a series of "proofs and refutations".

As an example of his critique, in connection with the origin of the concept of real function, *Lakatos* writes (p. 151):

"Some infallibist historians of mathematics use here the ahistorical technique of condensing a long development full of struggle and criticism into one single action of infallible insight and attribute to Dirichlet the maturity of later analysts."

It seems as if the idea of this quote could also be applied to the written history of Gauss' thesis.

Another line of questioning of the proof of the fundamental theorem of algebra was undertaken by the Constructivists. *Brouwer* and *Weyl* in 1924 both published articles which gave constructive proofs for yielding a zero of a complex polynomial. But as is emphasized by the computer scientist, what good is a constructive solution if it takes 10^{10} years with the fastest computers (say even fastest in principle). Thus a Constructivist approach to be satisfactory today should be paired with a theorem on the speed or cost of computation. In fact in some of the literature on root finding methods, there is a successful effort to measure the number of steps. See for example *Dejon* and *Henrici*, *Henrici*, and *Collins*. Also, the book by *Ostrowski* gives a very useful account of fast algorithms for solving systems of equations in general and for finding a zero of a polynomial in particular.

3. Here we give some of the background of our project related to economics and numerical analysis. Why economics?! There are a couple of related reasons. I was brought to the complexity questions through my work in economics. Also economic theory, besides being concerned with the existence of equilibria, has seriously considered the computational question of finding equilibria as well.

About 100 years ago the great early economic theorist L. Walras saw economic equilibrium for several markets as a solution to a system of (nonlinear) equations, supply equals demand. In this model, supply and demand functions are generated by the individual agents of the economy, consumers and producers.

The modern rigorous development of this theory is due especially to Arrow and Debreu and can be found in *Debreu*. Arrow and Debreu transform the equation of supply equals demand into a fixed point problem, and then apply a fixed point theorem (that of Kakutani, or in simpler versions, Brouwer's). In this way a coherent unifying structure is given to classical economic theory.

Scarf has developed a technique for computing these economic equilibria, and fixed points generally. Although Scarf is himself a mathematical economist, his method falls into the area of operations research (techniques of pivoting, etc.).

Working in equilibrium theory and following the work of Scarf, I perceived the problem of the existence of an equilibrium in terms of solving a system of equations (closer to Walras). In particular, one tries to find a price system $p = (p_1, \ldots, p_l) \in \mathbf{R}^l_+$ which makes the excess demand Z (supply minus demand) zero. Now the values of Z are commodities, so that $Z: \mathbf{R}^l_+ \to \mathbf{R}^l$ is a *morphism* whose source space consists of price systems and target space consists of commodity bundles. Fixed point theory deals with *endomorphisms* or maps of a space into itself. Thus I found it more natural to solve the equation $Z(p) = 0$ by directly constructing a solution in contrast to the method of artificially transforming it into a fixed point problem, and then either using a fixed point theorem or Scarf's method. In fact, eventually, the existence theorems of economic equilibrium theory could be proved directly and constructively this way (see *Smale* (to appear)).

I abstracted these ideas to the general problem of solving a system of nonlinear equations. The method was analogous to Scarf's method in some ways, but it emphasized the equation approach (versus fixed points) on one hand and differentiability on the other hand. I called it the "Global Newton Method", because locally it was essentially Newton's method and it worked globally (see *Smale* (1976)). It was developed further, in several respects including polynomial systems in *Hirsch-Smale*.

Having proved convergence theorems, it was natural to consider the question of speed of convergence and thus the problems of the present paper.

In some sense then there is a little of the Global Newton methodology in Newton, but what is more interesting is its connection to the thesis of Gauss.

To see that connection better, consider the basic idea for the case of a complex polynomial $f: \mathbf{C} \to \mathbf{C}$ (although, the same construction works for polynomial maps $f: \mathbf{C}^m \to \mathbf{C}^n$). To emphasize the "morphism" aspect we can write $f: \mathbf{C}_{\text{source}} \to \mathbf{C}_{\text{target}}$, and we want to solve $f(z) = 0_{\text{target}}$. Suppose $f(z_0)$ does not lie on one of the finite number of rays in $\mathbf{C}_{\text{target}}$ which contain the critical values $f(\theta)$ (i.e., $f'(\theta) = 0$). Of course such $z_0 \in \mathbf{C}_{\text{source}}$ exists. Now take the segment from $f(z_0)$ to 0 in $\mathbf{C}_{\text{target}}$ and lift it back by f^{-1} to a curve starting at z_0 in $\mathbf{C}_{\text{source}}$. This lifting cannot go to ∞ or to a critical point and therefore the lifted curve goes to z with $f(z) = 0$. This proves the fundamental theorem of algebra; it also is the basic idea of the "Global Newton Method" and the basic idea of this paper. For sufficiently differentiable maps $f: \mathbf{R}^m \to \mathbf{R}^m$ which are not polynomial maps, one uses Sard's theorem to choose the z_0, and the argument becomes a bit more subtle.

Now we can compare this construction with the passage in Gauss' thesis quoted earlier in §2. There is a reasonable sense in which Gauss attempts to find a zero of a polynomial by following an algebraic curve.

Pont (p. 32) is right in relating this footnote to the precursors of algebraic topology. More exactly, Pont refers to a sentence later in the same footnote where Gauss speaks of the principles of the geometry of position "which are no less valid than the principles of the geometry of magnitudes".

Besides the mentioned references, there are points of contact between this

Global Newton method I have described previously, and the work of Branin, Davidenko, *Hirsch*, *Kellog-Li-Yorke* and many others. See *Hirsch-Smale* for more information.

4. I would like to say a few words on the discrete mathematics of complexity theory of computer science. Even though the mathematics is somewhat removed from mine, it was an important element in my motivation here. Complexity theory made me aware of the problem of computational cost of algorithms that I had been working with (i.e., the "Global Newton" method).

It was the computer scientists who focused on the key question of polynomial versus exponential bounds on the cost. And of course that is important in formulating my theorem here. Also computer scientists have seen the importance of dealing with problems statistically and not just studying the (often very slow) worst case.

More particularly, let me refer to *Hartmanis* for a useful perspective and history with references. One can see *Garey* and *Johnson* for an account of the problem $P = NP$ associated to Cook and Karp. Traub and Wozniakowski have explicitly discussed complexity in the analytic framework; see *Traub*. In some of the work in numerical analysis and algebraic root finding, one finds results on the cost. See §2 of this paper.

5. We give some further details on the geometric background to the algorithm used in this paper. This geometry is based on an idea which is known, but not usually explicated. Namely, *Newton's method for solving* $f(z) = 0$ *is an Euler approximation to the ordinary differential equation*

$$dz/dt = -\tfrac{1}{2}\,\mathrm{grad}|f|^2. \qquad (*)$$

The prior idea can be restated to be true quite generally; see *Smale* (1976) and *Hirsch-Smale*. Here we restrict ourselves to the case of a single complex polynomial f.

In $(*)$ $\mathrm{grad}|f|^2$ denotes the gradient vector field of the real valued function $|f|^2$ on $\mathbf{R}^2 = \mathbf{C}$, $\mathrm{grad}|f|^2 = (\partial|f|^2/\partial x, \partial|f|^2/\partial y)$. Thus $dz/dt = -\tfrac{1}{2}\,\mathrm{grad}|f|^2$ is a real ordinary differential equation (1st order, autonomous) on $\mathbf{R}^2$. Let f' denote the (complex) derivative of f and $\bar{f}$ the complex conjugate. Then it is easily shown that

LEMMA. $\tfrac{1}{2}\,\mathrm{grad}|f|^2 = f\bar{f'}$.

As a consequence $-\tfrac{1}{2}\,\mathrm{grad}|f|^2 = \rho(z)(-f(z)/f'(z))$ where $\rho\colon \mathbf{R}^2 \to \mathbf{R}$ is the positive scalar function, $\rho(z) = f'(z)\overline{f'(z)}$.

Therefore the vector fields on $\mathbf{R}^2$ defined by $-f(z)/f'(z)$ and $-\tfrac{1}{2}\,\mathrm{grad}|f|^2$ differ only by a rescaling, which has the effect of desingularizing $-f(z)/f'(z)$ at the critical points z where $f'(z) = 0$.

An Euler approximation (see *Hurewicz*) of one of these two vector fields may thus be regarded as an Euler approximation of the other, by changing the step size by ρ.

In particular, if z_0 is fixed then

$$z_{n+1} = z_n - h_n f(z_n)/f'(z_n), \qquad n = 1, 2, \ldots, \tag{**}$$

with $h_n > 0$ is such an Euler approximation, and if $h_n = 1$, this is Newton's method. Generally speaking (**) describes an algorithm, and that is the algorithm studied in this paper where $h_n = h$ is independent of n; we will also assume $0 < h \leqslant 1$.

From these considerations some geometrical insight into the algorithm can be obtained by looking at the solution structure of (*) in $\mathbf{R}^2 = \mathbf{C}$. We will do this by stating some facts which can be proved without great difficulty, and by giving an example. First the example.

Let $f(z) = z^2 - a^2$, $a > 0$. Then the "phase portrait" of (*) is given by Figure 1.

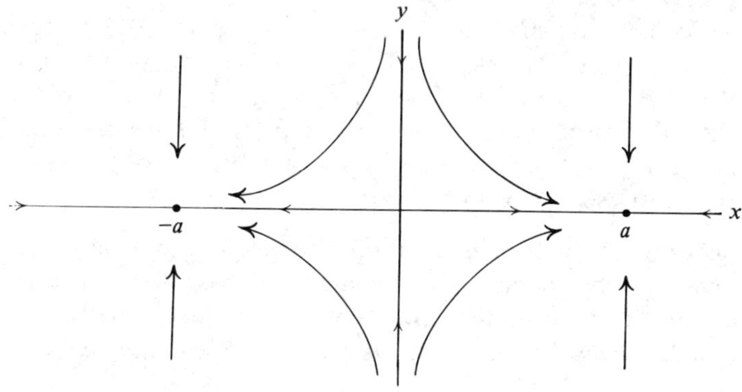

FIGURE 1

Thus the point at ∞ is a single source, and there are two sinks, a and $-a$, the zeros of f. There is one saddle point, namely the origin which is a critical point of f.

The solution curves, which are purely imaginary don't approach the zeros of f, while all others do. Thus one may expect that if the real part of z_0 is not zero and $h > 0$ is small enough, then (*) converges to a or $-a$. That is true and is the basis for the general success of (*).

For general complex polynomials f, ∞ is the single source, and the sinks are precisely the zeros of f. The zeros of the vector field, i.e., the equilibria of the equation (*) other than the zeros of f, are the zeros of f'.

These zeros of f' or the critical points (when distinct from the zeros of f) are saddle points when their multiplicity is one.

One other fact of geometrical importance for us here is that the solution curves of (*) are the same as the inverse image of rays in $\mathbf{C}$ under the map f: $\mathbf{C} \to \mathbf{C}$. This explains why, in the subsequent analysis, we are especially concerned with what happens in the "target" space. See also §3 of this paper.

I emphasize that the above considerations are not used in our proof of the main theorem, but may provide helpful background geometry.

PART II

1. Here is a simple consequence of Theorem 1. Suppose that the polynomial

$$f(z) = z + a_2 z^2 + \cdots + a_d z^d$$

has no critical values $f(\theta)$ in the unit disk in **C** (i.e., $|f(\theta)| \geqslant 1$ if $f'(\theta) = 0$). Then $|a_2| \leqslant 2$ and it is perhaps true that $|a_2|$ is always less than $\frac{1}{2}$. It is somewhat subtle even that $|a_2|$ has any bound (independent of d).

THEOREM 1. *Let* $f(z) = \sum_{i=0}^{d} a_i z^i$, *a complex polynomial with* $a_0 = 0$, $a_1 \neq 0$. *Then there is a critical point* $\theta \in \mathbf{C}$ *(i.e.,* $f'(\theta) = 0$*) such that for every* $k = 1$, $2, 3, \ldots$

$$|a_k/a_1|^{1/(k-1)}|f(\theta)|/|a_1| \leqslant 4.$$

REMARK 1. We will sometimes write instead

$$|a_k/a_1|^{1/(k-1)}|f(\theta)|/|a_1| \leqslant K.$$

Thus K could be 4. On the other hand 4 may not be the sharpest value. I suspect one could take $K = 1$, but, we will give examples which show that $K \geqslant 1$. See also problem number 1 of Part III.

REMARK 2. The proof yields sharper results for each k. More precisely

$$|a_k/a_1|^{1/(k-1)}|f(\theta)|/|a_1| \leqslant \beta_k$$

where $\beta_2 = 2$, $\beta_3 = \sqrt{5}$, $\beta_4 = (14)^{1/3}$, $\beta_5 = (42)^{1/4} \sim 2.55$, $\beta_6 \sim 2.61$, $\beta_7 \sim 2.65$, $\beta_{20} \sim 3.29$ and of course $\beta_k \leqslant 4$.

REMARK 3. The theorem is false in a quite extreme way for entire functions. There is no bound at all for $k = 2$ or any k. This is even true for entire functions which have no critical points in **C**. For example let $f(z) = (1/\alpha)e^{\alpha z} - 1/\alpha$, $\alpha > 1$.

The proof of Theorem 1 goes by a theorem of Loewner. See *Hayman Jenkins*, or *Schober*. Loewner proved this theorem at the same time that he showed that $|a_3| \leqslant 3$ in the Bieberbach conjecture (1923).

Define D_r to be the disk $\{z \in \mathbf{C}|\ |z| < r\}$ throughout.

THEOREM (LOEWNER). *Let* $g\colon D_1 \to \mathbf{C}$ *be a "schlicht" function, i.e.* $g(z)$ *can be expressed as a convergent power series,* $g(z) = \sum_{i=0}^{\infty} b_i z_i$, $|z| < 1$, *with* $b_0 = 0$, $b_1 = 1$ *and* g *is 1-1. Suppose that* $f\colon \Omega \to D_1$ *is an inverse to* g *with* $0 \in \Omega$. *Let* $f(z) = \sum_{i=0}^{\infty} a_i z^i$ *near* 0 *(therefore* $a_0 = 0$, $a_1 = 1$*). Then*

$$|a_k| \leqslant B_k, \qquad k = 1, 2, \ldots$$

where

$$B_k = 2^k \frac{1 \cdot 3 \ldots (2k-1)}{1 \cdot 2 \ldots (k+1)}.$$

We need to extend the theorem to the case where D_1 is replaced by D_R and $b_1 \neq 0$ is arbitrary.

EXTENDED LOEWNER'S THEOREM. *Let* $g\colon D_R \to \mathbf{C}$, $(R > 0)$ *be 1-1,* $g(z) = \sum_{i=0}^{\infty} b_i z^i$, $b_0 = 0$, $b_1 \neq 0$ *and* $f\colon \Omega \to D_R$ *be an inverse to* g *with* $0 \in \Omega$. *Let*

$f(z) = \sum_{i=0}^{\infty} a_i z^i$ *near* 0. *Then*

$$|a_k/a_1|^{1/(k-1)} R/|a_1| \leqslant B_k^{1/(k-1)}.$$

PROOF. First let f, g be as in the extended theorem with $R = 1$. Note that $a_1 \neq 0$ and in fact $a_1 = f'(0) = 1/b_1 = 1/g'(0)$. Let $g_0(\omega) = (1/b_1)g(\omega)$ and $f_0(z) = f(z/a_1)$. So

$$f_0(g_0(\omega)) = f\left(\frac{1}{a_1}\left(\frac{1}{b_1}g(\omega)\right)\right) = \omega \quad \text{for } |\omega| < 1.$$

Since

$$f_0(z) = z + \frac{a_2}{a_1^2}z^2 + \frac{a_3}{a_1^3}z^3 + \cdots,$$

and since Loewner's theorem applies to g_0, we have

$$|a_k/a_1^k| \leqslant B_k.$$

Thus we conclude

LEMMA 1. *If* $g: D_1 \to \mathbf{C}$ *is* 1-1, $g(0) = 0$, $g'(0) \neq 0$ *and* $f: \Omega \to D_1$, $0 \in \Omega$, $f(0) = 0$, $f(z) = a_1 z + a_2 z^2 + \cdots$ *is its inverse, then*

$$|a_k/a_1|^{1/(k-1)} 1/a_1 \leqslant B_k^{1/(k-1)}.$$

Next let $g: D_R \to \mathbf{C}$, and f be as in the extended theorem and let $g_1(w) = g(Rw)$. Then g_1 and f_1 satisfy Lemma 1 where $f_1(z) = (1/R)f(z)$, since $f_1(g_1(w)) = (1/R)f(g(Rw)) = w$. Also $f_1(z) = (a_1/R)z + (a_2/R)z^2 + \cdots$. Apply Lemma 1 to obtain

$$|a_k/a_1|^{1/(k-1)} R/|a_1| \leqslant B_k^{1/(k-1)}.$$

This proves the theorem.

LEMMA 2. $B_k^{1/(k-1)} \leqslant 4$.

PROOF.

$$B_k = 2^{k-1} \cdot \frac{2}{k+1} \cdot \frac{1 \cdot 3 \cdot 5 \cdots 2k-1}{1 \cdot 2 \cdot 3 \cdots k} \leqslant 2^{k-1}\left(\frac{2}{k+1}\right)2^{k-1} \leqslant 2^{k-1} \cdot 2^{k-1}.$$

It follows.

LEMMA 3. *Let* $f(z)$ *be a polynomial of exact degree* d, $f(0) = 0$ *and such that*

$$R = \min_{\substack{\theta \\ f'(\theta)=0}} |f(\theta)| > 0.$$

Then there exists an analytic function $g: D_R \to \mathbf{C}$ *with* $g(0) = 0$, *and* $f(g(w)) = w$ *for all* $w \in D_R$ *and* g *is* 1-1.

PROOF. First note that

$$f: \mathbf{C} - \bigcup_{\substack{\theta \\ f'(\theta)=0}} f^{-1}(f(\theta)) \to \mathbf{C} - \bigcup_{\substack{f'(\theta)=0}} f(\theta)$$

is proper (the inverse image of a compact set is compact) and a local homeomorphism. It is a covering map. Therefore we may lift it back by g over D_R, uniquely if $g(0) = 0$. This proves Lemma 3.

Combining Lemmas 2 and 3 with the Extended Loewner Theorem yields Theorem 1.

We give some lower bounds for the constant $K = K_{d,k}$ of Remark 1 of Theorem 1. Thus for each k, d as in Theorem 1, let $C_{d,k}$ be the lower bound exhibited by $f(z) = (z - 1)^d - (-1)^d$ if $k < d$ and $f(z) = z^d - dz$ if $k = d$. Then

PROPOSITION. $C_{d,d} = ((d - 1)/d)(1/d)^{d-1}$.

$$C_{d,k} = \left(\frac{(d - 1)!}{k!\,(d - k)!} \right)^{1/(k-1)} \frac{1}{d} \quad if \ k < d.$$

COROLLARY. K OF (REMARK 1). *Theorem* 1 *must be at least* 1 (*it is independent of* k, d).

The proof of the Proposition is a direct calculation which we omit.

Note that $C_{d,2} = \frac{1}{2}(d - 1)/d$ for all $d > 2$ and that $C_{d,k} \leq (d - 1)/d$ for all k, d. Note also that

$$\max_{d} C_{d,2} = \frac{1}{2}, \qquad \max_{d} C_{d,3} = \left(\frac{1}{6} \right)^{1/2},$$

$$\max_{d} C_{d,k} = \left(\frac{k - 1}{k} \right)\left(\frac{1}{k} \right)^{1/(k-1)}, \qquad k > 3.$$

This gives the Corollary.

We end this section by remarking that there is a large related literature on the critical points of polynomials; see *Marden*.

2. Here we make some estimate on the effect of one step of our algorithm. As usual we are taking a target space perspective, i.e., estimating values. In the next section the result is used to estimate the effect of many steps.

Let K be as in Remark 1 after Theorem 1, e.g., $K = 4$.

THEOREM 2. *Let* f *be a polynomial and* $z \in \mathbf{C}$ *such that* $f(z)$ *and* $f'(z)$ *are not zero. Let*

$$h_0 = \min_{\substack{\theta \\ f'(\theta)=0}} \frac{1}{K} \left| \frac{f(\theta) - f(z)}{f(z)} \right|$$

and $0 < h < h_0$. *If* $z' = z - hf(z)/f'(z)$, *then*

$$f(z')/f(z) = 1 - h + \alpha h^2/ (h_0 - h)$$

for some $\alpha \in \mathbf{C}, |\alpha| \leq 1$.

The proof uses the following version of Theorem 1.

THEOREM 1A. *Let* f *be a polynomial and* $z \in \mathbf{C}$ *such that* $f'(z) \neq 0$. *There is* *a critical point θ of f such that for all $k = 2, 3, \ldots ,$*

$$\left| \frac{f^{(k)}(z)}{k!f'(z)} \right|^{1/(k-1)} \leqslant \frac{K}{|f(z) - f(\theta)|} .$$

Furthermore, if $f(z) \neq 0$, and h_0 is as in Theorem 2, then for $k = 2, 3, \ldots$

$$\left| \frac{f^{(k)}(z)}{k!} \frac{f(z)^{k-1}}{f'(z)^k} \right|^{1/(k-1)} \leqslant \frac{1}{h_0} .$$

Note first that the last statement of Theorem 1A is a consequence of the first part. We now check the first part. For this let f, z be as in Theorem 1A and expand f about z so that $f(w) = \sum_{k=0}^{d}(f^k(z)/k!)(w - z)^k$. Define the polynomial g by $g(u) = \sum_{k=1}^{d}(f^k(z)/k!)u^k$. So $g(w - z) = f(w) - f(z)$ and $g'(w - z) = f'(w)$. Apply Theorem 1 to g to obtain

$$\left| \frac{f^k(z)}{k!f'(z)^k} \right|^{1/(k-1)} |g(\sigma)| \leqslant K \quad \text{for some } \sigma \text{ with } g'(\sigma) = 0.$$

Thus

$$\left| \frac{f^k(z)}{k!f'(z)^k} \right|^{1/(k-1)} |f(\theta) - f(z)| \leqslant K \quad \text{for } \theta = z + \sigma.$$

This proves 1A.

We now prove Theorem 2. Again expand f by Taylor's series about z, using $z' = z - hf(z)/f'(z)$, to obtain $f(z')/f(z) = 1 - h + h\gamma$ where

$$\gamma = \sum_{k=2}^{d} (-h)^{k-1} \frac{f^k(z)f(z)^{k-1}}{k!f'(z)^k} .$$

Now by Theorem 1A we obtain

$$|\gamma| \leqslant \sum_{k=2}^{d} \left| \frac{h}{h_0} \right|^{k-1} .$$

Thus

$$|\gamma| \leqslant \sum_{l=1}^{\infty} \left| \frac{h}{h_0} \right|^{l} = \sum_{l=0}^{\infty} \left| \frac{h}{h_0} \right|^{l} - 1 \leqslant \frac{1}{1 - h/h_0} - 1 = \frac{h}{h_0 - h} .$$

Let $\alpha = ((h_0 - h)/h)\gamma$; then $|\alpha| \leqslant |\gamma| |(h_0 - h)/h| \leqslant 1$.

This proves Theorem 2.

We apply Theorem 2 to obtain information about Newton's method.

COROLLARY A. *Let $C \geqslant 1$ and $\rho_f = \rho = \min_{\theta: f'(\theta) = 0}|f(\theta)|$ for a polynomial f. If $|f(z)| < \rho/CK + K + 1$ then $|f(z')/f(z)| < 1/C$ where $z' = z - f(z)/f'(z)$ (Newton's method).*

PROOF OF COROLLARY A. We first remark that for any critical point θ, $|f(\theta) - f(z)| + |f(z)| \geqslant |f(\theta)|$ and $|f(\theta) - f(z)| \geqslant |f(\theta)| - |f(z)| \geqslant \rho - |f(z)|$. By hypothesis, $|f(z)|(CK + K + 1) < \rho$, so $|f(z)|K(1 + C) < \rho -$

$|f(z)|$. Thus by the prior remark, for any critical point θ, $|f(z)|K(1 + C) < |f(\theta) - f(z)|$.

Thus we have shown that for any critical point θ,

$$|f(\theta) - f(z)| / K|f(z)| > 1 + C.$$

So for h_0 of Theorem 2, we have $h_0 > 1 + C$ and $1/(h_0 - 1) < 1/C$.

Apply Theorem 2 with $h = 1$ to obtain

$$|f(z')/f(z)| \leqslant 1/(h_0 - 1) < 1/C.$$

This proves Corollary A.

COROLLARY B. *Let K and ρ be as before with $|f(z)| < \rho/(2K + 1)$. Then Newton's method starting at z converges to z^*, a solution of $f(z^*) = 0$. (Recall we could take $K = 4$.)*

PROOF. Choose $C > 1$ such that $|f(z)| < \rho/(CK + K + 1)$. Use Corollary A. If $z_1 = z - f(z)/f'(z)$, then $|f(z_1)/f(z)| < 1/C$. Inductively let $z_{n+1} = z_n - f(z_n)/f'(z_n)$. Thus $|f(z_{n+1})| < (1/C)|f(z_n)| < \rho/(2K + 1)$, so $f'(z_n)$ is never zero and $|f(z_{n+1})/f(z)| < (1/C)^{n+1}$. This proves Corollary B.

Note that Corollaries A and B give a new criterion for Newton's method to work. This is used strongly in our main result.

3. Here we make estimates in the target space or on the values $f(z_0)$, $f(z_1)$, $f(z_2)$, ... as the algorithm constructs z_1, z_2, ... Under suitable initial conditions, it must be shown that $f(z_n)$ tends to zero sufficiently faster than it veers off toward some critical value $f(\theta)$. At the critical points, the algorithm is not defined, and near them goes very slowly. Thus this analysis is central. We use heavily Theorem 2 of the preceding section.

Throughout K will be the constant in the remark after Theorem 1. Certainly $K \geqslant 1$ and K could be taken as 4. As in §2, ρ_f is the number

$$\rho_f = \min_{\substack{\theta \\ f'(\theta)=0}} |f(\theta)|.$$

We now define for each polynomial f and $z_0 \in \mathbf{C}$ such that $f(z_0) \neq 0$ and $\rho_f > 0$,

$$\mathcal{H} = \mathcal{H}_{f,z_0} = \min_{\substack{\theta \\ f'(\theta)=0 \\ |f(\theta)| < 2|f(z_0)|}} \left\{ 1, \left| \arg \frac{f(\theta)}{f(z_0)} \right| \right\}.$$

Moreover, under the same conditions, define

$$\zeta = \zeta_{f,z_0} = \max \left\{ 2, \frac{(3K + 1)|f(z_0)|}{\rho_f} \right\}.$$

THEOREM 3. *There exist two universal functions $H: \mathbf{R}^+ \times \mathbf{R}^+ \to \mathbf{R}^+$, $S: \mathbf{R}^+ \times \mathbf{R}^+ \to \mathbf{Z}^+$, exhibited later, with these properties: If f, z_0 satisfy $\rho_f > 0$, $\mathcal{H} = \mathcal{H}_{f,z_0} > 0$ and $h = H(\mathcal{H}, \zeta)$, $s = S(\mathcal{H}, \zeta)$ then $z_k = z_{k-1} - hf(z_{k-1})/f'(z_{k-1})$ is well defined inductively for all k starting from $k = 1$ and*

$|f(z_s)| < \rho_f/(3K + 1)$. *Moreover H and S have the form*

$$H(\mathcal{H}, \zeta) = \frac{1}{K} \frac{\sin^2(\mathcal{H}/2)}{(3\sin(\mathcal{H}/2) + \log\zeta)}$$

$$S(\mathcal{H}, \zeta) = K\left(3 + \frac{\log\zeta}{\sin(\mathcal{H}/2)}\right)^2.$$

REMARK. According to Corollary A of Theorem 2, for $k > s$, one can change h to 1 (Newton's method) so that $z_k \to z^*$ as $k \to \infty$ with $f(z^*) = 0$. After s steps, we have found an approximate solution.

Define $h_* = (1/K)\sin(\mathcal{H}/2)$. The following is the main lemma.

LEMMA 1. *Suppose* $s \in Z^+$, $0 < h < h_*/2$, *depending on* ζ, $\mathcal{H}$ *of Theorem* 3 *satisfy* (1) *and* (2).

$$\left(1 - h + h^2/\left(h_* - h\right)\right)^s < 1/\zeta, \tag{1}$$

$$s\left(\frac{h^2}{h_* - h}\right)\left(\frac{1}{1 - h}\right) < \sin\frac{\mathcal{H}}{2}. \tag{2}$$

Then the conclusions of Theorem 3 *(except the last sentence) are satisfied with these choices of h and s.*

Note that once Lemma 1 has been proven, the rest of the proof of Theorem 3 amounts to solving equations (1) and (2) for h, s and exhibiting their dependence on $\mathcal{H}$ and ζ.

For the proof of Lemma 1, we use

LEMMA 2. (a) *For any* τ, $\beta \in \mathbf{C}$, $\beta \neq 0$, $|\sin\arg((\beta + \tau)/\beta)| \leq |\tau/\beta|$,
(b) $|\sin\arg(f(\theta)/f(z))| \leq |(f(z) - f(\theta))/f(z)|$ *if* $f(z) \neq 0$,
(c) *for* $0 < h < h_*/2$, $\alpha \in \mathbf{C}$, $|\alpha| \leq 1$,

$$\left|\sin\arg\left(1 - h + \frac{\alpha h^2}{h_* - h}\right)\right| \leq \left(\frac{h^2}{h_* - h}\right)\left(\frac{1}{1 - h}\right).$$

PROOF OF LEMMA 2. (a) can be checked via trigonometry. (b) follows from (a) with $\tau = f(\theta) - f(z)$, $\beta = f(z)$. For (c), take in (a), $\beta = 1 - h$, $\tau = \alpha h^2/(h_* - h)$.

LEMMA FROM TRIGONOMETRY.

$$\left|\sin\sum\alpha_i\right| \leq \sum|\sin\alpha_i|.$$

Let z_0 be given as in Theorem 3. We proceed by induction to prove Lemma 1. Let $k \leq s$ and suppose inductively for $i < k$ that $f'(z_i) \neq 0$, and hence $z_{i+1} = z_i - hf(z_i)/f'(z_i)$ is well defined, and moreover

$$h_i = \frac{1}{K}\min_{\substack{\theta \\ f'(\theta)=0}}\left|\frac{f(z_i) - f(\theta)}{f(z_i)}\right| > h_*.$$

Note that if ever $f(z_k)$ (or $f(z_i)$, $i < k$) is zero, we are finished trivially. To finish the induction, we will show that $f'(z_k) \neq 0$ and $h_k > h_*$.

LEMMA 3.

$$\left|\frac{f(z_k)}{f(z_0)}\right| < \left(1 - h + \frac{h^2}{h_* - h}\right)^k < 1, \tag{a}$$

$$\left|\arg\frac{f(z_k)}{f(z_0)}\right| < \frac{\mathcal{H}}{2}. \tag{b}$$

PROOF OF LEMMA 3. First write $f(z_k)/f(z_0) = f(z_k)/f(z_{k-1}) \cdots f(z_1)/f(z_0)$. So we may apply Theorem 2 to obtain

$$\frac{f(z_i)}{f(z_{i-1})} = 1 - h + \frac{\alpha_i h^2}{h_{i-1} - h}, \qquad i = 1, \ldots, k, |\alpha_i| \leq 1, 0 < h < h_{i-1}.$$

By the induction hypothesis $h_{i-1} > h_*$, $i = 1, \ldots, k$, and thus since $h < h_*/2$, we have

$$\left|1 - h + \frac{\alpha_i h^2}{h_{i-1} - h}\right| < \left|1 - h + \frac{h^2}{h_* - h}\right|.$$

Therefore

$$\left|\frac{f(z_k)}{f(z_0)}\right| \leq \left|1 - h + \frac{h^2}{h_* - h}\right|^k.$$

This proves (a) of Lemma 3.

For part (b) we have

$$\arg\frac{f(z_k)}{f(z_0)} = \sum_{i=1}^{k} \arg\left(\frac{f(z_i)}{f(z_{i-1})}\right)$$

and thus by the lemma from trigonometry,

$$\left|\sin \arg\frac{f(z_k)}{f(z_0)}\right| \leq \sum_{i=1}^{k} \left|\sin \arg\frac{f(z_i)}{f(z_{i-1})}\right|.$$

But by Lemma 2(c) we may estimate

$$\left|\sin \arg\frac{f(z_i)}{f(z_{i-1})}\right| \leq \left(\frac{h^2}{h_{i-1} - h}\right)\left(\frac{1}{1 - h}\right) \leq \left(\frac{h^2}{h_* - h}\right)\left(\frac{1}{1 - h}\right),$$

thus

$$\left|\sin \arg\frac{f(z_k)}{f(z_0)}\right| \leq k\left(\frac{h^2}{h_* - h}\right)\left(\frac{1}{1 - h}\right).$$

By (2), since $k \leq s$,

$$\left|\sin \arg\frac{f(z_k)}{f(z_0)}\right| < \sin\frac{\mathcal{H}}{2} < \frac{\mathcal{H}}{2},$$

the last since $|\arg(f(z_k)/f(z_0))| \leq \mathcal{H} \leq 1$. This proves Lemma 3.

Now $|\arg(f(\theta)/f(z_0))| \geqslant \mathcal{H}$ for any critical point $\theta \in C$ such that $|f(\theta)| \leqslant 2|f(z_0)|$. Since

$$\arg\frac{f(\theta)}{f(z_0)} = \arg\frac{f(\theta)}{f(z_k)} + \arg\frac{f(z_k)}{f(z_0)},$$

we conclude from Lemma 3 that $|\arg\{f(\theta)/f(z_k)\}| > \mathcal{H}/2$ for any critical point θ with $|f(\theta)| \leqslant 2|f(z_0)|$. In particular, $|f(z_k)| \leqslant 2|f(z_0)|$ (Lemma 3(a)) so that z_k is not a critical point, $f'(z_k) \neq 0$.

Next, $|\sin \arg(f(\theta)/f(z_k))| > \sin(\mathcal{H}/2)$, so using Lemma 2(b)

$$\left|\frac{f(z_k) - f(\theta)}{f(z_k)}\right| \geqslant \left|\sin \arg\frac{f(\theta)}{f(z_k)}\right| > \sin\frac{\mathcal{H}}{2}$$

for any critical point θ such that $|f(\theta)| \leqslant 2|f(z_0)|$. On the other hand, if $|f(\theta)| > 2|f(z_0)|$, since $|f(z_k)| < |f(z_0)|$,

$$\left|\frac{f(z_k) - f(\theta)}{f(z_k)}\right| > 1 > \sin\frac{\mathcal{H}}{2}.$$

Therefore

$$h_k = \min_{\substack{\theta \\ f'(\theta)=0}} \frac{1}{K}\left|\frac{f(z_k) - f(\theta)}{f(z_k)}\right| > \frac{1}{K}\sin\frac{\mathcal{H}}{2} = h_*.$$

This finishes the induction. To finish the proof of Lemma 1, one has to show that $|f(z_s)| < \rho_f/(3K + 1)$. Use Lemma 3(a) to get $|f(z_s)| < (1 - h + h^2/(h_* - h))^s |f(z_0)|$. So $|f(z_s)| < (1/\zeta)|f(z_0)|$ by (1) which in turn is less than $\rho_f/(3K + 1)$ by the definition of ζ. This proves Lemma 1.

Now we solve (1) and (2) for h and s to finish the proof of Theorem 3. Recall

$$\left(1 - h + \frac{h^2}{h_* - h}\right)^s < \frac{1}{\zeta}. \tag{1}$$

$$s\left(\frac{h^2}{h_* - h}\right)\left(\frac{1}{1 - h}\right) < \sin\frac{\mathcal{H}}{2} = Kh_*. \tag{2}$$

(1) is equivalent to

$$\frac{\log \zeta}{\log\left(1 - h + h^2/(h_* - h)\right)^{-1}} < s \tag{1'}$$

and (2) to

$$s < Kh_*\left[\frac{(1 - h)(h_* - h)}{h^2}\right]. \tag{2'}$$

LEMMA 4 (EASY). $\log(1/(1 - y)) > y$ if $0 < y < 1$.

In (1') let $y = h - h^2/(h_* - h) = h((h_* - 2h)/(h_* - h))$ and use Lemma

4 to see that (1″) implies (1′).

$$\frac{\log \zeta}{y} < s. \tag{1″}$$

Next use $h = h_*/C$ where $C > 2$ to replace (1″) and (2′) respectively by

$$(\log \zeta)\frac{C}{h_*}\left(\frac{C-1}{C-2}\right) < s, \tag{1‴}$$

$$s < K(C - h_*)(C - 1). \tag{2″}$$

Besides (1‴) and (2″), we have the condition that s be an integer.

TECHNICAL LEMMA. *If* $C \geqslant 3 + \log \zeta/Kh_*$, *then*

$$(\log \zeta)\frac{C}{h_*}\left(\frac{C-1}{C-2}\right) + 1 < K(C - h_*)(C - 1).$$

PROOF OF TECHNICAL LEMMA. Note first that if $C > 3$ then

$$\frac{C-2}{K}\frac{1}{C}\left(\frac{1}{C-1} + Kh_*\right) \leqslant 1$$

(using $h_* < \frac{1}{2}$, since $\mathcal{H} \leqslant 1$).

Therefore from the hypothesis of the Technical Lemma,

$$\frac{\log \zeta}{Kh_*} + \frac{(C-2)}{K}\left(\frac{1}{C}\right)\left(\frac{1}{C-1} + Kh_*\right) \leqslant C - 2.$$

Next divide and multiply through this expression to obtain the Technical Lemma.

To finish we choose $C = 3 + \log \zeta/Kh_*$ and $h = h_*/C$. Then choose s to be the greatest integer in $(\log \zeta)(C/h_*)((C-1)/(C-2)) + 1$.

In fact we may calculate h and s more explicitly as follows.

$$h = Kh_*^2/\left(\log \zeta + 3h_* K\right), \qquad h_* = (1/K)\sin^2(\mathcal{H}/2)$$

giving $H(\mathcal{H}, \zeta)$ as in Theorem 3.

Going back to (2″) we could take s as simply KC^2 since increased s (fixed h) will still satisfy our conditions in Theorem 3. Then

$$s = \left(\frac{(1/K)\sin(\mathcal{H}/2)}{h}\right)^2 = \frac{K(3\sin(\mathcal{H}/2) + \log \zeta)^2}{\sin^2(\mathcal{H}/2)}.$$

This finishes the proof of Theorem 3.

4. This section gives the main probability (or volume) estimates which are used in our statistics. Our approach is more geometric than measure-theoretic although we do make use of Fubini's theorem. There is an interesting historical point about a theorem of *Herman Weyl* on the volume of tubes that we use. The statistician (and economist) *Harold Hotelling* proved a result on the volume of a tubular neighborhood of curves (see Lemma 2 below). This is motivated by problems of regression equations in theoretical statistics. Subsequently *Hotelling's* theorem was extended by *Weyl* (see Lemma 5 below) and these two papers lie next to each other in the American Journal of 1939. *Weyl* writes: "In a lecture before the Mathematics Club at Princeton last year Professor Hotelling stated the following geometric problem as one of primary important for certain statistical investigations: . . . ." and Weyl goes on to state and prove his theorem on the volume of tubes.

In the intervening years Weyl's result has become well known to geometers; see *Griffith's* extensive discussion. Now the result of Weyl is found useful in our statistical analysis.

There is another amusing aside in this connection. In the use of Weyl's theorem, I had need to estimate the integral of the Gaussian curvature K of a complex algebraic curve γ in $\mathbf{C}^2$ in terms of the degree δ. I took this problem to Osserman and after a little thought, using his joint paper with *Hoffman* he came up with the neat estimate:

$$\frac{1}{2\pi} \int K \leqslant -\delta(\delta - 1).$$

This is nice because the volume of γ is infinite, and it might also have singularities where the curvature blows up. Although I believe this estimate is not explicit in the literature, it now follows easily from standard results. Griffiths subsequently showed me another proof.

Eventually, I bypassed the need for the estimate by noticing a simple sign argument which did the trick.

We proceed to the theorems on volume estimates. Let $\mathcal{P}_d$ be the space of polynomials of degree d with leading coefficient 1. Thus $\mathcal{P}_d$ can be identified with complex Cartesian space $\mathbf{C}^d = \{(a_0, \ldots, a_{d-1}) = a | a_i \in \mathbf{C}\}$ or we may write $f \in \mathbf{C}^d$ where $f(z) = \Sigma_{i=0}^d a_i z^i$, $a_d = 1$. Let $P(R)$ denote the polycylinder $\{a \in \mathcal{P}_d | |a_i| < R, i = 0, \ldots, d - 1\}$. Note that the volume of $P(R)$ or vol $P(R)$ is $(\pi R^2)^d$. Here we use the standard volume on $\mathbf{C}^d = \mathbf{R}^{2d}$ for $\mathcal{P}_d$.

Let $W_0 = \{f \in \mathcal{P}_d | f(\theta) = 0$ for some θ with $f'(\theta) = 0\}$. It is easily seen that W_0 is the set of f with multiple roots. Finally define $U_\rho(W_0) = \bigcup_{f \in W_0} U_\rho(f_0)$ where $U_\rho(f_0) = \{f \in \mathcal{P}_d | |f(\theta) - f_0(0)| < \rho$ and $f'(z) = f_0'(z)$ for all $z\}$. Then $f \in U_\rho(f_0)$ if and only if $|f_0(0) - f(0)| < \rho$ and the coefficients a_i in f and f_0, $i > 0$, coincide.

THEOREM 4A.

$$\frac{\mathrm{Vol}\left[\, U_\rho(W_0) \cap P(R)\,\right]}{\mathrm{Vol}\, P(R)} \leqslant \frac{d\rho^2}{R^2}.$$

This theorem is not used in the sequel; however it serves as a simple prototype of Theorems 4B and 4C which are important for our development.

The proof of Theorem 4A goes via

PROPOSITION 1. *The subset* $W_0 \subset \mathcal{P}_d \cong \mathbf{C}^d$ *has a representation as a complex algebraic hypersurface given by a single polynomial equation of degree* $2d - 1$ *which is of degree d in* a_0. *More precisely, there is a polynomial* $F \colon \mathbf{C}^d \to \mathbf{C}$,

$$F(a_0, \ldots, a_{d-1}) = \sum_{i=0}^{d} F_i(a_1, \ldots, a_{d-1}) a_0^i,$$

where the F_i are polynomials, F has total degree $2d - 1$ and

$$W_0 = \{a \in \mathbf{C}^d | F(a) = 0\}.$$

PROOF OF PROPOSITION 1. Note that

$$W_0 = \{f \in \mathcal{P}_d | f(\theta) = 0, f'(\theta) = 0 \text{ for some } \theta \in \mathbf{C}\}.$$

Then the proof goes by the description of the resultant (see *Van der Waerden*, Vol. I (p. 84) or *Lang*). Thus F is the resultant of f and f', $R(f, f')$. The proof of the proposition then follows from an examination of the form of $R(f, f')$ given in either reference.

For the proof of Theorem 4A, define χ: $\mathbf{C}^d \to \mathbf{R}$ to be the characteristic function of $U_\rho(W_0)$, i.e., χ is 1 on $U_\rho(W_0) \subset \mathbf{C}^d$ and zero off of it.

Now observe that for almost every (in the sense of Lebesgue measure) $(a_1, \ldots, a_{d-1})$, W_0 intersected with the one (complex) dimensional coordinate plane through $(a_1, \ldots, a_{d-1})$ consists of at most d points. This follows from the proposition, with the exceptional $(a_1, \ldots, a_{d-1})$ being those points where all the F_i vanish. From this, using the definitions, we may conclude that for almost every $(a_1, \ldots, a_{d-1})$,

$$\left| \int_{|a_0| < R} \chi(a_0, \ldots, a_{d-1}) \, da_0 \right| \leq d\pi\rho^2.$$

Thus by Fubini's theorem

$$\frac{\text{Vol } U_\rho(W_0) \cap P(R)}{\text{Vol } P(R)} = \frac{1}{(\pi R^2)^d} \int_{P(R)} \chi(a) \, da$$

$$= \frac{1}{(\pi R^2)^d} \int_{|a_1| < R} \left[\int_{|a_0| < R} \chi(a_0, \ldots, a_{d-1}) \, da_0 \right] da_1 \cdots da_{d-1}$$

$$\vdots$$
$$|a_{d-1}| < R$$

$$\leq \frac{1}{(\pi R^2)^d} \int_{|a_1| < R, \ldots, |a_{d-1}| < R} (d\pi\rho^2) \, da_1 \cdots da_{d-1}$$

$$\leq \frac{1}{(\pi R^2)^d} (d\pi\rho^2)(\pi R^2)^{d-1} \leq \frac{d\rho^2}{R^2}.$$

This proves Theorem 4A.

We will now use Re z and Im z systematically for the real and imaginary parts of a complex number z.

Consider the subspace W_* of $\mathcal{P}_d$, $W_* = \{f \in \mathcal{P}_d | \text{Im}(\overline{f(0)}f(\theta)) = 0$ for some critical point θ of $f\}$. So $f \in W_*$ provided Re f', Im f' and $\text{Im}(\overline{f(0)}f)$ have a common zero. Let $U_\rho(W_*) = \bigcup_{f_0 \in W_*} U_\rho(f_0)$, where $U_\rho(f_0)$ is defined earlier in this section.

THEOREM 4B.

$$\frac{\text{Vol}\left[U_\rho(W_*) \cap P(R)\right]}{\text{Vol } P(R)} \leqslant \frac{3\rho(d-1)^2}{R}.$$

The proof of this theorem uses the following Algebraic Lemma.

ALGEBRAIC LEMMA. *The set $W_* \subset \mathcal{P}_d$ is a closed subset of a real algebraic variety in $\mathbf{R}^{2d} = \mathbf{C}^d$. In fact this variety is the set of zeros of a real polynomial G in the $2d$ variables $(\text{Re } a_i, \text{Im } a_i)_{i=0}^{d-1}$ and G has degree $(d-1)^2$ in $(\text{Re } a_0, \text{Im } a_0)$.*

To prove the Algebraic Lemma, we use elimination theory which can be found in *Van der Waerden*, Vol. II. See especially p. 15.

THEOREM FROM ELIMINATION THEORY. *n generic homogeneous polynomials in n-variables have a resultant F which is an integral polynomial in their (indeterminate) coefficients. The vanishing of this resultant for particular $f_1, \ldots, f_n$ with coefficients in a field is a necessary and sufficient condition for the existence of a solution (perhaps in an extension field) of the system of equations $f_1 = 0, \ldots, f_n = 0$, distinct from the zero solution. The resultant is homogeneous in the coefficients of f_k of degree $D_k = \Pi_{i \neq k} d_i$ (the product) where degree $f_i = d_i$.*

We need to apply this result to the case of n inhomogeneous polynomials in $(n-1)$ variables. This is done as usual by adding a new variable to homogenize the polynomials. New solutions may be introduced, however. We obtain in this way

COROLLARY. *Let $\mathcal{F}_i$, $i = 1, \ldots, n$, be given spaces of inhomogeneous polynomials f_i of degree d_i in $(n-1)$ variables over a fixed field K. Thus $\mathcal{F}_i$ can be identified with the space of coefficients of f_i. Then there is a polynomial (the resultant) $F: \Pi_{i=1}^n \mathcal{F}_i \to K$ in the coefficients of these f_i with the property that if particular $f_1, \ldots, f_n$ have a common zero, then $F(f_1, \ldots, f_n) = 0$. Furthermore F is homogeneous in the coefficients of f of degree $D_k = \Pi_{i=1; i \neq k}^n d_i$.*

Now fix the field to be $\mathbf{R}$, the real number field. Let $\mathcal{F}_i$ and $\mathcal{F}_2$ each be the space of real (inhomogeneous) polynomials of 2 variables of degree $d - 1$. Let $\mathcal{F}_3$ be the space of real polynomials of 2 variables of degree d.

The above corollary applies to yield a polynomial $F: \mathcal{F}_1 \times \mathcal{F}_2 \times \mathcal{F}_3 \to \mathbf{R}$ such that $F(g_1, g_2, g_3) = 0$ whenever g_1, g_2, g_3 have a common zero $(x, y) \in \mathbf{R}^2$.

Next define a map $\Phi: \mathcal{P}_d \to \mathcal{F}_1 \times \mathcal{F}_2 \times \mathcal{F}_3$, $\Phi = (\Phi_1, \Phi_2, \Phi_3)$ by $\Phi(f) = (\text{Re } f', \text{Im } f', \text{Im}(\overline{f(0)}f))$. Here f', the derivative of f, may be naturally interpreted as the sum of 2 polynomials, $f' = \text{Re } f' + i \text{ Im } f'$, where $\text{Re } f'$ is a real polynomial of 2 real variables $(\mathbf{C} = \mathbf{R}^2)$ of degree $d - 1$. Similarly $\text{Im}(\overline{f(0)}f)$ has a natural interpretation as a real polynomial of 2 real variables of degree d. Explicitly $\text{Im } \overline{f(0)}f$ is the imaginary part of the polynomial

$$\sum_{k=0}^d (\overline{\alpha_0 + i\beta_0})(\alpha_k + i\beta_k)(x + iy)^k$$

where $a_k = \alpha_k + i\beta_k$, $k = 0, \ldots, d$ and $i = \sqrt{-1}$ (recall $a_d = 1$).

Thus Φ is well defined and moreover $\Phi_1 f$, $\Phi_2 f$ are independent of α_0, β_0 and Φ_3 is linear in α_0, β_0.

From the construction of $F: \mathcal{F}_1 \times \mathcal{F}_2 \times \mathcal{F}_3 \to \mathbf{R}$, if $f \in W_*$, then $F \circ \Phi(f) = 0$. Let $G = F \circ \Phi$. Then G is a polynomial in the $\alpha_0, \beta_0, \ldots, \alpha_{d-1}, \beta_{d-1}$ which has degree $(d - 1)^2$ in (α_0, β_0). This finishes the proof of the Algebraic Lemma.

Now let γ be a plane real algebraic curve of degree δ. Thus γ can be thought of as a locus of zeros in $\mathbf{R}^2$ of a real polynomial $F(x, y)$ of degree δ ($\delta = (d - 1)^2$ in our application). The *singular points* of γ are those points in $\mathbf{R}^2$ which are common zeros of $\partial F/\partial x$ and $\partial F/\partial y$. Thus by Bezout's theorem there are at most $(\delta - 1)^2$ singular points.

Let $N_\rho(\gamma)$ be the set of all points in $\mathbf{R}^2$ within ρ of some point of γ. Let $D_R \subset \mathbf{R}^2$ be the open disk of radius $R > 0$ about zero.

PROPOSITION 2. *If γ is a real plane algebraic curve of degree δ, then*

$$\text{area}\left[N_\rho(\gamma) \cap D_R \right] \leqslant 3\pi R\delta\rho.$$

For the proof introduce the set $T_\rho(\gamma) \subset \mathbf{R}^2$ of all points which lie on line segments of length $< \rho$ normal to a nonsingular point of γ.

LEMMA 1.

$$N_\rho(\gamma) \cap D_R \subset V_\rho \cup T_\rho(\gamma \cap D_{R+\rho})$$

where

$$V_\rho = \cup D_\rho(z), \qquad z \text{ a singular point of } \gamma.$$

Here $D_\rho(z)$ is the disk of radius ρ about z.

For the proof of Lemma 1, let $x \in N_\rho(\gamma)$, $x \in D_R$. Then there is $y \in \gamma$ such that $d(x, y) < \rho$ and we may suppose $d(x, y)$ is minimal. Therefore $y \in D_{R+\rho}$. If y is a singular point, then $x \in D_\rho(y) \subset V_\rho$. If y is not singular, then since $d(x, y)$ is minimal, x lies on a normal segment to γ at y of length less than ρ, i.e., $x \in T_\rho(\gamma \cap D_{R+\rho})$. This proves Lemma 1.

The following is a special case of a theorem of *Hotelling*.

LEMMA 2. *Let γ_0 be a smooth plane curve with nonvanishing tangent (but not necessarily connected) of total length l. Then*

$$\text{Area } T_\rho(\gamma_0) \leqslant 2\rho l.$$

LEMMA 3. *Let γ be a real algebraic plane curve of degree δ. Then the length $l(\gamma \cap D_R)$ is given by*

$$l(\gamma \cap D_R) \leqslant \pi\delta R.$$

For the proof of Lemma 3, we follow *Santalo* pp. 17–31 and especially pp. 30–31. Let dG be the standard measure on the space of lines in the plane. The length L of a curve C satisfies(*Santalo* p. 31)

$$2L = \int n \, dG$$

where n is the number of points of intersection of a line and the curve. The integral is over all lines. Now in our case, almost every line (in the sense of our measure on the space of lines in $\mathbf{R}^2$) meets C in at most δ points. This is a very special case of Bezout's theorem. Moreover, the integral is just over the lines which meet D_R. So

$$2L \leqslant \delta \int_{\text{lines meeting } D_R} dG \leqslant \delta \, 2\pi R.$$

The last evaluation, that

$$\int_{\text{lines meeting } D_R} dG = 2\pi R,$$

is shown on *Santalo* p. 30. This proves Lemma 3.

We now give the proof of Proposition 2. Use Lemma 1 to obtain

$$\text{Area}\big(N_\rho(\gamma) \cap D_R\big) \leqslant \text{Area } V_\rho + \text{Area } T_\rho(\gamma \cap D_{R+\rho}).$$

Clearly, Area $V_\rho \leqslant \pi\rho^2(\delta - 1)^2$ since there are at most $(\delta - 1)^2$ singularities (again we use Bezout's theorem). By Lemma 2, Area $T_\rho(\gamma \cap D_{R+\rho}) \leqslant 2\rho l(\gamma \cap D_{R+\rho})$ and by Lemma 3, this is less than $2\rho\pi\delta(R + \delta)$. Putting these estimates together yields

$$\begin{aligned}
\text{Area}\big(N_\rho(\gamma) \cap D_R\big) &\leqslant \pi\rho^2(\delta - 1)^2 + 2\rho(\delta\pi(R + \rho)) \\
&\leqslant \pi\big[2\rho\delta R + \rho^2[2\delta + (\delta - 1)^2]\big] \\
&\leqslant \pi\big[2\rho\delta R + \rho^2(\delta + 1)\big].
\end{aligned}$$

Now observe that if $\rho \geqslant R/3\delta$, then $3\pi\delta R\rho \geqslant \pi R^2$ and since

$$\text{area}\big[N_\rho(\gamma) \cap D_R\big] \leqslant \text{area } D_R \leqslant \pi R^2,$$

the proposition is true in this case.

Thus for the proof we may assume $\rho < R/3\delta$. Since $(R\rho/3\delta)(\delta^2 + 1) \leqslant \delta R\rho$ (easily checked), $\rho^2(\delta^2 + 1) \leqslant \delta R\rho$. Together with our previous estimate this yields the proposition.

Now the proof of Theorem 4B proceeds as the proof of Theorem 1A. Let $\chi\colon \mathbf{C}^d \cong \mathbf{R}^{2d} \to \mathbf{R}$ be the characteristic function of $U_\rho(W_*)$. Now represent W_* as in the algebraic lemma. The almost everywhere considerations used in the proof of Theorem 4A apply here as well. Thus by Proposition 2, for almost every $(a_1, \ldots, a_{d-1})$,

$$\left| \int_{|a_0| < R} \chi(a_0, a_1, \ldots, a_{d-1}) \, da_0 \right| \leqslant 3(d - 1)^2 \pi\rho R.$$

Then by Fubini's theorem,

$$\frac{\text{Vol } U_\rho(W_*) \cap P(R)}{\text{Vol } P(R)} = \frac{1}{(\pi R^2)^d} \int_{|a_1| < R} \cdots \int_{|a_{d-1}| < R} \int \chi(a_0, \ldots, a_{d-1}) \, da_0 da_1 \ldots da_{d-1}$$

$$\leqslant \frac{1}{(\pi R^2)^d} 3(d-1)^2 \pi \rho R \int_{|a_1| < R, \ldots, |a_{d-1}| < R} da_1 \ldots da_{d-1}$$

$$\leqslant \frac{3(d-1)^2}{(\pi R^2)^d} \pi \rho R (\pi R^2)^{d-1} = 3(d-1)^2 \frac{\rho}{R}.$$

This proves Theorem 4B.

Define

$$W_1 = \{ f \in \mathcal{P}_d | f(\theta) = f(0) \text{ for some critical point } \theta \}.$$

Let

$$L_\rho(f_0) = \left\{ f \in \mathcal{P}_d \middle| |f'(0) - f_0'(0)| < \rho \text{ and } \left| \frac{f''(0)}{2} - \frac{f_0''(0)}{2} \right| < \rho \right\}$$

and

$$L_\rho(W_1) = \cup L_\rho(f_0), \qquad f_0 \in W_1.$$

THEOREM 4C.

$$\frac{\text{Vol}[L_\rho(W_1) \cap P(R)]}{\text{Vol } P(R)} \leqslant 4\left(\frac{\rho}{R}\right)^2 (d+2).$$

As in the proof of Theorem 4A we use the resultant to describe W_1 as a variety in $\mathbf{C}^d$. The basic fact is that $f \in W_1$ if and only if f' and $f - f(0)$ have a common zero. As in *van der Waerdan*, Vol. I, and *Lang*, we consider the resultant $R(f', f - f(0))$ as a polynomial F in the coefficients $(a_0, a_1, \ldots, a_{d-1})$. Note here that of course $a_d = 1$, and also that F is independent of a_0. Furthermore it can be seen from the form of the resultant that F has degree $(d+2)$ in (a_1, a_2).

On $\mathbf{C}^2$ we will use the norm $\| \ \|_B$, defined by $\|z\|_B = \|(z_1, z_2)\|_B = \max(|z_1|, |z_2|)$.

PROPOSITION 3. *Let γ be a complex algebraic plane curve of degree $\delta > 3$ and define*

$$N_\rho(\gamma) = \{ z \in \mathbf{C}^2 | \ \|z - y\|_B < \rho \text{ for some } y \in \gamma \}.$$

Then

$$\text{Vol}[N_\rho(\gamma) \cap P(R)] \leqslant 4\delta(\pi \rho R)^2.$$

Here $P(R)$ is the polycylinder $D_R \times D_R$, D_R the disk of radius R in $\mathbf{C}$.

The proof of Proposition 3 is analogous to the proof of Proposition 2 with Lemmas 4, 5, 6 playing the role of Lemmas 1, 2, 3 respectively. Also Proposition 3 plays the same role in the proof of Theorem 4C that Proposition 2 did in the proof of Theorem 4B.

Define $T_\rho(\gamma)$ just as in Lemma 1, in terms of distance along line segments normal to the surface γ in $\mathbf{R}^4$. The same applies to $T_\rho(\gamma \cap P(R + \rho))$. Define

$$V_\rho = \bigcup_{z \in \Sigma} P_z(\rho)$$

where Σ is the set of singular points of γ and $P_z(\rho)$ is the polycylinder

$$P_z(\rho) = \left\{ y \in \mathbf{C}^2 \mid \|y - z\|_B < \rho \right\}.$$

LEMMA 4.

$$N_\rho(\gamma) \cap P(R) \subset V_\rho \cup T_\rho(\gamma \cap P(R + \rho)).$$

REMARK. It seems likely that this is true omitting V_ρ. However, this doesn't affect our estimates much and we leave it in. P. Griffiths, in conversation, has confirmed the possibility of omitting V_ρ.

PROOF OF LEMMA 4. Let $x \in P(R) \cap N_\rho(\gamma)$. Suppose $y \in \gamma$ minimizes $\|x - y\|_B$. Thus $\|x - y\|_B < \rho$ and if $x \notin V$, then y is nonsingular. Therefore the segment $\overline{xy}$ is normal to γ. Finally $\|x\|_B < R$, $\|x - y\|_B < \rho$ so $\|y\|_B < R + \rho$ and $y \in P(R + \rho)$. This proves Lemma 4.

See *Weyl* (1939) (and also *Griffiths*) for the following

LEMMA 5 (WEYL).

$$\text{Vol } T_\rho(\gamma) \leqslant \pi \left[\frac{\rho^2}{4} \text{ area } \gamma + \frac{\rho^4}{24} \int_\gamma K \right].$$

Here γ is supposed to be a smooth surface in $\mathbf{R}^2$ with equality in the case of no overlapping. Actually Weyl supposes γ is compact, no boundary (or singularities). But his analysis applies to our situation as long as we interpret T_ρ via normal segments. In the integral K is the Gaussian curvature of γ.

We first note that since γ is a complex algebraic curve, it is a minimal surface; therefore everywhere $K \leqslant 0$ and we may discard this term. See also the introduction to this section.

LEMMA 6. *If γ is a complex plane algebraic curve of degree δ, then*

$$\text{area}\left[\gamma \cap P(R) \right] \leqslant 6\pi\delta R^2.$$

PROOF OF LEMMA 6. Here we follow *Santalo* once more. We integrate functions over the space of 2 dimensional planes in $\mathbf{R}^4 = \mathbf{C}^2$.

This goes as follows. We write L_2 for a typical 2-plane in $\mathbf{R}^4$ and dL_2 for the measure over the space of such planes. Then

$$\frac{O_4 O_3}{O_2 O_1} \text{ area}(\gamma \cap P(R)) \leqslant \int_{L_2} \#(\gamma \cap L_2) \, dL_2$$

where $\#(\gamma \cap L_2)$ is the number of points of intersection of γ and L_2 (see

Santalo, 14.70, p. 245). Here O_k is the surface area of the unit k-sphere and in particular $O_0 = 2$, $O_1 = 2\pi$, $O_2 = 4\pi$, $O_2/O_4 = 3/2\pi$ (*Santalo*, p. 9).

Now $\#(\gamma \cap L_2) \leqslant \delta$ by Bezout's theorem except for a set of L_2 of measure zero. Then

$$\int_{L_2} \#(\gamma \cap L_2)\, dL_2 \leqslant \delta \int_{L_2 \cap P(R) \neq \varnothing} dL_2.$$

Now for any convex set K (*Santalo*, p. 233), $\int_{L_2 \cap K} dL_2$ is evaluated in terms of a certain *quermassintegrale* $W_r(K)$. Thus

$$\int_{L_2 \cap K} dL_2 = 2 \frac{O_2 O_1}{O_1 O_0} W_2(K).$$

I was not able to use *Santalo's* formula (p. 225) for $K = P(R)$; so crudely use the fact that $P(R) \subset D_{\sqrt{2R}}$, the ball, and *Santalo's* evaluation on p. 224

$$W_2(D_{\sqrt{2R}}) = \frac{O_3 R^2}{2}.$$

Thus

$$\int_{L_2 \cap K} dL_2 \leqslant \frac{O_3 O_2}{O_0} R^2.$$

Putting these estimates together we obtain

$$\text{area}(\gamma \cap P(R)) \leqslant \frac{O_2 O_1}{O_4 O_3} \delta \frac{O_3}{O_0} O_2 R^2 \leqslant 6\pi\delta R^2.$$

This proves Lemma 6.

We now are able to prove Proposition 3 in much the manner of Proposition 2. First use Lemma 4.

$$\text{Vol}(N_\rho(\gamma) \cap P(R)) \leqslant \text{Vol } V_\rho + \text{Vol } T_\rho(\gamma \cap P(R + \rho)),$$

where

$$\text{Vol } V_\rho \leqslant (\delta - 1)^2 \text{Vol } P(\rho) = (\delta - 1)^2(\pi\rho^2)^2.$$

By Lemma 5 (and comments),

$$\text{Vol } T_\rho(\gamma \cap P(R + \rho)) \leqslant \pi \frac{\rho^2}{4} \text{area}(\gamma \cap P(R + \rho));$$

then using Lemma 6,

$$\text{Vol } T_\rho(\gamma \cap P(R + \rho)) \leqslant \frac{\pi\rho^2}{4} \cdot 6\pi\delta(R + \rho)^2,$$

so

$$\text{Vol}\big[N_\rho(\gamma) \cap P(R) \big] \leqslant \pi^2\rho^4(\delta - 1)^2 + \tfrac{3}{2}\pi^2\rho^2\delta(R + \rho)^2. \qquad (*)$$

Now we prove the estimate of Proposition 3 in two cases. First suppose that

$\rho > (2/3\delta)^{1/2}R$. Then

$$4\delta\pi^2 R^2\rho^2 \geqslant (\pi R^2)^2 \geqslant \operatorname{Vol} P(R) \geqslant \operatorname{Vol}\big[\, N_\rho(\gamma) \cap P(R)\,\big].$$

On the other hand, suppose $\rho \leqslant (2/3\delta)^{1/2}R$. Use our previous estimate $(*)$. So we have to show that if $\rho \leqslant (2/3\delta)^{1/2}R$, then

$$\pi^2\rho^4(\delta - 1)^2 + \tfrac{3}{2}\pi^2\rho^2\delta(R + \rho)^2 \leqslant 4\delta\pi^2\rho^2 R^2$$

or

$$\rho^2(\delta - 1)^2 + \tfrac{3}{2}\delta(R + \rho)^2 \leqslant 4\delta R^2$$

or yet

$$\frac{2}{3\delta^2}(\delta - 1)^2 + \frac{3}{2}\left(1 + \left(\frac{2}{3\delta}\right)^{1/2}\right)^2 \leqslant 4.$$

But this last is easily checked using the fact that $\delta \geqslant 4$. This proves Proposition 3.

It remains to prove Theorem 4C, which will follow closely the proof of Theorem 4B. Thus let $\chi\colon \mathbf{C}^d \to \mathbf{R}$ be the characteristic function of $L_\rho(W_1)$.

Now we have shown that W_1 is a complex hypersurface in $\mathbf{C}^d$ with the property that for almost every $(a_0, a_3, a_4, \ldots, a_{d-1})$, it has degree $(d + 2)$ in a_1, a_2. Furthermore

$$L_\rho(W_1) = \bigcup \{(a_0, a_1', a_2', a_3, \ldots, a_{d-1})|$$
$$\times |a_1' - a_1| < \rho,\ |a_2' - a_2| < \rho\}, \qquad (a_0, a_1, \ldots, a_{d-1}) \in W_1.$$

By Proposition 3, for almost every $(a_0, a_3, \ldots, a_{d-1})$

$$\int_{\substack{|a_1| < R \\ |a_2| < R}} \chi(a_0, \ldots, a_{d-1})\, da_1 da_2 \leqslant 4(d + 2)(\pi\rho R)^2.$$

Then

$$\frac{\operatorname{Vol} L_\rho(W_1) \cap P(R)}{\operatorname{Vol} P(R)}$$

$$\leqslant \frac{1}{(\pi R^2)^d} \int_{\substack{|a_0| < R \\ |a_3| < R \\ \vdots \\ |a_{d-1}| < R}} \left[\int \chi(a_0, \ldots, a_{d-1})\, da_1 da_2\right] da_0 \ldots da_{d-1}$$

$$\leqslant \frac{1}{(\pi R^2)^d}\left[\int_{\substack{|a_0| < R \\ |a_3| < R \\ |a_{d-1}| < R}} da_0 da_3 \ldots da_{d-1}\right] 4(d + 2)(\pi\rho R)^2$$

or

$$\frac{\mathrm{Vol}(L_\rho(W_1) \cap P(R))}{\mathrm{Vol}\, P(R)} \leqslant \frac{(\pi R^2)^{d-2}}{(\pi R^2)^d} 4(d+2)\pi^2 R^2 \rho^2$$

$$\leqslant 4(d+2)\frac{p^2}{R^2}.$$

This proves Theorem 4c.

5. Here we will prove our main result, as stated in the introduction and in Theorem 6. Toward this end we combine Theorem 3 and Theorems 4B, 4C. This is done by first using Theorems 4B and 4C to prove Theorem 5. Then the main result will follow easily from Theorems 3 and 5.

We take $U_{\sigma R}(W_*)$ as defined in §4 and define

$$Q_\sigma = \{ f \in \mathcal{P}_d |\ |f(\theta) - f(0)| < \sigma \text{ for some critical point } \theta \},$$

$$Y_\sigma = Q_\sigma \cup U_{\sigma R}(W_*).$$

THEOREM 5. (1) *If* $R > \frac{1}{3}$,

$$\frac{\mathrm{Vol}\, Y_\sigma \cap P(R)}{\mathrm{Vol}\, P(R)} \leqslant 150(d+2)^{4/3}\sigma^{2/3},$$

(2) *Let* $\sigma < 1$ *and* $f \notin Y_\sigma$. *Then*
   (a) $\rho_f > \sigma R$,
   (b) $4R \sin \mathcal{K}_f \geqslant \sigma^2$.

Here (see also §3)

$$\rho_f = \min_{\substack{\theta \\ f'(\theta)=0}} |f(\theta)|, \qquad \mathcal{K} = \mathcal{K}_f = \min_{\substack{\theta \\ f'(\theta)=0}} \left| \arg\frac{f(\theta)}{f(0)} \right|.$$

For the proof of (1) we use Lemma 1.

LEMMA 1. *If* $\sigma < \frac{1}{8}$, $Q_\sigma \cap P(R) \subset L_\alpha(W_1)$,

$$\alpha = \max\{2\sigma^{2/3}, \sigma^{1/3}, 6R\sigma^{1/3}\}$$

where $L_\alpha(W_1)$ *is defined in Theorem* 4C.

PROOF. Let $f \in Q_\sigma \cap P(R)$, $f(z) = \sum_{i=0}^d a_i z^i$, $f(\theta) - f(0) = b$, $|b| < \sigma$, $f'(\theta) = 0$. We will construct $f_0 \in W_1$ such that $f \in L_\alpha(f_0)$. There are two cases to consider; first suppose that $|\theta| \geqslant \sigma^{1/3}$. Define in this case $f_0(z) = f(z) - c_1 z - c_2 z^2$ where $c_1 = 2b/\theta$, $c_2 = -b/\theta^2$. One can compute directly that $|c_1| < \alpha$, $|c_2| < \alpha$, so that it remains to show $f_0 \in W_1$. To do that it is sufficient to prove that $f_0'(\theta) = 0$ and $f_0(\theta) = f_0(0)$. But both of these are checked by direct calculations.

To finish the proof of Lemma 1, it remains to construct the desired f_0 in case $|\theta| < \sigma^{1/3}$. For that let $f_0(z) = f(z) - a_1 z$ and choose $\theta' = 0$. Then clearly $f_0'(\theta') = 0$ and $f_0(\theta') = f_0(0)$, so $f_0 \in W_1$. To show $f \in L_\alpha(f_0)$, it is sufficient to prove that $|a_1| < 6R\sigma^{1/3}$.

Since $f'(\theta) = 0$, $\Sigma_{k=0}^d ka_k\theta^{k-1} = 0$, so

$$|a_1| = \left|\sum_{k=2}^d ka_k\theta^{k-1}\right| \leqslant R\sum_{k=2}^d k|\theta|^{k-1}.$$

The last uses the fact that $|a_k| < R$ since $f \in P(R)$. Thus

$$|a_1| \leqslant R\cdot\sum_{k=2}^\infty k(\sigma^{1/3})^{k-1} \leqslant R\left\{\sum_{k=0}^\infty k(\sigma^{1/3})^{k-1} - 1\right\}.$$

But for $0 < \beta < 1$

$$\sum_{k=0}^\infty k\beta^{k-1} = \frac{d}{d\beta}\sum_{k=0}^\infty \beta^k = \frac{d}{d\beta}\left(\frac{1}{1-\beta}\right) = \frac{1}{(1-\beta)^2}.$$

We conclude for $\beta = \sigma^{1/3}$

$$|a_1| \leqslant R\left\{\frac{1}{(1-\beta)^2} - 1\right\} \leqslant R\frac{\beta(2-\beta)}{(1-\beta)^2}.$$

Now

$$\frac{2-\beta}{(1-\beta)^2} \leqslant 6 \quad \text{if } \beta \leqslant \frac{1}{2} \text{ or } \sigma \leqslant \frac{1}{8}.$$

Thus

$$|a_1| \leqslant R6\beta = 6R\sigma^{1/3}.$$

This finishes the proof of Lemma 1.

The proof of (1) of Theorem 5 now goes as follows. We may assume $\sigma < \frac{1}{8}$. Use Lemma 1 with $\alpha = 6R\sigma^{1/3}$ taking into account the limitations $R > \frac{1}{3}$, $\sigma < \frac{1}{8}$. First

$$\frac{\text{Vol}(Y_\sigma \cap P(R))}{\text{Vol } P(R)} \leqslant \frac{\text{Vol}\big(U_{\sigma R}(W_*) \cap P(R)\big)}{\text{Vol } P(R)} + \frac{\text{Vol}(Q_\sigma \cap P(R))}{\text{Vol } P(R)}.$$

By Theorem 4B,

$$\frac{\text{Vol}\big(U_{\sigma R}(W_*) \cap P(R)\big)}{\text{Vol } P(R)} \leqslant 3\sigma(d-1)^2.$$

By Lemma 1 and Theorem 4C,

$$\frac{\text{Vol}(Q_\sigma \cap P(R))}{\text{Vol } P(R)} \leqslant \frac{\text{Vol}(L_{6R\sigma^{1/3}}(W_1) \cap P(R))}{\text{Vol } P(R)} \leqslant 144(d+2)\sigma^{2/3}.$$

We may now suppose that $150(d+2)^{4/3}\sigma^{2/3} < 1$. This implies $d^2\sigma < (1/150)^3 < 2^{3/2}$ so that $3\sigma(d-1)^2 \leqslant 6(d+2)^{4/3}\sigma^{2/3}$. On the other hand

$$144(d+2)\sigma^{2/3} \leqslant 144(d+2)^{4/3}\sigma^{2/3};$$

thus

$$3\sigma(d-1)^2 + 144(d+2)\sigma^{2/3} \leqslant 150(d+2)^{4/3}\sigma^{2/3}.$$

With our previous estimates, this proves (1) of Theorem 5.

LEMMA 2. $f \in U_\rho(W_0)$ *if and only if* $\rho_f < \rho$ *(here* $U_\rho(W_0)$ *is defined in §4).*

PROOF OF LEMMA 2. We have to show $f \in U_\rho(W_0)$ if and only if $|f(\theta)| < \rho$ for some critical point θ. By definition, $f \in U_\rho(W_0)$ implies there is some $f_0 \in W_0$ with $f \in U_\rho(f_0)$. Thus $|f(z) - f_0(z)| < \rho$ for all z. But there is some critical point θ with $f_0(\theta) = 0$ since $f_0 \in W_0$. Also f and f_0 have the same critical points. Thus $f \in U_\rho(W_0)$ implies $|f(\theta)| < \rho$ some critical point θ. Conversely if $|f(\theta)| < \rho$ for some critical point θ, define $f_0(z) = f(z) - f(\theta)$. Then $f_0 \in W_0$ and $f \in U_\rho(f_0)$. This proves Lemma 2.

Since $W_0 \subset W_*$, $U_\rho(W_0) \subset U_\rho(W_*)$ and $U_{\rho R}(W_0) \subset U_{\rho R}(W_*)$. This yields (2)(a) in Theorem 5.

Incidentally (not used here), we have as a Corollary (use Theorem 4A),

PROPOSITION.

$$\frac{\mathrm{Vol}\{f \in \mathcal{P}_d | \rho_f < \rho\} \cap P(R)}{\mathrm{Vol}\, P(R)} \leqslant \frac{d\rho^2}{R^2}.$$

For the proof of part (2)(b) of Theorem 5, suppose that $f \notin Q_\sigma$. We will show that under the estimate $4R \sin \mathcal{K}_f < \sigma^2, f \in U_{\sigma R}(W_*)$.

Consider (see Figure 2) (a similar picture with β reversed if $|f(\theta)| < |f(0)|$)

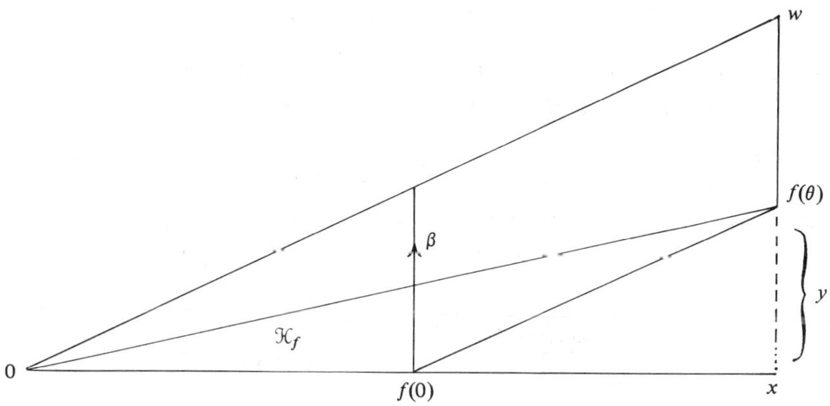

FIGURE 2

The segment $\overline{0w}$ is drawn parallel to $\overline{f(0)f(\theta)}$. We define β to be the vector starting at $f(0)$, perpendicular to $\overline{0f(0)}$ which meets the segment $\overline{0w}$, and $\beta = w - f(\theta)$ defines the magnitude of w.

Since $f(0) + \beta$ and $f(\theta) + \beta$ lie on the same ray,

$$\mathrm{Im}((f(0) + \beta)/(f(\theta) + \beta)) = 0.$$

By a similar triangle argument $|\beta|/|f(0)| = y/(x - |f(0)|)$ where (x, y) are as in the figure. If $f(0) > 0$, then (x, y) are the Cartesian coordinates of $f(\theta)$.

We obtain,

$$\frac{|\beta|}{R} \leqslant \frac{\beta}{|f(0)|} = \frac{y}{|x - f(0)|} \leqslant \frac{y}{(\sigma^2 - y^2)^{1/2}}.$$

The last inequality is true since $|f(\theta) - f(0)| \geqslant \sigma$ ($f \notin Q_o$) and thus

$$|x - f(0)|^2 + y^2 \geqslant \sigma^2.$$

So $|x - f(0)| \geqslant (\sigma^2 - y^2)^{1/2}$.

Furthermore

LEMMA 3.

$$\frac{y}{(\sigma^2 - y^2)^{1/2}} < \sigma.$$

PROOF OF LEMMA 3. By multiplying and squaring the inequality becomes

$$y^2 \leqslant \sigma^2(\sigma^2 - y^2) \quad \text{or} \quad y^2(1 + \sigma^2) \leqslant \sigma^4$$

or yet $y \leqslant \sigma^2/(1 + \sigma^2)^{1/2}$. Since $\sigma < 1$, $\frac{1}{2}\sigma^2 \leqslant \sigma^2/(1 + \sigma^2)^{1/2}$ and it is sufficient to prove that

$$2y \leqslant \sigma^2.$$

Now $y = x \sin \mathcal{H}_f$, $x \leqslant |f(\theta)| \leqslant 2R$, so by our hypothesis $2y = 2x \sin \mathcal{H}_f$ $\leqslant 2|f(\theta)|\sin \mathcal{H}_f \leqslant 4R \sin \mathcal{H}_f < \sigma^2$. This proves Lemma 3.

Now by Lemma 3 and the preceding inequality we have

$$\frac{|\beta|}{R} < \sigma \quad \text{or} \quad |\beta| < \sigma R.$$

Let $f_0(z) = f(z) + \beta$. Then $f_0 \in W_1, f \in \overline{U_{|\beta|}(W_1)} \subset U_{\sigma R}(W_1)$.

This finishes the proof of Theorem 5.

THEOREM 6. *Suppose* $0 < \mu < 1$, d *are given and* $R > \frac{1}{3}$. *Let* σ *be chosen so that* $\mathrm{Vol}(Y_\sigma \cap P(R))/\mathrm{Vol}\, P(R) < \mu$ *and let* $f \in P(R)$ *with* $f \notin Y_\sigma$. *Then for a suitable choice of* h *depending on* (μ, d), *and* $z_0 = 0$, $z_k = z_{k-1} - hf(z_{k-1})/f'(z_{k-1})$ *is well defined for all* k *and* z_s *is an approximate zero of* f *where*

$$s = 4\left\{3 + \frac{\log(15/\sigma)8R}{\sigma^2}\right\}^2, \qquad \sigma = \left(\frac{\mu}{150}\right)^{3/2}\frac{1}{(d + 2)^2}.$$

Thus an approximate zero is obtained after s steps with probability of at least $1 - \mu$.

Define $\sigma(\mu)$ by

$$\sigma(\mu) = \left(\frac{\mu}{150}\right)^{3/2}\frac{1}{(d + 2)^2}.$$

Therefore by Theorem 5(1),

$$\frac{\mathrm{Vol}(Y_{\sigma(\mu)} \cap P(R))}{\mathrm{Vol}\, P(R)} < \mu$$

for all μ, $0 < \mu < 1$.

This construction complements the second sentence of Theorem 6.

Now for the proof, suppose $f \notin Y_\sigma$, $f \in P(R)$. By Theorem 5(2), $\mathcal{H}_f$, $\rho_f > 0$, so Theorem 3 applies. In fact we obtain for a suitable h, the conclusion of Theorem 6 relative to the s defined by

$$s = K\left\{3 + \frac{\log \zeta}{\sin(\mathcal{H}/2)}\right\}^2.$$

Thus to finish the proof, we have to translate this function of ζ and $\mathcal{H}$ to our function of μ and d. For this we use the definition of ζ, the full strength of Theorem 5(2) and the explicit $\sigma = \sigma(\mu)$ stated previously.

Briefly then, since $|f(0)| < R$, $K = 4$, and $\rho_f > \sigma R$, we can take $\zeta = 15/\sigma$. Since, $4R \sin \mathcal{H} \geq \sigma^2$, take $\sin(\mathcal{H}/2) = \sigma^2/R$. This yields Theorem 6.

Note that crude estimates from Theorem 6 yield that if $R = 1$,

$$s = \frac{[100(d + 2)]^9}{\mu^7}$$

is sufficient. Thus the theorem stated in the introduction follows.

PART III

We devote this part to a discussion of open problems related especially to Part II.

PROBLEM 1A. Reduce the constant K of Theorem 1 (Remark 1) from 4. It is possible that $K = 1$. In the same body of ideas there are a number of other problems about polynomials, their critical points and values; we will go into this in more detail and suggest some way of looking at these problems.

In Theorem 1, the coefficients a_k are symmetric functions of the critical points $\theta_1, \ldots, \theta_{d-1}$ of $f(z) = \Sigma_{k=0}^d a_k z^k$ provided $f(0) = 0$ (as we have assumed) if we take $a_d = 1$, as we will now. To see this, just solve these equations for a_k,

$$\sum_1^d k a_k \theta_i^{k-1} = 0, \qquad i = 1, \ldots, d - 1$$

or better

$$\sum_1^d a_k z^k = \frac{1}{d} \int \prod_{i=1}^{d-1} (z - \theta_i) - \text{const.}$$

Recall the elementary symmetric functions which are defined by

$$S_k(z_1, \ldots, z_{d-1}) = \sum_{i_1 < i_2 < \cdots < i_k} z_{i_1} z_{i_2} \cdots z_{i_k}.$$

It follows that

$$\left|\frac{a_k}{a_1}\right|^{1/(k-1)} \frac{1}{|a_1|} = \left|\frac{1}{k} S_{k-1}\left(\frac{1}{\theta_1}, \ldots, \frac{1}{\theta_{d-1}}\right)\right|^{1/(k-1)} \frac{1}{|f'(0)|}$$

where $f'(0) = a_1 = d\theta_1 \cdots \theta_{d-1}$.

Thus for example we get from Theorem 1 for $k = 2$ (recall $K = 2$ if $k = 2$).

PROPOSITION. *If $f(0) = 0$, degree $f = d$, $a_d = 1$, then for some critical point θ,*

$$\left| \frac{1}{2} \sum_{i=1}^{d-1} \frac{1}{\theta_i} \right| \frac{|f(\theta)|}{|f'(0)|} \leqslant 2.$$

This brings our subject close to the theory of critical points of polynomials (see *Marden* for a survey). It also suggests an alternate approach to Theorem 1. One parametrizes all polynomials $f(z) = \sum a_i z^i$ of degree d with $f(0) = 0$, $a_d = 1$ by their critical points $\theta_1, \ldots, \theta_{d-1}$ (of course we list a multiple critical point repeatedly).

Thus each nontrivial $(d-1)$ tuple $(\theta_1, \ldots, \theta_{d-1})$ determines such a polynomial (and conversely up to an ordering of the θ_i).

Let $\theta = (\theta_1, \ldots, \theta_{d-1}) \in \mathbf{C}^{d-1} - 0$, and suppose now $d > 2$. The previous proposition motivates the definition of these functions,

$$\psi_m(\theta) = \left(\frac{1}{2} \sum_1^{d-1} \frac{1}{\theta_i} \right) \frac{f(\theta_m)}{f'(0)}, \qquad m = 1, \ldots, d-1.$$

It can be checked that each ψ_m is homogeneous of degree 0 in the θ_i (it is a rational function) and can be thought of as $\psi_m \colon P_{d-2}(\mathbf{C}) - \Sigma \to \mathbf{C}$. Here $P_{d-2}(\mathbf{C})$ is complex projective space and $\theta \in \Sigma$ if some $\theta_i = 0$.

The preceding proposition translates to

$$\min_m |\psi_m(\theta)| \leqslant 2.$$

Consistent with our preceding problem is

PROBLEM 1B. Can 2 be replaced by $\frac{1}{2}((d-1)/d)$ in the previous proposition?

Note that the function $\theta \to \min_m |\psi_m(\theta)|$ is a continuous function on $P_{d-2}(\mathbf{C})$ and this gives a proof that $\min_m |\psi_m(\theta)|$ is bounded. A bound can be computed directly to be approximately 2^d, but this is too weak for the main theorem.

We are naturally led to the problem of finding the maximum over $\theta \in P_{d-2}(\mathbf{C})$ of $\min_m |\psi_m|$. Finding this maximum could lead to the solution of several of the problems in this section.

PROBLEM 1C. Does the point $\theta = (\theta_1, \ldots, \theta_{d-1}) = (1, \ldots, 1)$ maximize $\min_m |\psi_m|$?

An affirmative answer to Problem 1C would yield an affirmative answer to Problem 1B. Note that $\theta = (1, \ldots, 1)$ corresponds to $f(z) = (z - 1)^d - (-1)^d$.

An easier question is

PROBLEM 1D. Is $\theta = (1, \ldots, 1)$ a *local* maximum for the function $\min_m |\psi_m|$?

Recall that Problems 1B, 1C, 1D all have to do with $k = 2$ in Theorem 1 and that for each k, there is a similar set of problems.

One should keep in mind for these questions, that very general maximum principles hold for complex analytic functions; see *Whitney*. Also I checked that these problems do have affirmative answers if $d = 3$ (and a number of other cases).

It is noteworthy that among the maxima (which might be unique) is one which is *Pareto optimal* as defined in mathematical economics (see *Smale* (1974) or *Wan*). One says that $\theta^* \in P_{d-2}(\mathbf{C})$ is Pareto Optimal for $|\psi|$, $m = 1, \ldots, d - 1$, provided there is no $\theta \in P_{d-2}(\mathbf{C})$ such that $|\psi_m(\theta)| \geqslant |\psi_m(\theta^*)|$ for all m with strict inequality for some m. Thus a Pareto optimality study of the $|\psi_m|$ might be in order.

Before we state the next problem, we prove a theorem which is an idealized version of the previous proposition, proved with crucial help from Mike Shub.

THEOREM. *If f is a polynomial with $f(0) = 0, f'(0) \neq 0$ then*

$$\min_{\substack{\theta \\ f'(\theta) = 0}} \left| \frac{f(\theta)}{\theta} \right| \frac{1}{|f'(0)|} \leqslant 4.$$

The proof goes via the Koebe constant. Note first that the expression on the left has numerator, denominator each linear in a_d and homogeneous of degree d in the critical points $\theta_1, \ldots, \theta_{d-1}$ of f. Here degree $f = d$.

We make the substitution $\theta_i \to \lambda\theta_i$, for all i, to reduce to the case $f'(0) = 1$. Next make a further substitution $f \to f(Rz)/R$ to reduce to the case $\min_\theta |f(\theta)| = 1$. Compare this to the proof of Theorem 1.

Now take the inverse of f to obtain a Schlict function g with $g(0) = 0$, $g'(0) = 1$, $w \to g(w)$ 1-1 for $|w| < 1$. Now the Koebe theorem (see *Hille*) asserts that the image of the unit disk D_1 under g contains a disk D_κ of radius $\kappa > 0$ (κ is the Koebe constant). Bieberbach evaluated $\kappa = \frac{1}{4}$ at the time he proved $|a_2| \leqslant 2$ and proposed the Bieberbach conjecture. Thus $|\theta| \geqslant \frac{1}{4}$ in our situation so $1/|\theta| \leqslant 4$ and our theorem follows.

PROBLEM 1E. Can one replace 4 by 1 in the theorem (or yet $(d - 1)/d$)?

One cannot do better because of the example $f(z) = z^d - dz$.

The following is an incidental problem which came up as I was working on some of the previous problems.

PROBLEM 1F. What is the set of polynomials f of degree d (for each d) such that $f(0) = 0$, the leading coefficient a_d is 1, and $f(\theta) = \theta$ for each critical point θ? That is to say, the critical points are fixed points.

If f' has simple zeros it must be $f(z) = z^d - (d/(1 - d))z$. Other examples are of the form $f(z) = (z - a)^d - (-a)^d$ or $(z^d - a)^l$-constant. But there are other more subtle examples if $d \geqslant 4$. I believe that I computed them for $d \leqslant 5$ and showed that for each d there were only finitely many.

PROBLEM 2. Extend the main result to polynomial maps $f: \mathbf{C}^n \to \mathbf{C}^n$ for each n. This is quite a nice problem. At first, I thought necessary estimates would involve some kind of Bieberback conjecture mathematics of several variables. But eventually I noticed the following counterexample to the Koebe theorem for $\mathbf{C}^2 \to \mathbf{C}^2$:

$$(z_1, z_2) \to (z_1, z_2 + \lambda z_1^2), \qquad \lambda > 0.$$

This map is globally invertible. But the image of the unit polydisk gets thin as $\lambda \to \infty$. On the other hand since there are no critical points in $\mathbf{C}^2$, the example is not bad from the point of view of computing by Newton type algorithms.

PROBLEM 3. Find a bound on cost for real polynomial maps $\mathbf{R}^n \to \mathbf{R}^n$. Here the natural algorithm for $n > 1$ is the "Global Newton". See *Hirsch-Smale*. One must take into account negative Jacobian determinants.

PROBLEM 4. Reduce the number of steps (and/or find a shorter proof!) of the main theorem. One might vary h in §3. Also §§4 and 5 might be developed in a different way to reduce the degree in d. Also §4 suggests various geometric problems.

PROBLEM 5. Analyse Part II in terms of round off error. The algorithm is robust, but still there is a question here. One can see Wilkinson on this subject. Then there are related purely discrete or algebraic problems.

PROBLEM 6. Find an analogue for our main theorem for the simplex method of linear programming. This is a very well-known problem. For example in discussing Khachian's recent work, *Wolfe* writes . . . "Dantzig's 'simplex method' has been shown not to run in polynomial, but in exponential time, in the worst case. 'Worst case' behaviour is always the easiest to study; a theory of 'average' behaviour, which would explain the fact that, in practice, the simplex method acts like a highly efficient polynomial time algorithm, does not exist".

PROBLEM 7. Estimate the probability of (strict) Newton's method converging. From results and with the notations in Part II, this could be accomplished by estimating

$$\frac{\text{Volume}\{\, f \in P(R)|\; |f(0)| < \rho_f/\,(2K + 1)\,\}}{\text{Volume } P(R)}.$$

PROBLEM 8. Prove an analogue of our theorem for the Scarf-Eaves algorithm (see *Eaves and Scarf*). In fact there are a large number of problems in operations research and numerical analysis that suggest themselves.

PROBLEM 9. It is a fact essentially due to *Barna* that for a polynomial f with all roots real, Newton's method itself converges to a zero starting with almost every real number. The exceptional set of starting points is homeomorphic to the Cantor set.

In fact $T\colon P_R \to P_R$ defined by $T(x) = x - f(x)/f'(x)$ is an Axiom A dynamical system where P_R is real 1-dimensional projective space. One can use a theorem of Bowen and Ruelle to prove the measure-theoretic statement. The problem is to find an analogue of our main result for this situation.

We summarize by noting some points about our main result. One might ask what about letting the initial point of the algorithm z_0 vary as well as the polynomial. I believe everything goes through directly with no problem. What about letting the leading coefficient a_d vary, not just stay at 1? I haven't checked this and don't know what happens.

One can also ask to what extent is it a reasonable computational problem to let the degree of a polynomial grow large. In fact there can be a certain amount of ill-posedness by taking high powers. On the other hand there has been much successful numerical work on this, cf. *Dejon-Henrici*. For example, in this book of conference proceedings, there is reported work by Dejon and Nickel on a rather random polynomial of degree 100. All the roots were found. Hirsch and I worked on a number of simple examples with a PDP11 with good success. I chose this problem of working with a polynomial equation because of the tradition associated to the fundamental theorem of algebra as well as its being a prototype of the fundamental and central problem of solving a (nonlinear) system of equations.

REFERENCES

B. Barna, 1956, *Über die Divergenzpunkte des Newtonsches Verfahrens zur Bestimmung von Wurzeln Algebraischen Gleichungen*. II, Publicatione Mathematicae, Debrecen, vol. 4, pp. 384–397.

L. E. J. Brouwer, 1924 (w. B. de Loor), *Intuitionischer Beweis des Fundamentalsatzes der Algebra*, Coll. Works, vol. 1 (1975), North-Holland, Amsterdam.

G. Collins, 1977, *Infallible calculation of polynomials to specified precision in mathematical software*. III, Academic Press, New York.

G. Debreu, 1959, *Theory of value*, Yale Univ. Press, New Haven, Conn.

B. Dejon and P. Henrici, 1969, *Constructive aspects of the fundamental theorem of algebra*, Wiley, New York.

C. Eaves and H. Scarf, 1976, *The solution of systems of piecewise linear equations*, Math. Operations Res., vol. 1, pp. 1–27.

H. Eves, 1976, *An introduction to the history of mathematics* (4th ed.), Holt, Rinehart and Winston, New York.

M. Garey and D. Johnson, 1979, *Computers and intractability*, Freeman, San Francisco.

C. F. Gauss, 1973, *Werke*, Band X, Georg Olms Verlag, New York.

P. Griffiths, 1978, *Complex differential and integral geometry and curvature integrals associated to singularities of complex analytic varieties*, Duke Math. J. **45**, pp. 427–512.

J. Hartmanis, 1979, *Observations about the development of theoretical computer science*, 20th Annual Sympos. on Foundations of Computer Science, IEEE, Long Beach, Calif.

W. Hayman, 1958, *Multivalent functions*, Cambridge Univ. Press, Cambridge, England.

P. Henrici, 1977, *Applied and computational complex analysis*, Wiley, New York.

E. Hille, 1962, *Analytic function theory*. II, Ginn, Boston.

M. Hirsch, 1963, *A proof of the non-retractability of a cell onto its boundary*, Proc. Amer. Math. Soc. **14**, pp. 364–365.

M. Hirsch and S. Smale, 1979, *On algorithms for solving $f(x) = 0$*, Comm. Pure Appl. Math. **32**, pp. 281–312.

D. Hoffman and R. Osserman, (to appear) *The geometry of the generalized Gauss map*.

H. Hotelling, 1939, *Tubes and spheres in n-space and a class of statistical problems*, Amer. J. Math. **61**, pp. 440–460.

W. Hurewicz, 1958, *Lectures on ordinary differential equations*, MIT Press, Cambridge, Mass.

J. Jenkins, 1965, *Univalent functions and conformal mapping*, Springer, New York.

R. Kellog, T. Li and J. Yorke, 1976, *A constructive proof of the Brouwer fixed point theorem and computational results*, SIAM J. Numer. Anal. **13**, pp. 473–483.

I. Lakatos, 1976, *Proofs and refutations*, Cambridge Univ. Press, Cambridge, England.

S. Lang, 1965, *Algebra*, Addison-Wesley, Reading, Mass.

M. Marden, 1966, *Geometry of polynomials*, Math. Surveys, no. 3, Amer Math. Soc., Providence, R. I.

A. Ostrowski, 1973, *Solutions of equations in Euclidean and Banach spaces*, Academic Press, New York.

Jean-Claude Pont, 1974, *La topologie algébrique, des origine à Poincaré*, Presses Universitaires de France, Paris.

L. Santalo, 1976, *Integral geometry and geometric probability*, Addison-Wesley, Reading, Mass.

H. Scarf, 1973, *The computation of economic equilibria* (in collaboration with T. Hansen) Yale Univ. Press, New Haven, Conn.

G. Schober, (to appear), *Coefficient estimates for inverses of Schlicht functions*, Proc. of NATO-LMS Conference on Aspects of Contemporary Complex Analysis, Academic Press, New York.

S. Smale, 1974, *Sufficient conditions for an optimum*, Dynamical Systems–Warwick 1974, Lecture Notes in Math., vol. 468, Springer-Verlag, Berlin and New York.

______, 1976, *A convergent process of price adjustment and global Newton methods*, J. Math. Econom., **3**, pp. 107–120.

______, (to appear), *Global analysis and economics*, Handbook of Mathematical Economics (Arrow and Intrilligator, eds.), North-Holland, Amsterdam.

D. Smith, 1953, *History of mathematics*, vol. II, Dover, New York.

D. Struik, 1969, *A source book in mathematics*, 1200–800, Harvard Univ. Press, Cambridge, Mass.

J. Traub, ed., 1976, *Analytic computational complexity*, Academic Press, New York.

B. Van der Waerden, 1953, *Modern algebra*, Vol. I, Ungar, New York.

______, 1950, *Modern algebra*, Vol. II, Ungar, New York.

Y.-H. Wan, 1975, *On local Pareto optima*, J. Math. Econom., **2**, pp. 35–42.

H. Weyl, 1924, *Randbemerkingen zu Hauptproblem der Mathematik*, Math. Z. **20**, pp. 131–150.

______, 1939, *On the volume of tubes*, Amer. J. Math. **61**, pp. 461–472.

H. Whitney, 1972, *Complex analytic varieties*, Addison-Wesley, Reading, Mass.

J. H. Wilkinson, 1963, *Rounding errors in algebraic processes*, Prentice-Hall, Englewood Cliffs, N. J.

P. Wolfe, 1980, *The ellipsoid algorithm* (letter to the editor), Science, **208**, pp. 240–242.

DEPARTMENT OF MATHEMATICS, UNIVERSITY OF CALIFORNIA, BERKELEY, CALIFORNIA 94720

Section 7
ERGODIC THEORY AND RECURRENCE

Proceedings of Symposia in Pure Mathematics
Volume 39 (1983), Part 2

Poincaré Recurrence and Number Theory

HARRY FURSTENBERG

Introduction. Poincaré is largely responsible for the transformation of celestial mechanics from the study of individual solutions of differential equations to the global analysis of phase space. A system of differential equations such as those which embody the laws of Newtonian mechanics generates a one-parameter group of transformation of the manifold that represents the set of states of a dynamical system. The evolution of the dynamical system in time corresponds to a particular solution of the system of differential equations; it also corresponds to an orbit of the group of transformations acting on a single state. The efforts of the classical analysts in celestial mechanics had been directed to extracting by analytical means as much information as possible about the individual solutions to the system of differential equations. Poincaré's work gave impetus to a global approach which studies the totality of solutions and shifts attention to the transformation group of phase space.

Two of Poincaré's achievements which can be traced to this point of view are his theory of periodic solutions and his recurrence theorem. In the former of these, the topological nature of the phase space plays a key role; Poincaré showed how an essentially topological analysis of orbits, under certain conditions, can be used to establish the existence of a periodic solution curve. In his recurrence theorem, Poincaré demonstrated how measure-theoretic ideas, particularly the idea of a measure-preserving group of transformations, lead to the existence of numerous "approximately periodic"—or, "recurrent" —solution curves. The impact of these ideas is felt today in the establishment of two new disciplines: topological dynamics and ergodic theory. In topological dynamics one abstracts from the classical setup the topological space representing the totality of states of a dynamical system, together with the group of homeomorphisms corresponding to the evolution of the system from its position at time 0 to its position at time t. For ergodic theory, the phase space is replaced by an abstract measure space and the "dynamics" come from the action of a group of measure-preserving transformations of the measure space.

Initially these abstract settings for dynamical theory were contemplated in order to shed light on classical dynamics by focusing on those aspects of dynamical systems that were pertinent to the phenomena being studied. At the same time, however, the scope of dynamics was considerably broadened by the generality of the new theory, and the groundwork was laid for

Reprinted from Bulletin Amer. Math. Soc. (N.S.) 5 (1981), 211–234.
1980 *Mathematics Subject Classification.* Primary 05A05, 28D05, 58F11; Secondary 10A99.

applying similar ideas to areas that, on the face of it, are quite unrelated to dynamics.

We shall be interested in two particular dynamical theorems and their ramifications when taken in the broadest possible context. One of these is the Poincaré recurrence theorem alluded to previously; the other is a topological analogue due to Birkhoff. Choosing a nonconventional model of a dynamical system rather than a classical model, we will obtain results of interest in number theory. For example we will see that van der Waerden's theorem on arithmetic progressions is a consequence of an appropriate generalization of Birkhoff's recurrence theorem. A more recent result is that of Szemerédi stating that a subset of the integers having positive upper density contains arbitrarily long arithmetic progressions. This will be seen to relate in a natural way to an extension of Poincaré's recurrence theorem.

For a comprehensive treatment of the results described here the reader is referred to [3]–[5].

1. Dynamical systems and measure-preserving systems. A dynamical system evolving in time is described by a one-parameter group $\{T_t, -\infty < t < \infty\}$, $T_{t+s} = T_t \circ T_s$, acting on a space X. Since we are interested primarily in asymptotic behavior as $t \to \infty$, the dynamical aspects are usually already reflected in the action of the subgroup $\{T_n, n \in \mathbf{Z}\}$, $\mathbf{Z} = $ integers. For our purposes, then, a *dynamical system* will consist of a space X on which a one-one transformation T acts, thereby generating a group $\{T^n, n \in \mathbf{Z}\}$ of transformations.

The space X will be either a topological space (in our discussion, a metric space) in which case we require T to be a homeomorphism, or a measure space in which case T will be required to be measure preserving. The central phenomenon to be studied is that of *recurrence*. For this phenomenon to occur one must assume some kind of boundedness of X. When X is a topological space, the appropriate assumption is that X is compact. When X is a measure space, the boundedness of X is expressed by requiring X to have finite measure.

Let us then define formally a *dynamical system* as a pair (X, T), X being a compact metric space, and T a homeomorphism of X. A *measure-preserving system* will be a quadruple $(X, \mathcal{B}, \mu, T)$ where X is an abstract space, $\mathcal{B}$ is a σ-algebra of subsets of X, μ is a probability measure in $\mathcal{B}$, and T is a measure-preserving transformation of $(X, \mathcal{B}, \mu)$. These last two conditions mean that $\mu(A) \geqslant 0$ for $A \in \mathcal{B}$, with $\mu(X) = 1$, and for $A \in \mathcal{B}$, $T^{-1}A \in \mathcal{B}$ with $\mu(T^{-1}A) = \mu(A)$.

In the topological context recurrence is a property of individual points of the space X. We shall be studying several versions of this notion, the most basic of which we shall simply call *recurrence*. We say a point $x \in X$ is a *recurrent* point of (X, T) if for some sequence $n_k \to \infty$, $T^{n_k}x \to x$. Equivalently we could say that a point x is recurrent if for each neighborhood V of x, $T^n x \in V$ for some $n > 0$.

Closely related to this is the notion of recurrence implicit in Poincaré's recurrence theorem.

THEOREM 1.1. *Let $(X, \mathcal{B}, \mu, T)$ be a measure-preserving system, and let $V \in \mathcal{B}$ with $\mu(V) > 0$. There exists some point $x \in V$ with $T^n x \in V$ for some $n > 0$.*

The proof is extremely simple. Assume no point $x \in V$ returned to V. Then $T^{-n}V \cap V = \varnothing$ for all $n > 0$, and so $T^{-n}V \cap T^{-m}V = \varnothing$ whenever $n \neq m$. But the sets $T^{-m}V$ all have the same measure $\mu(V) > 0$, and they cannot be disjoint since $\mu(\bigcup_{n=1}^{\infty} T^{-n}V) \leqslant \mu(X) = 1$.

Actually Theorem 1.1 implies the stronger statement that almost every point (excepting a set of measure 0) $x \in V$ returns sometimes to V. For this set of points is, in any case, measurable (i.e., in $\mathcal{B}$) and we can consider the subset $V' \subset V$ of points that never return to V. *A fortiori* they never return to V' and so, by Theorem 1.1, $\mu(V') = 0$.

Now take X to be a separable metric space as well as a measure space. Cover X by countably many balls of radius $\varepsilon/2$, and apply the foregoing to each ball. We conclude that almost every point X returns to within ε of itself. Since $\varepsilon > 0$ is arbitrary, we conclude that *almost every point of X is recurrent*.

It can be shown that any (compact) dynamical system (X, T) possesses some measure on Borel sets which is preserved by T. Poincaré's theorem then implies that *for any dynamical system (X, T) there exist points $x \in X$ which are recurrent*. It is Birkhoff who initiated the study of dynamical systems in the context of metric spaces [2], and he gave a purely topological proof of this statement. Let us present a streamlined proof of this result which we call the *Birkhoff recurrence theorem*.

THEOREM 1.2. *If X is a compact metric space, T a continuous map of X to itself, then there exists some point $x \in X$ with $T^{n_k}x \to x$ for some sequence $n_k \to \infty$.*

PROOF. Consider the family of closed sets $Y \subset X$ satisfying $TY \subset Y$. It is easy to see that Zorn's lemma applies to this family (on account of compactness) so that there exists a minimal closed invariant set. Let Y_0 be such a set and let $y \in Y_0$. Then $Y_1 = \overline{\{T^n y, n = 1, 2, 3, \ldots\}}$ is again a closed invariant set contained in Y_0. By minimality $Y_1 = Y_0$; hence $y \in \overline{\{T^n y, n = 1, 2, 3, \ldots\}}$. Thus y is a recurrent point.

Actually Birkhoff proved more than this. In fact Birkhoff uses the term recurrence for a stronger version of this notion that we shall call "uniform recurrence". Birkhoff's argument shows that every compact dynamical system possesses uniformly recurrent points.

DEFINITION. A point $x \in X$ is *uniformly recurrent* for the dynamical system (X, T) if, for any neighborhood V of x, there is a length $L < \infty$, such that in any interval (a, b) with $b - a > L$ there is an integer $n \in (a, b)$ with $T^n x \in V$.

For a uniformly recurrent point, any sufficiently long segment of its orbit comes arbitrarily close to the point. There is a close connection between uniform recurrence and the notion of a minimal dynamical system.

DEFINITION. A system (X, T) is *minimal* if there does not exist a nonempty closed T-invariant subset $Y \subsetneq X$.

Equivalently, a system (X, T) is minimal if every (forward) orbit $\{T^n x, n \geqslant 0\}$, $x \in X$, is dense in X. For if (X, T) is minimal, the closed T-invariant set $\overline{\{T^n x, n \geqslant 0\}}$ must coincide with X. Conversely, if every forward orbit is dense, so is every T-invariant set, and the system is minimal.

THEOREM 1.3. *If (X, T) is minimal, then each $x \in X$ is uniformly recurrent for (X, T). Conversely, if a point $x \in X$ is uniformly recurrent, then x belongs to a closed T-invariant set $Y \subset X$ for which the subsystem (Y, T) is minimal.*

According to the second part of the theorem, if every point of X is uniformly recurrent, then X is the union of *minimal sets*, i.e., minimal closed T-invariant subsets of X. Note that by definition, two different minimal sets are necessarily disjoint.

PROOF OF THE THEOREM. Assume (X, T) is minimal and let V be any open set in X. The set $\bigcup_{n=0}^{\infty} T^{-n} V$ is an open set whose complement is a T-invariant closed set. Since it is nonempty, we must have $\bigcup_{n=0}^{\infty} T^{-n} V = X$. Since X is compact, we will already have $\bigcup_{n=0}^{L} T^{-n} V = X$ for some L. Hence for any $x \in X$ and any n, one of the points $T^n x, T^{n+1} x, \ldots, T^{n+L} x$ belongs to V. This shows that x is uniformly recurrent.

Conversely, suppose that $x \in X$ is uniformly recurrent. We wish to show that x is contained in a minimal set of X. Suppose Z is a T-invariant closed subset of the forward orbit closure of x, $Y = \overline{\{T^n x, n \geqslant 0\}}$. If $x \in Z$ then Z coincides with Y. If $x \notin Z$, then $x \in V = X \setminus Z$. By uniform recurrence, for any n, one of the points $T^n x, T^{n+1} x, \ldots, T^{n+L} x$ belongs to V. For any $z \in Z$ there are points $T^n x$ arbitrarily close to z. It follows that one of the points $z, Tz, \ldots, T^L z$ belongs to V.

But this is a contradiction since these points are all in Z and $Z \cap V = \varnothing$. It follows that $Z = Y$ and so Y is minimal. This completes the proof.

The proof of the theorem shows that in the case of a minimal system, if V is an open set, there is an L so that for each $x \in X$, one of the points $x, Tx, \ldots, T^L x$ meets V. It follows that for any $\varepsilon > 0$ there is some M so that each orbit segment $x, Tx, \ldots, T^M x$ comes within ε of each point of the space. Using now the second assertion of the theorem, this applies in particular to the orbit closure of a uniformly recurrent point. This gives us the following characterization of uniform recurrence.

COROLLARY. *A point is uniformly recurrent iff every sufficiently long segment of its orbit comes arbitrarily close to every point in its orbit.*

The proof of Theorem 1.2 shows that any system (X, T) with X compact possesses a minimal subsystem. On the basis of the foregoing theorems we now conclude:

THEOREM 1.4. *If X is a compact metric space, T a homeomorphism of X, there exists a point $x \in X$ which is uniformly recurrent for (X, T).*

2. Symbolic systems. In the course of our discussion we shall extend and refine the recurrence theorems of the preceding section. To obtain results of a number-theoretical nature we shall apply these theorems to a particular kind of nonconventional dynamical system—*a symbolic system.* Let $\Lambda = \{a, b, c, \dots\}$ be a finite set and form the space

$$\Omega = \Lambda^{\mathbf{Z}} \text{ of all sequences with entries from } \Lambda:$$

$$x \in \Omega \iff x = \{\dots, x(-2), x(-1), x(0), x(1), x(2), \dots\}.$$

Ω can be made into a compact metric space, taking as metric, for example,

$$(2.1) \qquad d(x, x') = \inf\left\{\frac{1}{k+1} \middle| x(i) = x'(i) \text{ for } |i| < k\right\}.$$

Here $d(x, x') = 1$ if $x(0) \neq x'(0)$, and $d(x, x') <$ otherwise. We define the *shift* homeomorphism $T: \Omega \to \Omega$ by $Tx(n) = x(n + 1)$. If X is any closed T-invariant subset of Ω we call (X, T) a *symbolic dynamical system.*

Given a particular function $\xi: \mathbf{Z} \to \Lambda$ so that ξ is a point in Ω, we form the smallest shift invariant closed set $X \subset \Omega$ containing the point ξ. The idea which we shall use repeatedly is that the dynamic properties of (X, T) reflect certain aspects of the behavior of ξ. More specifically, recurrence properties for X will imply the existence of certain patterns in the function $\xi(n)$.

To illustrate the idea, consider the implication of the following extension of Theorem 1.2 which will be proved in §6.

THEOREM 2.1. *Let X be a compact metric space, and let $T: X \to X$ be a continuous map. Then for any integer $l \geqslant 1$, there exists a point $x \in X$ and a sequence $n_k \to \infty$ with $T^{n_k}x \to x$, $T^{2n_k}x \to x$, $\dots$, $T^{ln_k}x \to x$.*

Let (X, T) be any symbolic dynamical system, $X \subset \Omega = \Lambda^{\mathbf{Z}}$. According to Theorem 2.1, there exists $x \in X$ and an $n > 0$ with the points $x, T^n x, T^{2n} x, \dots, T^{ln} x$ within distance < 1 of one another. Now for two points of Ω, $d(x, x') < 1$ implies $x(0) = x'(0)$. It follows that $x(0) = T^n x(0) = T^{2n} x(0) = \cdots = T^{ln} x(0)$. Bearing in mind that $T^k x(0) = x(k)$ we have

PROPOSITION 2.2. *Let Λ be a finite set. If X is any shift invariant closed subset of $\Omega = \Lambda^{\mathbf{Z}}$ and $l \geqslant 1$, there exists a point $x \in X$ and an $n \geqslant 1$ with $x(0) = x(n) = x(2n) = \cdots = x(ln)$.*

If we take $\Lambda = \{1, 2, \dots, r\}$, then a point $\xi \in \Lambda^{\mathbf{Z}}$ corresponds to a partition of $\mathbf{Z}$ into r sets, $\mathbf{Z} = \cup\, C_i$, where $C_i = \{n: \xi(n) = i\}$. Let $X \subset \Omega$ be the closure of the set of all translates of ξ: $X = \overline{\{T^n \xi, n \in \mathbf{Z}\}}$. Apply Proposition 2.2 to X, and we find there is an $x \in X$ with $x(0) = x(n) = \cdots = (ln)$. Since $X = \overline{\{T^n \xi, n \in \mathbf{Z}\}}$, in view of the definition of the metric on Ω, we can find m so that $T^n \xi$ and x agree on the interval $(-ln, ln)$. It then follows that $\xi(m) = \xi(m + n) = \xi(m + 2n) = \cdots = \xi(m + ln)$. Say the common value is j. Then C_j contains the arithmetic progression $m, m + n, m + 2n, \dots, m + ln$. We have thus established van der Waerden's theorem on arithmetic progressions.

THEOREM 2.3. *Let $\mathbf{Z} = \cup_{i=1}^{r} C_i$ be a partition of the integers into r subsets. Then one of the sets C_j contains an arithmetic progression of length $l + 1$.*

Our next example shows how measure-theoretic ideas can be used. The following refinement of Poincaré's recurrence theorem will be proved in the next section.

THEOREM 2.4. *Let $(X, \mathscr{B}, \mu, T)$ be a measure-preserving system, and let $A \in \mathscr{B}$ with $\mu(A) > 0$. Then there exists an integer of the form $n = m^2$ such that $\mu(A \cap T^{-n}A) > 0$.*

Let us now return to our symbolic dynamical system, taking this time $\Lambda = \{0, 1\}$. Suppose that $X \subset \Lambda^{\mathbf{Z}}$ is T-invariant and closed and let $A_1 = \{x \in X; x(0) = 1\}$. Assume now that we can find a T-invariant Borel measure μ on X for which $\mu(A_1) > 0$. We can then form the measure-preserving system $(X, \mathscr{B}, \mu, T)$, with $\mathscr{B}$ the Borel sets of X and T the shift. Applying Theorem 2.4 to the system with $A = A_1$, we conclude that there exists some point $x \in A_1 \cap T^{-m^2}A_1$. Then $x(0) = 1$ and $x(m^2) = T^{m^2}x(0) = 1$.

Now let $S \subset \mathbf{Z}$ be a subset of *positive upper density*. For the present purposes this will be taken to mean that there is a sequence of intervals $[a_k, b_k] \subset \mathbf{Z}$ with $b_k - a_k \to \infty$ and

$$\frac{|S \cap [a_k, b_k]|}{b_k - a_k + 1} > \delta$$

for some $\delta > 0$ and all k. (For any finite set Q, we denote by $|Q|$ the number of elements in Q.) Now set $\xi(n) = 1$ for $n \in S$ and $\xi(n) = 0$. Again let $X = \overline{\{T^n\xi, n \in \mathbf{Z}\}}$. We claim that we can indeed find some T-invariant measure on X with $\mu(A_1) > 0$.

LEMMA 2.5. *Let $X \subset \{0, 1\}^{\mathbf{Z}}$ contain a point ξ satisfying $\xi(n) = 1$ for a set of n of positive upper density. Then there exists a T-invariant probability measure μ on the Borel sets of X satisfying $\mu(A_1) > 0$.*

PROOF. A Borel measure μ on the compact metric space X determines a linear functional on $C(X)$, the continuous functions in X, by

$$L(f) = \int f \, d\mu.$$

If μ is a probability measure then L satisfies
   (i) $L(f) \geqslant 0$ for $f \geqslant 0$,
   (ii) $L(1) = 1$.
Conversely, given L satisfying these conditions there corresponds to L a probability measure on Borel sets of X. If μ is a T-invariant Borel measure then the corresponding functional satisfies
   (iii) $L(f \circ T) = L(f)$.
Conversely, if L satisfies (i)–(iii) it defines a T-invariant measure μ.

To obtain a linear functional on $C(X)$ satisfying (i)–(iii) we take the point $\xi \in X$ and set

$$L_k(f) = \frac{1}{b_k - a_k + 1} \sum_{n=a_k}^{b_k} f(T^n\xi).$$

Let $\{f_1, f_2, \ldots, f_n, \ldots\}$ be a countable dense set of functions in $C(X)$. Since $\{L_k(f_l), k = 1, 2, 3, \ldots\}$ is bounded for each l, we can choose a subsequence that converges. By a diagonal procedure we can choose a subsequence $\{k_n\}$ so that

$$\lim_{n \to \infty} L_{k_n}(f_l)$$

exists for each l. Finally, since $\{f_l\}$ is dense in $C(X)$, we find that the corresponding limit exists for each $f \in C(X)$. Thus replacing $\{a_k, b_k\}$ by a subsequence we may assume that

$$L(f) = \lim_{k \to \infty} \frac{1}{b_k - a_k + 1} \sum_{n=a_k}^{b_k} f(T^n \xi)$$

exists for all $f \in C(X)$. It is readily verified that L satisfies (i)–(iii).

We next put to use the fact that $\xi(n) = 1$ on a set of integers having positive upper density. The function

$$\varphi(A) = x(0)$$

is a continuous function on X. We have

$$L(\varphi) = \lim_{k \to \infty} \frac{1}{b_k - a_k + 1} \sum_{n=a_k}^{b_k} T^n \xi(0)$$

$$= \lim \frac{|S \cap [a_k, b_k]|}{b_k - a_k + 1} \geqslant \delta > 0.$$

Now if L corresponds to the measure μ on X, then

$$\mu(A_1) = \int \varphi \, d\mu = L(\varphi) > 0.$$

This proves the lemma.

The converse of the lemma is also true. In fact, by the ergodic theorem,

$$\lim_{N \to \infty} \frac{1}{N + 1} \sum \varphi(T^n \eta) = g(\eta)$$

exists for almost all $\eta \in X$ with respect to μ, and $\int g(\eta) d\mu(\eta) = \int \varphi \, d\mu = \mu(A_1)$. So $g(\eta) > 0$ for some $\eta \in X$ and so $\eta(n) = 1$ on a set of positive density, where now the limit

$$\lim_{N \to \infty} \frac{|S \cap [0, N]|}{N + 1}$$

exists and is positive.

Note the following corollary of the proof of Lemma 2.5.

COROLLARY. *If X is a compact metric space and T is a continuous map of X to itself, there exists some probability measure μ on Borel sets of X which is invariant with respect to T.*

Combining Lemma 2.5 and Theorem 2.4 we obtain the following.

THEOREM 2.6. *Let S be a subset of $\mathbf{Z}$ having positive upper density. Then there exist $s, t \in S$ with $s - t = m^2$, m an integer.*

PROOF. As before let $\xi(n) = 1$ for $n \in S$, $\xi(n) = 0$ otherwise, and let $X = \overline{\{T^n, n \in \mathbf{Z}\}} \subset \{0, 1\}^{\mathbf{Z}}$. Let $A_1 = \{x \in X : x(0) = 1\}$ and let μ be a

T-invariant measure on X with $\mu(A_1) > 0$. Apply Theorem 2.4 to $(X, \mathscr{B}, \mu, T)$ and the set $A_1 \subset X$. If $\mu(A_1 \cap T^{-n}A_1) > 0$ with $n = m^2$, then, in particular, $A_1 \cap T^{-n}A_1 \neq \varnothing$, and so there exists $\eta \in A_1$ with $T^{m^2}\eta \in A_1$. In other words there is a point $\eta \in X$ with $\eta(0) = \eta(m^2) = 1$. Now η is in the orbit closure of ξ, and we can find a neighborhood of η so small that any point in that neighborhood will have the same coordinates as η in the interval between $-m^2$ and m. It follows that for some s,

$$T^s\xi(0) = T^s\xi(m^2) = 1$$

or

$$\xi(s) = \xi(s + m^2) = 1.$$

Hence $s \in S$ and $s + m^2 \in S$. This concludes the proof.

The next two theorems describe in a more general manner how recurrence results are used to prove the existence of patterns in subsets of integers. These theorems extend Theorems 2.3 and 2.6 respectively, and the proofs are entirely analogous. In the formulation of these two theorems the term *configuration* will be synonomous with a finite subset of $\mathbf{Z}$, except that two sets that are translates of one another are said to define the same configuration.

THEOREM 2.7. *Let Q be a family of configurations having the property that for any compact metric dynamical system (X, T) and any $\varepsilon > 0$, there exists a point $x_0 \in X$ and a configuration $(n_1, n_2, \ldots, n_l) \in Q$ such that $T^{n_1}x_0, T^{n_2}x_0, \ldots, T^{n_l}x_0$ are all within ε of one another. Then if $\mathbf{Z} = C_1 \cup C_2 \cup \cdots \cup C_r$ is any partition of $\mathbf{Z}$ into finitely many sets, one of the subsets C_j contains some configuration in Q.*

In particular the hypothesis is fulfilled when Q consists of arithmetic progressions of a given length, according to Theorem 2.1. Theorem 2.3 is thus a special case of the foregoing. It should be remarked that a slight modification of the argument in proving Theorem 2.7 shows that the same result is true for partitions of the natural numbers $\mathbf{N} = C_1 \cup C_2 \cup \cdots \cup C_r$.

THEOREM 2.8. *Let Q be a family of configurations having the property that for any measure-preserving system $(X, \mathscr{B}, \mu, T)$ and any set $A \in \mathscr{B}$ with $\mu(A) > 0$, there exists a subset $A' \in \mathscr{B}$ with $\mu(A') > 0$ and a configuration $(n_1, n_2, \ldots, n_l) \in Q$ such that $T^{n_1}A' \subset A, \ldots, T^{n_l}A' \subset A$. Then if S is any subset of $\mathbf{Z}$ of positive upper density, S necessarily contains some configuration in Q.*

Theorem 2.6 is the special case of this where $Q = \{(0, m^2), m \in \mathbf{Z}\}$. In §7 we will discuss the applicability of Theorem 2.8 to the case $Q = $ arithmetic progression of length l.

We remark that if Q is a family of configurations satisfying the condition of Theorem 2.8, it also satisfies the condition of Theorem 2.7. For by the corollary to Lemma 2.5, if (X, T) is a compact metric dynamical system, we can endow X with a T-invariant probability measure μ and thereby obtain a measure-preserving system. For any $\varepsilon > 0$ we can find a set of diameter ε

having positive measure. Let A be such a set and let $(n_1, n_2, \ldots, n_l)$ be a configuration in Q. If A' satisfies $T^{n_1}A' \subset A$, $T^{n_2}A' \subset A$, $\ldots$, $T^{n_l}A' \subset A$, then for any $x_0 \in A'$, the points $T^{n_1}x_0$, $T^{n_2}x_0$, $\ldots$, $T^{n_l}x_0$ are within ε of one another.

We can also see that the conclusion of Theorem 2.8 is stronger than that of Theorem 2.7. For if $\mathbf{Z} = \bigcup C_i$ is a finite partition, some C_j has positive upper density. So any configuration that occurs in any set of positive upper density also occurs in some subset of any finite partition of $\mathbf{Z}$.

3. Sets of recurrence. The simplest nontrivial configurations that one can look for in subsets of $\mathbf{Z}$ are two-point configurations. These may be assumed to have the form $(0, n)$. In this section we shall inquire which families Q of configurations $\{(0, n)\}$ satisfy the condition of Theorem 2.8. In other words, for which sets $R \subset \mathbf{N}$ can we assert that whenever $(X, \mathcal{B}, \mu, T)$ is a measure-preserving system, A a subset in $\mathcal{B}$ with $\mu(A) > 0$, then there exists $A' \subset A$, $\mu(A') > 0$, with $T^r A' \subset A$ for some $r \in R$? A set R with this property will be called a *set of recurrence*. Our considerations in this section are measure-theoretic.

We shall say that a set $R \subset \mathbf{N}$ is an *infinite difference set* if R consists of all differences $s_j - s_i$, $i < j$ where $\{s_1 < s_2 < \cdots < s_n < \cdots\}$ is some sequence in $\mathbf{N}$. Our first general result is the following.

THEOREM 3.1. *Every infinite difference set is a set of recurrence.*

The proof is identical to the proof of Poincaré's recurrence theorem. Let $R = \{s_j - s_i, i < j\}$ and let $(X, \mathcal{B}, \mu, T)$ be a measure-preserving system, $A \in \mathcal{B}$ with $\mu(A) > 0$. If $\mu(T^{-(s_j - s_i)}A \cap A) = 0$ for all $i < j$. Then

$$\mu(T^{-s_j}A \cap T^{-s_i}A) = 0,$$

and the sets $T^{-s_i}A$ are essentially disjoint sets all having the same positive measure $\mu(A)$. So we must have $\mu(T^{-(s_j - s_i)}A \cap A) > 0$ for some $i < j$ and we can take $A' = T^{-(s_j - s_i)}A \cap A$.

Choosing the sequence $\{s_n\}$ rapidly increasing shows that sets of recurrence can be extremely sparse. On the other hand a set of recurrence cannot be too sparse.

THEOREM 3.2. *If $R = \{r_1 < r_2 < \cdots < r_n < \cdots\}$ is lacunary, i.e., if $r_{n+1}/r_n > q > 1$, then R is not a set of recurrence.*

Before proving Theorem 3.2 we prove the following general lemma.

LEMMA 3.3. *If R_1, $R_2 \subset \mathbf{N}$ and neither R_1 and R_2 is a set of recurrence, then $R = R_1 \cup R_2$ is not a set of recurrence.*

PROOF. Since R_1 and R_2 are not sets of recurrence we can find systems $(X_i, \mathcal{B}_i, \mu_i, T_i)$ and sets $A_i \in \mathcal{B}_i$ with $\mu_i(A_i) > 0$ satisfying $\mu_i(A_i \cap T_i^{-r}A_i) = 0$ for $r \in R_i$, $i = 1, 2$. Form $X = X_1 \times X_2$, $\mathcal{B} = \mathcal{B}_1 \times \mathcal{B}_2$, $\mu = \mu_1 \times \mu_2$ and let $T = T_1 \times T_2$. It is now clear that

$$\mu((A_1 \times A_2) \cap T^{-r}(A_1 \times A_2)) = 0$$

for $r \in R$. Thus R is not a set of recurrence.

On the basis of this lemma it suffices to establish Theorem 3.2 for arbitrarily large q.

LEMMA 3.4. *If $R = \{r_1 < r_2 < \cdots < r_n < \cdots \}$ satisfies $r_{n+1}/r_n \geqslant q > 4$, then R is not a set of recurrence.*

PROOF. We shall determine a number α so that the measure-preserving system consisting of translation by α in the group $\mathbf{R}/\mathbf{Z}$ of reals modulo 1 provides a counterexample to the recurrence of R. Consider the sets $\Lambda_n \subset \mathbf{R}/\mathbf{Z}$ where $\Lambda_n = \{t: r_n t \in \Delta\}$ and Δ consists of numbers whose fractional part is between $1/3$ and $2/3$. Λ_n consists of intervals of length $(3r_n)^{-1}$ repeated periodically with period r_n^{-1}. Since $r_{n+1} > 4r_n$ it follows that each Λ_n interval contains some Λ_{n+1} interval. Hence $\cap \Lambda_n \neq \varnothing$. Take $\alpha \in \cap \Lambda_n$. We let $X = \mathbf{R}/\mathbf{Z}$ with Lebesgue measure and let T be a translation by α. Finally let A be any interval of length $< 1/6$. $T^{r_n}A = A + r_n\alpha$, and since the fractional part of the $r_n\alpha$ is between $1/3$ and $2/3$, $(A + r_n\alpha) \cap A = \varnothing$. So $T^{-r_n}A \cap A = \varnothing$ and R is not a set of recurrence. This completes the proof.

The next theorem describes sets of recurrence of polynomial growth. Theorem 2.4 is a special case of this theorem.

THEOREM 3.5. *Let $p(t)$ be a polynomial with integer coefficients and with $p(0) = 0$. The set $\{p(n), n = 1, 2, 3, \ldots \}$ is a set of recurrence.*

PROOF. We consider the Hilbert space $L^2(X, \mathscr{B}, \mu)$ and the unitary operator defined on it by

$$(Uf)(x) = f(Tx).$$

We use $\langle f, g \rangle$ to denote the inner product in $L^2(X, \mathscr{B}, \mu)$. According to the spectral theorem, if $f \in L^2(X, \mathscr{B}, \mu)$, there is a nonnegative measure ω on $[0, 1]$ such that

$$\langle U^n f, f \rangle = \int_0^1 e^{2\pi i n \theta} d\omega(\theta).$$

Now let $A \in \mathscr{B}$ with $\mu(A) > 0$ and let $f = 1_A$ denote the characteristic function of the set A. Suppose that we had $\mu(T^{-n}A \cap A) = 0$ whenever $n = p(m)$, $m \in \mathbf{N}$. Then

$$(3.1) \qquad \int_0^1 e^{2\pi i p(m)\theta} d\omega(\theta) = 0$$

for every $m \in \mathbf{N}$. We now use Weyl's theorem on equidistribution [9] which implies that

$$(3.2) \qquad \frac{1}{N} \sum_{m=1}^{N} e^{2\pi i p(m)\theta} \to 0$$

as $N \to \infty$ for every irrational θ. Write $\omega = \omega_i + \omega_r$, where ω_r is the atomic part of ω supported on rational points of $[0, 1]$, and ω_i is what is left of ω. Then (3.1) and (3.2) imply

$$(3.3) \qquad \int_0^1 \lim_{N \to \infty} \frac{1}{N} \sum_{m=1}^{N} e^{2\pi i p(m)\theta} d\omega_r(\theta) = 0,$$

or, ω_r being atomic,

$$(3.4) \qquad \sum_{\theta \in \mathbf{Q}/\mathbf{Z}} \left\{ \lim_{N \to \infty} \frac{1}{N} \sum_{m=1}^{n} e^{2\pi i p(m)\theta} \right\} \omega_r\{\theta\} = 0.$$

Now, since the polynomial $p(t)$ has no constant term, $a|p(am)$ for integers a, m. Since (3.1) is equally valid if m is replaced by a multiple of m, we can rewrite (3.4) replacing $p(m)$ by $p(k!m)$, k an arbitrary number. However,

$$e^{2\pi i p(k!m)\theta} = 1$$

for each rational θ whose denominator is $\leqslant k$. Hence

$$\sum_{\theta} \omega_r\{\theta\} = \lim_{k \to \infty} \sum_{\theta \in \mathbf{Q}/\mathbf{Z}} \left\{ \lim_{N \to \infty} \frac{1}{N} \sum_{m=1}^{N} e^{2\pi i p(k!m)\theta} \right\} \omega_r\{\theta\} = 0$$

by (3.4). Since ω_r is nonnegative, $\omega_r = 0$, and, in particular $\omega_r\{0\} = 0$. But then

$$(3.5)$$

$$\lim_{N \to \infty} \frac{1}{N} \sum_{n=1}^{N} \langle U^n f, f \rangle = \int_0^1 \lim_{N \to \infty} \frac{1}{N} \sum_{n=1}^{N} e^{2\pi i n\theta} \, d\omega(\theta) = \omega\{0\} = 0.$$

On the other hand, by the mean ergodic theorem,

$$\lim_{N \to \infty} \frac{1}{N} \sum_{n=1}^{N} U^n f = \bar{f}$$

exists in $L^2(X, \mathcal{B}, \mu)$, $\bar{f}$ is U-invariant and $\int \bar{f} \, d\mu = \int f \, d\mu$. By (3.5), $\langle \bar{f}, f \rangle = 0$. Hence $\langle \bar{f}, U^n f \rangle = \langle U^n \bar{f}, U^n f \rangle = \langle \bar{f}, f \rangle = 0$, and by averaging again over n, $\langle \bar{f}, \bar{f} \rangle = 0$, or $\bar{f} = 0$. But then

$$\mu(A) = \int 1_A \, d\mu = \int \bar{f} \, d\mu = 0,$$

a contradiction. This proves the theorem.

We may now combine this result with Theorem 2.8. Let Q be the family of configurations $\{(0, p(m))\}$, $p(t)$ being a polynomial with integer coefficients and with no constant term. We obtain the following number-theoretic result.

THEOREM 3.6. *Let S be a subset of $\mathbf{Z}$ of positive upper density. Let $p(t)$ be a polynomial with integer coefficients and with no constant term. Then one can solve the equation*

$$x - y = p(z)$$

in integers x, y, z with $x, y \in S$ and $z \geqslant 1$.

This result has been obtained independently by Conze and Sárközy.

4. Proximality and a lemma of Schur. We now turn to a more elaborate study of recurrence in the topological framework. In this section the notion of uniform recurrence (§1) plays a central role. According to Theorem 1.4, every compact dynamical system possesses uniformly recurrent points. In a sense to be made explicit, every orbit of a compact dynamical system keeps coming closer and closer to the orbit of some uniformly recurrent point. To form a picture of what happens, imagine that X contains a unique minimal closed T-invariant subset which reduces to a point $\{y\}$. The point y is a fixed point and certainly uniformly recurrent. Since every orbit closure in X contains some minimal subset and in our case there is a unique such set, we conclude that every orbit comes arbitrarily close to y. Once the orbit is close to y, by continuity, it will spend a certain interval within a small neighborhood of y. In particular, for each $\varepsilon > 0$, there will be arbitrarily long intervals of n for which $d(T^n x, y) < \varepsilon$. This is a special case of what is called *proximality*.

DEFINITION. Two points $x, y \in X$ are *proximal* for a dynamical system (X, T) if

$$\liminf_{n \to \infty} d(T^n x, T^n y) = 0.$$

Note that in this definition $n \to \infty$ and not $|n| \to \infty$. We might have referred to this as "forward" proximality. It is this form which will be useful and we call attention to the lack of symmetry of past and future.

A far reaching extension of Theorem 1.4 is the following result of J. Auslander and R. Ellis.

THEOREM 4.1. *If (X, T) is a dynamical system, X a compact metric space, and if x is any point in X, there exists a uniformly recurrent point $y \in X$ such that x and y are proximal.*

We refer the reader to [3] or [5] for the proof of this result. We remark that it would be easy to show that any point x is proximal to some minimal set Y. What is more delicate is showing that x is proximal to a particular point $y \in Y$. Recall that by Theorem 1.3, every point in a minimal set is uniformly recurrent.

We now consider what the implications of Theorem 4.1 are for symbolic systems. Suppose then that $X \subset \Lambda^{\mathbf{Z}}$, Λ a finite set, with X a closed translation invariant subset, and let T be defined on X as the shift $Tx(n) = x(n + 1)$. We first inquire what uniform recurrence means for a point $x \in X$. Call a finite sequence of elements of Λ a *word*. We say that a *word occurs in a point* $x \in \Lambda^{\mathbf{Z}}$ if for some n, the sequence $x(n), x(n + 1), \ldots, x(n + l - 1)$ coincides with the given word. Similarly we speak of a short word occurring in a long word.

PROPOSITION 4.2. *A point $x_0 \in X \subset \Lambda^{\mathbf{Z}}$ is uniformly recurrent for (X, T) iff every word that occurs in x_0 occurs in every sufficiently large word occurring in x_0.*

The proof is a straightforward adaptation of uniform recurrence to the present situation. The metric on X being given by (2.1), a neighborhood of a particular point $x_0 \in X$ consists of all points for which a given word occurs as $(x(-k), x(-k + 1), \ldots, x(k - 1), x(k))$ for a particular k. A translate of x_0 appears in this neighborhood if the word in question occurs later on in x_0. The assertion of the proposition is then quite clear.

We now treat proximality.

PROPOSITION 4.3. *Two points $x_0, y_0 \in X$ are proximal iff arbitrarily long words occur at the same position arbitrarily far along the sequences $\{x_0(n)\}$ and $\{y_0(n)\}$.*

In other words, for some $n_l \to \infty$, we have

$$x(n_l) = y(n_l), x(n_l + 1) = y(n_l + 1), \ldots, x(n_l + l - 1) = y(n_l + l - 1).$$

Once again, this follows directly from the definition of proximality. If the condition in the proposition holds then

$$d(T^{n_{2l+1}+l}x, T^{n_{2l+1}+l}y) < \frac{1}{l+1}.$$

The converse direction is also clear.

Let us now apply Theorem 4.1 to the full symbolic system $(\Lambda^{\mathbf{Z}}, T)$. Every $x \in \Lambda^{\mathbf{Z}}$ is therefore proximal to a point $y \in \Lambda^{\mathbf{Z}}$ which is uniformly recurrent. We will use this to conclude the following.

PROPOSITION 4.4. *If x is any point of $\Lambda^{\mathbf{Z}}$, there exist integers $p_1, p_2 > 0$ with $x(p_1 + p_2) = x(p_1) = x(p_2)$.*

PROOF. Let $y \in \Lambda^{\mathbf{Z}}$ be a uniformly recurrent point of $\Lambda^{\mathbf{Z}}$ proximal to x and let $a = y(0)$. Since y is uniformly recurrent, the word consisting of the single symbol a occurs inside of every sufficiently long word of y. Since y and x agree on arbitrarily long blocks of positive values of n, we can find $p_1 > 0$ with $x(p_1) = y(p_1) = a$. Consider next the word $(y(0), y(1), \ldots, y(p_1))$. This word also occurs inside every sufficiently long word in y, and there will be a word of this length occurring in x at the same position. Then for appropriate $p_2 > 0$ we will have

$$(x(p_2), x(p_2 + 1), \ldots, x(p_2 + p_1)) = (y(0), y(1), \ldots, y(p_1)).$$

Hence $x(p_2) = y(0) = y(p_1) = x(p_2 + p_1)$. This proves the proposition.

Proposition 4.4 gives us the following result of Schur [6].

THEOREM 4.5. *If the natural numbers $\{1, 2, 3, \ldots\}$ are partitioned into finitely many sets, then one of these sets contains two numbers p_1, p_2 together with their sum $p_1 + p_2$.*

To see this suppose $\mathbf{N} = C_1 \cup C_2 \cup \cdots \cup C_r$. Define a point $x \in \{1, 2, \ldots, r\}^{\mathbf{Z}}$ by setting

$$x(n) = \begin{cases} i & \text{if } n > 0 \text{ and } n \in C_i, \\ \text{anything} & \text{if } n \leqslant 0. \end{cases}$$

According to the proposition we can find $p_1, p_2 > 0$ with $x(p_1) = x(p_2) = x(p_1 + p_2)$. If j is this common value then $p_1, p_2, p_1 + p_2 \in C_j$.

Note that unlike the configurations described in §2, the configuration $(p_1, p_2, p_1 + p_2)$ is not translation invariant. It thus happens that if we decompose $\mathbf{N}$ into odd and even numbers, one set contains solutions to $a + b = c$ whereas the other set does not. In particular the kind of configuration described here need not occur in every subset $S \subset \mathbf{N}$ of positive upper density.

Schur was interested in this combinatorial result because it relates to Fermat's last theorem. Theorem 4.5 leads readily to the following "finite" version.

THEOREM 4.5′. *There exists an integer-valued function $N(k) \geqslant 1$, such that if the interval of integers $[1, N(k)]$ is partitioned into k sets $C_1 \cup C_2 \cup \cdots \cup C_k$, then some C_j contains a solution a, b, c to $a + b = c$.*

Now let q be any natural number and let p be a prime with $p > N(q)$ and with $q | p - 1$. The numbers $\{1, 2, \ldots, p - 1\}$ may be partitioned into q cosets modulo the multiplicative subgroup of qth powers mod p. Theorem 4.5′ asserts that we can find a, b, c inside the same coset satisfying $a + b = c$. This means that the congruence

$$aX^q + aY^q \equiv aZ^q \pmod{p}$$

has a nontrivial solution and therefore the congruence

$$(4.1) \qquad\qquad X^q + Y^q \equiv Z^q \pmod{p}$$

has a solution with $XYZ \not\equiv 0 \pmod{p}$. It follows that for any q, Fermat's assertion of the unsolvability of $X^q + Y^q = Z^q$ is unlikely to be established by showing that the corresponding congruence has no solution, since (4.1) always has solutions for p sufficiently large. This is Schur's result.

To obtain further consequences of Theorem 4.1 we introduce the notion of an *IP-set* of integers.

DEFINITION. A subset of **N** is called an *IP-set* if it consists of a sequence of integers (not necessarily distinct) $p_1, p_2, p_3, \ldots$, together with all sums of these for distinct indices, i.e., all $p_{i_1} + p_{i_2} + \cdots + p_{i_k}$ with $i_1 < i_2 < \cdots < i_k$.

The notion of an IP-set is somewhat weaker than that of a semigroup. We shall find that IP-sets occur in connection with recurrence.

PROPOSITION 4.6. *Let (X, T) be a dynamical system, X compact metric, let $x, y \in X$ with x proximal to y and y a uniformly recurrent point. Then for any $\varepsilon > 0$, there exists an IP-set $S = \{p_{i_1} + p_{i_2} + \cdots + p_{i_k}\}$ with $d(T^n x, y) < \varepsilon$ for $n \in S$.*

PROOF. We prove first that for any $\delta > 0$ we can find p such that

$$(4.2) \qquad\qquad d(T^p x, y) < \delta \quad \text{and} \quad d(T^p y, y) < \delta.$$

By uniform recurrence of y there is an N so that for any $n \in \mathbf{Z}$, $d(T^{n+i} y, y) < \delta/2$ for some $i = 0, 1, \ldots, N$. Let δ' be so small that

$$d(y_1, y_2) < \delta' \Rightarrow d(T^i y_1, T^i y_2) < \delta/2 \quad \text{for } i = 0, 1, \ldots, N.$$

By proximality of x and y there is an n with $d(T^n x, T^n y) < \delta'$. For some $i = 0, 1, \ldots, N$ we will have $d(T^{n+i} y, y) < \delta/2$ and at the same time $d(T^{n+i} x, T^{n+i} y) < \delta/2$. Then for $p = n + i$ we have both inequalities of (4.2).

Now suppose we have found $p_1, p_2, \ldots, p_n$ so that each $p = p_{i_1} + p_{i_2} + \cdots + p_{i_k}, i_1 < i_2 < \cdots < i_k$, satisfies

$$(4.3) \qquad\qquad d(T^p x, y) < \varepsilon, \quad d(T^p y, y) < \varepsilon.$$

In these inequalities we can replace y by anything sufficiently close to it. Let $\delta > 0$ be chosen so that for any $z \in X$ with $d(z, y) < \delta$ we have $d(T^p z, y) < \varepsilon$ for the aforementioned finite set of p. Determine p_{n+1} so that

$$d(T^{p_{n+1}} x, y) < \delta, \quad d(T^{p_{n+1}} y, y) < \delta.$$

We then have

$$d(T^{p + p_{n+1}} x, y) < \delta, \quad d(T^{p + p_{n+1}} y, y) < \delta.$$

We can then establish (4.3) for an IP-set of p and this proves the proposition.

Note than an IP-set in $\mathbf{N}$ always contains a solution to $a + b = c$. It was conjectured by Graham and Rothschild that Schur's lemma could be extended to state that in any finite partition of the natural numbers, one of the subsets contains an IP-set. This result was proved by N. Hindman. We now show how this follows from Proposition 4.6.

THEOREM 4.7. *If the natural numbers $\{1, 2, 3, \dots\}$ are partitioned into finitely many sets, then one of these sets contains an IP-set.*

PROOF. Let $\mathbf{N} = C_1 \cup C_2 \cup \cdots \cup C_r$ be the partition in question and define $x \in \{1, 2, \dots, r\}^{\mathbf{Z}}$ by setting

$$x(n) = \begin{cases} j & \text{if } n > 0 \text{ and } n \in C_j, \\ \text{anything} & \text{for } n \leqslant 0. \end{cases}$$

Let $y \in \{1, 2, \dots, r\}^{\mathbf{Z}}$ be a uniformly recurrent point proximal to x. Apply Proposition 4.6 to x and y taking $\varepsilon = 1$. There exists accordingly an IP-set S with $d(T^p x, y) < 1$ for $p \in S$. Let $j = y(0)$. We have $x(p) = y(0)$ for $p \in S$; i.e., $S \subset C_j$. This proves the theorem.

A still more general result is given in the next theorem which was also proved by Hindman. In [3] and [5] it is shown how the more general result follows from the special case.

THEOREM 4.8. *If an IP-set of integers is partitioned into finitely many subsets, then one of these subsets contains an IP-set.*

5. Recurrent sets and strong recurrence. Let (X, T) be a dynamical system with X a compact metric space and let A be a closed subset of X. We shall say that *the set A is recurrent* if for any $\varepsilon > 0$ there exist points $x, y \in A$ and an integer $n > 0$ with $d(T^n x, y) < \varepsilon$. With some additional assumptions on A we can conclude that *some* point of A is recurrent for (X, T) and possibly even that *every* point of A is recurrent.

DEFINITION. A subset $A \subset X$ is *homogeneous* for the system (X, T) if there exists a group G of homeomorphisms of X commuting with T, $ST = TS$ for $S \in G$, such that for each $S \in G$, $SA = A$, and moreover that G acts transitively on A; i.e., for $x, x' \in A$ there exists $S \in G$ with $Sx = x'$. A is *weakly homogeneous* if the condition that G is transitive on A is replaced by the condition that G acts minimally (so that every G-orbit in A is dense in A).

We shall prove the following theorem.

THEOREM 5.1. *If A is a homogeneous recurrent set for (X, T) then every point of A is recurrent. If A is weakly homogeneous and recurrent, then there exists a dense set of recurrent points in A.*

It is obvious that an *automorphism* of (X, T), i.e., a homeomorphism $S: X \to X$ which commutes with T, takes recurrent points to recurrent points:

$$T^{n_k}x \to x \Rightarrow T^{n_k}Sx \to Sx.$$

Hence both statements of the theorem will follow once it is established that at least one point of A is recurrent. We prove this under the hypothesis of weak homogeneity of A.

LEMMA 5.2. *Suppose A is a weakly homogeneous recurrent set. Then for any $\varepsilon > 0$ and for any $y \in A$, there exists $x \in A$ with $d(T^n x, y) < \varepsilon$ for some $n > 0$.*

PROOF. By hypothesis we can find sequences $\{x_n\}$, $\{y_n\}$ with $d(T^{r_n}x_n, y_n) \to 0$. Passing to a subsequence we can assume that we have $y_n \to y'$. Then y' has the property that for every $\varepsilon > 0$ there exists $x \in A$ and $n > 0$ with $d(T^n x, y') < \varepsilon$. Now the set of points $y \in A$ with this property is clearly closed. It is also invariant under G. Since G acts minimally on A and the property holds for some point of A, it holds for every point of A.

LEMMA 5.3. *Suppose A has the property that for each $y \in A$ and $\varepsilon > 0$ there exists $x \in A$ and $n > 0$ with $d(T^n x, y) < \varepsilon$. Then for any $\varepsilon > 0$ there exists a point $z \in A$ with $d(T^n z, z) < \varepsilon$ for some $n > 0$.*

PROOF. Let $\varepsilon > 0$ be given. We shall define inductively a sequence of points $z_0, z_1, z_2, \ldots$ in A, one of which will satisfy $d(T^n z, z) < \varepsilon$. Set $\varepsilon_1 = \varepsilon/2$. Choose z_0 arbitrarily in A and let $z_1 \in A$ and $n_1 > 0$ be such that

$$(5.1) \qquad\qquad d(T^{n_1}z_1, z_0) < \varepsilon_1.$$

Now choose $\varepsilon_2 < \varepsilon_1$ so that $d(z, z_1) < \varepsilon_2$ implies that the inequality (5.1) remains valid when z_1 is replaced by z. Then if we find $z_2 \in A$ and $n_2 > 0$ with

$$d(T^{n_2}z_1, z_1) < \varepsilon_2,$$

we will also have

$$d(T^{n_1 + n_2}z_2, z_0) < \varepsilon_1.$$

Proceed inductively in this way, obtaining an array of inequalities

$$(5.2) \qquad\qquad d(T^{n_j + n_{j-1} + \cdots + n_i}z_j, z_i) < \varepsilon_{i+1}$$

whenever $i < j$. The successive ε_k are chosen so that $d(z, z_j) < \varepsilon_{j+1}$ implies that all the inequalities of (5.2) for $i < j$, j fixed, are still valid when z_j is replaced by z. Having determined ε_{j+1} we find $z_{j+1} \in A$ and $n_{j+1} > 0$ so that

$$d(T^{n_{j+1}}z_{j+1}, z_j) < \varepsilon_{j+1}.$$

We then obtain the inequalities of (5.2) for $j + 1$ instead of j. The hypothesis of the lemma enables us to proceed indefinitely. So we can suppose that (5.2) is valid for all $i < j < \infty$. Now for some i, j we shall have $d(z_i, z_j) < \varepsilon_1$, so that

$$d\left(T^{n_j + n_{j+1} + \cdots + n_i}z_j, z_j\right) < \varepsilon_{i+1} + \varepsilon_1 < \varepsilon,$$

and this proves the lemma.

PROOF OF THE THEOREM. As we have seen it suffices to show that if A is a weakly homogeneous recurrent set, it contains a recurrent point. Form the function

$$F(x) = \inf_{n > 1} d(T^n x, x).$$

Clearly a point is recurrent if $F(x) = 0$. Now $F(x)$ is readily seen to be upper semicontinuous. It follows that it has a point of continuity when restricted to A. Let z be such a point. Assume $F(z) > 0$. We can find a relatively open set $V \subset A$ with $F(x) > \delta > 0$ for all $x \in V$. Now $\bigcup_{S \in G} S(V)$ is an open G-invariant subset of A and by minimality, this must be all of A. By compactness of A some finite set of $S(V)$ covers A: $A = \bigcup S_i(V)$. Now for each $S \in G$, $F(Sx) = \inf d(ST^n x, Sx)$, and it follows that for $\delta > 0$ there exists $\varepsilon > 0$ so that $F(x) < \varepsilon \Rightarrow F(Sx) < \delta$. Letting $S = S_i^{-1}$ and taking ε' as the minimum of the corresponding ε, we will have $F(x) \geqslant \varepsilon'$ for all $x \in A$. For if $F(x) < \varepsilon'$ with $x \in S_i(V)$ then $F(S_i^{-1}x) < \delta$ contradicting the choice of δ. This shows that $F(z) = 0$ at a point of continuity z of F. This proves that A possesses, in fact, a residual set of recurrent points. This completes the proof.

We illustrate the theorem by considering "group extensions" of a given dynamical system.

DEFINITION. Let (X, T) be a dynamical system and let G be a compact group. Let $\psi: X \to G$ be a continuous function and define $\tilde{T}$ on $\tilde{X} = X \times G$ by

$$\tilde{T}(x, g) = (Tx, \psi(x)g).$$

Then $(\tilde{X}, \tilde{T})$ is said to be a *group extension* of (X, T).

Note that the group G acts on $\tilde{X}$ by setting

$$(x, g)^{g'} = (x, gg'),$$

and these transformations commute with T. The fibers $x \times G \subset \tilde{X}$ are invariant under the actions of G and they form homogeneous closed sets. Finally $x \times G$ is a recurrent set of $(\tilde{X}, \tilde{T})$ iff x is a recurrent point of (X, T). Applying Theorem 5.1 we have

THEOREM 5.3. *If $(\tilde{X}, \tilde{T})$ is a group extension of (X, T) then a point $(x, g) \in \tilde{X}$ is recurrent for $(\tilde{X}, \tilde{T})$ iff x is recurrent for (X, T).*

We illustrate this in the following proposition.

PROPOSITION 5.4. *Let $\mathbf{T}$ denote the circle group $\mathbf{T} = \mathbf{R}/\mathbf{Z}$. Let $h_0 \in \mathbf{T}$ and for $i = 1, 2, \ldots, d - 1$, let h_i be a continuous map from $\mathbf{T}^i$ into $\mathbf{T}$. Define a* *transformation of* $\mathbf{T}^d \to \mathbf{T}^d$ *by*

$$T(\theta_1, \theta_2, \ldots, \theta_d)$$

(5.3)
$$= (\theta_1 + h_0, \theta_2 + h_1(\theta_1), \ldots, \theta_d + h_{d-1}(\theta_1, \theta_2, \ldots, \theta_{d-1})).$$

Then every point of $\mathbf{T}^d$ *is a recurrent point.*

PROOF. By induction on d. Notice that for $d + 1$ the system in question is a group extension of the system for d. For $d = 0$ we take the system to be a single point which is naturally recurrent.

The toral systems described by (5.3) have another property which we shall proceed to study. We call this property *strong recurrence*.

To define this notion we recall that the product of two systems (X, T) and (X', T') is the system $(X \times X', T \times T')$ where $(T \times T')(x, x') = (Tx, T'x')$.

DEFINITION. A point $x \in X$ is *strongly recurrent* for (X, T) if for any system (X', T') and any point $x' \in X'$ which is recurrent for it, the pair (x, x') is recurrent for the product system $(X \times X', T \times T')$.

We will see presently what are the implications of strong recurrence. In particular it will be seen to imply uniform recurrence. In the meantime we have the following example.

PROPOSITION 5.5. *Each point of the system described in Proposition 5.4 is strongly recurrent.*

PROOF. For in this case $(X \times X', T \times T')$ is obtained by a succession of group extensions of (X', T'), so by Theorem 5.3, (x, x') is recurrent for $(X \times X', T \times T')$ iff x' is recurrent for (X', T').

For the remainder of this section we will be concerned with the notion of strong recurrence. In the next section we return to the notion of recurrent sets.

The next result relates strong recurrence to IP-sets.

THEOREM 5.6. *A point x is a strongly recurrent point of a dynamical system (X, T) iff for any neighborhood V of x and for any IP-set $S \subset \mathbf{N}$, there exists $q \in S$ with $T^q x \in V$.*

For the proof of the theorem we will use the following proposition.

PROPOSITION 5.7. *If $x_0 \in X$ is a recurrent point for (X, T) and V is a neighborhood of x_0, then the set of $n \in \mathbf{N}$ for which $T^n x_0 \in V$ contains an IP-set. Conversely, if S is an IP-set in $\mathbf{N}$, there is a system (X, T), a recurrent point $x_0 \in X$, and a neighborhood V of x_0 such that $S \supset \{n: T^n x_0 \in V\}$.*

PROOF. Assume $x_0 \in X$ is a recurrent point for (X, T) and V is a neighborhood of x_0. For some number, say p_1, we have $T^{p_1} x \in V$. Then $x \in V \cap T^{-p_1} V = V_1$, and V_1 constitutes a neighborhood of x. Let p_2 be such that $T^{p_2} x \in V_1$; then $x \in V_1 \cap T^{-p_2} V_1 = V_2$. Continue in this way obtaining a succession of neighborhoods V_k of x_0 with $V_{k+1} = V_k \cap T^{-p_{k+1}} V_k$. At each stage the corresponding power p_{k+1} is defined by the condition $T^{p_{k+1}} x_0 \in V_k$.

We then have

$$V_k = \bigcap_{i_1 < i_2 < \cdots \leq n} T^{-(p_{i_1} + p_{i_2} + \cdots + p_{i_k})} V.$$

It follows immediately that $\{n: T^n x_0 \in V\}$ contains the IP-set generated by $\{p_i\}$.

For the converse direction suppose $S \subset \mathbf{N}$ is an IP-set; say, $S = \{p_{i_1} + p_{i_2} + \cdots + p_{i_k}; i_1 < i_2 < \cdots < i_k\}$ with $p_1, p_2, \ldots$ a sequence in $\mathbf{N}$. If we set

$$q_1 = p_{r_1} + p_{r_1+1} + \cdots + p_{s_1}, \quad q_2 = p_{r_2} + p_{r_2+1} + \cdots + p_{s_2}$$

$$\ldots, q_n = p_{r_n} + p_{r_n+1} + \cdots + p_{s_n}, \ldots$$

where $\{r_n\}$ and $\{s_n\}$ are two sequences satisfying $r_1 < s_1 < r_2 < s_2 < \cdots$, then the IP-set generated by the q_n: $\{q_{i_1} + q_{i_2} + \cdots + q_{i_l}; i_1 < i_2 < \cdots < i_l\}$ is a subset of S. In this manner we can arrive at an IP-set whose generators satisfy $q_n > 3q_{n-1}$. Let R be this set together with 0. Now let $X = \{0, 1\}^{\mathbf{Z}}$ and let $x_0 = 1_R \in X$. T will be the shift. Let $V = \{x \in X: x(0) = 1\}$. Clearly $R = \{n: T^n x_0 \in V\}$. To complete the proof we show that x_0 is a recurrent point. We claim that $T^{q_n} x_0 \to x_0$. Namely, let $m \in \mathbf{Z}$. We show that if $m \notin R$, then for n large, $m + q_n \notin R$ and if $m \in R$ then for n large $m + q_n \in R$. The second statement is evident. For the first statement, notice that on account of the rate of growth of q_n,

$$q_{i_k} < q_{i_1} + \cdots + q_{i_k} < \frac{3}{2} q_{i_k}.$$

So for any m, if $m + q_n \in R$ for large n, then since $m + q_n \sim q_n$, $m + q_n = q_{i_1} + q_{i_2} + \cdots + q_{i_k}$ for large n implies $n = i_k$, so that $m = q_{i_1} + \cdots + q_{i_{k-1}}$. This completes the proof.

To prove the theorem we suppose x is strongly recurrent, that V is a neighborhood of x and that S is an IP-set. Let x' be a recurrent point for a system (X', T') and V' a neighborhood of x for which $\{n: T'^n x' \in V'\} \subset S$. Then (x, x') is recurrent for $(X \times X', T \times T')$, and for some n, $(T \times T')^n(x, x') \in V \times V'$. But then $T'^n x \in V'$ so that $n \in S$; hence $T^n x \in V$ for some $n \in S$. Conversely, suppose x has the property in question and that x' is recurrent for (X', T'). Let W be a neighborhood of (x, x') in $X \times X'$. $W \subset V \times V'$ where V, V' are neighborhoods of x, x' respectively. Since $T'^n x' \in V'$ along some IP-set of n, and for one of these $T^n x \in V$, by hypothesis, we will have $(T \times T')^n(x, x') \in W$ so that (x, x') is recurrent.

We shall use this characterization of strong recurrence in order to prove that strong recurrence implies uniform recurrence. First a lemma.

LEMMA 5.8. *If a subset $S \subset \mathbf{N}$ contains arbitrarily long intervals, then S contains an IP-set.*

PROOF. We have $[a_k, b_k] \subset S$ with $b_k - a_k \to \infty$. Suppose $p_1, p_2, \ldots, p_n$ have been found in S so that every sum of the form $p_{i_1} + p_{i_2} + \cdots + p_{i_k}$, $i_1 < i_2 < \cdots < i_k \leq n$, belongs to S. We choose $p_{n+1} = a_j$ for some j so large that $b_j - a_j > p_1 + p_2 + \cdots + p_n$. Clearly every sum $p_{i_1} + p_{i_2} + \cdots + p_{i_k}$, $i_1 < i_2 < \cdots < i_k \leq n + 1$, now belongs to S.

THEOREM 5.9. *Strong recurrence implies uniform recurrence.*

PROOF. Suppose $x \in X$ is not uniformly recurrent for the system (X, T). Then for some neighborhood V of x, the set $\{n; T^n x \in V\}$ has arbitrarily long gaps. By Lemma 5.8, there is an IP-set of n for which $T^n x \notin V$. So by Theorem 5.6, x is not strongly recurrent.

The converse of this is not true. In demonstrating this we shall use the following lemma.

LEMMA 5.10. *If $x_1 \in X_1$ is a strongly recurrent point for (X_1, T_1) and $x_2 \in X_2$ is strongly recurrent for (X_2, T_2), then (x_1, x_2) is strongly recurrent for $(X_1 \times X_2, T_1 \times T_2)$.*

PROOF. Let x' be recurrent for some third system (X', T'). By strong recurrence (x_2, x') is recurrent for $(X_2 \times X', T_2 \times T')$ and by strong recurrence of x_1, (x_1, x_2, x') is recurrent for $(X_1 \times X_2 \times X', T_1 \times T_2 \times T')$. But this shows that (x_1, x_2) is strongly recurrent.

The horocycle flow for a compact 2-dimensional surface of constant negative curvature is known to be minimal as a one-parameter flow $(X, T_t; -\infty < t < \infty)$. It follows that every point of X is uniformly recurrent for the flow and so it is uniformly recurrent for (X, T_1). It is also known that the horocycle flow is weakly mixing ([1]), so that the product system $(X \times X, T_1 \times T_1)$ possesses dense orbits. Now a point $(x, y) \in X \times X$ having a dense orbit cannot be uniformly recurrent, for that would imply, by Theorem 1.3, that $(X \times X, T_1 \times T_1)$ is minimal, and this cannot be the case since the diagonal $\{(x, x); x \in X\}$ is an invariant subset. Hence (x, y) is not uniformly recurrent, and so it is not strongly recurrent. But by Lemma 5.10 this means that one of the points x and y is not strongly recurrent. We thus see that uniform recurrence does not imply strong recurrence.

The same example shows that Lemma 5.10 is not valid if strong recurrence is replaced by uniform recurrence. For each point is uniformly recurrent for the horocycle system and yet there are pairs (x, y) which are not uniformly recurrent for the product system.

The next theorem gives two more characterizations of strong recurrence. The proof can be found in [3].

THEOREM 5.11. *Each of the following conditions for a point $x \in X$ is equivalent to strong recurrence for (X, T).*

(a) *x is uniformly recurrent and it is not proximal to any point $y \neq x$ in its orbit closure.*

(b) *For any uniformly recurrent point x' for a system (X', T') the point (x, x') is uniformly recurrent for $(X \times X', T \times T')$.*

The characterization (a) of strong recurrence relates the notion to that of *point distality*. Two points of a system are *distal* if they are not proximal. A minimal system with a point that is distal from every other point is *point distal*. There is a rather deep structure theorem of Veech and Ellis which shows how one can construct all point distal metric dynamical systems [8]. The idea of the theorem is that certain extensions of point distal systems preserve this property. For example, the considerations leading to Theorem 5.3 show that if x is strongly recurrent for (X, T) and $(\tilde{X}, \tilde{T})$ is a group extension of (X, T), then each (x, g) is strongly recurrent for $(\tilde{X}, \tilde{T})$. This means that the passage from (X, T) to $(\tilde{X}, \tilde{T})$ preserves point distality. The structure theorem in question asserts that by a possibly transfinite sequence of "allowable" extensions of this kind one can arrive at an arbitrary point distal system starting with the trivial one-point system. The implication of this is that strong recurrence is a relatively rare phenomenon.

We conclude this section with some observations comparing recurrence properties for the systems (X, T) and (X, T^m). First note that as a special case of Proposition 5.5, the periodic system (Y_m, T_m): $Y_m = \{0, 1, \ldots, m - 1\}$, $T_m y = y + 1 \pmod{m}$, has every point strongly recurrent. From this it follows that if a point x is recurrent for (X, T), $(x, 0)$ is recurrent for $(X \times Y_m, T \times T_m)$, and from this it follows that x is recurrent for (X, T^m).

By a similar argument, using Theorem 5.11(b), we deduce that if x is a uniformly recurrent point for (X, T) then it is also uniformly recurrent for (X, T^m).

We can also show that a strongly recurrent point of (X, T) is strongly recurrent for (X, T^m). For by Theorem 5.6, x is strongly recurrent for (X, T^m) iff for every neighborhood V of x and every IP-set $S \subset \mathbf{N}$, some $T^n x \in V$ for $n \in S$. But if S is an IP-set so is mS; so it follows that some $T^{mn} x \in V$ for $n \in S$. Now this implies that x is strongly recurrent for (X, T^m).

One can also show that for each of the notions of recurrence a recurrent point for (X, T^m) is also recurrent for (X, T). We leave it to the reader to verify this.

Combining the foregoing remarks with Lemma 5.10 we obtain the following.

PROPOSITION 5.11. *If x is a strongly recurrent point of (X, T) then $(x, x, \ldots, x) \in X^r$ is a strongly recurrent point of $(X^r, T \times T^2 \times \cdots \times T^r)$. In particular it is recurrent and so there exists a sequence $n_k \to \infty$ with $T^{n_k} x \to x$, $T^{2n_k} x \to x$, \ldots, $T^{rn_k} x \to x$.*

We will see in the next section that the latter property always holds for some point of any system (X, T). On the other hand we do not know if there always exists a point x such that $(x, x, \ldots, x)$ is a uniformly recurrent point for $T \times T^2 \times \cdots \times T^r$.

6. Multiple recurrence. We turn now to the proof of Theorem 2.1 which, as we saw in §2, implies the van der Waerden theorem on arithmetic progressions. We prove the following more general theorem.

THEOREM 6.1. *Suppose X is a compact metric space and $T_1, T_2, \ldots, T_k$ are commuting maps of X into itself. Then there exists a point $x \in X$ and a sequence $n_k \to \infty$ with $T_1^{n_k} x \to x$, $T_2^{n_k} x \to x$, \ldots, $T_k^{n_k} x \to x$.*

The idea of the proof is to show that some point of the diagonal $\Delta = \{(x, x, \ldots, x) \in X^l; x \in X\}$ is recurrent for $T_1 \times T_2 \times \cdots \times T_l$. In order to do this we make use of the notion of recurrent sets discussed in the preceding section.

We shall assume that the maps T_i are invertible. It is easy to pass from this case to the general case. The proof of the theorem will proceed by induction on l. The case $l = 1$ is Birkhoff's theorem (Theorem 1.2). Suppose then that the result has been established for $l - 1$ and we have l invertible maps $T_1, T_2, \ldots, T_l$ on X. Let G_1 be the group of transformations of X generated by $T_1, T_2, \ldots, T_l$, and let G_1 consist of the transformations of $X^l = X \times X \times \cdots \times X$ of the form $\tilde{S} = S \times S \times \cdots \times S, S \in G_1$. Note that G_l takes the diagonal Δ into itself, and in fact that action of G_l on Δ simply mirrors the action of G_1 on X. Now we can assume that G_1 acts minimally on X; otherwise simply restrict the discussion to a minimal closed G_1-invariant subset of X. If G_1 acts minimally on X, G_l acts minimally on Δ. Let $\tilde{T} = T_1 \times T_2 \times \cdots \times T_l$. $\tilde{T}$ commutes with all the transformations of G_l; hence Δ is a weakly homogeneous subset for the system $(X^l, \tilde{T})$. If we show that Δ is a recurrent set for $\tilde{T}$, then by Theorem 5.1, it contains a recurrent point and this will prove the theorem.

Now use the induction hypothesis applied to the $l - 1$ transformations $S_1 = T_1 T_l^{-1}, S_2 = T_2 T_l^{-1}, \ldots, S_{l-1} = T_{l-1} T_l^{-1}$. Accordingly, there exists a point $x \in X$ and a sequence $n_k \to \infty$ with

$$S_1^{n_k} x \to x, \; S_2^{n_k} x \to x, \ldots, \; S_{l-1}^{n_k} x \to x.$$

This means that for any $\varepsilon > 0$ we can find n with the two points

$$(x, x, \ldots, x) \in X^l, \quad (T_1 \times T_2 \times \cdots \times T_l)^n (T_l^{-n} x, T_l^{-n} x, \ldots, T_l^{-n} x) \in X^l$$

within distance ε of one another. Since the points $(x, x, \ldots, x)$ and $(T_l^{-n} x, T_l^{-n} x, \ldots, T_l^{-n} x)$ are in Δ, it follows that Δ is recurrent for $\tilde{T} = T_1 \times T_2 \times \cdots \times T_l$. This now completes the proof of the theorem.

Theorem 6.1 leads to a multidimensional extension of van der Waerden's theorem. Namely suppose $\mathbf{Z}^h = C_1 \cup C_2 \cup \cdots \cup C_r$ is a partition of the h-dimensional lattice $\mathbf{Z}^h$. Consider the space $\Omega = \{1, 2, \ldots, r\}^{\mathbf{Z}^h}$ and let $\xi \in \Omega$ be the point $\xi(v) = i \Leftrightarrow v \in C_i$. Here v denotes a typical point of $\mathbf{Z}^h$. Let $\{e_1, \ldots, e_l\}$ be any finite subset of $\mathbf{Z}^h$, and let $T_i \colon \Omega \to \Omega$ be defined by $T_i \omega(v) = \omega(v + e_i), i = 1, \ldots, l$. Let $X \subset \Omega$ be the smallest closed subset of Ω containing ξ and invariant under $T_1, T_2, \ldots, T_l$. The T_i commute and so by Theorem 6.1, there exists $\eta \in X$ and $n \in \mathbf{N}$ with the points $T_1^n \eta, T_2^n \eta, \ldots, T_l^n \eta$ within distance ε of one another. In particular ε can be chosen so that this implies that the values of $T_1^n \eta(0), T_2^n \eta(0), \ldots, T_l^n \eta(0)$ coincide. This means that $\eta(ne_1) = \eta(ne_2) = \cdots = \eta(ne_l)$. Using the fact that η belongs to the closure of translates of ξ we find that there is a vector $v \in \mathbf{Z}^h$ with

$$\xi(v + ne_1) = \xi(v + ne_2) = \cdots = \xi(v + ne_l).$$

If j is this common value, then $v + ne_i \in C_j$ for $i = 1, 2, \ldots, l$. We have thereby proved the following result, due to Gruenwald.

THEOREM 6.2. *Let F be any finite configuration in $\mathbf{Z}^h$ and suppose $\mathbf{Z}^h = C_1 \cup C_2 \cup \cdots \cup C_r$ is a finite partition. Then some C_j contains a configuration "similar" to F, i.e., a set of the form $v + nF$ for some $v \in \mathbf{Z}^h$ and $n \in \mathbf{N}$.*

7. Szemerédi's theorem. It is natural to expect that just as Theorem 6.1 extends Birkhoff's theorem (Theorem 1.2) to the case of commuting transformations, there should exist a theorem extending the Poincaré recurrence theorem to "multiple recurrence". This is indeed the case and one has the following result.

THEOREM 7.1. *Let $(X, \mathcal{B}, \mu)$ be a measure space and let $T_1, T_2, \ldots, T_l$ be commuting measure-preserving transformations of $(X, \mathcal{B}, \mu)$. Let $A \in \mathcal{B}$ with $\mu(A) > 0$. Then there exists $n \in N$ with*

$$\mu(A \cap T_1^{-n}A \cap T_2^{-n}A \cap \cdots \cap T_l^{-n}A) > 0.$$

A special case of Theorem 7.1 consists of setting $T_i = T^i$, $i = 1, 2, \ldots, l$ where $(X, \mathcal{B}, T)$ is a measure-preserving system. It follows that the set of arithmetic progressions of length l form a set Q of configurations satisfying the hypothesis of Theorem 2.8. According to Theorem 2.8 we can conclude that any subset of $\mathbf{Z}$ of positive upper density contains arithmetic progressions of length l. This is Szemerédi's theorem [7]:

THEOREM 7.2. *If $S \subset \mathbf{Z}$ is a subset of positive upper density, then S contains arbitrarily long arithmetic progressions.*

As in the case of van der Waerden's theorem, the multiple recurrence theorem for *arbitrary* commuting transformation of a measure space implies (indeed, is equivalent to) a multidimensional extension of Szemerédi's theorem. Namely we have

THEOREM 7.3. *Let S be a subset of $\mathbf{Z}^h$ of positive upper density. If F is any finite configuration in $\mathbf{Z}^h$, S contains a configuration "similar" to F, i.e., $S \supset v + nF$ for some $v \in \mathbf{Z}^h$ and $n \in \mathbf{N}$.*

Theorem 7.3 is an example of a combinatorial result which was first proved (and, so far, this is the only proof) by ergodic-theoretic means. For a complete discussion of this as well as the proof of Theorem 7.1 the reader is referred to [3] and [4].

In contrast to the proof of Theorem 6.1 which is the topological analogue of Theorem 7.1, the proof of the measure-theoretic result involves a careful study of the structure of the measure space with regard to the action of $T_1, T_2, \ldots, T_l$. In proving Theorem 6.1 we presented an argument that covers all cases. In proving Theorem 7.1 it seems necessary to distinguish between *mixing actions* of the T_i and some form of *almost periodic* action. At one extreme the expression

$$\mu(A \cap T_1^{-n}A \cap \cdots \cap T_l^{-n}A)$$

tends to a constant, $\mu(A)^{l+1}$. At the other extreme this expression behaves like an almost periodic function returning "almost periodically" to the value of the expressions for $n = 0$, namely $\mu(A)$. More precisely one forms the group Γ

of transformations generated by $T_1, T_2, \ldots, T_l$. The measure space $(X, \mathcal{B}, \mu)$ is then decomposed into a succession of factors such that at each stage the group Γ operates and Γ decomposes into two subgroups, one operating in a *relatively mixing* manner, and the other in a *relatively compact* manner. It is rather curious, and not yet understood, why in the topological situation there is no need to carry out such a careful analysis of the structure of the action, whereas this appears, so far, unavoidable in the measure-theoretic case.

References

1. L. Auslander, L. Green and F. Hahn, *Flows on homogeneous spaces*, Ann. of Math. Studies, No. 53, Princeton Univ. Press, Princeton, N. J., 1963.

2. G. D. Birkhoff, *Dynamical systems*, Amer. Math. Soc. Colloq. Publ., vol. 9, Amer. Math. Soc., Providence, R. I., 1927.

3. H. Furstenberg, *Recurrence in ergodic theory and combinatorial number theory*, Princeton Univ. Press, Princeton, N. J., 1981.

4. H. Furstenberg and Y. Katznelson, *An ergodic Szemerédi theorem for commuting transformations*, J. d'Analyse Math. **34** (1978), 275–291.

5. H. Furstenberg and B. Weiss, *Topological dynamics and combinatorial number theory*, J. d'Analyse Math. **34** (1978), 61–85.

6. I. Schur, *Über die Kongruenz $x^m + y^m \equiv z^m \pmod p$*, Jahresber. Deutsch. Math.-Verein. **25** (1916), 114–117.

7. E. Szemerédi, *On sets of integers containing no k elements in arithmetic progression*, Acta Arith. **27** (1975), 199–245.

8. W. A. Veech, *Topological dynamics*, Bull. Amer. Math. Soc. **85** (1977), 775–830.

9. H. Weyl, *Über die Gleichverteilung von Zahlen mod Eins*, Math. Ann. **77** (1916), 313–352.

INSTITUTE OF MATHEMATICS, HEBREW UNIVERSITY OF JERUSALEM, JERUSALEM, ISRAEL

Proceedings of Symposia in Pure Mathematics
Volume 39 (1983), Part 2

The Ergodic Theoretical Proof of Szemerédi's Theorem

H. FURSTENBERG, Y. KATZNELSON AND D. ORNSTEIN[1]

Introduction. In 1975, E. Szemerédi proved the following theorem conjectured some forty years earlier by Erdös and Turan:

THEOREM I. *Let $\Lambda \subset Z$ be a subset of the integers of positive upper density, then Λ contains arbitrarily long arithmetic progressions.*

Partial results were obtained previously by K. F. Roth (1952) who established the existence of arithmetic progressions of length three in subsets of Z of positive upper density, and by E. Szemerédi (1969) who proved the existence of progressions of length four.

In 1976 Furstenberg noticed that the statement of Theorem I is equivalent to a statement about "multiple recurrence" of measure-preserving transformations, namely

THEOREM II. *Let $(X, \mathscr{B}, \mu)$ be a probability measure space, let T be an invertible, measure-preserving transformation on $(X, \mathscr{B}, \mu)$, and let $A \in \mathscr{B}$ be a set of positive measure. Then for any positive integer k, there exists a subset $B \subset A$ with $\mu(B) > 0$ and an integer $n \geqslant 1$ with*

$$T^n B \subset A, \, T^{2n} B \subset A, \ldots, T^{(k-1)n} B \subset A$$

or what amounts to the same,

$$\mu\left(\bigcap_{j=0}^{k-1} T^{-jn} A\right) > 0.$$

It turned out to be possible to give an ergodic theoretic proof of Theorem II, thereby providing a new proof of Szemerédi's theorem.

Various elements of Furstenberg's original proof were simplified by Katznelson and Ornstein, and making use of this it became possible to prove a generalization of Theorem II with $T, T^2, \ldots, T^k$ replaced by any commuting set of measure-preserving transformations (cf. [FK]). This result leads to an analogue of Szemerédi's theorem for Z^r and, in fact, this proof proceeding by way of ergodic theory is the only one available so far for this analogue.

Reprinted from Bulletin Amer. Math. Soc. (N. S.) 7 (1982), 527–552.

1980 *Mathematics Subject Classification.* Primary 05, 10, 28, 60.

[1] Partial support of the second and third authors was given by National Science Foundation grant MCS81-07092.

217

Our purpose here is to give an exposition, as widely accessible as possible, of the ergodic theoretic proof of Theorem I. For a detailed account of the interrelation between dynamics and combinatorial number theory the reader is referred to [**F**].

Theorem 2 is valid for all measure-preserving systems but not for the same reason. There are two distinct phenomena, mutually exclusive, which account for the existence of positive measure intersections of the form $\bigcap_{j=0}^{k-1} T^{-jn}A$. One, compactness, is seen in the case of group rotations (T being a translation by a group element on the Haar measure space of a compact group) where for appropriate values of n, T^n is "close" to the identity so that $T^{jn}A$ differs from A by very little, for $0 < j < k$, and $\mu(\bigcap_{j=0}^{k-1} T^{-jn}A)$ is very close to $\mu(A)$. The other phenomenon is that of weak mixing, defined by the condition that for every set A, $\mu(A \cap T^{-n}A) \sim \mu(A)^2$ for most n. It can then be proved (cf. §3) that $\mu(\bigcap_{j=0}^{k-1} T^{-jn}A) \sim \mu(A)^k$ for most values of n.

It is not true that these two phenomena are complementary; there exist systems $(X, \mathcal{B}, \mu, T)$ which are neither weakly mixing nor group rotations. However, if $(X, \mathcal{B}, \mu, T)$ is not weakly mixing then there exists a nontrivial T-invariant sigma-algebra $\mathcal{B}_1 \subset \mathcal{B}$ such that T restricted to $\mathcal{B}_1$ acts like a group rotation, so that in any case the assertion of Theorem II is valid for all the sets A in some T-invariant sigma algebra of $\mathcal{B}$. The strategy of the proof of Theorem II is (a) to show that there exists a T-invariant subsigma-algebra $\mathcal{B}_1 \subset \mathcal{B}$ which is maximal, with respect to inclusion, in the class of T-invariant subalgebras of $\mathcal{B}$ for which the statement of Theorem II is valid. (b) Assuming $\mathcal{B}_1 \neq \mathcal{B}$, study the behavior of sets $A \subset \mathcal{B}$ under T "relative to $\mathcal{B}_1$," and show that either we have "relative weak mixing" or else there exist $\mathcal{B}_2 \supset \mathcal{B}_1$ for which the action of T is "relatively compact". In either case we show that there exists a bigger subalgebra for which the statement of Theorem II is valid, contradicting the maximality of $\mathcal{B}_1$. This implies $\mathcal{B}_1 = \mathcal{B}$ and completes the proof.

In §1 we shall show that Theorems I and II are equivalent. In §2 we verify Theorem II in two special cases. The next three sections give a limited version of Theorem II, but the ideas used there form a basis for the general arguments used subsequently. The formal proof of Theorem II begins in §6 in which the notion of a factor system is described, and is carried out in a series of four steps which take up the last four sections.

1. We inherit the translation from Z as the "shift" homeomorphism $T\{\omega_n\} = \{\omega_{n+1}\}$ and with it the possibility to check whether an arithmetic progression, say $\{a + jb\}_{j=0}^{k-1}$, is contained in a set Λ. Formally, if we denote by A_0 the subset of 2^Z defined by the condition $\omega_0 = 1$, and by $\bar{\omega}$ the indicator function of Λ, then

$$(1.1) \qquad \{a + jb\}_{j=0}^{k-1} \subset \Lambda \Leftrightarrow \bar{\omega} \in \bigcap_{j=0}^{k-1} T^{-(a+jb)}A_0.$$

It is not enough at this point to just check that the intersection in (1.1) is not empty; it certainly is not empty, containing all the elements of 2^Z which contain $\{a + jb\}_{j=0}^{k-1}$. Most of these points are irrelevant, the only relevant ones being $\bar{\omega}$ itself and any ω which agrees with $\bar{\omega}$ on a set containing our given arithmetic progression. If we only want to know that Λ contains an arithmetic progression of length k and step b it is enough to know that $\bigcap_{j=0}^{k-1} T^{-jb}A_0$ contains some translate of $\bar{\omega}$. We remove the irrelevant points from 2^Z and consider the subspace $X =$ the closure in 2^Z of $\{T^n\bar{\omega}\}_{n=-\infty}^{\infty}$. X is clearly invariant under T and, writing $A = A_0 \cap X$, we see that Λ contains arithmetic progressions of length k if, and only if, for some $b \neq 0$, $\bigcap_{j=0}^{k-1} T^{-jb}A$ is nonempty (since, being open, it must contain translates of $\bar{\omega}$, the set of which is dense in X).

If we assume now that Λ has positive upper density we can construct a probability measure μ on X which is T-invariant and such that $\mu(A) > 0$. Once we have this, Theorem II applies and gives us nonempty intersections of the form $\bigcap_{j=0}^{k-1} T^{-jb}A$ and Theorem I follows.

Let $\{[a_n, b_n)\}$ be a sequence of intervals such that $b_n - a_n \to \infty$ and

$$\lim_{n \to \infty} |\Lambda \cap [a_n, b_n)|/(b_n - a_n) \to d > 0.$$

Put $\mu_n = (b_n - a_n)^{-1}\sum_{j=a_n}^{b_n-1} \delta_{T^j\bar{\omega}}$, where δ_x denotes the unit point mass at x. μ_n is a probability measure on X which, for $n \to \infty$, becomes more and more T-invariant. Specifically,

$$T\mu_n - \mu_n = (b_n - a_n)^{-1}\left(\delta_{T^{b_n}\bar{\omega}} - \delta_{T^{a_n}\bar{\omega}}\right)$$

and its total mass is bounded by $2(b_n - a_n)^{-1}$. If μ is any ω^*-limit point of μ_n then μ is clearly T-invariant and, as $\mu_n(A) = |\Lambda \cap [a_n, b_n)|/(b_n - a_n)$, we obtain $\mu(A) = d > 0$. We have now proved that Theorem II implies Theorem I.

To prove that Theorem I implies Theorem II we first deduce a finite version of Theorem I.

THEOREM $\mathrm{I_F}$. *For every $\varepsilon > 0$ and positive integer k, there exists $N = N(\varepsilon, k)$ such that if Λ is a set of integers contained in some interval $[a, b]$ such that $b - a > N$ and $|\Lambda| \geqslant \varepsilon(b - a)$, then Λ contains an arithmetic progression of length k.*

The implication $\mathrm{I_F} \Rightarrow \mathrm{I}$ is obvious and we claim that we also have $\mathrm{I} \Rightarrow \mathrm{I_F}$. In fact if $\mathrm{I_F}$ were false we would have an $\varepsilon > 0$ and a positive integer k such that for every N there exists a sequence Λ_N carried by an interval $[a_N, b_N]$, $b_N - a_N > N$, $|\Lambda_N| > \varepsilon(b_N - a_N)$ and Λ_N does not contain an arithmetic progression of length k. The properties listed for Λ_N are translation invariant so we may assume that the Λ_N's are well separated, say $a_{N+1} > b_N + (b_{N+1} - a_{N+1})$. Writing $\Lambda = \cup \Lambda_N$ we see that Λ has positive upper density ($\geqslant \varepsilon$) and contains no arithmetic progression of length k since, because of the separation, such progression will have to be contained in one of the Λ_N's. This would contradict Theorem I.

Theorem $\mathrm{I_F}$ makes it clear that the existence of arbitrarily long arithmetic progressions is insured also for a class of sequences of density zero; diminishing density can be compensated by the size of the intervals on which the density is checked. For precise statements in this direction we need an estimate of $N(\varepsilon, k)$ of Theorem $\mathrm{I_F}$. Szemerédi's proof consists in giving bounds for $N(\varepsilon, k)$. The "ergodic proof", at least so far, gives no estimates of $N(\varepsilon, k)$.

We now deduce the following consequence of $\mathrm{I_F}$.

THEOREM III. *Let $\varepsilon > 0$ and k be given and write $N_1 = N(\varepsilon/2, k)$ (of $\mathrm{I_F}$). Let $(X, \mathfrak{B}, \mu)$ be a probability measure space and $B_l \in \mathfrak{B}$, $\mu(B_l) \geqslant \varepsilon$ for $l = 1, 2, \ldots, N_1$. Then there exists an arithmetic progression of length k in $[1, \ldots, N_1]$, say $\{a + mb\}_{m=0}^{k-1}$, such that*

$$(1.2) \qquad \mu\left(\bigcap_{m=0}^{k-1} B_{a+mb} \right) > \frac{\varepsilon}{2} N_1^{-2}.$$

PROOF. For $x \in X$ write $\Lambda(x) = \{l;\ 1 \leqslant l \leqslant N_1,\ x \in B_l\}$. We have

$$(1.3) \qquad \int |\Lambda(x)|\, d\mu = \sum_{l=1}^{N_1} \mu(B_l) \geqslant N_1 \varepsilon$$

and consequently

$$(1.4) \qquad \mu(\{x;\ |\Lambda(x)| \geqslant \varepsilon N_1/2\}) \geqslant \varepsilon/2.$$

By $\mathrm{I_F}$ and the choice of N_1, for each point x of the set appearing in (1.4), $\Lambda(x)$ contains an arithmetic progression of length k, say $\{a(x) + mb(x)\}_{m=0}^{k-1}$. There are fewer than N_1 choices for either $a(x)$ or $b(x)$ which implies, in view of (1.4), that for some pair (a, b) we have (1.2).

The proof of Theorem II, assuming we know I (and hence III), is done by writing $B_l = T^{-l}A$ and noticing that

$$\mu\left(\bigcap_{m=0}^{k-1} T^{-(a+bm)}A \right) = \mu\left(T^{-a} \bigcap_{m=0}^{k-1} T^{-bm}A \right) = \mu\left(\bigcap_{m=0}^{k-1} T^{-bm}A \right).$$

With a and b chosen so that the measure in question is positive and with $B = \bigcap_{m=0}^{k-1} T^{-bm}A$, we obtain the conclusion of Theorem II.

2. Two special cases. In this section we discuss two examples of measure-preserving systems for which the assertion of Theorem II is readily established. By appropriately generalizing these two examples we shall obtain a strategy for proving Theorem II "step by step".

Our first example is that of a *Bernoulli system*. A Bernoulli system is the dynamical system that corresponds to a stochastic process of infinitely many independent, identically distributed, "Bernoulli" trials. To be precise, a Bernoulli system consists of a space Ω which is the space of all sequences $\{\omega_n\}_{n \in \mathbf{Z}}$ with values in a finite set, say $\Gamma = \{1, 2, \ldots, r\}$. A σ-algebra of sets $\mathfrak{B}$ is obtained in Ω by letting $\mathfrak{B}$ be the smallest σ-algebra for which each $\omega \to \omega_n$ is measurable. The probability measure μ on $\mathfrak{B}$ is the product measure

$$\mu\{\omega_{i_1} = j_1,\ \omega_{i_2} = j_2, \ldots, \omega_{i_n} = j_n\} = p_{j_1} p_{j_2} \cdots p_{j_n}$$

where $p_1, p_2, \ldots, p_r$ is a probability distribution on Γ: $p_i \geqslant 0$, $\Sigma_{i=1}^{r} p_i = 1$. Finally the measure-preserving transformation T in this system is the *shift* $T\{\omega_n\} = \{\omega_{n+1}\}$.

In the case of a Bernoulli system, Theorem II follows from the following stronger assertion.

PROPOSITION 2.1. *If $(X, \mathcal{B}, \mu, T)$ is a Bernoulli system and $A_0, A_1, \ldots, A_k$ are $k + 1$ sets in $\mathcal{B}$, then as $n \to \infty$*

$$(2.1) \quad \mu\left(A_0 \cap T^{-n}A_1 \cap T^{-2n}A_2 \cap \cdots \cap T^{-kn}A_k\right) \to \mu(A_0)\mu(A_1) \cdots \mu(A_k).$$

Thus if $A \in \mathcal{B}$, $\mu(A) > 0$, we will have for n sufficiently large,

$$\mu\left(A \cap T^{-n}A \cap \cdots \cap T^{-kn}A\right) > 0$$

which, of course, implies Theorem II.

The proof of Proposition 2.1 is seen easily once it is noticed that the assertion of the proposition will follow for all $k + 1$-tuples of sets if it is known for A_i in a "dense" subfamily of $\mathcal{B}$. Elementary considerations show that if $\mathcal{B}_0$ is an algebra of sets spanning $\mathcal{B}$ as a σ-algebra, then $\mathcal{B}_0$ is dense in $\mathcal{B}$ in the sense that for $A \in \mathcal{B}$ and $\varepsilon > 0$ there exists $A' \in \mathcal{B}_0$ with $\mu(A \bigtriangleup A') < \varepsilon$, where $A \bigtriangleup A'$ denotes the symmetric difference

$$(A \setminus A') \cup (A' \setminus A).$$

In our case we may take $\mathcal{B}_0$ to be the algebra of *cylinder sets*, i.e., sets in Ω defined by conditions on finitely many coordinates: $A = \{\omega: (\omega_{i_1}, \omega_{i_2}, \ldots, \omega_{i_m}) \in \bar{A} \subset \Delta^m\}$. Now if $A_0, A_n, \ldots, A_k$ are cylinder sets we have, in fact,

$$\mu\left(A_0 \cap T^{-n}A_1 \cap \cdots \cap T^{-kn}\right) = \mu(A_0)\mu(A_1) \cdots \mu(A_k)$$

as soon as n is so large that the sets of defining coordinates for $T^{-nl}A_l$, $l = 0, \ldots, k$, are disjoint. This establishes the proposition.

Our second example is of a rather different nature. Perhaps the most trivial case of Theorem II occurs for T periodic, i.e., $T^p = $ identity for some p. Not quite so trivial is the case of T "almost periodic", a phenomenon that is exemplified by $T = $ irrational rotation of the circle. To be precise, let X be the circle which we represent as the reals modulo the integers, $X = R/Z$, $\mathcal{B}$ the σ-algebra of Borel subsets of X, μ Lebesgue measure and $T: X \to X$ defined by $Tx = x + \alpha$ for any fixed α. We now have

PROPOSITION 2.2. *With $(X, \mathcal{B}, \mu, T)$ as above, $A \in \mathcal{B}$ with $\mu(A) > 0$, we have for each $k = 1, 2, \ldots,$*

$$\liminf_{N \to \infty} \frac{1}{N} \sum_{1}^{N} \mu\left(A \cap T^{-n}A \cap \cdots \cap T^{-kn}A\right) > 0.$$

PROOF. From the fact that $\int_A 1_A(x + y) \, d\mu(x)$ is a continuous function of y we see that for any $\varepsilon > 0$ there exists $\delta > 0$ so that if $|y| < \delta$, $\mu(A \cap (A - y)) > \mu(A) - \varepsilon$. From this it follows that

$$\mu(A \cap (A - y) \cap (A - 2y) \cap \cdots \cap (A - ky)) > \mu(A) - (k + 1)\varepsilon.$$

Choose $\varepsilon < (k + 1)^{-1}\mu(A)$ and taking the corresponding δ, set $S_\delta = \{n: n\alpha \in (-\delta, \delta) \pmod 1, n \geq 1\}$. We see that if $n \in S_\delta$,

$$\mu\left(A \cap T^{-n}A \cap T^{-2n}A \cap \cdots \cap T^{-kn}A\right) > \mu(A) - (k + 1)\varepsilon > 0.$$

The proposition now follows from the fact that for any α and $\delta > 0$, the set S has positive density. This latter fact is trivial if α is rational, and for α irrational it is also easily deduced since for large l, $\{0, \alpha, 2\alpha, \ldots, l\alpha\}$ is 2δ-dense in R/Z and so the gap between successive numbers in S_δ cannot exceed $l + 1$.

Let us notice that in both of these cases we establish a sharper version of Theorem II. It is this version which we shall in fact obtain in general. Namely, we shall prove

THEOREM IV. *For any measure-preserving system* $(X, \mathcal{B}, \mu, T)$ *and* $A \in \mathcal{B}$ *with* $\mu(A) > 0$, *and for any* $k = 1, 2, \cdots$

$$(2.2) \qquad \liminf_{N \to \infty} \frac{1}{N} \sum_{n=1}^{N} \mu(A \cap T^{-n}A \cap \cdots \cap T^{-kn}) > 0.$$

REMARK 1. We do not know if the limit in (2.2) exists in general. In the two examples under discussion the lim inf may be replaced by lim.

REMARK 2. In the Bernoulli case the intersection $A_0 \cap T^{-n}A_1 \cap \cdots \cap T^{-kn}A_k$ is eventually nonempty for any sets $A_0, A_1, \ldots, A_k$ with positive measure. In the "almost periodic" case we may have $A_0 \cap T^{-n}A_1 \cap T^{-2n}A_2 = \varnothing$ for all n. The assertion (2.2) holds only for multiple intersections involving the same set.

In the next two sections we shall extend the phenomenon encountered in the examples of this section to as wide a class of systems as possible. The Bernoulli example will appear as a special case of *weak mixing systems* and the almost periodic example a special case of *compact systems*. These two notions play a central role in the sequel. The manner in which (2.2) will be established in these two classes of cases will be quite different. For weak mixing systems (2.2) will be a consequence of "mixing" which takes place for any $k + 1$-tuple of sets; for any $A_0, A_1, \ldots, A_k \in \mathcal{B}$, the measure $\mu(A_0 \cap T^{-n}A_1 \cap \cdots \cap T^{-kn}A_k)$ is close to $\mu(A_0)\mu(A_1) \cdots \mu(A_k)$ for most n, $n \to \infty$. In the case of compact systems, the translate $T^{-n}A$ of a set $A \in \mathcal{B}$ returns sufficiently closely to A, that the iterated translates $T^{-2n}A, \ldots, T^{-kn}A$ all overlap, and this for a set of n of positive density.

3. Weak mixing systems. A system $(X, \mathcal{B}, \mu, T)$ is *mixing* if for any two sets $A, B \in \mathcal{B}$, $\lim_{n \to \infty} \mu(A \cap T^{-n}B) = \mu(A)\mu(B)$, so that asymptotically the portion of A that arrives in B after n steps is porportional to the measure of B. The system is *weak mixing* if, instead, one only has

$$(3.1) \qquad \lim_{n \to \infty} \frac{1}{N} \sum_{n=1}^{N} \left(\mu(A \cap T^{-n}B) - \mu(A)\mu(B)\right)^2 = 0.$$

A weak mixing system is clearly ergodic, since for any two sets A, B of positive measure some intersection $A \cap T^{-n}B$ is nonempty, and so an invariant set must have either measure 0 or its complement will have measure 0. (3.1) is a special case of

$$(3.2) \qquad \lim_{N \to \infty} \frac{1}{N} \sum_{1}^{N} \left(\int fT^n g \, d\mu - \int f \, d\mu \int g \, d\mu\right)^2 = 0$$

where $f, g \in L^2(X, \mathcal{B}, \mu)$ and $T^n g$ is, by definition, the function $g(T^n x)$, (3.1) is obtained from (3.2) by restricting the latter to functions taking on the values 0, 1 (indicator functions). On the other hand, (3.2) follows for all functions in $L^2(X, \mathcal{B}, \mu)$ once it is known for functions spanning linearly a dense subset; hence (3.1) $\Rightarrow$ (3.2).

We also remark that if $(X, \mathcal{B}, \mu, T)$ is weak mixing, so is the product system $(X \times X, \mathcal{B} \times \mathcal{B}, \mu \times \mu, T \times T)$. For, once again, it suffices to check (3.2) for a spanning system of functions in $L^2(X \times X)$. Such a spanning system consists of functions $f \otimes g$ where $f \cap g(x_1, x_2) = f(x_1)g(x_2)$. Now (3.2) is equivalent to the assertion that for any $\varepsilon > 0$

$$\left| \int f T^n g \, d\mu - \int f \, d\mu \int g \, d\mu \right| < \varepsilon$$

but for a set of n of zero density. The same assertion now follows for the tensor products, since

$$\int f_1 \otimes f_2 (T \times T)^n g_1 \otimes g_2 \, d(\mu \times \mu) = \int f_1 T^n g_1 \, d\mu \int f_2 T^n g_2 \, d\mu,$$

$$\int f_1 \otimes f_2 \, d(\mu \times \mu) = \int f_1 \, d\mu \int f_2 \, d\mu,$$

$$\int g_1 \otimes g_2 \, d(\mu \times \mu) = \int g_1 \, d\mu \int g_2 \, d\mu.$$

The proof of Theorem IV in the case of weak mixing systems will follow from the next theorem which, in effect, states that a weak mixing system is "weak mixing of all orders".

THEOREM 3.1. *If* $(X, \mathcal{B}, \mu, T)$ *is a weak mixing system and* $A_0, A_1, \ldots, A_k$ *are sets in* $\mathcal{B}$, *then*

$$(3.3) \qquad \lim_{N \to \infty} \frac{1}{N} \sum_{n=1}^{N} \left(\mu\left(A_0 \cap T^{-n}A_1 \cap T^{-2n}A_2 \cap \cdots \cap T^{-kn}A_k \right) \right.$$

$$\left. - \mu(A_0)\mu(A_1) \cdots \mu(A_k) \right)^2 = 0.$$

The property described in this theorem is one of several possible analogues of the notion of (strong) mixing of all orders. If $(X, \mathcal{B}, \mu, T)$ is mixing of all orders then for all $A_0, A_1, \ldots, A_n \in \mathcal{B}$,

$$\lim_{\substack{n_j \to \infty \\ |n_j - n_i| \to \infty}} \mu\left(A_0 \cap T^{-n_1}A_1 \cap T^{-n_2}A_2 \cap \cdots \cap T^{-n_k} \right) = \mu(A_0)\mu(A_1) \cdots \mu(A_k).$$

It is still unknown if mixing implies mixing of all orders, or even if mixing implies

$$\lim_{n \to \infty} \mu\left(A_0 \cap T^{-n}A_1 \cap T^{-2n}A_2 \right) = \mu(A_0)\mu(A_1)\mu(A_2).$$

Before giving the proof we will briefly sketch the strategy. Let $k = 2$. $T^n f_1 \cdot T^{2n} f_2$ will have the right product with f_0 for most n [(3.4)$_2$] if $(1/N)\sum_{n=1}^{N} T^n f_1 \cdot T^{2n} f_2$ is close to a constant in the L_2 norm [(3.5)$_2$]. To check the above we need to show that $\int T^i f_1 T^{2i} f_2 T^j f_1 T^{2j} f_2$ is close on the average to $\int T^i f_1 \cdot \int T^{2i} f_2 \cdot \int T^j f_1 \cdot \int T^{2j} f_2$. Write those products where $j - i$ is fixed as

$$\int T^i\big(f_1 \cdot T^{(j-i)} f_1\big) \cdot T^{2i}\big(f_2 \cdot T^{2(j-i)} f_2\big) = \int\big(f_1 \cdot T^{j-i} f_1\big) \cdot T^i\big(f_2 T^{2(j-i)} f_2\big).$$

Weak mixing says that if i varies through enough values (depending on $j - i$) then the average value of the above integral will be $\int f_1 T^{(j-i)} f_1 \cdot \int f_2 T^{2(j-i)} f_2$ and for most $j - i$ this will equal $(\int f_1)^2 \cdot (\int f_2)^2$. In order to make the above argument work we need to have i vary through a large number of values compared to $j - i$. Therefore we check the average of the L_2 norms of $(1/H)\Sigma_{n=j}^{j+H} T^n f_1 \cdot T^{2n} f_2$ (3.9).

PROOF. To prove Theorem 3.1 we prove the following two variants of (3.3) by induction on k. Here $f_0, f_1, \ldots, f_k \in L^\infty(X, \mathcal{B}, \mu)$.

$$(3.4)_k \qquad \lim_{N \to \infty} \frac{1}{N} \sum_{n=1}^{N} \left[\int \prod_{l=0}^{k} T^{ln} f_l \, d\mu - \prod_{l=0}^{k} \int f_l \, d\mu \right]^2 = 0,$$

$$(3.5)_k \qquad \lim_{N \to \infty} \left\| \frac{1}{N} \sum_{n=1}^{N} \prod_{l=1}^{k} T^{ln} f_l - \prod_{l=1}^{k} \int f_l \, d\mu \right\|_{L^2(X)} = 0.$$

Clearly (3.3) is a special case of $(3.4)_k$. $(3.5)_k$ refers to convergence in $L^2(X, \mathcal{B}, \mu)$, and constitutes a generalized mean ergodic theorem for weak mixing systems.

Recall that if $(X, \mathcal{B}, \mu, T)$ is weak mixing so is the product system (which we abbreviate $X \times X$). The strong convergence in $(3.5)_k$ implies weak convergence, so that $(3.5)_k$ implies

$$(3.6) \qquad \frac{1}{N} \sum_{n=1}^{N} \int f_0 \prod_{l=1}^{k} T^{ln} f_l \, d\mu \to \prod_{l=0}^{k} \int f_l \, d\mu.$$

Since $X \times X$ is also weak mixing, we obtain the analogue of (3.6) replacing f_l by $f_l \otimes f_l$ and T by $T \times T$. The integrals on $X \times X$ become products of integrals on X and we obtain

$$(3.7) \qquad \frac{1}{N} \sum_{n=1}^{N} \left[\int f_0 \prod_{l=1}^{k} T^{ln} f_l \, d\mu \right]^2 \to \prod_{l=0}^{k} \left[\int f_l \, d\mu \right]^2.$$

Now it is an elementary exercise to see that if $(1/N)\Sigma_1^N a_n \to \alpha$ and $(1/N)\Sigma_1^N a_n^2 \to \alpha^2$, then $(1/N)\Sigma_1^N (a_n - \alpha)^2 \to 0$. Thus (3.6) and (3.7) give us $(3.4)_k$, i.e., $(3.5)_k \Rightarrow (3.4)_k$ (more precisely, the validity of $(3.5)_k$ for all weak mixing systems implies the same for $(3.4)_k$). Now $(3.4)_1$ is simply (3.2) and so our theorem will be proved if we show that $(3.4)_{k-1} \Rightarrow (3.5)_k$.

We claim that to prove $(3.5)_k$ in general it suffices to consider the case where some $\int f_{l_0} \, d\mu = 0$ so that $(3.5)_k$ can be replaced by

$$(3.5)_k' \qquad \frac{1}{N} \sum_{n=1}^{N} \prod_{l=1}^{k} T^{ln} f_l \to 0$$

in $L^2(X, \mathcal{B}, \mu)$. The reason for this is the identity

$$(3.8) \qquad \prod_{l=1}^{k} a_l - \prod_{l=1}^{k} b_l = \sum_{j=1}^{k} \left(\prod_{l=1}^{j-1} a_l \right)(a_j - b_j)\left(\prod_{l=j+1}^{k} b_l \right);$$

alternatively: the reason for this is that the identity allows us to replace $\prod_{l=1}^{k} T^{ln}f_l - \prod_{l=1}^{k} \int f_l \, d\mu$ by a sum of products satisfying the above condition.

Now set

$$\psi_N = \frac{1}{N} \sum_{n=1}^{N} \prod_{l=1}^{k} T^{ln}f_l$$

and let $N \to \infty$. We fix a large number H to be determined presently and rewrite

$$(3.9) \qquad \psi_N = \frac{1}{N} \sum_{j=1}^{N} \left[\frac{1}{N} \sum_{n=j}^{j+H-1} \prod_{l=1}^{k} T^{ln}f_l \right] + \psi_N'' = \psi_N' + \psi_N''.$$

Since the f_l are bounded, $\psi_N'' \to 0$ uniformly as $O(H/N)$ as $N \to \infty$, and it will suffice to show that for appropriate H, $\limsup_{N \to \infty} \|\psi_N'\|_{L^2(X)} < \varepsilon$, where ψ_N' denotes the first summand in (3.9). By convexity,

$$\psi_N'^2 \leqslant \frac{1}{N} \sum_{j=1}^{N} \left[\frac{1}{H} \sum_{n=j}^{j+H-1} \prod_{l=1}^{k} T^{ln}f_l \right]^2,$$

and so

$$\|\psi_N'\|_{L^2(X)}^2 \leqslant \frac{1}{N} \sum_{j=1}^{N} \frac{1}{H^2} \sum_{n,m=j}^{j+H-1} \int \prod_{l=1}^{k} T^{ln}f_l T^{lm}f_l \, d\mu$$

$$= \frac{1}{NH^2} \sum_{j=1}^{N} \sum_{n,m=j}^{j+H-1} \int \prod_{l=1}^{k} T^{ln}\left(f_l T^{l(m-n)}f_l \right) d\mu.$$

Since T, and therefore T^n, is measure preserving, we can replace each $T^{ln}(\ )$ by $T^{(l-1)n}(\ )$ in the foregoing expression. Thus

$$(3.10) \quad \|\psi_N'\|_{L^2(X)}^2 \leqslant \frac{1}{NH^2} \sum_{j=1}^{N} \sum_{n,m=j}^{j+H-1} \int \prod_{l=0}^{k-1} T^{ln}\left(f_{l+1} T^{(l+1)(m-n)}f_{l+1} \right) d\mu.$$

But now notice that the integrals occurring are those that occur in $(3.4)_{k-1}$ with the functions f_l replaced by $g_{l,m-n} = f_{l+1}T^{(l+1)(m-n)}f_{l+1}$. If we rewrite (3.10) as

$$(3.11) \quad \|\psi_N'\|_{L^2(X)}^2 \leqslant \frac{1}{H} \sum_{r=1-H}^{H-1} \left(1 - \frac{|r|}{H} \right)\left(\frac{1}{N} \sum_{n=1}^{N} \int \prod_{l=0}^{k-1} T^{ln}g_{l,r} \, d\mu \right)$$

we can use $(3.4)_{k-1}$ to replace the right-hand side by

$$\frac{1}{H} \sum_{r=1-H}^{H-1} \left(1 - \frac{|r|}{H} \right)\left(\prod_{l=0}^{k-1} \int g_{l,r} \, d\mu \right) + \frac{\varepsilon}{2}$$

for N large. On the other hand, some $\int f_{l_0}\,d\mu = 0$ and by (3.2) most of the terms $\int g_{l_0-1,r}\,d\mu$ are small for H large, since

$$\int g_{l_0-1,r}\,d\mu = \int f_{l_0} T^{l_0 r} f_{l_0}\,d\mu$$

which is, on the average, close to $(\int f_{l_0}\,d\mu)^2 = 0$. Since everything in sight is bounded, we can choose H so large that

$$\left| \frac{1}{H} \sum_{r=1-H}^{H-1} \left(1 - \frac{|r|}{H}\right) \left(\prod_{l=0}^{k-1} \int g_{l,r}\,d\mu\right) \right| < \frac{\varepsilon}{2}$$

and so

$$\limsup_{N\to\infty} \|\psi_N'\|_{L^2(X)}^2 < \frac{\varepsilon}{2} + \frac{\varepsilon}{2} = \varepsilon.$$

This completes the induction and proves Theorem 3.1.

4. Compact systems. We now abstract from the example of the irrational rotation of the circle the property that enables one to obtain a simple verification of

$$(2.2\,\text{bis}) \qquad \liminf_{N\to\infty} \frac{1}{N} \sum_{1}^{N} \mu(A \cap T^{-n}A \cap \cdots \cap T^{-kn}A) > 0$$

for $\mu(A) > 0$.

DEFINITION 4.1. The system $(X, \mathcal{B}, \mu, T)$ is *compact* if for every $f \in L^2(X, \mathcal{B}, \mu)$ the closure in $L^2(X, \mathcal{B}, \mu)$ of the orbit $\{f, Tf, T^2f, \ldots, T^nf, \ldots\}$ is compact.

The topology of $L^2(X, \mathcal{B}, \mu)$ to which compactness and closure refer in this definition is the strong, or norm, topology. It is easily shown that if the unitary operator $f \to Tf$ has discrete spectrum then the system $(X, \mathcal{B}, \mu, T)$ is compact. The converse is also true but somewhat more difficult to prove.

THEOREM 4.1. *If $(X, \mathcal{B}, \mu, T)$ is compact then for any $f \in L^\infty(X, \mathcal{B}, \mu), f \geq 0$ but f not a.e. 0,*

$$(4.1) \qquad \liminf_{N\to\infty} \frac{1}{N} \sum_{1}^{N} \int f T^n f\, T^{2n}f \cdots T^{kn}f\,d\mu > 0.$$

Specializing to the case $f = 1_A$, the characteristic function of a set $A \in \mathcal{B}$ with $\mu(A) > 0$, we obtain (2.2).

PROOF. Let $a = \int f^{k+1}\,d\mu$. We have $a > 0$. We can assume without loss of generality that $0 \leq f \leq 1$. Choose $\varepsilon < a/(k+1)$. Let $g_0, \ldots, g_k$ be measurable functions with $0 \leq g_i \leq 1$ and with $\|f - g_i\| < \varepsilon$, $i = 0, 1, \ldots, k$. Then

$$\left| \int \prod_{l=0}^{k} g_l\,d\mu - \int f^{k+1}\,d\mu \right| \leq \sum_{j=0}^{k} \int \prod_{l=0}^{j-1} g_l\, |g_j - f| f^{k-j}\,d\mu \leq (k+1)\varepsilon < a$$

using the identity (3.8). If we set $a' = a - (k+1)\varepsilon$ then we shall have $\int \prod_{l=0}^{k} g_l\,d\mu \geq a'$. This observation will be used to prove (4.1) since we shall show that for a set of n of positive lower density, $\|T^{ln} - f\| < \varepsilon$ for $l = 0, 1, \ldots, k$. In fact, if we show that $\|T^n f - f\| < \varepsilon/k$ for a set of n of positive lower density, then since T is measure preserving it will follow that $\|T^{(j+1)n}f - T^{jn}f\| < \varepsilon/k$ for these n, and by the triangle inequality, $\|T^{ln}f - f\| < \varepsilon$ for $l = 0, 1, \ldots, k$.

The property in question follows, however, directly from the compactness of the orbit closure $\overline{\{T^n f\}} \subset L^2(X, \mathcal{B}, \mu)$. For $\{T^n f, n = 0, 1, 2, \ldots\}$ being totally bounded we can find a finite subset $\{T^{n_1}f, T^{n_2}f, \ldots, T^{n_r}f\}$ which is ε/k-separated: $\|T^{n_i}f - T^{n_j}f\| \geqslant \varepsilon/k$, and which has the maximum cardinality r of such a subset. Now for any n, $\{T^{n+n_1}f, T^{n+n_2}f, \ldots, T^{n+n_r}f\}$ is again ε/k-separated and has the same cardinality. Thus for some i, $\|T^{n+n_i}f - f\| < \varepsilon/k$. This shows that $\|T^n f - f\| < \varepsilon/k$ for a set of n of positive lower density (in fact, a set of n with bounded gaps), and this completes the proof of the theorem.

5. Weak mixing and compact factors. The two special cases of Theorem IV dealt with in the foregoing sections are clearly very different from one another, and, as is easily shown, are mutually exclusive.

In terms of the spectral decomposition of the unitary operator defined by T on $L^2(X, \mathcal{B}, \mu)$, these cases correspond respectively to purely continuous spectrum (on the subspace of functions with 0 integral) and to pure point spectrum.

These two cases together are far from exhausting all possibilities; nevertheless, we shall show in this section that taking into account both weak mixing and compact systems, one can establish Theorem IV for *some* "factor" of an *arbitrary* system. This will be the first step of our proof of Theorem IV. The proof will then proceed by showing that in a like manner one can establish Theorem IV for larger and larger factors until one arrives at the given system. First, let us explain what is a factor.

Let $(X, \mathcal{B}, \mu, T)$ be a measure-preserving system. $\mathcal{B}$ is a σ-algebra of subsets of X, and T is measurable with respect to $\mathcal{B}$, so that $T^{-1}\mathcal{B} \subset \mathcal{B}$. For sets of $\mathcal{B}$ we have $\mu(T^{-1}A) = \mu(A)$. Now suppose $\mathcal{B}_1 \subset \mathcal{B}$ is another σ-algebra which is T-invariant: $T^{-1}\mathcal{B}_1 \subset \mathcal{B}_1$. We can form a new measure-preserving system $(X, \mathcal{B}_1, \mu, T)$ and we refer to this as a *factor* of $(X, \mathcal{B}, \mu, T)$. We shall presently explain why the new system is called a factor of the original system. A factor $(X, \mathcal{B}_1, \mu, T)$ will be said to be *nontrivial* if $\mathcal{B}_1$ contains sets of measure strictly between 0 and 1. The purpose of this section is to prove the following.

THEOREM 5.1. *A system* $(X, \mathcal{B}, \mu, T)$ *is weak mixing iff it has no nontrivial compact factors.*

As a result, if a system is not weak mixing, some nontrivial factor is compact. Thus in any case either Theorem 3.1 or Theorem 4.1 may be used to establish Theorem IV for *some* nontrivial factor of an arbitrary system.

We have already noted in §3 that if a system $(X, \mathcal{B}, \mu, T)$ is weak mixing it is ergodic. We now prove a converse.

PROPOSITION 5.2. *If the system $(X \times X, \mathcal{B} \times \mathcal{B}, \mu \times \mu, T \times T)$ is ergodic then $(X, \mathcal{B}, \mu, T)$ is weak mixing.*

PROOF. Here we shall use the weakest version of the ergodic theorem for ergodic systems. Namely, if a system $(X, \mathcal{B}, \mu, T)$ is ergodic, then for any $g \in L^2(X, \mathcal{B}, \mu)$, the averages

$$\frac{1}{N+1}\big(g(x) + g(Tx) + \cdots + g(T^N x)\big)$$

converge to $\int g \, d\mu$ in the weak topology of $L^2(X, \mathcal{B}, \mu)$. (See [**H, RN**].) In other words, for $f, g \in L^2(X, \mathcal{B}, \mu)$,

$$(5.1) \qquad \frac{1}{N+1} \sum_{n=0}^{N} \int f T^n g \, d\mu \to \int f \, d\mu \int g \, d\mu.$$

If we assume $T \times T$ is ergodic then so is T and we have (5.1) as well as the corresponding assertion for $f \otimes f$ and $g \otimes g$. (Recall that $f \otimes f(x_1, x_2) = f(x_1) f(x_2)$.) This gives us

$$(5.2) \qquad \frac{1}{N+1} \sum_{n=0}^{N} \left(\int f T^n g \, d\mu \right)^2 \to \left(\int \int f \, d\mu \right)^2 \left(\int g \, d\mu \right)^2$$

since, e.g., $\int f \otimes f \, d\mu \times \mu = (\int \int f \, d\mu)^2$. But now (5.1) and (5.2) together give us (by the remark following (3.7), i.e., a sequence is almost constant if the L_1 and L_2 norms are almost the same)

$$(5.3) \qquad \frac{1}{N+1} \sum_{n=0}^{N} \left(\int f T^n g \, d\mu - \int f \, d\mu \int g \, d\mu \right)^2 \to 0$$

which proves the proposition.

Recall now the definition of compact systems. A system $(X, \mathcal{B}, \mu, T)$ was said to be compact if for every $f \in L^2(X, \mathcal{B}, \mu)$ the orbit closure $\overline{\{T^n f\}} \subset L^2(X, \mathcal{B}, \mu)$ is compact. Now this may happen for some f and not for others. Let us say $f \in L^2(X, \mathcal{B}, \mu)$ is AP (almost periodic) if its orbit closure is compact.

PROPOSITION 5.3. *If for a system $(X, \mathcal{B}, \mu, T)$ the square $T \times T$ is not ergodic, then there is a nonconstant $f \in L^2(X, \mathcal{B}, \mu)$ which is AP.*

PROOF. Let $H(x, x')$ be a nonconstant $T \times T$-invariant function in $L^2(X \times X)$. We can suppose T itself to be ergodic; otherwise a T-invariant function on X would provide the desired AP function. The function $\int H(x, x') \, d\mu(x)$ is seen to be T-invariant, hence constant, and adding a constant to H we may suppose this vanishes. Since H is not identically 0, there must be some $\varphi \in L^2(X, \mathcal{B}, \mu)$ with $\int H(x, x') \varphi(x') \, d\mu(x') \neq 0$ for a set, of x, of positive measure. It follows that

$$(5.4) \qquad f(x) = \int H(x, x') \varphi(x') \, d\mu(x')$$

is also nonconstant, since $\int f(x)\,d\mu(x) = \int \varphi(x')\int H(x, x')\,d\mu(x)\,d\mu(x') = 0$. Now the function in (5.4) is AP. For

$$T^n f(x) = \int H(T^n x, x')\varphi(x')\,d\mu(x') = \int H(T^n x, T^n x')\varphi(T^n x')\,d\mu(x')$$

by the invariance of μ, or

$$T^n f(x) = \int H(x, x')T^n \varphi(x')\,d\mu(x').$$

If $\tilde{H}\colon L^2(X, \mathcal{B}, \mu) \to L^2(X, \mathcal{B}, \mu)$ denotes the integral operator

$$\tilde{H}\psi(x) = \int H(x, x')\psi(x')\,d\mu(x')$$

then $\overline{\{T^n f\}} = \overline{\{\tilde{H}(T^n \varphi)\}}$. However it is well known that the operator $\tilde{H}$ is a compact operator [**RN**] and since the norms of $T^n \varphi$ are constant $\overline{\{\tilde{H}(T^n \varphi)\}}$ is compact. This completes the proof.

REMARK. We will sketch a more complete picture that comes out of the above argument. (However, it is very helpful when we relativize later, and necessary for several commuting transformations, that we prove only what we actually need.) First, note that we get an invariant distance between points of X, $d(x_1, x_2) = \int |H(x_1, x') - H(x_2, x')|\,dx'$. If we identify points that are 0 distance apart and consider only sets that respect this identification we get an invariant σ-algebra or factor. (Factors will be discussed later and this remark may be clearer after that discussion.) T acting on this factor is an isometry of a metric space. We would then argue that this metric space is totally bounded. (For any ε there are at most a finite number, $N(\varepsilon)$, of points ε apart. This is so because the set of points at distance $< \frac{1}{2}\varepsilon$ from some x has measure $> \alpha > 0$ and by ergodicity this is true for a.e. x. If T is an isometry of a totally bounded metric space and f any function defined on this space then the orbit closure $\overline{\{T^n f\}}$ is compact.)

We turn now the the proof of Theorem 5.1.

PROOF OF THEOREM 5.1. Assume first that $(X, \mathcal{B}, \mu, T)$ is weak mixing. We shall show that there can be no nonconstant functions $f \in L^2(X, \mathcal{B}, \mu)$ with compact orbit closure $\overline{\{T^n f\}} \subset L^2(X, \mathcal{B}, \mu)$. Indeed, the proof of Theorem 4.1 shows that if f has compact orbit closure, for any $\varepsilon > 0$, there exists a subset $S \subset N$ of positive lower density so that for $n \in S$, $\|f - T^n f\|_{L^2(X)} < \varepsilon$. On the other hand, (3.2) implies that for any $\delta > 0$, $f, g \in L^2(X, \mathcal{B}, \mu)$

$$\left| \int fT^n g\,d\mu - \int f\,d\mu \int g\,d\mu \right| < \delta$$

but for a set n of density 0. In particular,

$$\left| \int fT^n f\,d\mu - \left(\int f\,d\mu\right)^2 \right| < \delta$$

but for a set of n of density 0. For some $n \in S$ we will have

$$\left| \int f^2\,d\mu - \left(\int f\,d\mu\right)^2 \right| < \delta + \varepsilon\|f\|_{L^2(X)},$$

and since ε, δ are arbitrary $\int f^2 \, d\mu = (\int f \, d\mu)^2$, which for real f implies $f = $ constant a.e.

Now suppose that $(X, \mathcal{B}, \mu, T)$ is not weak mixing. By Proposition 5.2, there is a nontrivial invariant function on $X \times X$. By Proposition 5.3, there exists a nonconstant $f \in L^2(X, \mathcal{B}, \mu, T)$ which is AP. We now construct a nontrivial σ-algebra $\mathcal{B}_1$, invariant with respect to T, so that the factor $(X, \mathcal{B}_1, \mu, T)$ is compact. First, recall that a subset of a complete metric space has compact closure iff for every $\varepsilon > 0$, the subset may be covered by finitely many balls of radius $\leq \varepsilon$. Using this it is easy to verify that the set of $\varphi \in L^2(X, \mathcal{B}, \mu)$ which are AP is a closed linear subspace of $L^2(X, \mathcal{B}, \mu)$. It can also be seen that it is closed under lattice operations $\varphi_1, \varphi_2 \to \max(\varphi_1, \varphi_2)$, $\min(\varphi_1, \varphi_2)$. From this it follows that if f is AP and $\mathcal{B}_0$ is the smallest σ-algebra of sets with respect to which f is measurable then each 1_A, $A \in \mathcal{B}_0$ is AP. Since φ is AP iff $T\varphi$ is AP, the same is true for $\mathcal{B}_1 = $ the smallest σ-algebra of sets with respect to which $f, Tf, T^2f, \ldots$ are measurable. Finally, if each 1_A, $A \in \mathcal{B}_1$ is AP, so is each $\varphi \in L^2(X, \mathcal{B}_1, \mu)$. It follows that this factor $(X, \mathcal{B}_1, \mu, T)$ is compact. This completes the proof of Theorem 5.1.

We now summarize what has been done in the last three sections. Our goal is to prove Theorem IV which asserts that for any measure-preserving system $(X, \mathcal{B}, \mu, T)$ and $A \in \mathcal{B}$ with $\mu(A) > 0$, and for any k

$$(2.2 \text{ bis}) \qquad \liminf_{N \to \infty} \frac{1}{N} \sum_{n=1}^{N} \mu(A \cap T^{-n}A \cap \cdots \cap T^{-kn}) > 0.$$

If the system $(X, \mathcal{B}, \mu, T)$ is weak mixing, then we are done by Theorem 3.1. If the system $(X, \mathcal{B}, \mu, T)$ is not weak mixing, it possesses a factor $(X, \mathcal{B}_1, \mu, T)$ which is nontrivial and which is compact. Finally by Theorem 4.1, for this system the assertion of Theorem IV is valid. Thus, in any case, Theorem IV is valid for a nontrivial factor of the given system.

What we have done so far will appear as the first step in the proof of Theorem IV. Let us now make the strategy of proof more precise. Fix the measure-preserving system $(X, \mathcal{B}, \mu, T)$. For any T-invariant σ-algebra $\mathcal{B}_1 \subset \mathcal{B}$ (i.e., for any factor $(X, \mathcal{B}_1, \mu, T)$), we can ask if the conclusion of Theorem IV is valid for elements of $\mathcal{B}_1$; that is, do we have

$$\liminf_{N \to \infty} \frac{1}{N} \sum_{n=1}^{N} \mu(A \cap T^{-n}A \cap \cdots \cap T^{-kn}A) > 0$$

for every $k \in Z^+$ and $A \in \mathcal{B}_1$ such that $\mu(A) > 0$? If the answer is "yes" we say that the action of T on $\mathcal{B}_1$ is "Szemerédi" (shortened to SZ). We have shown that for some nontrivial factor, the action of T is SZ. Naturally, if $\mathcal{B}_1 \supset \mathcal{B}_2$ and the action of T on $\mathcal{B}_1$ is SZ, so is the action of T on $\mathcal{B}_2$.

Consider the family of all factors (T-invariant sub-σ-algebras) of $\mathcal{B}$ ordered by inclusion. We shall show in the sequel that

(a) the set $\mathcal{F}$ of factors for which T is SZ contains a maximal element.

(b) No *proper* factor can be maximal in $\mathcal{F}$.

Clearly (a) and (b) together imply that the action of T on the full σ-algebra $\mathcal{B}$ is SZ, and with that the proof of Theorem IV will be complete.

The proof of (a) is not hard and we shall dispose of it rather quickly. We shall then turn to the proof of (b). The proof of (b) will be carried out by relativizing the ideas in §§3–5. Namely we shall define the notion of a factor $(X, \mathcal{B}_2, \mu, T)$ being *relatively weak mixing* with respect to another factor $(X, \mathcal{B}_1, \mu, T)$, $\mathcal{B}_2 \supset \mathcal{B}_1$. We shall also speak of *relative compactness* of factors. In both of these cases we shall show that if the action of T is SZ for the smaller factor, it is so also for the larger factor. Finally we show that if for some factor $(X, \mathcal{B}_1, \mu, T)$, the original system $(X, \mathcal{B}, \mu, T)$ fails to be a relatively weak mixing extension, then there exists $\mathcal{B}_1' \supsetneqq \mathcal{B}_1$ with $(X, \mathcal{B}_1', \mu, T)$, a relatively compact extension of $(X, \mathcal{B}_1, \mu, T)$. This implies that if $\mathcal{B}_1 \neq \mathcal{B}$, $(X, \mathcal{B}_1, \mu, T)$ cannot be a maximal member of $\mathcal{F}$, proving (b).

6. Factors. In this section we shall elaborate on the notion of a factor of a measure-preserving system. The following is an important example of a factor of a system. Let $(Y, \mathcal{D}, \nu, S)$ be a measure-preserving system, and suppose $(Z, \mathcal{E}, \theta)$ is a measure space. Suppose we have a map $y \to \sigma(y)$ of Y into the measure-preserving maps of $(Z, \mathcal{E}, \theta)$ such that $(y, z) \to \sigma(y)z$ is a measurable function from $Y \times Z$ to Z with respect to the σ-algebra $\mathcal{D} \times \mathcal{E}$ on $Y \times Z$. It is easy to see that if we set

$$T(y, z) = (Sy, \sigma(y)z)$$

then T is measure preserving on $(Y \times Z, \mathcal{D} \times \mathcal{E}, \nu \times \theta)$. Setting $X = Y \times Z$, $\mathcal{B} = \mathcal{D} \times \mathcal{E}$, $\mu = \nu \times \theta$, we obtain a measure-preserving system $(X, \mathcal{B}, \mu, T)$. We shall say that $(X, \mathcal{B}, \mu, T)$ is a *skew product* of $(Y, \mathcal{D}, \nu, S)$ with $(Z, \mathcal{D}, \theta)$. Now let $\pi: X \to Y$ be the projection $\pi(y, z) = y$ and set $\mathcal{B}_1 = \pi^{-1}(\mathcal{D}) \subset \mathcal{B}$. There is a one-one correspondence between sets of $\mathcal{D}$ and sets of $\mathcal{B}_1$ and for $D \in \mathcal{D}$, $\pi^{-1}(SD) = T\pi^{-1}(D)$. Thus we can identify the "factor" $(X, \mathcal{B}_1, \mu, T)$ with $(Y, \mathcal{D}, \nu, S)$, the latter being the image of $(X, \mathcal{B}, \mu, T)$ under the map π. It will often be helpful to think of this case as the typical example of a factor. Indeed, if $(X, \mathcal{B}, \mu, T)$ is an ergodic system, a theorem of Rokhlin [**R**] asserts that for any T-invariant σ-algebra $\mathcal{B}_1 \subset \mathcal{B}$ we can find a system $(Y, \mathcal{D}, \nu, S)$ so that $(X, \mathcal{B}, \mu, T)$ can be identified as a skew product over $(Y, \mathcal{D}, \nu, S)$ and $\mathcal{B}_1$ arises in the manner described.

We shall not need the full strength of Rokhlin's theorem, but rather a consequence of it which is valid quite generally (even without ergodicity). In the case of a skew product, Fubini's theorem gives

$$\mu(A) = \int \theta(A_y) \, d\nu(y)$$

for $A \in \mathcal{D} \times \mathcal{E}$, where $A_y = \{z: (y, z) \in A\}$. We can also write

$$\theta(A_y) = (\delta_y \times \theta)(A)$$

where δ_y is the point mass concentrated at $y \in Y$. Thus we can "disintegrate" the measure μ into a family of measures $\mu_y = \delta_y \times \theta$ for which

$$(6.1) \qquad \mu(A) = \int \mu_y(A) \, d\nu(y),$$

the measure μ_y being concentrated on the fiber $\pi^{-1}(y) \subset X$. Moreover we have

$$(6.2) \qquad \mu_y(T^{-1}A) = \mu_{Sy}(A).$$

For the sequel we shall not need to retain the full details of the skew product construction. We shall however make use of the following description which can be applied without real loss of generality to any system and a factor of it. Namely, given a system $(X, \mathcal{B}, \mu, T)$ and an invariant sub-σ-algebra $\mathcal{B}_1 \subset \mathcal{B}$, we can suppose that there exists a system $(Y, \mathcal{D}, \nu, S)$ and a measurable, measure-preserving map $\pi: X \to Y$ such that $\mathcal{B}_1 = \pi^{-1}(\mathcal{D}) = \{\pi^{-1}(D): D \in \mathcal{D}\}$. π maps the action of T into that of S: $\pi(Tx) = S\pi(x)$. Moreover we can find a measurable map $y \to \mu_y$ of Y into measures on $(X, \mathcal{B})$ with μ_y supported on the inverse image $\pi^{-1}(y)$, and such that (6.1) and (6.2) are valid. We can identify the factor $(X, \mathcal{B}_1, \mu, T)$ with the system $(Y, \mathcal{D}, \nu, S)$, since in terms of the measures of sets and the respective actions of S and T on sets, $D \to \pi^{-1}(D)$ gives an isomophism of the two systems.

Assume now $g \in L^2(X, \mathcal{B}, \mu)$ and let $(Y, \mathcal{D}, \nu, S)$ be a factor of $(X, \mathcal{B}, \mu, T)$. We denote by $E(g \mid \mathcal{Y})$ the function "g conditioned on $\mathcal{Y}$" defined by

$$(6.3) \qquad E(g \mid \mathcal{Y})(y) = \int g \, d\mu_y.$$

Another way of obtaining the same function—or rather the corresponding function in $L^2(X, \mathcal{B}_1, \mu)$ is to regard $L^2(X, \mathcal{B}_1, \mu)$ as a closed subspace of $L^2(X, \mathcal{B}, \mu)$, and to let $g \to E(g \mid \mathcal{Y})$ represent the orthogonal projection of $L^2(X, \mathcal{B}, \mu) \to L^2(X, \mathcal{B}_1, \mu) \simeq L^2(Y, \mathcal{D}, \nu)$. It will be convenient to identify the latter two spaces, and so we shall identify functions on Y with the corresponding functions on X, measurable with respect to $\mathcal{B}_1$. Having said this we can formulate two facts which follow from (6.2) and (6.3).

$$(6.4) \qquad E(gf \mid \mathcal{Y}) = gE(f \mid \mathcal{Y}) \quad \text{if } g \text{ is measurable } \mathcal{B}_1.$$

$$(6.5) \qquad SE(f \mid \mathcal{Y}) = E(Tf \mid \mathcal{Y}).$$

Finally we shall define the *fiber-square* or the *relative-square* of a system $(X, \mathcal{B}, \mu, T)$ with respect to a factor $\mathcal{Y} = (Y, \mathcal{D}, \nu, S)$.

Here it will be convenient to use the picture of $(X, \mathcal{B}, \mu, T)$ as a skew product of $(Y, \mathcal{D}, \nu, S)$ with a space $(Z, \mathcal{E}, \theta)$. We then set $\tilde{X} = Y \times Z \times Z$ and let $\tilde{\mathcal{B}} = \mathcal{D} \times \mathcal{E} \times \mathcal{E}$ with $\tilde{\mu} = \nu \times \theta \times \theta$. If $T(y, z) = (Sy, \sigma(y)z)$ then $\tilde{T}(y, z, z') = (Sy, \sigma(y)z, \sigma(y)z')$. In general if π denotes the measure-preserving map from X to Y we can set $\tilde{X} = \bigcup_{y \in Y} \pi^{-1}(y) \times \pi^{-1}(y) \subset X \times X$, with $\tilde{\mu} = \int \tilde{\mu}_y \, d\nu(y)$ where $\tilde{\mu}_y = \mu_y \times \mu_y$. $\tilde{\mathcal{B}}$ is then the restriction of $\mathcal{B} \times \mathcal{B}$ to $\tilde{X} \subset X \times X$.

We denote the system $(\tilde{X}, \tilde{\mathcal{B}}, \tilde{\mu}, \tilde{T})$ by $X \times_Y X$.

An identity that will be used repeatedly is

$$(6.6)$$

$$\int_{X \times_Y X} f(y, z) f(y, z') \, d\tilde{\mu}(y, z, z')$$

$$= \int \int \int f(y, z) f(y, z') \, d\mu_y(z) \, d\mu_y(z') \, d\nu(y)$$

$$= \int E(f \mid \mathcal{Y})^2 \, d\nu.$$

7. Maximal SZ factors. Let us now return to assertion (a) at the end of §5. Namely it is claimed that the family $\mathcal{F}$ of factors of $(X, \mathcal{B}, \mu, T)$ which are SZ has a maximal element. Let $\{\mathcal{B}_\alpha\}$ be a totally ordered (by inclusion) family of factors and recall that $\sup_\alpha \mathcal{B}_\alpha$ is, by definition, the σ-algebra spanned by $\cup \mathcal{B}_\alpha$. More concretely, a set $A \in \mathcal{B}$ belongs to $\sup \mathcal{B}_\alpha$ if for every $\varepsilon > 0$ there exist some $A_0 \in \cup_\alpha \mathcal{B}_\alpha$ such that

$$\mu(A \setminus A_0) + \mu(A_0 \setminus A) < \varepsilon.$$

It is clear that if $\mathcal{B}_\alpha$ if T-invariant for every α, so is $\sup \mathcal{B}$. The key to the proof of (a) is the following

PROPOSITION 7.1. *Let $\{\mathcal{B}_\alpha\}$ be a totally ordered family of factors and assume $\mathcal{B}_\alpha \in \mathcal{F}$ (that is, the action of T on $\mathcal{B}_\alpha$ is SZ) for every α. Then $\sup \mathcal{B}_\alpha \in \mathcal{F}$.*

PROOF. Let $A \in \sup \mathcal{B}_\alpha$, $\mu(A) > 0$, and k be fixed. Take $\eta = \frac{1}{2}(k + 1)^{-1}$ and $A_0' \in \mathcal{B}_{\alpha_0}$ such that

$$(7.1) \qquad \mu(A \setminus A_0') + \mu(A_0' \setminus A) < \tfrac{1}{4}\eta\mu(A).$$

We now apply to the factor $(X, \mathcal{B}_{\alpha_0}, \mu, T)$ the description given above. Namely we suppose given a system $(Y, \mathcal{D}_0, \nu, T_0)$ and a map $\pi: X \to Y$ as before so that $\mathcal{B}_{\alpha_0} = \pi^{-1}(\mathcal{D}_0)$. $A_0' \in \mathcal{B}_{\alpha_0}$ corresponds to a set $A_0'' \in \mathcal{D}_0$, $A_0' = \pi^{-1}(A_0'')$. By (7.1) $\mu(A_0') \geqslant \mu(A) - \tfrac{1}{4}\eta\mu(A) > 0$. We claim that the set of $y \in A_0''$ such that $\mu_y(A) < 1 - \eta$ has measure less than $\tfrac{1}{4}\mu(A)$. For otherwise

$$\mu(A_0' \setminus A) = \int_{A_0''} \mu_y(A_0' \setminus A)\, d\nu(y)$$

$$= \int_{A_0''} \left(1 - \mu_y(A)\right) d\nu(y) \geqslant \tfrac{1}{4}\eta\mu(A)$$

since for $y \in A_0''$, $\mu_y(A_0') = 1$, and this inequality contradicts (7.1). Denote by A_0 the subset of points $y \in A_0''$ for which $\mu_y(A) > 1 - \eta$. $A_0 \in \mathcal{D}_0$ and $\nu(A_0) > \nu(A_0'') - \tfrac{1}{4}\mu(A) = \mu(A_0') - \tfrac{1}{4}\mu(A) > \tfrac{1}{2}\mu(A)$. Since the action of T on $\mathcal{B}_{\alpha_0}$ is SZ, or equivalently the action of T_0 on $\mathcal{D}_0$ is SZ, we have

$$(7.2) \qquad \liminf_{N \to \infty} \frac{1}{N} \sum_{j=1}^{N} \nu\left(A_0 \cap T_0^{-j}A_0 \cap \cdots \cap T_0^{-kj}A_0 \right) = a > 0.$$

We claim now that for every j

$$(7.3) \qquad \tfrac{1}{2}\nu\left(A_0 \cap T_0^{-j}A_0 \cap \cdots \cap T_0^{-kj}A_0 \right) < \mu\left(A \cap T^{-j}A \cap \cdots \cap T^{-kj}A \right).$$

In part, on account of (6.1), the latter will follow if we show that for $y \in A_0 \cap T_0^{-j}A_0 \cap \cdots \cap T_0^{-kj}A_0$,

$$(7.4) \qquad \mu_y\left(A \cap T^{-j}A \cap \cdots \cap T^{-kj}A \right) > \tfrac{1}{2}.$$

But if $y \in T_0^{-lj}A_0$, $l = 0, 1, \ldots, k$, we obtain from the definition of A_0 and (6.2) that $\mu_y(T_0^{-lj}A) > 1 - \eta$. The intersection of $k + 1$ sets each having probability $> 1 - \eta$ has itself probability $> 1 - (k + 1)\eta = 1/2$ and (7.4) follows. This proves (7.3) which together with (7.2) implies

$$\liminf_{N \to \infty} \frac{1}{N} \sum_{j=1}^{N} \left(A \cap T^{-j}A \cap \cdots \cap T^{-kj}A \right) \geqslant \frac{a}{2} > 0,$$

and this completes the proof of Proposition 7.1.

The proof of part (a) now follows immediately from Zorn's lemma and Proposition 7.1.

8. Relative weak mixing. Let $(X, \mathscr{B}, \mu, T)$ be a measure-preserving system and let $\mathscr{Y} = (Y, \mathscr{D}, \nu, S) \simeq (X, \mathscr{B}_1, \mu, T)$ be a factor. It will be convenient to assume that X is ergodic.

DEFINITION. X is *weakly mixing relative to* Y if $X \times_Y X$ is ergodic.

LEMMA 8.1. *Let* $(X, \mathscr{B}, \mu, T)$ *be a relatively weak mixing extension of* $(Y, \mathscr{D}, \nu, S)$ *and let* $f, g \in L^{\infty}(X, \mathscr{B}, \mu)$. *Then*

$$(8.1) \quad \lim_{N \to \infty} \frac{1}{N} \sum_{n=1}^{N} \int \left[E(fT^n g \mid \mathscr{Y}) - E(f \mid \mathscr{Y}) S^n E(g \mid \mathscr{Y}) \right]^2 d\nu = 0.$$

PROOF. By (6.4) and (6.5) we may assume with no loss of generality that $E(f \mid \mathscr{Y}) = 0$ (write $f = (f - E(f \mid \mathscr{Y})) + E(f \mid \mathscr{Y})$ and note that (8.1) is trivial if f is measurable $\mathscr{B}_1$). Now we change the scene to $X \times_Y X$. Put $f \otimes f(y, z, z') = f(y, z)f(y, z')$ and $g \otimes g(y, z, z') = g(y, z)g(y, z')$. We have to prove

(8.2)

$$\lim_{N \to \infty} \frac{1}{N} \sum_{1}^{N} \int \left[E(fT^n g \mid \mathscr{Y}) \right]^2 d\nu = \lim_{N \to \infty} \frac{1}{N} \int \sum f \otimes f \tilde{T}^n(g \otimes g) \, d\tilde{\mu}$$

$$= \lim_{N \to \infty} \int f \otimes f \left(\frac{1}{N} \sum_{1}^{N} \tilde{T}^n(g \otimes g) \right) d\tilde{\mu} = 0.$$

But since $\tilde{T}$ is ergodic (the relative weak mixing) we have by the ergodic theorem

$$\lim_{N \to \infty} \frac{1}{N} \sum \tilde{T}^n(g \otimes g) = \int g \otimes g \, d\tilde{\mu} = \text{constant}$$

and

$$\int f \otimes f \, d\tilde{\mu} = \int E(f \mid \mathscr{Y})^2 \, d\nu = 0,$$

which proves (8.2) and hence (8.1).

If f and g are the characteristic functions of A and B, (8.1) can be rewritten as

$$(8.3) \quad \int \frac{1}{N} \sum_{n=1}^{N} |\mu_y(A \cap T^{-n}B) - \mu_y(A)\mu_y(T^{-n}B)|^2 \, d\nu(y) \to 0.$$

This shows that in case of relative weak mixing, for most n and most $y \in Y$, the sets A and $T^n B$ are almost independent with respect to μ_y.

LEMMA 8.2. *Let* $(X, \mathcal{B}, \mu, T)$ *be a relative weak mixing extension of* $\mathcal{Y} = (Y, \mathcal{D}, \nu, S)$. *Then* $(X \times_Y X, \tilde{\mathcal{B}}, \tilde{\mu}, \tilde{T})$ *is also a relatively weak mixing extension of* $\mathcal{Y}$.

PROOF. We denote $X \times_Y X$ by $\tilde{X}$ and $\tilde{X} \times_Y \tilde{X}$ by $\hat{X}$. We want to show that $(\hat{X}, \hat{\mathcal{B}}, \hat{\mu}, \hat{T})$ is ergodic. For this it suffices to show that for a dense set of functions $F, G \in L^2(\hat{X}, \hat{\mathcal{B}}, \hat{\mu})$

$$(8.4) \qquad \frac{1}{N} \sum_{n=1}^{N} \int F \hat{T}^n G \, d\hat{\mu} \to \int F \, d\hat{\mu} \int G \, d\hat{\mu}.$$

It therefore suffices to verify (8.4) for F and G of the form

$$F(y, z_1, z_2, z_3, z_4) = f_1(y, z_1) f_2(y, z_2) f_3(y, z_3) f_4(y, z_4),$$
$$G(y, z_1, z_2, z_3, z_4) = g_1(y, z_1) g_2(y, z_2) g_3(y, z_3) g_4(y, z_4).$$

We then have

$$\int F \hat{T}^n G \, d\hat{\mu} = \int \left[\int f_1 T^n g_1 \, d\mu_y \int f_2 T^n g_2 \, d\mu_y \int f_3 T^n g_3 \, d\mu_y \int f_4 T^n g_4 \, d\mu_y \right] d\nu$$

$$= \int E(f_1 T^n g_1 \mid \mathcal{Y}) E(f_2 T^n g_2 \mid \mathcal{Y}) E(f_3 T^n g_3 \mid \mathcal{Y}) E(f_4 T^n g_4 \mid \mathcal{Y}) \, d\nu.$$

But now in light of Lemma 8.1, each expression $E(f_i T^n g_i \mid \mathcal{Y})$ can be replaced by $E(f_i \mid \mathcal{Y}) S^n E(g_i \mid \mathcal{Y})$ and so the left-hand side of (8.4) can be replaced by

$$\frac{1}{N} \sum_{n=1}^{N} \int E(f_1 \mid \mathcal{Y}) E(f_2 \mid \mathcal{Y}) E(f_3 \mid \mathcal{Y}) E(f_4 \mid \mathcal{Y})$$

$$\cdot S^n \left[E(g_1 \mid \mathcal{Y}) E(g_2 \mid \mathcal{Y}) E(g_3 \mid \mathcal{Y}) E(g_4 \mid \mathcal{Y}) \right] d\nu.$$

Since $(Y, \mathcal{D}, \nu, S)$ is ergodic

$$\frac{1}{N} \sum_{n=1}^{N} S^n \left[E(g_1 \mid \mathcal{Y}) E(g_2 \mid \mathcal{Y}) E(g_3 \mid \mathcal{Y}) E(g_4 \mid \mathcal{Y}) \right]$$

$$\to \left[\int g_1 \, d\mu_y \int g_2 \, d\mu_y \int g_3 \, d\mu_y \int g_4 \, d\mu_y \right] d\nu = \int G \, d\hat{\mu},$$

and (8.4) follows.

We now show how relative weak mixing implies "relative weak mixing of all orders".

THEOREM 8.3. *Let* $(X, \mathcal{B}, \mu, T)$ *be a relative weak mixing extension of* $(Y, \mathcal{D}, \nu, S)$. *Then if* $f_l \in L^\infty(X, \mathcal{B}, \mu)$, $l = 0, 1, \ldots, k$, *we have*

$$(8.5)_k \qquad \lim_{N \to \infty} \frac{1}{N} \sum_{n=1}^{N} \left[E\left(\prod_{l=0}^{k} T^{ln} f_l \mid \mathcal{Y} \right) - \prod_{l=0}^{k} S^{ln} E(f_l \mid \mathcal{Y}) \right]^2 d\nu = 0,$$

and

$$(8.6)_k \qquad \lim_{N \to \infty} \left\| \frac{1}{N} \sum_{n=1}^{N} \left(\prod_{l=1}^{k} T^{ln} f_l - \prod_{l=1}^{k} T^{ln} E(f_l \mid \mathcal{Y}) \right) \right\|_{L^2} = 0.$$

Note that $E(f_l \mid \mathcal{Y})$ is regarded both as a function in $L^\infty(X, \mathcal{B}, \mu)$ and in $L^\infty(Y, \mathcal{D}, \nu)$.

PROOF. We proceed by induction on k. Lemma 8.1 gives us $(8.5)_1$ and $(8.6)_1$ follows from the (norm) ergodic theorem. We now assume that $(8.5)_{k-1}$ is valid in full generality, that is, for all relatively weak mixing extensions of $\mathcal{Y}$ (in particular, by 3.2, for $X \times_Y X$) and prove

(i) $(8.5)_{k-1} \Rightarrow (8.6)_k$,

(ii) $(8.6)_k$ (for $X \times_Y X$) $\Rightarrow (8.5)_k$ (for X).

We begin with (ii) which is easier and follows the line of proof of Lemma 8.1.

If f_0 is measurable $\mathcal{B}_1 = \pi^{-1}(\mathcal{D})$, the integrals in $(8.5)_k$ have the form

$$(8.7) \quad \int f_0^2 \left[E\left(\prod_1^k T^{ln}f_l \mid \mathcal{Y} \right) - \prod_{l=1}^k S^{ln}E(f_l \mid \mathcal{Y}) \right]^2 d\nu$$

$$\leqslant \sup |f_0^2| \int \left[E\left(\prod_{l=0}^{k-1} T^{ln}f_{l+1} \mid \mathcal{Y} \right) - \prod_{l=0}^{k-1} S^{ln}E(f_{l+1} \mid \mathcal{Y}) \right]^2 d\nu$$

(use (6.4), (6.5) and the fact that S^n is measure preserving) and $(8.5)_k$ is reduced to the case $(8.5)_{k-1}$. This enables us to assume, as in the proof of Lemma 8.1, that $E(f_0 \mid \mathcal{Y}) = 0$. With this, $(8.5)_k$ takes the form (see (6.6))

$$(8.8) \quad \lim_{N \to \infty} \int f_0 \otimes f_0 \left(\frac{1}{N} \sum_{n=1}^N \prod_{l=1}^k \tilde{T}^{ln}f_l \otimes f_l \right) d\tilde{\mu}.$$

By $(8.6)_k$, applied to $X \times_Y X$, the limit in (8.8) is the same as

$$\lim_{N \to \infty} \int f_0 \otimes f_0 \left(\frac{1}{N} \sum_{n=1}^N \prod_{l=1}^k \tilde{T}^{ln}E(f_l \otimes f_l \mid \mathcal{Y}) \right) d\tilde{\mu}$$

which is 0 (for all N) since the sum appearing there is constant on fibers and $E(f_0 \mid \mathcal{Y}) = 0$.

Now to the proof of implication (i). First, write

$$(8.9) \quad \prod_{l=1}^k T^{ln}f_l - \prod_{l=1}^k T^{ln}E(f_l \mid \mathcal{Y})$$

$$= \sum_{j=1}^k \left(\prod_{l=1}^{j-1} T^{ln}f_l \right) T^{jn} \left(f_j - E(f_j \mid \mathcal{Y}) \prod_{j+1}^k T^{ln}E(f_l \mid \mathcal{Y}) \right)$$

and notice that this enables us to prove $(8.6)_k$ under the additional assumption that for some l_0, $1 \leqslant l_0 \leqslant k$, $E(f_{l_0} \mid \mathcal{Y}) = 0$. We now have to show that under this assumption, $\lim \|\psi_N\|_{L^2} = 0$ where $\psi_N = (1/N)\Sigma_{n=1}^N \Pi_{l=1}^k T^{ln}f_l$.

Rewrite

$$\psi_N = \frac{1}{N} \sum_{j=1}^N \left(\frac{1}{H} \sum_{n=j}^{j+H-1} \prod_{l=1}^k T^{ln}f_l \right) + O(H/N)$$

(where H will be chosen large but much smaller than N). By the convexity of the function $\varphi(x) = x^2$, we have (up to $O(H/N)$)

$$\psi_N^2 \leqslant \frac{1}{N} \sum_{j=1}^{N} \left(\frac{1}{H} \prod_{n=j}^{j+H-1} \prod_{l=1}^{k} T^{ln} f_l \right)^2.$$

By integration and the fact that T is measure preserving,

$$(8.10) \quad \|\psi_N\|_{L^2}^2 \leqslant \frac{1}{N} \sum_{j=1}^{N} \frac{1}{H^2} \sum_{n,m=j}^{j+H-1} \int \prod_{l=1}^{k} T^{ln} f_l T^{lm} f_l \, d\mu$$

$$= \frac{1}{NH^2} \sum_{j=1}^{N} \sum_{n,m=j}^{j+H-1} \int \prod_{l=1}^{k} T^{(l-1)n} \left(f_l T^{l(m-n)} f_l \right) d\mu.$$

Set $r = m - n$, notice that a pair (n, m) appears in (8.10) only if $|r| = |m - n| < H$, and then for $H - |r|$ values of j, and rewrite (8.10) as

$$(8.11) \quad \|\psi_N\|_{L^2}^2 \leqslant \frac{1}{H} \sum_{r=1-H}^{H-1} \left(1 - \frac{|r|}{H} \right) \left[\frac{1}{N} \sum_{n=1}^{N} \int \prod_{l=1}^{k} T^{(l-1)n} \left(f_l T^{lr} f_l \right) d\mu \right]$$

$$+ O(H/N).$$

By $(8.5)_{k-1}$, for a fixed H and every r such that $|r| < H$ we can replace the integrated terms in (8.11), provided N is large enough, by $\prod T^{(l-1)n} E(f_l T^{lr} f_l \mid \mathcal{Y})$ and obtain
(8.12)

$$\|\psi_N\|_{L^2}^2 \leqslant \frac{1}{H} \sum_{r=1-H}^{H-1} \left(1 - \frac{|r|}{N} \right) \left[\frac{1}{N} \sum_{n=1}^{N} \int \prod_{l=1}^{k} T^{(l-1)n} E\left(f_l T^{lr} f_l \mid \mathcal{Y} \right) d\mu \right]$$

$$+ O(H/N).$$

Now estimate the integrals appearing in (8.12) by

$$\| E\left(f_{l_0} T^{l_0 r} f_{l_0} \mid \mathcal{Y} \right) \|_{L^2} \prod_{l \neq l_0} \| f_l \|_\infty^2$$

and recall that $E(f_{l_0} \mid \mathcal{Y}) = 0$ so that by Lemma 8.1 most of the terms appearing in (8.12) are small provided H is large enough. Since all the terms are bounded by $\prod \| f_l \|_\infty^2$ and most of them are small, their average (8.12) is small, arbitrarily for large H and N, and the proof is complete.

THEOREM 8.4. *Let $(X, \mathcal{B}, \mu, T)$ be a relatively weak mixing extension of $(Y, \mathcal{D}, \nu, S)$. If the action of S on $\mathcal{D}$ is SZ, then so is the action of T on $\mathcal{B}$.*

PROOF. Let $A \in \mathcal{B}$, $\mu(A) > 0$ and write f for the indicator function of A. Let $a > 0$ be small enough to insure that, for $A_1 = \{y : E(f \mid \mathcal{Y}) \geqslant a\}$, $\nu(A_1) > 0$. By (6.3) and Theorem 8.3 (and $E(f \mid \mathcal{Y}) \geqslant a I_{A_1}$) we have for all k

$$\frac{1}{N} \sum_{1}^{N} \mu\left(\bigcap_{l=0}^{k} T^{-lj} A \right) > \frac{1}{2} a^{k+1} \cdot \frac{1}{N} \sum_{1}^{N} \nu\left(\bigcap_{l=0}^{k} S^{-lj} A_1 \right),$$

provided N is large enough.

9. Compact extensions. The notion of *compact extension* which we discuss in this section, is a natural generalization of group translations to the relative case. We shall prove the analogue of Theorem 8.4 for compact extensions and later show that whenever X is a nonrelatively weak mixing extension of $\mathcal{Y}$, there exists an intermediate extension which is compact. (Thus, in either case, a proper factor $\mathcal{Y}$ cannot be maximal with respect to the property that the action of T is SZ, which completes the proof of Theorem IV.)

A function $f \in L^2(X, \mathcal{B}, \mu)$ is said to be almost periodic (AP) relative to the factor $\mathcal{Y}$ if for every $\delta > 0$ there exist functions $g_1, \ldots, g_n \in L^2(X, \mathcal{B}, \mu)$ such that for every $j \in Z$, $\inf_{1 \leqslant s \leqslant n} \|T^j f - g_s\|_{L^2(\mu_y)} < \delta$ for almost all $y \in \mathcal{Y}$. We denote the subset of AP functions in $L^2(X, \mathcal{B}, \mu)$ by AP.

DEFINITION. $(X, \mathcal{B}, \mu, T)$ is a *compact extension* of $(Y, \mathcal{D}, \nu, S)$ if AP is dense in $L^2(X, \mathcal{B}, \mu)$.

We now show that the property SZ lifts to compact extensions.

THEOREM 9.1. *If $(X, \mathcal{B}, \mu, T)$ is a compact extension of $(Y, \mathcal{D}, \nu, S)$ and if the action of S on $(Y, \mathcal{D}, \nu)$ is SZ, then so is the action of T on $(X, \mathcal{B}, \mu)$.*

PROOF. Let $A \in \mathcal{B}$, $\mu(A) > 0$, and k be given. We have to prove

$$(*) \qquad \lim_{N \to \infty} \frac{1}{N} \sum_{j=1}^{N} \mu\left(\bigcap_{l=1}^{k} T^{-jl}A \right) > 0$$

which clearly follows from the same inequality holding for a subset of A. We remove from A its portions sitting on fibers for which $\mu_y(A) \leqslant \frac{1}{2}\mu(A)$. This removes less than half the measure of A and we may therefore assume without loss of generality that $\mu_y(A) \geqslant \alpha = \frac{1}{2}\mu(A)$ for $y \in A_1$, $\nu(A_1) > \frac{1}{2}\mu(A)$, and $\mu_y(A) = 0$ for $y \notin A_1$.

Denote $f = 1_A$, the indicator function of A. Our next step is to show that there is no loss of generality in assuming that f is AP. By our assumption there exist, for every $\varepsilon > 0$, an AP function f' such that $\|f - f'\|_{L^2(\mu)} < \varepsilon^2$. This implies $\|f - f'\|_{L^2(\mu_y)} < \varepsilon$ for all y outside a set E_ε such that $\nu(E_\varepsilon) < \varepsilon^2$. If we denote by A_ε the set obtained from A by removing the part of A included in the fibers above points in E_ε and by f_ε the corresponding indicator function, we have that on every fiber and for every $j \in Z$, either

$$\|T^j f_\varepsilon - T^j f'\|_{L^2(\mu_y)} < \varepsilon \quad \text{or} \quad \|T^j f_\varepsilon\|_{L^2(\mu_y)} = 0.$$

For $\delta > 0$ let $g_1, \ldots, g_m$ be functions as assured by Definition 4.1 for f', write $g_0 = 0$ and notice that

$$(9.1) \qquad \inf_{0 \leqslant s \leqslant m} \|T^j f_\varepsilon - g_s\|_{L^2(\mu_y)} < \delta + \varepsilon, \quad \text{for a.e. } y, j \in Z.$$

Notice that (4.1) remains valid if we replace A_ε, and correspondingly f_ε, by its intersection with sets in $\pi^{-1}(\mathcal{D})$ (i.e., replace f_ε by zero on some fibers). We repeat the procedure for a sequence $\{\varepsilon_j\}$ going to zero fast enough to insure $\Sigma \varepsilon_j^2 < \frac{1}{2}\mu(A)$, thus removing from A less than half of its measure and obtaining a set whose indicator function is AP.

The condition $f \in$ AP is equivalent to requiring that the sequence $\{T^j f\}_{j \in Z}$ be totally bounded, or relatively compact, in $L^2(\mu_y)$ for almost all y. Since T

maps μ_y onto μ_{S_y} it is clear that $\{T^j f\} \subset L^2(\mu_y)$ is isometric with $\{T^j f\} \subset L^2(\mu_{S_y})$. We could thus require the total boundedness only for a set of positive measure on Y and the ergodicity of S on Y would give it almost everywhere in a uniform manner.

We shall be considering vectors of the form $(f, T^n f, T^{2n} f, \ldots, T^{kn} f)$ on a fiber above $y \in Y$. Denote by $\bigoplus_{l=0}^{k} L^2(\mu_y)$ the direct sum of $k + 1$ copies of $L^2(\mu_y)$ endowed with the norm $\|(f_0, f_1, \ldots, f_k)\|_y = \max \|f_j\|_{L^2(\mu_y)}$. It is clear that if $f \in \mathrm{AP}$, the set $\mathcal{L}(k, f) = \{(f, T^n f, \ldots, T^{kn} f)\}_{n \in Z}$ is totally bounded in $\bigoplus_{l=0}^{k} L^2(\mu_y)$ for almost all $y \in Y$, in fact uniformly in $y \in Y$.

So we assume now that $f = 1_A$ is AP and keep the notation $A_1 = \{y; \mu_y(A) > \mu(A)/2\} = \{y; \mu_y(A) > 0\}$. Let $k > 0$ be given; we write

$$\mathcal{L}(k, f, y) = \{(f, T^n f, \ldots, T^{kn} f)_y\}_{n \in Z} \subset \bigoplus_{l=0}^{k} L^2(\mu_y).$$

As we mentioned before, $\mathcal{L}(k, f, y)$ are totally bounded uniformly in y. We are interested only in $y \in A_1$ and for these points, in the elements of $\mathcal{L}(k, f, y)$ for which all the components are nonzero (hence have norm $\geqslant \sqrt{\tfrac{1}{2}\mu(A)}$ in $L^2(\mu_y)$). We denote this subset of $\mathcal{L}(k, f, y)$ by $\mathcal{L}^*(k, f, y)$, and remark that these are still uniformly totally bounded. For $y \in A_1$ and $\varepsilon > 0$ let $M(\varepsilon, y)$ denote the maximum cardinality of an ε-separated subset of $\mathcal{L}^*(k, f, y)$, that is, a subset such that the $\| \ \|_y$ distance between any two of its members exceeds ε. The uniform total boundedness of $\mathcal{L}^*(k, f, y)$ clearly implies that $M(\varepsilon, y)$ is bounded on A_1. For every y, $M(\varepsilon, y)$ is a monotone decreasing function of ε, integer valued, hence locally constant except at the countable set of "critical" ε's where it has jump discontinuity. As a function of y, $M(\varepsilon, y)$ is clearly measurable and we can find some $\varepsilon_0 < \mu(A)/10k$, $\eta > 0$, and $A_2 \subset A_1$, $\nu(A_2) > 0$, so that $M(\varepsilon, y)$ is constant, say M, for $\varepsilon_0 - \eta \leqslant \varepsilon \leqslant \varepsilon_0$ and $y \in A_2$.

Take $y_0 \in A_2$ and find integers $m_1, \ldots, m_M$ so that $\{(f, T^{m_j} f, \ldots, T^{km_j} f)_{y_0}\}_{j=1}^{M}$ is a maximal ε_0-separated set in $\mathcal{L}^*(k, f, y_0)$. Consider the function $\|T^{lm_i} f - T^{lm_j} f\|_{L^2(\mu_y)}$, for $1 \leqslant i < j \leqslant M$ and $l = 0, 1, \ldots, k$, as functions on Y; these are measurable and we can suppose that y_0 has been chosen so that each neighborhood of the values of these functions at y_0 occurs with positive measure in the set A_2. We now let A_3 be the subset of A_2 of points y for which, for each i, j, l, $1 \leqslant i \leqslant j \leqslant M$, $0 \leqslant l \leqslant k$,

$$(9.2) \qquad \|T^{lm_i} f - T^{lm_j} f\|_{L^2(\mu_y)} > \|T^{lm_i} f - T^{lm_j} f\|_{L^2(\mu_{y_0})} - \eta.$$

This will be a subset with $\nu(A_3) > 0$.

We use now the assumption that the action of S on Y is SZ, applying it to A_3. Let $n \in Z$ such that $\nu(\bigcap_{l=0}^{k} S^{-nl} A_3) > 0$ and let $y \in \bigcap_{l=0}^{k} S^{-nl} A_3$. We have $S^{nl} y \in A_3$ for $l = 0, \ldots, k$, and on the other hand, by the definition of $\mathcal{L}^*(k, f, y)$, $A_3 \subset \bigcap_{l=0}^{k} S^{-lm_j} A_1$ for $j = 1, \ldots, M$. Thus $S^{l(n+m_j)} y \in A_1$ for $l = 0, \ldots, k$ and $j = 1, \ldots, M$. We claim that the vectors $\{(f, T^{n+m_j} f, \ldots, T^{k(n+m_j)} f)_y\}_{j=1}^{M}$ are $\varepsilon_0 - \eta$ separated in $\mathcal{L}^*(k, f, y)$, hence form a maximal such set which is therefore $\varepsilon_0 - \eta$ dense in $\mathcal{L}^*(k, f, y)$. To prove the separation take $j \neq i$; by the definition of the norm $\| \ \|$ there exists some l, $0 < l \leqslant k$, such that

$$\| T^{lm_j}f - T^{lm_i}f \|_{L^2(\mu_{y_0})} \geqslant \varepsilon_0;$$

therefore, $\| T^{lm_j}f - T^{lm_i}f \|_{L^2(\mu_{Sl^{n_y}})} \geqslant \varepsilon_0 - \eta$ by (9.2), since the points $S^{ln}y$ are all inside A_3. We have $(f, f, \ldots, f)_y \in \mathcal{L}^*(k, f, y)$, and hence, for an appropriate j, $(f, T^{n+m_j}f, \ldots, T^{k(n+m_j)}f)_y$ is ε_0-close to it. By the choice of ε_0, this implies that

$$\mu_y\left(\bigcap_{l=0}^{k} T^{-l(n+m_j)}A \right) = \int \prod_{l=0}^{k} T^{l(n+m_j)}f \, d\mu_y > \frac{9}{10}\mu_y(A) > \frac{1}{3}\mu(A).$$

The index j depends on y, but if we sum over j we will have for each $y \in \bigcap_{l=0}^{k} S^{-ln}A_3$

$$\sum_{j=1}^{M} \mu_y\left(\bigcap_{l=0}^{k} T^{-l(n+m_j)}A \right) > \frac{1}{3}\mu(A).$$

Integrating over $\bigcap_{l=0}^{k} S^{-ln}A_3$ we obtain

$$\sum_{j=1}^{M} \mu\left(\bigcap_{l=0}^{k} T^{-l(n+m_j)}A \right) \geqslant \frac{\mu(A)}{3} v\left(\bigcap_{l=0}^{k} S^{-ln}A_3 \right).$$

Finally averaging for $1 \leqslant n \leqslant N$ and passing to the limit $N \to \infty$, we obtain

$$M \liminf_{N \to \infty} \frac{1}{N} \sum_{p=1}^{N} \mu\left(\bigcap_{l=0}^{k} T^{-lp}A \right) \geqslant \frac{\mu(A)}{3} \liminf_{N \to \infty} \frac{1}{N} \sum_{p=1}^{N} \left(\bigcap S^{-lp}A_3 \right).$$

This completes the proof of Theorem 9.1.

10. Existence of compact extensions. Let us take account of where we stand. By Proposition 7.1 we know that for any $(X, \mathcal{B}, \mu, T)$ there exists a maximal factor for which the action of T is SZ. By Theorem 8.4 we know that no proper extension of this factor could be relatively weak mixing, for otherwise this extension would enjoy the property SZ. By Theorem 9.1 no proper extension of this factor may be relatively compact. We complete the argument proving Theorem IV by showing in this section that if $(X, \mathcal{B}, \mu, T)$ is a proper extension of $(Y, \mathcal{D}, v, S)$ and is not relatively weak mixing, there necessarily exists a proper subextension of $(Y, \mathcal{D}, v, S)$ which is relatively compact. It follows that $(X, \mathcal{B}, \mu, T)$ already has the property SZ.

THEOREM 10.1. *If $\mathcal{X} = (X, \mathcal{B}, \mu, T)$ is an extension of $\mathcal{Y} = (Y, \mathcal{D}, v, S)$ which is not relatively weak mixing, then there exists an intermediate factor $\mathcal{X}^*$ between $\mathcal{Y}$ and $\mathcal{X}$ so that $\mathcal{X}^*$ is a compact extension of $\mathcal{Y}$.*

PROOF. It will be convenient to use the Rokhlin picture and represent $\mathcal{X}$ as a skew product: $(X, \mathcal{B}, \mu) = (Y, \mathcal{D}, v) \times (Z, \mathcal{E}, \theta)$ with $T(y, z) = (Sy, \sigma(y)z)$. We also suppose, again for convenience, that $\mathcal{X}$ is ergodic. Since $\tilde{T}$ on $X \times_Y X$ is not ergodic, there exists a bounded function $H(x, x')$ on $X \times_Y X$ which is invariant under $\tilde{T}$, but is not a function of x alone or x' alone. We set

$X \times_Y X = Y \times Z \times Z$ and write $H(x, x') = H(y, z, z')$. There exists a function $\varphi \in L(X, \mathscr{B}, \mu)$, so that the convolution $H * \varphi$ defined by

$$(10.1) \qquad H * \varphi(y, z) = \int H(y, z, z')\varphi(y, z') \, d\theta(z')$$

is not a function of y alone. We have
(10.2)

$$T(H*\varphi)(y, z) = H * \varphi(Sy, \sigma(y)z) = \int H(Sy, \sigma(y)z, z')\varphi(Sy, z') \, d\theta(z')$$

$$= \int H(Sy, \sigma(y)z, \sigma(y)z')\varphi(Sy, \sigma(y)z') \, d\theta(z')$$

$$= \int H(y, z, z')\varphi(Sy, \sigma(y)z') \, d\theta(z') = H * T\varphi$$

using the fact that $\sigma(y)$ is measure preserving and that H is $\tilde{T}$-invariant.

For each y the integral operator (10.1) is compact and it follows that for every $\delta > 0$, there exists an integer $M = M(y, \delta)$ such that $\{T^j(H * \varphi)\}_{j=-M}^{M}$ $= \{H * T^j\varphi\}_{j=-M}^{M}$ is δ-dense in $\{T^j(H * \varphi)\}_{j \in Z}$ in the $L^2(\mu_y)$ norm. For every $\varepsilon > 0$ we can now take $M_{\varepsilon, \delta}$ big enough to insure $M(y, \delta) < M_{\varepsilon, \delta}$ for all y outside some set $E(\delta, \varepsilon)$ such that $v(E(\delta, \varepsilon)) < \varepsilon$. We repeat this argument for a sequence $\{\delta_j\}$, $\delta_j \to 0$ and $\{\varepsilon_j\}$, $\Sigma_1^\infty \varepsilon_j$ arbitrarily small and write

$$(10.3) \qquad f(y, z) = \begin{cases} 0 & \text{if } y \in \bigcup E(\delta_j, \varepsilon_j), \\ H * \varphi & \text{otherwise.} \end{cases}$$

Clearly $\| f - H * \varphi \|_{L^2} \leqslant \| H \|_{L^\infty} \| \varphi \|_{L^\infty} \Sigma_{j=1}^{\infty} \varepsilon_j$ (which is arbitrarily small) and for every δ the family $\{0\} \cup \{T^j(H * \varphi)\}_{j=-M}^{M}$ is δ-dense, for M large enough, in $\{T^j f\}_{j \in Z}$ in the $L^2(\mu_y)$ norm for every y. If we denote by $\mathscr{F}$ the algebra spanned by $\{H * \varphi; H \in L^\infty(X \times_Y X), \tilde{T}H = H, \varphi \in L^\infty(X)\}$ then by (10.2) $\mathscr{F}$ is T-invariant and the AP functions in $\mathscr{F}$ are dense in $\mathscr{F}$. Let $\mathscr{B}^*$ be the smallest sub-σ-algebra of $\mathscr{B}$ with respect to which all the elements of $\mathscr{F}$ are measurable. $\mathscr{B}^*$ clearly contains (properly) $\mathscr{B}_1 = \pi^{-1}(\mathscr{D})$ and the T-invariance of $\mathscr{F}$ implies the same for $\mathscr{B}^*$. $\mathscr{F}$ is dense in $L^2(X, \mathscr{B}^*, \mu)$ and so the set of AP functions is dense in $L^2(X, \mathscr{B}^*, \mu)$. If we denote the factor corresponding to $(X, \mathscr{B}^*, \mu, T)$ by $\mathscr{X}^*$ then $\mathscr{X}^*$ is a compact extension of $\mathscr{Y}$. This completes the proof of Theorem 10.1.

We remark that throughout our use of ergodicity has been merely for convenience and essentially the same arguments will work in general without the use of the Rokhlin theorem representing an extension as a skew product. In any case we have achieved a proof of Theorem IV in the case that $(X, \mathscr{B}, \mu, T)$ is ergodic. Now in fact using the decomposition of an arbitrary measure into ergodic components it is easy to see that Theorem IV, once established for ergodic systems, is proved for arbitrary measure-preserving systems.

References

[F] H. Furstenberg, *Recurrence in ergodic theory and combinatorial number theory*, Princeton Univ. Press, Princeton, N. J., 1981.

[FK] H. Furstenberg and Y. Katznelson, *An ergodic Szemerédi theorem for commuting transformations*, J. d'Analyse Math. **34** (1978), 275–291.

[H]. P. R. Halmos, *Lectures on ergodic theory*, Chelsea, New York, 1960.

[R] V. A. Rokhlin, *Selected topics from the metric theory of dynamical systems*, Amer. Math. Soc. Transl. (2) **49** (1966), 171–209.

[RN] F. Riesz and B. Sz.-Nagy, *Functional analysis*, Ungar, New York, 1955.

INSTITUTE OF MATHEMATICS, HEBREW UNIVERSITY OF JERUSALEM, JERUSALEM, ISRAEL

DEPARTMENT OF MATHEMATICS, STANFORD UNIVERSITY, STANFORD, CALIFORNIA 94305

Section 8
HISTORICAL MATERIAL

Proceedings of Symposia in Pure Mathematics
Volume **39** (1983), Part 2

Poincaré and Topology

P. S. ALEKSANDROV [1]

To the question of what is Poincaré's relationship to topology, one can reply in a single sentence: he created it; but it is also possible to reply with a course of lectures in which Poincaré's fundamental topological results would be discussed in greater or lesser detail. The first of these two approaches to my present problem, I can regard as exhausted; for the second approach, naturally, I do not have the time. So I have to look for some intermediate compromise solution—unsatisfactory like all compromises, but in any case closer to the first alternative than to the second; this solution can only be an attempt to renew and illuminate the emotion of delight in the act of the grand scientific creation accomplished by the great French geometer in a field whose effect upon all mathematical knowledge not only has surpassed all expectations of his contemporaries, but is still increasing every year.

Poincaré lived in the romantic era of the history of the mathematical sciences when the consistency of non-Euclidean geometry was first proved (by himself, and by Klein) as a result of which our views on geometry and the very concept of a geometrical space expanded beyond description, when the new geometrical ideas found their application (including, once again, in the works of Poincaré himself) to the special theory of relativity—a theory already sufficient to shake our concepts of the universe which had seemed stable since the time of Galileo and Newton; at a time when in the abstract depths of mathematics itself a theory had arisen, which in the opinion of many outstanding mathematicians lay outside their science, perhaps outside science in general—set theory, which caused a revolution in mathematics just as significant as the revolution produced in physics by the theory of relativity.

By his own mathematical tastes and by the tradition he had inherited, Poincaré was a representative of classical mathematics—of the great French school of mathematical analysis created by Lagrange, Laplace and Cauchy. Poincaré represented mathematical analysis in the universal concept of the word, including the theory of functions and all aspects of differential equations and "mathematical physics" in the widest sense. And Poincaré's universality as a mathematician was reflected precisely in the way in which he created a new field of mathematics—topology. For Poincaré topology was from the beginning and above all a powerful instrument for solving problems arising in the classical branches of mathematics. These were in the first place: the theory

---

[1] This article is a free transcription of the speech given by the author, under the same title, at the celebration session (in the Hague) of the International Congress of Mathematicians in honour of the centenary of Poincaré's birth. (Amsterdam 1954)

Reprinted from Russian Math. Surveys **27** (1972), 157–168 with the permission of the London Mathematical Society and the British Library.

of functions of a complex variable, whose close connections with geometry, which Riemann had observed in embryonic form, was first understood in all its depth by Poincaré; the theory of differential equations—a theory inseparable for Poincaré from celestial mechanics; and geometry itself. But by understanding the power of topological methods in "classical mathematics" and often predicting them, where in his time these methods could not be applied yet in all their force, Poincaré opened up for mathematics a whole world of new problems—problems of a "qualitative" that is, of a truly topological nature, a whole world in its essence unattainable not only to the methods, but, if I may so express it, to the universe of "classical" mathematics, at the centre of which lie formula and computation (that is, the technique of operating with formulae). Thus, Poincaré, the supreme representative of classical mathematics, like no one else, "exploded" its traditions from within and opened it not only to new methods of investigation, but also, which is possibly more important, to new means of seeing things of interest to them.

Let us clarify this a little. All mathematical creativity in the last analysis has as its basis our (mathematical) intuition. Prolonged, persistent and concentrated thought finally leads to a (more or less sudden) assessment of the essence of these laws, towards the search for which we have been directing our labour and thought. The purpose of subsequent investigations, which are often also very tedious, consists of checking this assessement, of checking our intuition which (if it is confirmed by this checking) is the true germ of the result obtained. No one has described this mechanism of mathematical creativity better than Poincaré in his books, "Science and method", "Science and hypothesis". But the character of mathematical intuition is by no means the same in all cases and in all mathematicians: Jacobi's intuition was not similar to that of Hilbert, nor the intuition of Weierstrass to that of Poincaré.

Probably there exists what may be called "formula intuition"; the capacity of predicting the result of a complicated transformation (for example, in tensor analysis). There exists also intuition of an algebraic-logical character; the vision (and prediction) of complex logical relationships (for example, in set theory and in abstract algebra). And, finally (or rather, first and foremost) there exists geometrical intuition sometimes spoken of as the only "genuine" intuition in mathematics. I think that this is untrue and that there actually exist different forms of mathematical intuition, even beyond the limits of the few examples I have named.

"Topological" intuition is generally regarded to be a special case of ordinary geometical intuition; however, this special case itself displays such richness and diversity of different possibilities and, on the other hand, differs so strongly from other forms of geometrical intuition that probably it deserves to be put into a special category. The intuition of topology is not connected with straight lines, perspective transformations and other images, however fundamental they may be for projective geometry, for example. Topological intuition is the intuition of the form and position of figures in a pure aspect, a pure "Freude an der Gestalt" as Klein said. This is the most geometrical of all the varieties of geometrical intuition.

Poincaré was a master of it as no other mathematician of his time and preceding eras; maybe only Riemann could be compared to him in this respect, but he did not succeed in developing it with such breadth and diversity of applications as Poincaré.

Topological intuition penetrates the majority of Poincaré's most significant works—the theory of automorphic functions and uniformization (this is the supreme triumph of the "Riemann" approach to the theory of functions of a complex variable); the qualitative theory of differential equations, which is perhaps the best illustration of Poincaré's ability of seeing in a completely new manner the most classical objects of mathematics and to what unprecedented problems he was able to subject them. And finally, of course, there is the whole series of his purely topological works.

Nothing less than the appearance of a genius of topological intuition was the fact that Poincaré saw in the concept of homology the fundamental pivot for all future development of topology. The first formulation of this concept (in the fundamental memoir of 1895 "Analysis Situs") appeals to direct geometrical visualization, which was put on a strict logical basis only several years later.

Just as intuitive, although entirely strictly formulated, was the second fundamental topological concept introduced by Poincaré—the concept of the fundamental group. By introducing this concept, Poincaré initiated the whole enormous trend of homotopic topology, with the further development with which we associate first and foremost the names of Brouwer, then H. Hopf, W. Hurevicz and a long list of later mathematicians. Here it must be noted that the definition of homotopy groups, that is, groups which generalize the concept of the fundamental group to any number of dimensions, was first given in 1932 by the famous Czech topologist E. Cech, who did not, however, subject them to further investigation; this went later, as is well known, to the credit of Hurevicz.

To the most remarkable and the earliest developed parts of homotopic topology there belong the theory of vector (and polyvector) fields and their singularities, which is closely connected with the theory of fixed points of continuous mappings. The founder of this theory was, once again, Poincaré himself; back in the 'eighties he established the first relevant definitions and facts, in particular, the fundamental concept of the singularity index of a vector field, in his works on the qualitative theory of differential equations, that is, before the creation of his proper topological works.

At the present time it is difficult to overestimate the fundamental significance of these ideas and results of Poincaré for the whole later development not only of the theory of differential equations, but also for that of modern mathematical analysis.

In particular, as regards theorems on the existence of fixed points under this or that continuous mapping, Poincaré already understood the significance of these theorems as a means of proving existence theorems in analysis. This is evident if only from the enormous efforts which he expended on the proof of his "last geometrical theorem" on the existence of a fixed point for a given class of continuous mappings of a plane annular ring onto itself. This last work of Poincaré's makes, I would say, a tragic impression on the reader. In a short introduction to it the author writes that he had never published so incomplete a work—in fact, he had been unable to find a proof of the fundamental result (Poincaré's last geometrical theorem) to which the work is devoted. Nevertheless Poincaré considered it possible and necessary to publish the partial results he had obtained, in view of the difficulty of the subject and the fact that, as he said, at his age he could no longer hope to obtain a complete solution of the problem. At that time Poincaré was only 57 years old, and the fact of the matter lay, of course, not in his age but in the serious illness which had already set in (at that time virtually beyond surgical intervention) from which he died a year later.

In general form "Poincaré's last geometrical theorem" was proved shortly after his death by the then young American mathematician J. G. Birkhoff, who at once became famous from this result. But today it is important for us to state how profoundly Poincaré could foresee the significance of topological theorems of the type of the 'fixed point" theorems for analysis and for celestial mechanics and to record him as the founder of the "fixed point method".

The force of Poincaré's geometrical intuition sometimes led him to ignore the pedantic strictness of proofs. Here there is still another side; finding himself under constant influx of a set of ideas in the most diverse fields of mathematics, Poincaré "did not have time to be rigorous", he was often satisfied when his intuition gave him the confidence that the proof of such and such a theorem could be carried through to complete logical rigour and then assigned the completion of the proof to others. Among the "others" were mathematicians of the highest rank. I shall quote a letter of Poincaré to Brouwer (which belongs to the last year of Poincaré's life and, as far as I know, has never been published before), which seems to me to be a good illustration of how the idea was put into practice.

Letter of Poincaré to Brouwer

Mon cher Collègue,

je vous remercie beaucoup de votre lettre; je ne vois pas pourquoi vous doutez que la correspondance entre les deux variétés soit analytique; les modules des surfaces de Riemann peuvent s'exprimer analytiquement en fonctions des constantes des groupes fuchsiens; il est vrai que l'on ne doit donner á certaines variables que des valeurs réelles, mais les fonctions de ces variables réelles n'en conservent pas moins le caratère analytique.

Ou bien la difficulté provient elle à vos yeux de ce que l'une de ces variétés dépend non des constantes du groupe mais bien des invariants. Si je me rappelle bien, j'envisageais une variété dépendant des constantes des substitutions fondamentales du groupe; à un groupe correspondera alors une infinité discrète de points de cette variété; je subdivisais ensuite cette variété en variétés partielles, de telle façon, qu'à un groupe corresponde un seul point de chaque variété partielle (de la même façon que l'on décompose le plan en parallelogrammes des périodes, ou bien le cercle fondamental en polygones fuchsiens). Le caractère analytique de la correspondance ne m'en semble pas altéré.

En ce qui concerne la variété des surfaces de Riemann on peut être embarassé si l'on considère ces surfaces de la façon de Riemann; on pourrait se demander par exemple si l'ensemble de ces surfaces ne forme pas *deux* variétés séparées. La difficulté disparaît dès que l'on envisage ces surfaces *au point de vue de M. Klein*; la continuité, l'absence de singularités, la possibilité de passer d'une surface à l'autre d'une manière continue deviennent alors des vérités presque intuitives.

Je vous demande pardon de la façon découpée et du desordre de ces explications; je n'espère pas qu'elle vous satisfassent par ce que je vous les ai très mal présentées; mais je pense qu'elles vous ameneront à préciser les points que vous embarassent de façon que je puisse ensuite vous donner entière satisfaction. Je suis heureux d'avoir cette occasion d'entrer en rapport avec un homme de votre valeur.

Votre bien dévoué collègue
POINCARÉ

My dear colleague,

Thank you very much for your letter; I do not see why you doubt that the correspondence between the two manifolds is analytic, the moduli of Riemann surfaces can be expressed analytically as functions of the constants of Fuchsian groups; true, one must only give real values to certain variables, but the functions of these real variables preserve no less their analytic nature.

Or maybe you see the difficulty in the fact that one of the manifolds depends not on the constants of the group but on invariants. If I remember correctly, I envisaged a manifold dependent on the constants of the fundamental substitutions of the group; to one group there will then correspond a discrete infinity of points of this manifold; I then subdivided this manifold into submanifolds in such a way that to one group there corresponds a single point of each submanifold (in the same manner in which one divides up the plane into parallelograms of periods or the fundamental circle into Fuchsian polygons). The analytic nature of the correspondence does not seem to me to have been changed.

As for the manifold of Riemann surfaces, one can find oneself in difficulties if one considers the surfaces in Riemann's manner; one could ask oneself, for example, whether the set of these surfaces does not form *two* separate manifolds. The difficulty disappears if one considers these surfaces *from the point of view of Herr Klein*; continuity, the absence of singularities, the possibility of passing from one surface to another in a continuous manner, then become almost intuitive truths.

I apologize for the disjointed and disordered manner of these explanations; I do not hope that they will satisfy you, since I have presented them to you very poorly, but I think that they will help you to clarify the points which were causing you difficulty so that later I can give you complete satisfaction. I am happy to have this opportunity to be in touch with a man of your worth.

Your most sincere colleague,
Poincaré.

Date (from postmark) 10 December 1911.

This letter is interesting not only as an illustration of some features of Poincaré's creative manner, it shows also that Poincaré had a high opinion of the mathematical, that is, the topological works of Brouwer. This can concern only those works of Brouwer belonging to the two years 1901–1911. Clearly, Poincaré not only knew these works well by the end of 1911 (when the letter was written), but he appreciated their depth. Moreover, Brouwer's topological papers were written with a very high degree of difficulty and were not in the least in the classical style. Consequently, Poincaré, even in the last year of his life, found sufficient energy and curiosity to master mathematical results and methods due to a completely different manner of mathematical creativity than his own—a feature that is characteristic only of very great scholars! Poincaré possessed this trait throughout his entire life. In 1883 Cantor constructed a set that is completely discontinuous (on an interval) which bears his name (the "Cantor discontinuum"). This was a discovery of genius not only on account of the significance which the Cantor set acquired in the whole of mathematics, but also because it introduced into mathematics an entirely new construction, unlike anything known to science up to then. Cantor envisaged a geometrical image taken beyond the limit of what can be regarded as dependent on geometrical intuition, thus extending its horizons and the very possibilities of our concept of space in an unprecedented fashion. He showed, first of all, that these possibilities can be extended to transformations concerning that very theory of sets, of which some famous and respected mathematicians (for example, Kronecker) even doubted whether it belongs to mathematics at all. Poincaré was not only one of the first mathematicians to accept Cantor's discovery; he was positively the first to apply it to concrete analytical investigations, as the chemists say, in *statu nascendi*, at the very moment of birth of this new mathematical "nature", so unlike anything in the whole old science.

Many outstanding mathematicians made various remarkable special constructions, proceeding along the new path of goemetrical intuition laid down by Cantor—Brouwer constructed his first examples of non-decomposable continua, Antoine his remarkable arcs for which the fundamental group of the residual space is non-trivial, Alexander his "horned" spheres—and many others. But the first step was taken by Cantor, and Poincaré was the first who understood not only the significance of this first step, but also its fruitfulness for mathematical analysis, and hence for all mathematics. Finally, we should note that, as Poincaré's last paper shows, to the end of his life to a remarkable degree he was master of the technique of geometrical set theory as far as it had been developed at the time.

Let us now turn to the concept of homology, which was introduced by Poincaré. As we have already mentioned, the concept was introduced, in intuitive form, in Poincaré's first topological paper—the famous "Analysis Situs". However, in this case his insufficiently rigorous approach had, so to speak, practical consequences, because it served as basis for the valid criticism of the Norwegian mathematician Heegaard. The fact is that in his first paper Poincaré did not pay sufficient attention to the phenomenon of torsion, which is limited essentially by the Betti numbers. But in his next publication on topology (in "Supplementary notes to 'Analysis Situs'") he filled the gap brilliantly. Here Poincaré took his stand on the combinatorial point of view, introducing the concept of a simplicial decomposition (triangulation) of a manifold, that is, the concept of a simplicial complex, and so created the fundamental method of combinatorial topology. Probably Poincaré regarded it as intuitively clear that the homological characterization of a manifold (and, in general, a polyhedron[1]), which he introduced into topology, cannot depend on the choice of the triangulation of the polyhedron. However, as we know, this fact is a deep and difficult theorem of topology. For its proof, in addition to the concept of an arbitrarily small subdivision of a given triangulation, which Poincaré possessed, of course one also needs the concept (based on that of a subdivision) of a simplicial (that is, piecewise-linear) approximation to a continuous mapping (which is a generalization of the approximation of a continuous curve by a broken line inscribed in it), and one concept or another equivalent to that of the degree of a mapping (that is, the multiplicity with which, under a continuous mapping say, of a simplex X onto a simplex Y or of a manifold X onto another manifold Y of the same dimension—Y is covered by the image of X). Both these fundamental concepts were introduced by Brouwer in 1911, just before the death of Poincaré; with their aid Brouwer proved his famous theorems on the topological invariance of the number of dimensions of a n-dimensional manifold, and on the invariance of internal points for sets lying in them; Jordan's general theorem (in the n-dimensional case), theorems on fixed points etc. However, Brouwer did not prove the theorem on the invariance on the homological characteristics of a polyhedron itself, although he had all the necessary tools to do this; this was first done in 1915 by the famous American topologist Alexander.

The proof of the invariance theorem was the first essential step in the further development of the homology theory created by Poincaré. The next step, unlike the first, was not connected with overcoming specific mathematical difficulties, but it had, nevertheless, a great theoretical significance. It was made by the famous algebraist Emmy Noether (in 1925–26) and consisted

---

[1]I use the word "polyhedron" here in its modern sense, namely as a set permitting decomposition into simplexes; Poincaré himself used the word "polyhedron" in the sense in which we now use "complex".

in replacing the numerical homological characteristics given by Poincaré, the Betti numbers and torsion coefficients, by the single concept of the Betti group (or, as we now prefer to say, homology group). Some famous topologists, for example Lefschetz, at first had a sceptical attitude to the innovations introduced by Emmy Noether, regarding them as purely formal (in fact, it seemed that there was no essential difference whether one spoke directly of the Betti group of a polyhedron or of its completely defined set of numerical characteristics, namely its rank, that is, the Betti number, and its torsion coefficients. However, closer investigations showed that the matter is not merely one of words.

In particular, and above all, according to the old approach, without the concept of homology groups, the development of one of the most significant of topological theories would be impossible, the theory of topological duality, whose foundations were laid by Poincaré himself, and which was later developed in new directions and aspects by Alexander, and then, in all its depths, by Pontryagin and other mathematicians.

Without the concept of the Betti group it is impossible to imagine two further essential advances in homology theory. The first consists in the transfer of homological concepts to more general geometrical objects than polyhedra, and, first of all, to compacta. It became possible due to the common apparatus of approximation of the most complex topological formations (compacta, bicompacta and even more general topological spaces) by the combinatorial-topological constructions by complexes, which I started in 1926 to carry out by means of the so-called projections spectra (afterwards subjected to various generalizations and variations). This approximation process is based on my introduction of the concept of the nerve of a covering of a given space and makes it possible to transfer to practically all topological spaces the fundamental concepts of combinatorial topology.[1]

The second fundamental progress in the theory of homology was the introduction by J. Alexander and A. N. Kolmogorov in 1934–35 of "higher" homology, now called cohomology. From homology and cohomology groups, whose domain of definition and applicability was always expanding, there arose in the end a new mathematical discipline—homological algebra, which essentially determines the face of a considerable part of modern mathematics.

Even in the briefest dicussion of the theme of my speech today, I could not pass by in silence Poincaré's remarkable popular article "Why does space have three dimensions?" ("Pourquoi l'espace a trois dimensions?"), which was published in the famous French journal "Revue de Métaphysique et de Morale".

This article is notable in that, more in a literary than a strictly scientific form, it proposes the problem and discusses the idea of one of the fundamental concepts of set-theoretical topology—the problem and idea of the general inductive definition of dimension; Poincaré's idea consisted in the fact that if a space has dimension n, then it can be divided into parts (as small as we

---

[1]Of course, the first sources of my concept of nerve lie in what Poincaré called the "reciprocal polyhedron" (polyhèdre réciproque).

please) by means of subspaces of dimension $n-1$. The first mathematican who gave a strict and well-founded form to these very indefinite obvious opinions of Poincaré was Brouwer (in a paper of 1913). Thus, Brouwer was the founder of the broad field of set-theoretical topology which was developed by Menger and (principally) by P. S. Uryson, and is now known as the general theory of dimension. At this moment, however, it is important to stress that the idea of the general concept of dimension goes back to Poincaré and gives us a new proof of the exceptional power of his geometrical intuition, which includes on this occasion the field of set-theoretical topology. In conclusion, let us note that the general theory of dimension received its full development after I had constructed 1928–32 what is known as the homological theory of dimension which subordinates the concept of dimension to that of homology and thus includes the theory of dimension in general homological topology.

I began my discussion with the remark that Poincaré lived in an era when ideas were being born in mathematics that astonished our minds by the force of their applicability to our perception of the world, and also by their capacity as it were suddenly to expand the horizons of mathematics itself, and finally by their beauty and internal perfection.

The mathematical ideas being born in our time are just as powerful (if not more so) and perhaps just as beautiful. They might never have been developed, however, had not Poincaré's discoveries stood by their cradle.

In his famous "Exposition du système du monde", Laplace once said that astronomy by the magnitude of its subject and the perfection of its theory is the best monument raised by the understanding of man and the finest manifestation of his intellect.

Such mathematicians as Poincaré prompt us to extend Laplace's words to mathematics and give it the right to be counted as a rival to astronomy as regards the magnitude of its subject and certainly the perfection of its theory.

Appendix
ON POINCARÉ'S LETTER TO BROUWER

V. K. Zorin

·Unfortunately we do not have available the letter of Brouwer to which Poincaré wrote his reply, but one can nevertheless, with a certain degree of evidence and hence confidence, judge on the one hand from Poincaré's letter and on the other hand from the subjects of Brouwer's papers [1–4] judge what its contents were.

Brouwer's paper [3] may serve as a key here. He gave it on 27 September 1911, at a meeting of the German Mathematical Society, and it is devoted to giving a basis to the "method of continuity" (Kontinuitäts-methode) in the theory of automorphic functions. Brouwer writes: "... Poincaré ([7], 368-370) gives a proof of the existence of a linearly-polymorphic function on a Riemann surface by the method of continuity, accepting without discussion the following two assertions:

THEOREM 1 [CONFORMAL EQUIVALENCE—V. Z.]. *Classes of a Riemann surface of genus g form a* $(6g - 6)$-*dimensional manifold without singularities.*

THEOREM 2. *A one-to-one and continuous image of an n-dimensional domain in an n-dimensional manifold again forms a domain.*

Due to a small change in the method we can avoid applying Theorem 1... and, thus, it all reduces to a justification of Theorem 2—the theorem of the invariance of domains, a proof of which I shall publish in the near future. [4]"

Thus, evidently, Brouwer approached Poincaré for explanations about the method of continuity which he used in [7].

To explain Poincaré's answer, it is advisable first to say a few words about the method of continuity.

This method, which is outstanding by its geometrical obviousness and at the same time by its great generality, was born on the peaks of the theory of automorphic functions of a single complex variable. Klein and Poincaré came to it simultaneously.

In [5], Fricke wrote on this subject: "Klein developed his ideas on the proof of continuity in his paper ([6], 704). On the other hand, Poincaré (in [7], 329) discusses the same question, and in his first close consideration of the essence of the matter discovers the exceptional significance depth and difficulty of this kind of proof, and at the same time indicates ideas for overcrowding these difficulties."

Here is an explicit account of the idea of the method of continuity from the pen of Poincaré ([7], 330):

"Suppose that any point m of (a manifold) S is put in correspondence with a point m' of (a manifold) S' in such a way that the coordinates of m' are analytic functions of the coordinates of m as long as m does not lie on the boundary of S, provided that S has one. Let us assume that to each point of S' there corresponds not more than one point of S. If S is a *closed manifold*, then we can be certain that to each point of S' there corresponds a point of S. If, however, S is an open manifold with a boundary, it is impossible to assert anything... It is this fact which Klein ignores. Here there are difficulties, which cannot be smoothed out in a few words."

As one of the difficulties in applying the method of continuity in the theory of automorphic functions, Klein ([6], 704) mentions the proof that the correspondence constructed by him is *analytic*. He writes ([6], 704): "For the following discussion I need a proposition which I do not doubt to be correct although I do not have a concise proof. The matter is that the connection between the manifolds M_1 and M_2 is analytic. I have no doubt that this will be proved with the further development of such existence proofs... If difficulties should arise here, one can use the formulae established by Poincaré for the connection between M_1 and M_2".

Comparing the above quotations from Brouwer's lecture with the views of Poincaré and Klein, one may understand that the fundamental question with which Brouwer approached Poincaré was that of the analyticity of the mapping; the first part of Poincaré's letter is devoted to answering this question.

From the point of view of topology this, in essence, is the question of whether the mapping used in the method of continuity is open.

As for the second part of Poincaré's letter (which is, indeed, closely connected with the first) where he explains Brouwer's Theorem 1 (above) according to Klein and Poincaré himself, any Riemann surface S may be mapped conformally onto a factor space K/Γ, where K is a circle, plane, or sphere, and Γ is a group of linear fractional transformations of K, isomorphic to $\pi_1(S)$. Here the spaces K/Γ_1 and K/Γ_2 are conformally equivalent if and only if Γ_1 and Γ_2 are conjugate subgroups of the group of linear fractional transformations of K. Using this representation of a Riemann surface, it is not difficult to work out the number of different classes of conformally non-equivalent compact Riemann surfaces of genus g depends on $6g - 6$ real parameters.

Instead of the term "substitution" used by Poincaré we often say nowadays "linear" or "linear fractional" transformation. Poincaré's "fundamental substitutions" are generators of the given group Γ of linear fractional transformations by which the factorization K/Γ proceeds.

Let us note in conclusion that the purely topological problem of the invariance of a domain and of the number of dimensions, which was only solved rigorously by Brouwer in 1911–12 [4], was born within the framework of the geometrical theory of analytic functions, in attempts to give a logically complete foundation to the method of continuity.

BIBLIOGRAPHY

[1] L. E. J. Brouwer, Über die topologischen Schwierigkeiten des Kontinuitätsbeweises der Existenz-Theoreme eindeutig umkehrbar polymorpher Funktionen auf Riemannschen Flächen, Nachr. Akad. Wiss. Göttingen No. 5 (1912), 603–606.

[2] L. E. J. Brouwer, Über die Singularitätenfreiheit der Modulmannigfaltigkeit, Nachr. Akad. Wiss. Göttingen No. 7 (1912), 803–806.

[3] L. E. J. Brouwer, Über den Kontinuitätsbeweis für das Fundamentaltheorem der automorphen Funktionen in Grenzkreisfalle, Jber. Deutsch. Math.-Verein 21 (1912), 154–157.

[4] L. E. J. Brouwer, Beweis der Invarianz des n-dimensionalen Gebiets, Math. Ann. 71 (1911), 305–313.

[5] R. Fricke, Beiträge zum Kontinuitätsbeweise der Existenz linear-polymorpher Funktionen auf Riemannschen Flächen, Math. Ann. 59 (1904), 449–513.

[6] F. Klein, Neue Beiträge zur Riemannschen Funktionentheorie, Gesammelte Mathematische Abhandlungen, vol. III, Springer-Verlag, Berlin, 1923, 531–710.

[7] H. Poincaré, Sur les groupes des équations linéaires, Oeuvres, vol. II, Gauthier-Villars, Paris, 1916, 300-401.

Translated by V. Rich

Proceedings of Symposia in Pure Mathematics
Volume 39 (1983), Part 2

Résumé Analytique [1,2]

HENRI POINCARÉ

INTRODUCTION.

J'ai classé les travaux que j'ai à résumer sous les sept rubriques suivantes:

1°. Equations Différentielles.

2°. Théorie générale des Fonctions.

3°. Questions diverses de Mathématiques pures (Algèbre, Arithmétique, Théorie des Groupes, Analysis Situs).

4°. Mécanique Céleste.

5°. Physique Mathématique.

6°. Philosophie des Sciences.

7°. Enseignement, vulgarisation, divers (Bibliographie, rapports divers).

Inutile d'ajouter que je n'ai pas poursuivi tous ces buts différents indépendamment les uns des autres et qu'il y a entre eux plus d'une connexion imprévue. On s'apercevra en effet que quelques mémoires ont dû figurer plusieurs fois, sous deux ou trois rubriques différentes.

Les chiffres entre parenthèses, en caractères gras, renvoient aux numéros de la bibliographie.

---

[1]Reprinted from H. Poincaré, *Analyse de ses travaux scientifiques*, Acta Math. 38 (1921), 36–135.

[2]Editorial Note: The extract consists of pages 36–135 of the original article and omits the Bibliography which was listed by the Journals in which the articles were published. In its present form, the references are to the bibliography given in this volume. The summary was written in 1901 by Poincaré at the request of Mittag-Leffler.

PREMIÈRE PARTIE.

ÉQUATIONS DIFFÉRENTIELLES.

—

I. Généralités. (1, 2, 47, 49, 52, 54, 189 chap. II.)

Dès que les principes du Calcul infinitésimal furent établis, l'analyste se trouva en face de trois problèmes:

Résolution des équations algébriques;

Intégration des différentielles algébriques;

Intégration des équations différentielles.

L'histoire de ces trois problèmes est la même. Après de longs et vains efforts pour les ramener à des problèmes plus simples, les géomètres se sont enfin résignés à les étudier pour eux-mêmes, et ils en ont été récompensés par le succès.

Longtemps on a pu espérer que l'on pourrait résoudre toutes les équations par radicaux. On y a renoncé, et aujourd'hui les fonctions algébriques nous sont aussi bien connues que les radicaux auxquel on voulait les ramener. De même les intégrales de différentielles algébriques, que l'on a cherché longtemps à ramener aux fonctions logarithmiques ou trigonométriques, s'expriment aujourd'hui à l'aide de transcendantes nouvelles.

Il devait en être à peu près de même des équations différentielles. Le nombre des équations intégrables par quadratures est extrêmement restreint, et tant qu'on ne s'est pas décidé à étudier les propriétés des intégrales en elles-mêmes, tout ce domaine analytique n'a été qu'une vaste *terra incognita* qui semblait à jamais interdite au géomètre.

C'est CAUCHY qui y a pénétré le premier, grâce à l'invention d'une méthode ingénieuse qu'il a appelée *calcul des limites*. A sa suite, MM. FUCHS, BRIOT et BOUQUET, et Mme KOWALÉVSKI ont employé avec succès la même méthode.

Ce sont donc les travaux de ces géomètres qui m'ont servi de point de départ.

En présence d'un problème si compliqué, ces divers savants, au lieu d'étudier la manière d'être des intégrales des équations différentielles ou des équations aux dérivées partielles *pour toutes les valeurs de la variable*, c'est-à-dire dans tout le plan, se sont d'abord occupés de déterminer les propriétés de ces intégrales *dans le voisinage d'un point donné*. Ils avaient ainsi reconnu que ces propriétés sont très différentes selon qu'il s'agit d'un point ordinaire ou d'un point singulier. Dans le voisinage d'un point ordinaire, l'équation différentielle peut se mettre sous la forme

$$(\mathrm{1}) \qquad \frac{dy}{dx} = f(x, y),$$

et y peut se développer suivant les puissances de $x - x_0$.

Dans le voisinage d'un point singulier, l'équation différentielle peut se mettre sous l'une des deux formes

$$(2) \qquad (x - x_0)\frac{dy}{dx} = f(x, y),$$

$$(3) \qquad (x - x_0)^m \frac{dy}{dx} = f(x, y)$$

si elle est du premier ordre ou, sous des formes analogues, si elle est d'un ordre supérieur ou aux dérivées partielles. Dans le cas où l'équation différentielle se met sous la forme (2) (ou sous des formes analogues pour le second ordre ou les ordres supérieurs), MM. Briot et Bouquet avaient signalé certaines propriétés des intégrales, et M. Fuchs en avait donné le développement en séries dans le cas particulier des équations linéaires.

J'ai complété les résultats de Cauchy relatifs aux points ordinaires (189, Ch. II) en montrant dans quelles conditions la solution peut être développée non seulement suivant les puissances de la variable indépendante, mais suivant celles des valeurs initiales, ou celle d'un petit paramètre arbitraire, dans le cas où les équations différentielles dépendent d'un pareil paramètre. J'ai montré comment ces séries procédant suivant les puissances de ce paramètre ou des valeurs initiales peuvent encore rester convergentes non seulement pour les petites valeurs de la variable indépendante, mais pour des valeurs quelconques de cette variable. Mais je me suis surtout occupé d'étudier ce qui se passe dans le voisinage d'un point singulier.

J'ai cherché d'abord à étudier l'équation (2), suppossée non linéaire et à trouver le développement en séries de ses intégrales. J'ai reconnu (78)[1] que ces intégrales peuvent se développer suivant les puissances de $(x - x_0)$ et de $(x - x_0)^\lambda$, λ étant une constante facile à déterminer, ou, dans un cas particulier, suivant les puissances de $(x - x_0)$ et de $\log (x - x_0)$. Le résultat peut d'ailleurs s'étendre aux équations d'ordre supérieur.

J'ai voulu ensuite (2) étudier du même point de vue les équations aux dérivées partielles du premier ordre. CAUCHY et Mme KOWALEVSKI nous avaient appris comment on peut développer en séries les intégrales de ces équations dans le voisinage d'un point ordinaire. Il restait à étudier ces intégrales dans le voisinage d'un point singulier, comme l'avaient fait MM. BRIOT et BOUQUET pour les équations différentielles. En abordant ce problème, je rencontrai deux sortes de singularités: les premières accidentelles et spéciales à l'intégrale particulière que l'on envisage, les secondes essentielles et provenant de l'équation aux différences partielles elles-mêmes. Dans le premier cas, je vis aisément que les intégrales satisfont à des équations algébriques, dont les coefficients sont holomorphes par rapport aux variables. Dans le second cas, les difficultés à surmonter étaient plus grandes.

J'ai envisagé d'abord l'équation

$$(4) \qquad X_1 \frac{dz}{dx_1} + X_2 \frac{dz}{dx_2} + \cdots + X_n \frac{dz}{dx_n} = \lambda z,$$

où les x sont des fonctions holomorphes de $x_1, x_2, \ldots, x_n$ (quand ces variables sont suffisamment voisines de zéro) et s'annulent avec ces variables.

Pour que cette équation admette une intégrale holomorphe, il faut d'abord que λ satisfasse à une certaine équation algébrique de degré n; mais cette condition n'est pas suffisante; les racines de cette équation doivent de plus être assujetties à une condition spéciale: le polygone convexe qui contient tous les points du plan qui représentent ces racines ne doit pas contenir l'origine. Si cette condition est remplie, il y a toujours une intégrale holomorphe, et il n'y en a pas, en général, dans le cas contraire. Nous verrons plus loin quelles sont les conséquences de ce fait dans la théorie générale des fonctions.

Considérant ensuite les équations

$$\frac{dx_1}{X_1} = \frac{dx_2}{X_2} = \cdots = \frac{dx_n}{X_n},$$

---

[1] Les chiffres placés entre parenthèses renvoient aux numéros de la Bibliographie.

je reconnus que les intégrales de ce système sont de la forme

$$\frac{T_1^{\lambda_1}}{K_1} = \frac{T_2^{\lambda_2}}{K_2} = \cdots = \frac{T_n^{\lambda_n}}{K_n},$$

où les T sont des fonctions holomorphes par rapport aux x, où les λ sont les racines de l'équation algébrique dont nous venons de parler et les K des constantes d'intégration.

Cela n'est vrai d'ailleurs que si les λ satisfont à la condition énoncée plus haut, et, dans ce cas, il est possible de trouver le développement des diverses intégrales particulières de l'équation (4).

Tout ce que je viens de dire ne s'applique qu'aux points singuliers les plus simples, analogues à celui de l'équation (2). Pour les singularités d'ordre plus élevé, telles que celle que présente l'équation (3), on ne sait presque rien. Ces singularités d'ordre supérieur se présentent en particulier dans l'étude des équations linéaires, dont les intégrales sont dites alors *irrégulières;* mais, même dans ce cas spécial, nous ne savons à leur sujet que fort peu de chose.

M. Thomé, qui les a étudiés, a montré que les équations sont alors satisfaites formellement par des séries de la forme suivante

$$e^{P(x)}\, \varphi\left(\frac{1}{x}\right),$$

P_x étant un polynôme entier en x et $\varphi\left(\frac{1}{x}\right)$ étant une série ordonnée suivant les puissances décroissantes de x. (Je suppose ici, pour fixer les idées, qu'on a rejeté le point singulier à l'infini.) Mais, pour que ces séries représentassent les intégrales cherchées, il faudrait qu'elles fussent convergentes, ce qui n'a lieu que dans des cas très particuliers. J'eus l'idée d'appliquer à ces intégrales irrégulières la transformation de Laplace (3, 47, 52 et 54), et j'obtins ainsi sous une forme nouvelle et simple la condition de convergence de ces séries; mais le cas de la convergence n'était qu'exceptionnel, et il semblait que, dans le cas général, on ne pût rien tirer des développements de M. Thomé. Il n'en était rien. On connaît depuis longtemps une série, celle de Stirling, qui, bien que divergente, peut être légitimement employée pour représenter la fonction

$$\frac{\Gamma'(x)}{\Gamma(x)};$$

car, si x est très grand, l'erreur commise sur cette fonction en s'arrêtant dans la série à un terme de rang convenable est extrêmement petite. J'ai montré que les séries de M. THOMÉ jouissent de la même propriété. Alors même qu'elles sont divergentes, elles représentent les intégrales des équations proposées de la même manière que la série de STIRLING représente la fonction $\dfrac{\Gamma'(x)}{\Gamma(x)}$.

J'ai trouvé en outre, en passant, un certain nombre de propriétés des équations linéaires, entre autres celle-ci:

Si une équation linéaire d'ordre n a pour coefficients des polynômes entiers d'ordre $m\,(m < n)$, elle admettra $n - m$ intégrales holomorphes dans tout le plan.

Mais l'étude des intégrales des équations différentielles *dans le voisinage d'un point donné*, quelle que soit son utilité au point de vue du calcul numérique, ne saurait être regardée que comme un premier pas. Ces développements, qui ne sont valables que dans un domaine très limité, ne nous apprennent pas, au sujet de ces équations, ce que nous apprennent les fonctions Θ au sujet des intégrales des différentielles algébriques: ils ne peuvent pas être considérés comme une véritable intégration.

Il faut donc les prendre comme point de départ dans une étude plus approfondie des intégrales des équations différentielles où l'on se proposera de sortir des domaines limités, où l'on s'était systématiquement cantonné, pour suivre les intégrales dans toute l'étendue du plan.

Mais cette étude peut être faite à deux points de vue différents:

1°. On peut se proposer d'exprimer les intégrales à l'aide de développements *toujours* valables et non plus limités à un domaine particulier. On est conduit ainsi à introduire dans la Science de nouvelles transcendantes; mais cette introduction est nécessaire, car les fonctions anciennement connues ne permettent d'intégrer qu'un très petit nombre d'équations différentielles.

2°. Mais ce mode d'intégration, qui nous fait connaître les propriétés des équations au point de vue de la théorie des fonctions, ne saurait suffire à lui seul si l'on veut appliquer les équations différentielles, par exemple, à des questions de Mécanique ou de Physique. Nos développements ne nous apprendraient pas, à moins d'un travail considérable, si par exemple la fonction va constamment en croissant, ou si elle oscille entre certaines limites, si elle peut croître au delà de toute limite. En d'autres termes, si l'on considère la fonction comme définissant une courbe plane, on ne saurait pas quelle est la forme générale de cette courbe. Dans certaines applications, toutes ces questions ont autant d'importance que le calcul numérique, et il y avait là un nouveau problème à résoudre.

Dans les paragraphes qui vont suivre, je vais exposer les efforts que j'ai faits pour trouver la solution de ces deux problèmes.

II. Fonctions fuchsiennes. (5, 6, 8, 10, 11, 12, 13, 14, 15, 16, 17, 23, 24, 25, 26, 27, 28, 36, 37, 38, 88, 131, 32, 33 34, 43, 44, 29, 3, 58, 53, 59).

Désirant, comme je l'ai expliqué plus haut, exprimer les intégrales des équations différentielles à l'aide de séries *toujours* convergentes, j'étais naturellement conduit à m'attaquer d'abord aux équations linéaires. Ces équations, en effet, qui ont été dans ces derniers temps l'objet des travaux de MM. Fuchs, Thomé, Frobenius, Schwarz, Klein et Halphen, étaient les mieux connues de toutes; on possédait depuis longtemps les développements de leurs intégrales *dans le voisinage d'un point donné* et, dans un assez grand nombre de cas, on était parvenu à les intégrer complètement à l'aide des fonctions anciennement connues. C'était donc en en abordant l'étude que j'avais le plus de chances d'arriver à un résultat.

Mais il était nécessaire de plus de faire une hypothèse au sujet des coefficients des équations que je voulais étudier. Si j'avais pris, en effet, pour coefficients des *fonctions quelconques*, j'aurais obtenu également pour les intégrales des *fonctions quelconques* et, par conséquent, je n'aurais pu dire quelque chose de précis au sujet de la nature de ces intégrales, ce qui était mon but. J'étais donc conduit à examiner les équations linéaires à coefficients rationnels et algébriques. Je supposerai, pour simplifier un peu l'exposé qui va suivre, que les coefficients sont rationnels.

Voici maintenant la classification que j'ai adoptée pour ces équations linéaires et qui est la plus naturelle au point de vue du problème que nous voulons résoudre (27, 44). Soit γ une intégrale d'une équation linéaire d'ordre n à coefficient rationnels. Posons

$$(5) \qquad z = e^{\int \lambda\, dx}\left(F_{n-1}\frac{d^{n-1}\gamma}{dx^{n-1}} + F_{n-2}\frac{d_{n-2}\gamma}{dx^{n-2}} + \cdots + F_1\frac{d\gamma}{dx} + F_0\,\gamma\right),$$

λ et les F étant des fonctions rationnelles de x. Il est clair que z satisfera comme γ à une équation linéaire d'ordre n à coefficients rationnels. Je dirai que ces deux équations appartiennent à la même *famille*. On voit aisément, en effet, que la connaissance des propriétés de la fonction γ entraîne celle des propriétés de la fonction z.

Il y a dans chaque famille une infinité d'équations différentes, mais certaines fonctions des coefficients ont même valeur pour les équations d'une même famille; en d'autres termes, il y a, comme je l'ai montré dans ma Note du 22 mai 1882, des *invariants* qui demeurent inaltérés par la substitution représentée par l'équation (5). Ces invariants ne sont pas les mêmes que ceux de **M.**

HALPHEN. Ce savant géomètre envisage la transformation qui consiste à remplacer x par une fonction *quelconque* de x' et à multiplier y par une autre fonction *quelconque* de x'. Au contraire, les fonctions qui entrent dans ma substitution (5) ne sont pas *quelconques*, mais *rationnelles*. Rien ne saurait mieux faire comprendre la différence du point de vue de M. HALPHEN et du mien. M. HALPHEN cherche avant tout des relations entre diverses intégrales, et il peut impunément introduire dans ses calculs des fonctions quelconques; au contraire, mon but étant d'étudier la *nature* de l'intégrale elle-même, cette nature serait évidemment altérée, si je multipliais l'intégrale par une fonction quelconque, comme le fait M. HALPHEN.

Mais cette étude intime de la nature des fonctions intégrales ne peut se faire que par l'introduction de transcendantes nouvelles, dont je vais maintenant dire quelques mots. Ces transcendantes ont une grande analogie avec les fonctions elliptiques, et l'on ne doit pas s'en étonner, car si j'imaginais ces fonctions nouvelles, c'était afin de faire pour les équations différentielles linéaires ce qu'on avait à l'aide des séries Θ elliptiques et abéliennes, pour les intégrales des différentielles algébriques.

C'est donc l'analogie avec les fonctions elliptiques qui m'a servi de guide dans toutes mes recherches. Les fonctions elliptiques sont des fonctions uniformes qui ne sont pas altérées quand on augmente la variable de certaines périodes. Cette notion est tellement utile dans l'Analyse mathématique, que tous les géomètres ont dû penser depuis longtemps qu'il conviendrait de la généraliser en cherchant des fonctions uniformes d'une variable x qui demeurent inaltérées, quand on fait subir à cette variable certaines transformations, mais ces transformations ne peuvent pas être choisies d'une manière quelconque. Elles doivent évidemment former un groupe, et, de plus, on ne doit pas pouvoir trouver dans ce groupe une transformation infinitésimale, c'est-à-dire qui ne fasse varier x que d'un infiniment petit. Sans cela, en répétant indéfiniment cette transformation, on ferait varier x d'une façon continue, et notre fonction uniforme, qui ne serait pas altérée quand la variable augmenterait d'une manière continue, se réduirait à une constante. En d'autres termes, notre groupe doit être *discontinu* (3, 6, 32). Le premier problème à résoudre est donc de trouver tous les groupes discontinus que l'on peut former.

Dans le cas des fonctions elliptiques, les transformations du groupe (qui est évidemment discontinu) consistent à ajouter à x certaines constantes. Ici encore une nouvelle analogie avec les fonctions elliptiques peut nous venir en aide. Pour étudier ces fonctions, on divise le plan en une infinité de parallélogrammes connus sous le nom de *parallélogrammes des périodes*. On peut obtenir tous les parallélogrammes en transformant l'un d'eux par les diverses substitutions du groupe, de sorte que la connaissance de la fonction à l'intérieur de l'un des parallélogrammes entraîne sa connaissance dans tout le plan. De même, si nous envisageons un groupe discontinu plus compliqué, engendrant une transcendante d'ordre plus élevé, nous pourrons partager le plan (ou la région du plan où la fonction existe) en une infinité de régions ou de polygones curvilignes, de telle façon qu'on puisse obtenir toutes ces régions en appliquant à l'une d'elles toutes les transformations du groupe. La connaissance de la fonction à l'intérieur d'un de ces polygones curvilignes entraîne sa connaissance pour toutes les valeurs possibles de la variable.

Il est aisé de voir quelle est l'espèce particulière de groupes discontinus qu'il convient d'introduire. On se rapelle quel est le mode de génération des fonctions elliptiques: on considère certaines intégrales appelées *de première espèce*, ensuite, par un procédé connu sous le nom d'*inversion*, on regarde la variable x comme fonction de l'intégrale; la fonction ainsi définie est uniforme et doublement périodique.

De même ici, nous envisagerons une équation linéaire du second ordre et, par une sorte d'inversion, nous regarderons la variable x comme fonction, non plus d'une intégrale, mais du *rapport z* des deux intégrales de notre équation. Dans certains cas, la fonction ainsi définie sera uniforme, et alors elle demeurera inaltérée par une infinité de substitutions linéaires changeant z en $\dfrac{\alpha z + \beta}{\gamma z + \delta}$.

Pour cela, le groupe formé par ces substitutions doit être discontinu, et il est aisé de voir que les polygones curvilignes dont il a été question plus haut sont limités par des arcs de cercle. J'ai supposé d'abord que les coefficients des substitutions $\left(z, \dfrac{\alpha z + \beta}{\gamma z + \delta}\right)$ étaient réels ou, ce qui revient au même, que ces substitutions n'altéraient pas un certain cercle appelé *fondamental*. Dans ce cas, les arcs de cercle qui servent de côtés à nos polygones curvilignes sont orthogonaux à ce cercle fondamental.

Quelle est alors la condition pour que le groupe engendré par un polygone curviligne donné soit discontinu? Pour résoudre ce problème, il y avait à surmonter une difficulté spéciale que je veux expliquer en quelques mots. Partant du polygone curviligne générateur, on construit aisément les polygones voisins, puis les polygones voisins de ceux-ci, et ainsi de suite. On a ainsi une sorte de surface qui va sans cesse en s'accroissant, et ce qu'il s'agit de faire voir, c'est que cette surface ne va pas se recouvrir elle-même partiellement ou totalement, c'est-à-dire qu'un polygone nouvellement annexé à notre surface ne va pas recouvrir en partie un polygone anciennement construit. Pour cela, il ne suffit pas de remarquer que notre surface est simplement connexe et sans point de ramification (*unverzweigt*). Cette façon de raisonner n'est qu'un paralogisme qui a déjà entraîné quelques savants dans diverses erreurs et qui, dans le problème qui nous occupe, nous égarerait certainement. Il faut encore faire voir que la surface recouvre une partie du plan qui est elle-même simplement connexe (le contraire pourrait avoir lieu et une surface simplement connexe pourrait, en se recouvrant plusieurs fois elle-même, couvrir une région plane à connexion multiple). Ici la région simplement connexe, recouverte une fois et une seule par notre surface, est la superficie du cercle fondamental.

Il s'agit donc de démontrer qu'en construisant successivement tous nos polygones, comme je l'ai dit plus haut, on ne sortira jamais de ce cercle et qu'on atteindra forcément un point *quelconque* du cercle. La seconde de ces propositions m'aurait peut-être arrêté longtemps sans l'aide que j'ai trouvée dans une théorie fort différente: je veux parler de la *Géométrie non euclidienne*. Cette Géométrie, fondée sur l'hypothèse que la somme des angles d'un triangle est plus petite que deux droits, ne semble d'abord qu'un simple jeu de l'esprit qui n'a d'intérêt que pour le philosophe, sans pouvoir être d'aucune utilité au mathématicien. Il n'en est rien; les théorèmes de la géométrie de LOWATSCHEVSKI sont aussi vrais que ceux de la géométrie d'Euclide, à la condition qu'on les interprète comme ils doivent l'être. Ainsi, par exemple, ces théorèmes ne sont pas vrais de la ligne droite, telle que nous la concevons, mais ils le deviennent si, partout où LOWATSCHEVSKI dit une droite», nous disons un cercle qui coupe orthogonalement le cercle fondamental». Je me trouvais donc en présence de toute une théorie, imaginée, il est vrai, dans un but métaphysique, mais dont chaque proposition, convenablement interprétée, me fournissait un théorème applicable à la Géométrie ordinaire. Il se trouva qu'en combinant tous ces théorèmes, j'obtins aisément la solution de la difficulté dont j'ai parlé plus haut.

Je pus ainsi construire tous les groupes discontinus formés de substitutions, n'altérant pas le cercle fondamental, et je les appelai *groupes fuchsiens*.

Mais un problème important se posait: étant donné un groupe fuchsien, existe-t-il des fonctions uniformes inaltérées par les substitutions de ce groupe (33)? C'est ce que j'ai démontré et j'ai donné à ces fonctions le nom de M. FUCHS. Pour arriver à ce résultat, il eût été possible, dans certains cas particuliers, d'appliquer la proposition connue sous le nom de *principe de Dirichlet*, si souvent appliquée par RIEMANN et démontrée plus récemment par M. SCHWARZ. Je ne connaissais pas ce principe à cette époque, mais l'eussé-je connu, que je ne m'en serais pas servi; car il ne pouvait me donner la solution du problème

tion: je veux parler des fonctions hyperfuchsiennes imaginées par M. Picard. Mais, comme elles ne peuvent guère s'appliquer aux équations différentielles proprement dites, je me réserve d'y revenir dans la deuxième Partie, consacrée à la théorie générale des fonctions.

Les résultats déjà obtenus faisaient dès lors pressentir quel intérêt il y aurait à déterminer les coefficients du groupe d'une équation linéaire en fonction des coefficients de l'équation elle-même (36, 37, 43). Ce problème n'était pas nouveau et il avait déjà fait l'objet des travaux de divers mathématiciens allemands, entre autres de MM. Fuchs et Hamburger. J'ai imaginé de nouvelles méthodes de calcul numérique, analogues à celles de ces savants, et j'ai reconnu qu'on pouvait varier ces procédés à l'infini. Mais ces méthodes ne nous apprennent rien, au point de vue de la théorie générale des fonctions, sur les propriétés des transcendantes, dont elles donnent seulement la valeur numérique. Il fallait chercher aussi à résoudre le problème à ce nouveau point de vue. J'ai obtenu dans cette voie divers résultats qui peuvent présenter quelque intérêt. Ainsi les coefficients du groupe considérés comme fonctions de certains coefficients de l'équation (les autres coefficients étant regardés comme constants) en sont des fonctions entières. J'ai étudié également les fonctions inverses qui, dans certains cas, sont uniformes.

Les résultats ainsi obtenus ne donnaient encore qu'une solution bien incomplète du problème que je m'étais proposé, c'est-à-dire l'intégration des équations différentielles linéaires. Les équations que j'ai appelées plus haut *fuchsiennes*, et qu'on peut intégrer par une simple inversion, ne sont que des cas très particuliers des équations linéaires du second ordre. On ne doit pas s'en étonner si l'on réfléchit un peu à l'analogie avec les fonctions elliptiques. Le procédé de l'inversion ne permet de calculer que les intégrales elliptiques de première espèce. Pour les intégrales de deuxième et troisième espèce, il faut procéder d'une autre manière.

Envisageons, par exemple, l'intégrale de deuxième espèce

$$u = \int_0^x \frac{x^2\,dx}{V(1-x^2)(1-k^2x^2)}.$$

Pour l'obtenir, nous considérerons comme équation auxiliaire celle qui donne l'intégrale de première espèce

$$z = \int_0^x \frac{dx}{V(1-x^2)(1-k^2x^2)};$$

d'où par inversion

$$x = \operatorname{sn} z .$$

Remplaçant x par sn z, on trouve que u est égal à une fonction uniforme de z, $Z(z)$, qui augmente d'une constante quand z augmente d'une période. On est donc conduit à employer ici un procédé analogue: étant donnée une équation linéaire E d'ordre quelconque, à coefficients aigébriques en x, on se sert d'une équation auxiliaire E' du second ordre, et cette équation auxiliaire doit être choisie de telle façon que x soit fonction fuchsienne du rapport z des intégrales de E' et que les intégrales de E soient des fonctions uniformes de z.

Est-il toujours possible de faire ce choix de manière à satisfaire à toutes ces conditions? Telle est la question qui se pose naturellement. Cela revient d'ailleurs à demander si, parmi les équations linéaires qui satisfont à certaines conditions, qu'il est inutile d'énoncer ici, il y a toujours une équation fuchsienne. Je suis parvenu à démontrer qu'on devait répondre affirmativement à cette question. Je ne puis expliquer ici en quoi consiste la méthode dont nous nous sommes servis d'abord, M. KLEIN et moi, dans l'étude de divers exemples particuliers; comment M. KLEIN a cherché à appliquer cette méthode dans le cas général, ni comment j'ai comblé les lacunes qui subsistaient encore dans la démonstration du géomètre allemand, en introduisant une théorie qui a les plus grandes analogies avec celle de la réduction des formes quadratiques (16, 17, 25, 26, 43).

On peut arriver au même résultat par une voie entièrement différente, comme l'ont reconnu divers savants. Il suffit de démontrer que l'équation

$$\Delta u = e^u$$

admet toujours sur une surface de RIEMANN donnée une solution présentant des singularités données. M. PICARD a donné le premier une démonstration de ce théorème; j'en ai donné une (58, 59) qui est entièrement différente et qui permet de compléter le résultat de M. PICARD en l'étendant à un cas que ce géomètre avait laissé de côté et qui est important au point de vue de la théorie des fonctions fuchsiennes. C'est celui où l'un des sommets du polygone générateur se trouve sur le cercle fondamental.

Ainsi l'équation auxiliaire E' existera toujours; mais il ne suffit pas de pouvoir démontrer son existence, il faut encore savoir la former. C'est là l'objet de la dernière Partie de mon Mémoire *Sur les groupes des équations linéaires*. J'ai donné, dans cette dernière Partie, des procédés pour calculer les coefficients de l'équation E', non pas exactement, ce qui est impossible, mais avec une approximation aussi grande que l'on veut.

Si maintenant on considère le rapport z des intégrales de cette équation auxiliaire, x est une fonction fuchsienne de z que j'appelle $f(z)$, et les intégrales de l'équation E sont des fonctions uniformes de z, qui subissent des transformations linéaires lorsque z subit une transformation du groupe, de la même manière que la fonction $Z(z)$ augmente d'une constante quand z augmente d'une période (44). Ces fonctions uniformes jouent pour l'intégration de l'équation E le même rôle que la fonction $Z(z)$ joue pour le calcul des intégrales elliptiques de seconde espèce. C'est pour cette raison que je les ai appelées *zétafuchsiennes*.

Ces fonctions zêtafuchsiennes sont évidemment susceptibles d'être mises sous la forme du quotient de deux séries ordonnées suivant les puissances croissantes de z. Ces deux séries sont convergentes à l'intérieur du cercle fondamental. Si la fonction $f(z)$ n'existe qu'à l'intérieur du cercle fondamental (ce que nous supposons), la variable z ne peut *jamais* sortir de ce cercle, en sorte que nos deux séries sont toujours convergentes. D'ailleurs, les coefficients de ces séries se calculent aisément par récurrence. A ce point de vue, on peut donc déjà dire que ces développements nous donnent l'intégration complète de l'équation E, puisqu'ils sont toujours valables au lieu d'être limités à un domaine particulier. Je ne me suis cependant pas contenté de ce résultat, car il est possible de donner des fonctions zêtafuchsiennes des développements beaucoup plus satisfaisants pour l'esprit, parce que les termes sont liés les uns aux autres par une loi simple, et que, par conséquent, le développement met en évidence les propriétés caractéristiques de ces fonctions. C'est ainsi que l'expression de $\operatorname{sn} z$ par les séries d'EISENSTEIN est beaucoup plus satisfaisante pour l'esprit (quoique moins convergente) que le développement de cette fonction suivant les puissances de z et de k^2. C'est dans ce but que j'ai exprimé les fonctions zêtafuchsiennes, par le quotient de deux séries; le dénominateur est une série thêtafuchsienne et le numérateur est une série à termes rationnels, où l'expression du terme général est fort simple.

Ainsi, il est possible d'exprimer les intégrales des équations linéaires à coefficients algébriques, à l'aide des transcendantes nouvelles, de la même manière que l'on a exprimé, à l'aide des fonctions abéliennes, les intégrales des différentielles algébriques. D'ailleurs ces dernières intégrales elles-mêmes sont susceptibles d'être obtenues aussi par l'intermédiaire des fonctions fuchsiennes, et l'on en a ainsi une expression nouvelle, entièrement différente de celle où entrent les séries Θ à plusieurs variables.

III. Équations non linéaires. (22, 45, 46).

Il resterait à faire pour les équations non linéaires ce que j'ai fait pour les équations linéaires, c'est-à-dire trouver des développements des intégrales qui soient *toujours* convergents. Je n'ai pu y parvenir; j'ai seulement reconnu que l'on peut, d'une infinité de manières, exprimer ces intégrales par des séries qui convergent pour toutes les valeurs réelles de la variable. Voici comment j'ai opéré (24, 77).

Je mets les équations différentielles sous la forme

$$\frac{d x_1}{X_1} = \frac{d x_2}{X_2} = \cdots = \frac{d x_n}{X_n},$$

les X_n étant des polynômes entiers par rapport aux variables x. Cela est toujours possible. J'introduis ensuite une variable auxiliaire s définie par l'équation

$$\frac{dx_1}{X_1} = \frac{dx_2}{X_2} = \cdots = \frac{d x_n}{X_n} = \frac{ds}{X_1^2 + X_2^2 + \cdots + X_n^2 + 1}.$$

Je puis alors démontrer que si α est convenablement choisi, les variables x peuvent se développer suivant les puissances croissantes de

$$\frac{e^{\alpha s} - 1}{e^{\alpha s} + 1},$$

et que les développements restent valables pour toutes les valeurs réelles de s.

Si l'on applique ce qui précède au problème des trois corps, on verra que, quand s varie de $-\infty$ à $+\infty$, t varie de $-\infty$ à $+\infty$, de sorte que les développements restent convergents pour toutes les valeurs réelles du temps. Il n'y aurait d'exception que dans l'hypothèse, assez peu vraisemblable d'ailleurs, où deux corps viendraient se choquer à l'époque t_0, et les développements ne nous apprendraient rien sur ce qui se passerait après l'époque du choc; le problème d'ailleurs ne se pose même pas. Si de plus on *suppose* que les éléments initiaux aient été choisis de telle sorte que les distances mutuelles restent constamment supérieures à une limite donnée, on peut remplacer la variable auxiliaire s par le temps lui-même et développer suivant les puissances de

$$\frac{e^{\alpha t} - 1}{e^{\alpha t} + 1}.$$

Ainsi que je l'ai dit plus haut, je n'ai donné cette solution qu'à titre d'exemple.

Une pareille intégration est d'un caractère bien différent et évidemment beaucoup moins satisfaisante pour l'esprit que l'intégration des équations linéaires par les fonctions fuchsiennes. Aussi y avait-il lieu de se demander si les méthodes qui avaient réussi pour les équations linéaires n'étaient pas applicables à d'autres classes d'équations, quoiqu'elles ne le fussent pas dans le cas général.

Un peu de réflexion fait tout de suite comprendre quelle est la différence essentielle entre le cas général et celui des équations linéaires. Les équations linéaires n'ont qu'un nombre fini de points singuliers, tandis que les équations non linéaires en ont en général une infinité. On est donc amené à rechercher s'il n'existe pas d'autres classes d'équations dont les points singuliers soient en nombre fini.

M. Fuchs a publié, dans les *Sitzungsberichte* de l'Académie de Berlin, un Mémoire où il expose les conditions nécessaires et suffisantes pour qu'une équation différentielle et, en particulier, pour qu'une équation du premier ordre n'ait qu'un nombre fini de points singuliers. On put croire un instant que l'on était sur la voie d'une nouvelle catégorie de transcendantes uniformes et d'une nouvelle classe d'équations intégrables.

Je fus donc amené à faire un examen plus approfondi de la question (45, 46); mais cet examen m'obligea à renoncer à l'espoir que j'avais conçu. Les équations du premier ordre qui satisfont aux conditions de M. Fuchs, ou bien se ramènent à l'équation de Riccati et par elle aux équations linéaires, ou bien sont intégrables par les fonctions elliptiques ou algébriques. On n'est donc jamais conduit à une classe réellement nouvelle d'équations intégrables. M. Painlevé a été plus heureux en passant aux équations d'ordre supérieur.

Quoi qu'il en soit, le résultat de M. Fuchs conserve encore son intérêt, puisqu'il nous fait connaître une catégorie d'équations différentielles intégrables algébriquement. Mais, en tout cas, le problème de l'intégration des équations non linéaires ne peut être regardé comme résolu.

IV. Intégration des équations par les fonctions algébriques et abéliennes.
(7, 9, 39, 40, 55, 56, 57).

Bien que le problème de l'intégration des équations linéaires soit résolu dans le cas général par l'emploi de nos transcendantes nouvelles, ce résultat laisse subsister tout entier l'intérêt qui s'attache aux cas particuliers où l'intégration peut se faire au moyen de fonctions plus simples, telles que les fonctions algébriques, elliptiques et abéliennes. D'ailleurs les procédés d'intégration par les fonctions algébriques et elliptiques rentrent facilement dans la méthode générale qui comprend ainsi comme cas particuliers les procédés déjà connus. Il en résulte que cette méthode jette quelque lumière sur les difficultés qui se rapportent à l'emploi des procédés particuliers. En ce qui concerne la recherche des cas d'intégrabilité algébrique, le premier problème à résoudre était de former les groupes d'ordre fini contenus dans le groupe linéaire. Ce résultat a été obtenu par M. JORDAN il y a quelques années; mais je ne crois pas que ce savant ait démontré qu'à tout groupe d'ordre fini correspond une équation linéaire intégrable algébriquement. L'emploi des fonctions fuchsiennes m'a fait voir aisément (39, 40) qu'à tout groupe d'ordre fini correspond, non pas une, mais une infinité d'équations dont les intégrales sont algébriques. Pénétrant ensuite plus profondément dans la question, j'ai cherché à quelles conditions une fonction algébrique dont on se donne le groupe de GALOIS satisfait à une équation linéaire d'ordre p. J'ai trouvé que certains déterminants dont les éléments s'expriment tantôt à l'aide des racines de l'unité, tantôt à l'aide des périodes des intégrales abéliennes de première espèce correspondant à la fonction algébrique considérée, devaient être nuls à la fois. D'autre part, on peut, sauf dans certains cas exceptionnels, trouver un système fondamental d'intégrales de première espèce, tel que les périodes normales de l'une quelconque d'entre elles soient des fonctions linéaires à coefficients entiers des périodes normales de la première. Je fus ainsi conduit à exprimer la condition cherchée sous la forme de certaines relations entre les périodes normales des intégrales de première espèce qu'on peut former avec la fonction algébrique considérée.

Au contraire, les procédés d'intégration par les fonctions abéliennes ne rentrent pas dans la méthode générale. On y est conduit en cherchant à généraliser les méthodes d'intégration par les fonctions elliptiques (9). On sait que la théorie des fonctions elliptiques permet de calculer les intégrales des équations linéaires du second ordre dans trois cas entièrement différents:

1º. Lorsque, les coefficients étant rationnels, il y a trois points singuliers tels que la différence des racines des trois équations déterminantes soit respectivement $\frac{1}{2}$, $\frac{1}{3}$ et $\frac{1}{6}$, ou bien $\frac{1}{2}$, $\frac{1}{4}$ et $\frac{1}{4}$, ou bien encore $\frac{1}{2}$, $\frac{1}{3}$ et $\frac{1}{3}$.

2º. Lorsque, les coefficients étant rationnels, il y a quatre points singuliers tels que la différence des racines de chaque équation déterminante soit $\frac{1}{2}$;

3º. Lorsque, les coefficients étant doublement périodiques, les intégrales n'offrent d'autre singularité que des pôles.

M. APPELL a généralisé le troisième cas en montrant que, lorsque le groupe de l'équation linéaire se réduit à un faisceau, la dérivée logarithmique de certaines intégrales est algébrique et que l'intégration peut s'effectuer par les fonctions abéliennes. J'ai voulu de même généraliser le premier et le second cas.

Je suis arrivé ainsi à une infinité d'équations linéaires du troisième ordre à coefficients algébriques dont les intégrales s'expriment à l'aide des fonctions abéliennes de deux variables. De même, les fonctions abéliennes à p variables permettent d'intégrer une infinité d'équations linéaires d'ordre $p + 1$.

J'ai indiqué ensuite succinctement les principales propriétés des groupes de ces équations.

Je me suis préoccupé aussi de rechercher des cas où les équations non linéaires sont susceptibles d'intégration algébrique, mais je me suis restreint aux équations du 1er ordre et du 1er degré.

La voie avait été ouverte par M. DARBOUX. J'e l'ai suivie à mon tour dans deux mémoires insérés aux *Rendiconti del Circolo Matematico di Palermo* (56, 57). Le problème doit se poser ainsi: étant donnée une équation différentielle, reconnaître par un nombre fini d'opérations, si elle est ou non intégrable algébriquement. Ce problème pourrait évidemment être regardé comme résolu si l'on savait déterminer une limite supérieure du degré de l'intégrale générale, si on la suppose algébrique.

Après avoir donné un certain nombre de relations numériques entre le degré de l'intégrale générale, son genre, le nombre des points singuliers des diverses espèces, le nombre des valeurs remarquables pour lesquelles la courbe algébrique qui représente l'intégrale générale se décompose et les degrés des composantes, je résous le problème dans un cas particulier, celui où les deux entiers caractéristiques relatifs à tous les cols sont égaux à 1.

Dans le cas général, je n'ai obtenu que des résultats partiels; j'ai par exemple limité le nombre des valeurs remarquables pour lesquelles notre courbe se décompose et le nombre des composantes; mais il y a encore deux entiers qui jouent un rôle dans le problème et qui restent inconnus, ce qui m'empêche de limiter le degré, sauf dans certains cas particuliers.

Quand on veut aller plus loin, les inégalités algébriques dont je me suis servi ne peuvent plus suffire et les difficultés à vaincre sont d'une nature pour ainsi dire arithmétique; c'est ce que je montre sur un exemple simple où ces difficultés peuvent être surmontées grâce à l'emploi des fonctions elliptiques.

Je termine par une étude de ce qui se passe dans le voisinage de certains points singuliers. Dans le voisinage d'un noeud quelconque, on peut trouver deux séries infinies X_1 et X_2 procédant suivant les puissances des variables et qui égalées à zéro, fournissent deux solutions particulières de l'équation différentielle. On voit alors que l'intégrale générale si elle est algébrique se réduit à une fonction rationnelle homogène de deux puissances entières de X_1 et X_2. Or à chaque nœud correspondra une semblable expression de notre intégrale générale. Si nous égalons deux de ces expressions, la discussion de l'égalité ainsi obtenue, discussion où s'introduisent les fonctions fuchsiennes, conduit à plusieurs résultats importants.

V. Courbes définies par les équations différentielles. (4, 19, 21, 18, 20, 50, 51).

Alors même qu'on parviendrait à faire pour une équation quelconque ce que j'ai fait pour les équations linéaires, c'est-à-dire à trouver des développements des intégrales valables dans toute l'étendue du plan, ce ne serait pas une raison pour laisser de côté les résultats que l'on peut obtenir par d'autres méthodes, car il peut arriver que ces méthodes nous fassent découvrir certaines particularités que les développements ne mettraient pas immédiatement en évidence. C'est ce qui m'a décidé à me placer à un point de vue nouveau et je ne saurais mieux le faire comprendre qu'en reproduisant ce que j'écrivais au moment où je commençais ces recherches[1]:

»Il est donc nécessaire d'étudier les fonctions définies par des équations différentielles en elles-mêmes et sans chercher à les ramener à des fonctions plus simples, ainsi qu'on a fait pour les fonctions algébriques, qu'on avait cherché à ramener à des radicaux et qu'on étudie maintenant directement, ainsi qu'on a fait pour les intégrales de différentielles algébriques, qu'on s'est efforcé longtemps d'exprimer en termes finis.

Rechercher quelles sont les propriétés des équations différentielles est donc une question du plus haut intérêt. On a déjà fait un premier pas dans cette voie en étudiant la fonction proposée *dans le voisinage d'un des points du plan.* Il s'agit aujourd'hui d'aller plus loin et d'étudier cette fonction *dans toute l'étendue du plan.* Dans cette recherche, notre point de départ sera évidemment ce que l'on sait déjà de la fonction étudiée *dans une certaine région du plan.*

L'étude complète d'une fonction comprend deux parties: 1º partie qualitative (pour ainsi dire), ou étude géométrique de la courbe définie par la fonction; 2º partie quantitative, ou calcul numérique des valeurs de la fonction.

Ainsi, par exemple, pour étudier une équation algébrique, on commence par rechercher, à l'aide du théorème de STURM, quel est le nombre des racines réelles: c'est la partie qualitative; puis on calcule la valeur numérique de ces

---

[1] *Journal de Liouville,* 3ᵉ série, t. VII.

racines, ce qui constitue l'étude quantitative de l'équation. De même, pour étudier une courbe algébrique, on commence par *construire* cette courbe, comme on dit dans les cours de Mathématiques spéciales, c'est-à-dire qu'on cherche quelles sont les branches de courbes fermées, les branches infinies, etc. Après cette étude qualitative de la courbe, on peut en déterminer exactement un certain nombre de points.

C'est naturellement par la partie qualitative qu'on doit aborder la théorie de toute fonction et c'est pourquoi le problème qui se présente en premier lieu est le suivant: *Construire les courbes définies par des équations différentielles.*

Cette étude qualitative, quand elle sera faite complètement, sera de la plus grande utilité pour le calcul numérique de la fonction, et elle y conduira d'autant plus facilement que l'on connaît déjà des séries convergentes qui représentent la fonction cherchée dans une certaine région du plan, et que la principale difficulté qui se présente est de trouver un guide sûr pour passer d'une région ou la fonction est représentée par une série à une autre région du plan où elle est exprimable par une série différente.[1]

D'ailleurs cette étude qualitative aura par elle-même un intérêt de premier ordre. Diverses questions fort importantes d'Analyse et de Mécanique peuvent en effet s'y ramener. Prenons par exemple le problème des trois corps: ne peut-on pas se demander si l'un des corps restera toujours dans une certaine région du ciel ou bien s'il pourra s'éloigner indéfiniment; si la distance de deux corps augmentera, ou diminuera à l'infini, ou bien si elle restera comprise entre certaines limites? Ne peut-on pas se poser mille questions de ce genre, qui seront toutes résolues quand on saura construire qualitativement les trajectoires des trois corps? Et, si l'on considère un nombre plus grand de corps, qu'est-ce que la question de l'invariabilité des éléments des planètes, sinon une véritable question de géométrie qualitative, puisque faire voir que le grand axe n'a pas de variations séculaires, c'est montrer qu'il oscille constamment entre certaines limites?

Tel est le vaste champ de découvertes qui s'ouvre devant les géomètres. Je n'ai pas eu la prétention de le parcourir tout entier, mais j'ai voulu du moins en franchir les frontières, et je me suis restreint à un cas très particulier, celui qui se présente d'abord tout naturellement, c'est-à-dire à l'étude des équations différentielles du premier ordre et du premier degré.»

---

[1] Ces considérations m'ont effectivement servi de guide dans des recherches relatives au calcul numérique de la fonction (22).

Je commençai donc mes recherches (4, 18) par l'étude des courbes définies par les équations différentielles de la forme

$$(1) \qquad \frac{dx}{X} = \frac{dy}{Y},$$

où X et Y sont des polynômes entiers en x et y, et je reconnus d'abord que ces courbes pouvaient représenter la forme de courbes fermées ou celle de spirales. Je démontrai également le théorème suivant:

Si une courbe définie par une équation de la forme (1) n'a pas de point d'arrêt et ne coupe aucune courbe algébrique qu'en un nombre fini de points réels, elle est une courbe fermée.

Pour pousser plus loin l'étude de la forme de ces courbes, j'ai dû commencer par rechercher ce qui se passe dans le voisinage d'un point singulier quelconque. En réalité, le problème était résolu par les travaux antérieurs de MM. Briot et Bouquet et par les miens (*Journal de l'École Polytechnique*, XLVᵉ Cahier, et Thèse inaugurale), mais j'avais à approprier la solution à mon nouveau but; dans les Mémoires que je viens de citer, et où je me plaçais au point de vue de la théorie des fonctions, j'attachais une égale importance au réel et à l'imaginaire. Pour mon but nouveau de géométrie qualitative, le réel seul m'intéressait et je devais faire une discussion spéciale qui me conduisît à distinguer quatre sortes de points singuliers (sans parler de points singuliers plus compliqués qui ne se présentent que dans certains cas particuliers et qui peuvent être regardés comme composés de plusieurs points singuliers simples confondus).

J'ai donné à ces quatre sortes les noms suivants:

1º. Les *cols*, par lesquels passent deux courbes définies par l'équation et deux seulement;

2º. Les *nœuds*, où viennent se croiser une infinité de courbes définies par l'équation;

3º. Les *foyers*, autour desquels ces courbes tournent en s'en rapprochant sans cesse à la façon d'une spirale logarithmique;

4º. Les *centres*, autour desquels ces courbes se présentent sous la forme de cycles fermés s'enveloppant mutuellement et enveloppant le centre. (On ne rencontre les centres que dans des cas très exceptionnels.)

J'ai étudié ensuite la distribution de ces divers points singuliers dans le plan. J'ai montré ainsi qu'il y en avait toujours (à distance finie ou infinie) et qu'il y avait toujours une relation simple entre le nombre des cols, des

que dans certains cas particuliers et, même dans ces cas, il pouvait servir à démontrer l'existence de la fonction, mais il n'en donnait pas le développement analytique.

C'est encore à l'analogie avec les fonctions elliptiques que j'ai dû faire appel. On sait que ces fonctions peuvent être regardées comme le quotient de deux transcendantes, non plus seulement uniformes, mais encore entières, et que l'on appelle les séries Θ. Les fonctions ne sont plus doublement périodiques, mais elles sont multipliées par une exponentielle quand la variable augmente d'une période. De même ici, je devais chercher à exprimer les fonctions fuchsiennes par le quotient de deux trancendantes finies et uniformes, tout à fait analogues aux fonctions Θ, et se reproduisant multipliées par un facteur simple, quand la variable z subit une des transformations du groupe.

Je trouvai aisément des séries satisfaisant à ces conditions et je les appelai *thêtafuchsiennes*. Le quotient de deux pareilles séries était évidemment une fonction fuchsienne: j'avais donc du même coup démontré l'existence de ces fonctions et trouvé leur expression analytique. Le quotient de l'unité par une série thêtafuchsienne est susceptible aussi d'un développement simple, et c'est la considération de ces développements nouveaux qui m'a permis de démontrer réciproquement que toute fonction fuchsienne peut être regardée comme le quotient de deux séries thêtafuchsiennes.

Ces fonctions fuchsiennes sont de deux sortes, les unes existant dans tout le plan, les autres n'existant qu'à l'intérieur du cercle fondamental. Dans les deux cas, il y a entre deux fonctions fuchsiennes qui ont même groupe une relation algébrique. La détermination du *genre* de cette relation est d'une importance capitale; je l'ai obtenue d'abord par des procédés analytiques, et plus simplement ensuite par la géométrie de situation.

Grâce à ces relations algébriques, il est possible d'utiliser les fonctions fuchsiennes pour l'étude des fonctions et des courbes algébriques. Ainsi l'*on peut exprimer les coordonnées des points d'une courbe algébrique par des fonctions fuchsiennes, c'est-à dire uniformes, d'un même paramètre.* On peut alors se servir de ces expressions des coordonnées pour arriver à un certain nombre de théorèmes sur ces courbes. On peut s'en servir également pour exposer d'une façon plus simple la théorie des fonctions abéliennes.

Si, dans une intégrale abélienne de première espèce, on remplace la variable par une fonction fuchsienne de z, cette intégrale devient à son tour une fonction uniforme de z dont on trouve aisément le développement analytique. Ainsi ces intégrales, qu'on savait déjà obtenir à l'aide des fonctions Θ, sont susceptibles d'une expression analytique entièrement différente, où entrent des transcendantes ne dépendant que d'une seule variable.

Mais ce n'est pas tout. Toute fonction fuchsienne peut être regardée comme provenant de l'inversion d'une équation du second ordre à coefficients algébriques, c'est-à-dire qu'on peut l'obtenir en regardant la variable x comme fonction du rapport z des intégrales de cette équation. Nos transcendantes nous fournissent donc immédiatement l'intégration d'une infinité d'équations linéaires que l'on peut appeler *fuchsiennes*.

Pour que l'analogie avec les fonctions elliptiques fût complète, il faudrait que les autres propriétés de ces fonctions, telles que les lois d'addition, de multiplication et de transformation pussent s'étendre aux nouvelles transcendantes.

La théorie de la transformation se généralise immédiatement, avec cette différence toutefois que le groupe des fonctions fuchsiennes étant beaucoup plus compliqué que celui des fonctions elliptiques, les cas à considérer sont beaucoup plus nombreux et variés. . Ce qui en fait surtout l'intérêt, c'est qu'on peut s'en servir pour jeter quelque lumière sur la question de la réduction des intégrales abéliennes (88). J'y reviendrai plus loin.

Au contraire, le théorème d'addition ne peut pas s'étendre à toutes les fonctions fuchsiennes. Cela n'est possible que dans un. cas particulier et pour une classe spéciale de ces transcendantes (131, 53). Je veux parler de ces fonctions fuchsiennes qui tirent leur origine de la considération des formes quadratiques ternaires indéfinies et sur lesquelles je reviendrai dans le paragraphe relatif à l'Arithmétique.

Les substitutions linéaires dont les coefficients ne sont plus réels, mais quelconques, peuvent aussi former des groupes discontinus que j'ai appelés *kleinéens* (13, 14, 34). Pour démontrer l'existence de ces groupes, je rencontrais la même difficulté que pour les groupes fuchsiens, et il semblait au premier abord impossible d'appliquer la géométrie non-euclidienne. Dans certains cas particuliers la difficulté était facile à surmonter; mais, dant le cas général, elle subsistait tout entière. J'imaginai alors un artifice qui me permît de me servir de la géométrie non euclidienne, non plus à deux, mais à trois dimensions, et je démontrai aisément l'existence des groupes kleinéens. Je n'avais plus qu'à appliquer les méthodes qui m'avaient réussi une première fois pour trouver une nouvelle catégorie de fonctions tout à fait analogues aux fonctions fuchsiennes. La seule différence digne d'être signalée est celle qui résulte de la forme du domaine à l'intérieur duquel ces fonctions existent. Ce domaine, au lieu d'être un cercle, est limité par une courbe non analytique qui n'a pas de rayon de courbure déterminé. Dans d'autres cas, ce domaine est limité par une infinité de circonférences.

Les fonctions fuchsiennes sont susceptibles d'une autre mode de généralisa-

foyers et des centres, et que, sur la courbe $X = 0$, les cols ou les nœuds et foyers se succédaient alternativement.

Ces problèmes résolus, je me suis occupé des contacts que peut avoir une courbe algébrique donnée avec les courbes définies par l'équation (1) et j'ai vu que, dans un très grand nombre de cas, il existe des branches de courbes fermées qui ne touchent en aucun point aucune des courbes qui satisfont à notre équation différentielle. Je les ai appelées *cycles sans contact* (20).

Il est facile de comprendre l'importance de la détermination des cycles sans contact; on voit aisément en effet qu'une courbe définie par l'équation (1) ne peut rencontrer un pareil cycle en plus d'un point. Si donc on imagine un point mobile décrivant notre courbe, dès qu'il sera sorti d'un cycle sans contact, il n'y pourra plus rentrer. En d'autres termes, si ce point a occupé une fois une position donnée, il ne pourra plus jamais y revenir, ni même revenir dans le voisinage immédiat de cette position. Les coordonnées du point n'oscilleront pas entre certaines limites et ne pourront être représentées par des séries trigonométriques, de sorte que, si l'on voulait appliquer à la trajectoire de ce point mobile le même langage qu'emploient les astronomes pour les orbites des planètes, il faudrait dire que l'orbite de ce point est *instable*.

Outre les cycles sans contact, il y a un autre genre de courbes fermées qui jouent un rôle capital dans cette théorie: ce sont les *cycles limites*. J'appelle ainsi les courbes fermées qui satisfont à notre équation différentielle et dont les autres courbes définies par la même équation se rapprochent asymptotiquement sans jamais les atteindre. Cette seconde notion n'est pas moins importante que la première. Supposons en effet que l'on ait tracé un cycle limite; il est clair que le point mobile dont nous parlions plus haut ne pourra jamais le franchir et qu'il restera toujours à l'intérieur de ce cycle, ou toujours à l'extérieur. Il est vrai que, les cycles limites sont en général des courbes transcendantes qu'on ne saurait tracer exactement. Mais on peut souvent tracer deux courbes algébriques fermées, concentriques l'une à l'autre, déterminant une sorte d'anneau, de telle façon qu'on peut distinguer dans le plan trois régions, l'intérieur de l'anneau, la région annulaire et l'extérieur de l'anneau. Supposons que l'on ait démontré d'une manière quelconque que le cycle limite se trouve dans la région annulaire; on sera certain alors que, si notre point mobile est à l'intérieur de l'anneau, il ne pourra jamais aller à l'extérieur de cet anneau. On peut donc, malgré l'*instabilité* de ce point mobile, assigner des limites supérieures à ses coordonnées.

Je reconnus ensuite qu'on pouvait dans tous les cas sillonner le plan par une infinité de courbes fermées, s'enveloppant mutuellement et rappelant par leur forme et leur disposition les courbes de niveau d'un plan topographique. Pour poursuivre cette comparaison, je dirai que, dans ce plan topographique, les sommets et les fonds seraient représentés par les nœuds et les foyers, et les cols par les points singuliers que j'ai appelés plus haut de ce nom. Parmi ces courbes fermées, les unes sont des cycles sans contact, les autres sont des cycles limites. A part ces cycles limites, les courbes définies par notre équation différentielle sont des spirales se rapprochant asymptotiquement des points singuliers et des cycles limites.

Après avoir démontré que le nombre des cycles limites est fini, sauf dans certains cas exceptionnels, j'ai donné une méthode générale pour déterminer ce nombre et pour tracer des régions annulaires dans lesquelles se trouve un cycle limite, et un seul.

A la fin du Mémoire, j'ai donné plusieurs exemples d'applications de cette méthode. Je citerai seulement le dernier de ces exemples, celui de l'équation

$$\frac{dx}{-y + x(x^2+y^2-2x-3)(x^2+y^2-2x-8)} = \frac{dy}{x + y(x^2+y^2-2x-3)(x^2+y^2-2x-8)}.$$

J'ai divisé le plan en quatre régions, limitées par les trois cercles[1]

$$(2) \qquad x^2 + y^2 = 1, \qquad x^2 + y^2 = 2x + 5.5, \qquad x^2 + y^2 = 16,$$

qui s'enveloppent mutuellement. De ces quatre régions, la deuxième et la troisième contiennent un cycle limite et n'en contiennent qu'un, les deux autres n'en contiennent pas. Il suit de là que si, à l'origine des temps, notre point mobile est à l'intérieur du premier des cercles (2), il ne pourra jamais sortir du second et que, s'il est à l'intérieur du second, il ne pourra jamais sortir du troisième.

Il y a un cas particulier qui mérite de fixer l'attention, bien qu'il ne se présente que très exceptionnellement: c'est celui où toutes les courbes définies par l'équation (1) sont des courbes fermées qui s'enveloppent mutuellement à la façon des courbes de niveau d'un plan topographique. C'est là le seul cas où, pour employer de nouveau une comparaison empruntée à l'Astronomie, le point mobile dont il a été question plus haut a une orbite *stable*. C'est le seul cas, en effet, où l'on ne puisse pas sillonner le plan de cycles sans contact (50).

---

[1] De telle façon que la première région soit intérieure au premier des cercles (2), la deuxième comprise entre le premier et le deuxième de ces cercles, la troisième comprise entre le deuxième et le troisième, et la quatrième région extérieure au troisième cercle.

Pour que ce cas particulier se présente, il faut une infinité de conditions, et l'on pourrait croire d'abord qu'il est impossible de reconnaître si elles sont toutes remplies à la fois. Cela est, au contraire, le plus souvent très facile, et l'on démontre *a priori* que ces conditions doivent être toutes satisfaites, dans un certain nombre de cas, et, en particulier, quand on a

$$\frac{dX}{dx} + \frac{dY}{dy} = 0.$$

J'ai appliqué ces principes à une équation différentielle rencontrée par DELAUNAY dans la théorie de la Lune.

J'abordai ensuite (19, 50) l'étude des équations du premier ordre et de degré supérieur de la forme suivante

$$(3) \qquad F\left(x, y, \frac{dy}{dx}\right) = 0,$$

F désignant un polynôme entier en x, y et $\frac{dy}{dx}$. Pour étudier plus facilement cette équation, j'emploie trois variables auxiliaires ξ, η, ζ, liées aux variables primitives, de telle façon que x, y et $\frac{dy}{dx}$ soient des fonctions rationnelles de ξ, η et ζ, et je considère ces trois variables comme les coordonnées d'un point dans l'espace. L'équation (3) signifie alors que ce point est situé sur une certaine surface algébrique. J'ai soin de choisir mes nouvelles variables, de telle façon que cette surface n'ait pas de nappes infinies et se réduise à un certain nombre de nappes fermées. J'envisage en particulier une de ces nappes, que j'appelle S. Grâce aux conventions faites, par chaque point non singulier de S passera une courbe définie par l'équation (3) et une seule. Quant aux points singuliers, ils se subdivisent en cols, en foyers, ne nœuds et en centres et jouissent des mêmes propriétés que les points que j'ai appelés plus haut de ces noms.

Une notion qui joue ici un rôle capital, c'est le *genre* de la nappe S. Je dirai que cette nappe est de genre 0, si elle est convexe à la façon d'une sphère; de genre 1, si elle présente un *trou* à la façon d'un tore; de genre 2, si elle présente deux trous, etc.

J'ai démontré une relation très simple entre le genre de cette nappe et le nombre des cols, des foyers et des nœuds qui s'y trouvent. C'est la généralisation d'une relation dont j'ai parlé plus haut et qui s'applique aux équations du premier ordre et du premier degré.

La suite de la discussion est d'ailleurs tout à fait la même que pour les courbes définies par l'équation (I), c'est-à-dire par une équation du premier degré. La nappe S est sillonnée d'une infinité de courbes fermées, qui sont des cycles sans contact ou des cycles limites; il y a toutefois une différence essentielle sur laquelle je désirerais appeler l'attention. Supposons, par exemple, que la nappe S soit un tore et qu'un cercle méridien de ce tore soit un cycle sans contact; contrairement à ce que nous avons remarqué dans le cas des équations du premier degré, rien ne s'oppose à ce qu'une courbe définie par notre équation différentielle vienne couper ce cercle méridien en plusieurs points et même en une infinité de points. Si cela arrive et qu'un point mobile décrive cette courbe en partant d'une position initiale donnée, il finira toujours par revenir dans une position aussi voisine qu'on le voudra de cette position initiale. On pourra donc dire que ce point mobile décrit une trajectoire *stable*.

Ainsi la stabilité qui, lorsqu'il s'agissait des équations du premier degré, ne se présentait que dans des cas très particuliers, n'est plus une exception quand il s'agit d'équations de degré supérieur.

D'ailleurs les points, en nombre infini, où le point mobile vient successivement rencontrer le cercle méridien, jouissent d'une propriété arithmétique inattendue.

Appelons μ un certain nombre incommensurable; appelons M_i le point où le point mobile vient rencontrer pour la $i^{\text{ième}}$ fois le cercle méridien. Cherchons dans quel ordre circulaire ces points M_i se rencontrent sur ce cercle. *Cet ordre sera le même que celui des nombres* $\mu i - E(\mu i)$.

Passons maintenant (21, 51) aux équations du second ordre, que j'écrirai sous la forme suivante

$$(4) \qquad \frac{dx}{X} = \frac{dy}{Y} = \frac{dz}{Z},$$

X, Y et Z désignant des polynômes entiers en x, y et z, et les variables x, y et z étant regardées comme les coordonnées d'un point dans l'espace. Nous pouvons alors étudier les courbes qui satisfont à ces équations et que j'appellerai les courbes C, et nous verrons que par chaque point de l'espace vient passer une courbe C et une seule, si toutefois on excepte les points singuliers, c'est-à-dire les points d'intersection des trois surfaces

$$(5) \qquad X = 0, \qquad Y = 0, \qquad Z = 0.$$

L'étude de ces points singuliers s'imposait tout d'abord. Je reconnus qu'il y en a de quatre sortes (sans parler des points singuliers que ne se rencontrent que très exceptionnellement, par exemple, les centres):

1°. Les *nœuds*, où viennent converger toutes celles des courbes C qui passent assez près du point singulier;

2°. Les *cols*, où viennent converger une infinité de ces courbes dont l'ensemble forme une surface et où passe, en outre, une autre courbe satisfaisant à l'équation et non située sur cette surface;

3°. Les *foyers*, où passe une courbe C et une seule, pendant que les autres courbes se rapprochent asymptotiquement du point singulier à la façon des spirales;

4°. Les *cols foyers*, par lesquels passe une courbe C et une seule, pendant qu'une infinité d'autres, dont l'ensemble forme une surface, se rapprochent asymptotiquement du point singulier.

J'ai étudié également le cas où les trois surfaces (5) ont une courbe commune qui devient alors une *ligne singulière*. J'ai reconnu que les différents points d'une ligne singulière ont des propriétés analogues à celles des points singuliers ordinaires dont nous venons de parler.

Dans le cas des équations du premier ordre, nous avons trouvé une relation entre les nombres des points singuliers des diverses espèces. Il n'en existe pas de pareille pour les équations du second ordre. Une analyse approfondie montre qu'il doit y en avoir pour toutes les équations d'ordre impair, et qu'au contraire les équations d'ordre pair n'en possèdent pas.

Néanmoins un assez grand nombre de propriétés des équations du premier ordre s'étendent à celles du second. Les surfaces sans contact sont tout à fait analogues aux cycles sans contact, et l'on peut démontrer, par exemple, qu'à l'intérieur de toute surface sans contact (si elles ne sont pas triplement connexes) il y a toujours des points singuliers.

On a vu, plus haut, que c'est l'étude des points singuliers des équations du premier ordre qui nous a fait connaître les principales propriétés des courbes définies par ces équations; au contraire, la théorie des points singuliers des équations du second ordre ne saurait suffire à elle seule pour nous faire pénétrer aussi profondément dans la connaissance des courbes C. Il faut introduire, en outre, une notion nouvelle qui joue, dans une certaine mesure, le même rôle que les points singuliers. Soient C_0 une courbe *fermée* quelconque satisfaisant à notre équation, et D un domaine comprenant tous les points suffisamment voisins de C_0; nous pouvons étudier la forme et la disposition générale des courbes C à l'intérieur de ce domaine. On reconnaîtra ainsi, indépendamment d'un grand nombre de cas moins importants, quatre cas principaux, qui sont les suivants:

1°. On peut faire passer par la courbe C_0 deux surfaces que l'on peut sillonner par une infinité de courbes C satisfaisant aux équations (4). Les autres courbes C, après être entrées dans le domaine D et s'être rapprochées de C_0, s'en éloignent ensuite et finissent par sortir du domaine. Je n'ai rien à ajouter sur ce premier cas, qui nous apprend peu de chose sur les propriétés de nos courbes.

2°. On peut construire une surface S présentant une forme annulaire analogue à celle du tore, et à l'intérieur de laquelle se trouve la courbe C_0, de la même façon que le cercle, lieu des centres des cercles méridiens, se trouve à l'intérieur d'un tore. De plus, cette surface S n'est tangente en aucun point, à aucune des courbes C: c'est une surface sans contact. Considérons un point mobile décrivant une courbe C; dès qu'il sera sorti de la surface S, il n'y pourra plus rentrer; nous avons donc *instabilité*, et cela semble être ici le cas général.

3°. On peut construire une surface S analogue à celle dont nous venons de parler; mais elle ne sera pas une surface sans contact, elle sera au contraire sillonnée par une infinité de courbes C. Alors, si notre point mobile est situé sur la surface S, il y restera toujours; de plus, s'il part d'une position initiale quelconque, il finira toujours par revenir aussi près que l'on veut de cette position. Son orbite est donc *stable*.

4°. Dans le quatrième cas enfin, le point mobile peut aller aussi près que l'on veut d'un point *quelconque* du domaine D, et, s'il part d'une position initiale donnée, il finira toujours par revenir aussi près que l'on veut de cette position. Dans ce sens, il y a donc *stabilité*, et la démonstration de cette stabilité serait complète, si l'on savait assigner des limites aux coordonnées du point mobile.

Malheureusement, mes méthodes ne me permettent presque jamais de distinguer le troisème cas du quatrième, ni, dans le quatrième, de trouver les limites entre lesquelles les coordonnées du point mobile restent comprises. C'est là une lacune importante que jusqu'ici j'ai vainement essayé de combler.

Ce troisième et ce quatrième cas ne se présentent que si X, Y et Z satisfont à une infinité de conditions, de sorte qu'ils semblent d'abord très exceptionnels. Ils ont néanmoins une grande importance pratique. On peut d'ailleurs démontrer qu'ils se présenteront toujours si le dernier multiplicateur M, défini par l'équation

$$\frac{d(MX)}{dx} + \frac{d(MY)}{dy} + \frac{d(MZ)}{dz} = 0,$$

est toujours uniforme et positif dans le domaine considéré. Or cette circonstance se rencontre précisément dans la plupart des applications.

Pour étendre les résultats précédents aux équations d'ordre supérieur au second, il faut renoncer à la représentation géométrique qui nous a été si commode, à moins d'employer le langage de l'hypergéométrie à n dimensions. Mais ce langage est si peu familier à la plupart des géomètres qu'on perdrait ainsi les principaux avantages que l'on peut attendre de la représentation en question. Les résultats n'en subsistent pas moins, et l'on retrouve les quatre cas dont nous avons parlé plus haut. Ce qu'il y a de remarquable, c'est que le troisième et le quatrième cas, c'est-à-dire ceux qui correspondent à la stabilité, se rencontrent précisément dans les équations générales de la Dynamique. Cette circonstance doit nous faire d'autant plus désirer de voir se combler la lacune que j'ai signalée plus haut.

Pour aller plus loin, il me fallait créer un instrument destiné à remplacer l'instrument géométrique qui me faisait défaut quand je voulais pénétrer dans l'espace à plus de trois dimensions. C'est la principale raison qui m'a engagé à aborder l'étude de l'Analysis Situs; mes travaux à ce sujet seront exposés plus loins dans un paragraphe spécial.

J'ai poursuivi ensuite mes recherches sur les courbes définies pas les équations différentielles, mais les résultats nouveaux que j'ai obtenus se rapportant avant tout à la Mécanique Céleste seront exposés dans la Quatrième Partie de cette Notice.

DEUXIÈME PARTIE.

THÉORIE DES FONCTIONS.

VI. Théorie générale des fonctions d'une variable.
(71, 72, 35, 78, 41, 73, 75, 79, 95, 93).

La théorie des fonctions d'une seule variable complexe a fait dans ces derniers temps des progrès considérables, grâce aux travaux de M. WEIERSTRASS et de M. MITTAG-LEFFLER.

Ces fonctions peuvent se répartir en trois classes: 1° fonctions uniformes existant dans toute l'étendue du plan; 2° fonctions uniformes à espaces lacunaires, c'est-à-dire n'existant pas dans toute l'étendue du plan; 3° fonctions non uniformes.

Parmi les fonctions de la première classe, les plus importantes sont les fonctions entières, c'est-à-dire celles qui peuvent se développer suivant les puissances de x, en séries toujours convergentes. M. WEIERSTRASS a fait voir qu'une pareille fonction peut toujours se décomposer en un produit d'une infinité de *facteurs primaires*. Un facteur primaire de genre n est le produit $\left(1 - \dfrac{x}{a}\right) e^{P(x)}$, $P(x)$ étant un polynôme entier de degré n. Une fonction de genre n est une fonction entière dont tous les facteurs primaires sont de genre inférieur.

Cette classification des fonctions entières en genres soulève un grand nombre de problèmes intéressants. J'ai voulu contribuer (75) à la solution de ces problèmes en étudiant la manière dont une fonction de genre n se comporte à l'infini et la rapidité avec laquelle décroissent les coefficients de son développement suivant les puissances de x. Je suis arrivé ainsi aux résultats suivants:

1°. Si $F(x)$ est une fonction de genre n et si le module de x croît indéfiniment avec un argument tel que $e^{ax^{n+1}}$ tende vers 0, le produit $F(x) e^{ax^{n+1}}$ tend aussi vers 0.

2°.   L'intégrale

$$\int_0^\infty e^{(xz)^{n+1}} F(z)\,dz$$

représente une fonction entière de $\dfrac{1}{x}$.

3°.   Si A_p est le coefficient de x^p dans le développement de $F(x)$, on a

$$\lim A_p \sqrt[n+1]{p!} = 0 \quad (\text{pour } p = \infty).$$

4°.   Si $F(x)$ est une fonction de genre 0, elle est susceptible d'être représentée par la série d'Abel étudiée par M. HALPHEN dans le tome X du *Bulletin de la Société mathématique de France*, et cela quelle que soit la constante β.

Malheureusement les réciproques de ces propositions ne sont pas toujours vraies. Il est aisé de voir la raison pour laquelle il est impossible de trouver un critère infaillible, donnant les conditions nécessaires et suffisantes pour qu'une fonction soit du genre n. En effet, la classification des fonctions en genres se rattache très étroitement à la théorie de la convergence des séries. Il y a donc toujours, ainsi que l'a montré M. HADAMARD des cas douteux où l'on peut hésiter entre le genre n et $n + 1$. Mais M. HADAMARD a montré quel parti on peut tirer de ces réciproques malgré les restrictions auxquelles elles sont soumises.

Pour qu'une fonction dont les zéros sont, par ordre de module croissant,

$$a_1, \quad a_2, \quad \ldots, \quad a_p, \quad \ldots,$$

soit de genre n, la première condition et la plus importante, c'est que la série

$$\frac{1}{a_1^{n+1}} + \frac{1}{a_2^{n+1}} + \cdots + \frac{1}{a_p^{n+1}} + \cdots$$

soit convergente. Or il n'y a point de critère de la convergence d'une série pouvant s'appliquer à tous les cas. C'est pour cela qu'il n'y a pas non plus de critère permettant de reconnaître dans tous les cas si une fonction est du genre n.

Outre les fonctions entières, la première classe comprend: 1° les fonctions qui ont des pôles; 2° celles qui ont un nombre fini de points singuliers essentiels; 3° celles qui en ont un nombre infini parmi lesquels on peut trouver des points singuliers *isolés* (j'appelle ainsi les points singuliers autour desquels on peut tracer un cercle assez petit pour ne contenir aucun autre point singulier);

4° celles qui ont une ligne singulière; 5° celles qui, ayant un nombre infini de points singuliers, mais n'ayant pas de lignes singulières, n'ont cependant pas de points singuliers isolés. (Les Allemands disent alors que les points singuliers forment *eine perfecte Menge*.) J'ai donné, pour la première fois,[1] un exemple de fonctions de cette dernière catégorie; ce sont les fonctions fuchsiennes qui existent dans toute l'étendue du plan. En appliquant un théorème de M. Picard, on peut voir en effet que ces fonctions ne peuvent avoir des points singuliers isolés.

Passons maintenant à la deuxième classe, celle des fonctions à espaces lacunaires signalées pour la première fois par M. Weiestrass. J'ai été conduit par deux voies différentes (79) à m'occuper de ces fonctions. En premier lieu les fonctions fuchsiennes et kleinéennes n'existent en général qu'à l'intérieur d'un cercle ou d'un domaine plus compliqué; elles me fournissaient donc un exemple de fonctions à espaces lacunaires. Les résultats de ma thèse inaugurale me conduisaient également à des fonctions présentant des lacunes. Si l'on veut bien en effet se reporter au paragraphe que j'ai intitulé *Généralités sur les équations différentielles* et à l'équation (4) de ce paragraphe, on verra que cette équation (4) n'a d'intégrale holomorphe que si le polygone convexe, qui contient tous les points représentatifs des différentes racines d'une certaine équation algébrique, ne contient pas l'origine. Cela ne pourrait pas arriver, si l'intégrale holomorphe de l'équation (4). *considérée comme fonction des racines de cette équation algébrique,* n'était une fonction à espace lacunaire.

Cette remarque m'a fait découvrir toute une classe de fonctions présentant des lacunes. Voici quel est leur mode de génération. On pose

$$\varphi(x) = \sum \frac{A_n}{x - b_n},$$

en supposant que la série ΣA_n soit absolument convergente et que les points b_n soient intérieurs à un certain domaine D ou situés sur le contour de ce domaine, et cela de telle façon que, si l'on prend sur ce contour un arc quelconque et aussi petit qu'on voudra, il y ait toujours sur cet arc une infinité de points b_n.

La fonction $\varphi(x)$ est alors une fonction uniforme admettant le domaine D comme espace lacunaire. Comme exemple particulier, j'ai cité la série

---

[1] *Fonctions fuchsiennes (passim).*

$$\varphi(x) = \sum \frac{u^m\, v^n\, w^p}{x - \dfrac{m\alpha + n\beta + p\gamma}{m + n + p}},$$

où u, v, w sont des constantes données, de module inférieur à 1, où α, β, γ sont des constantes imaginaires quelconques et où m, n et p peuvent prendre sous le signe Σ tous les systèmes de valeurs entières et positives.

La fonction $\varphi(x)$ a alors pour espace lacunaire le triangle $\alpha\beta\gamma$.

Il importe de se rendre compte de la véritable nature de ces fonctions à espaces lacunaires. Il arrive souvent que les développements en séries, à termes rationnels par exemple, sont convergents à la fois à l'intérieur et à l'extérieur de ce domaine et ne divergent que sur le contour même du domaine. Les deux parties du plan où la série converge sont alors complètement séparées par une ligne le long de laquelle le développement cesse d'être valable. Doit-on cependant considérer les deux fonctions représentées par le développement à l'intérieur et à l'extérieur du domaine comme le prolongement analytique l'une de l'autre? Plusieurs géomètres étaient autrefois tentés de le croire. M. WEIERSTRASS a montré pour la première fois que leur point de vue était faux, en donnant des exemples de séries qui représentent dans des domaines différents des fonctions manifestement différentes. J'en ai moi-même rencontré un exemple dont je veux ici dire un mot. Certains développements qui représentent à l'intérieur du cercle fondamental une de ces fonctions que j'ai appelées plus haut *thétafuchsiennes* représentent o à l'extérieur de ce cercle.

J'ai voulu donner (78) un argument nouveau à l'appui de la manière de voir de WEIERSTRASS. Considérons une fonction $F(x)$ admettant un domaine D comme espace lacunaire, et une autre fonction $F_1(x)$ n'existant au contraire qu'à l'intérieur de ce domaine et admettant par conséquent tout le reste du plan comme espace lacunaire. Divisons le contour du domaine D en deux arcs A et B. J'ai démontré qu'on pouvait trouver deux fonctions uniformes $\Phi(x)$ et $\Phi_1(x)$ existant dans tout le plan et admettant seulement, la première A, la seconde B comme ligne singulière; et cela de telle sorte que

$$\Phi + \Phi_1 = F \quad \text{à l'extérieur de } D,$$

$$\Phi + \Phi_1 = F_1 \quad \text{à l'intérieur de } D.$$

Si la fonction F avait un prolongement analytique naturel à l'intérieur de D, ce prolongement devrait être F_1; mais nous avons choisi cette fonction F_1 *d'une manière tout à fait arbitraire*, en l'assujettissant seulement à n'exister qu'à l'intérieur de D. Il est donc dénué de sens de parler du prolongement naturel d'une fonction à l'intérieur d'un de ses espaces lacunaires. J'avais en même temps ramené l'étude des fonctions à espaces lacunaires à celle des transcendantes uniformes à "ligne singulière essentielle. Je suis revenu sur la même question dans un mémoire plus étendu (95).

La théorie des fonctions non uniformes est loin d'être aussi avancée que celle des fonctions uniformes. J'ai montré d'abord (93) que le nombre de leurs déterminations s'il est infini est de la $1^{\text{ère}}$ puissance au sens de CANTOR.

Quoiqu'on connaisse assez bien la manière d'être de ces fonctions non uniformes *dans le voisinage d'un point donné*, quoique l'introduction des surfaces de RIEMANN ait jeté beaucoup de lumière sur les parties encore obscures de leur théorie, il y a encore bien de progrès à faire avant de connaître leurs principales propriétés. J'étais donc animé du désir de ramener leur étude à celle des transcendantes uniformes. La théorie des fonctions fuchsiennes me rapprochait déjà du but; j'avais démontré, en effet, que si $f(x, y) = 0$ est l'équation d'une courbe algébrique quelconque, on peut choisir un paramètre z de telle façon que x et y soient des fonctions uniformes de ce paramètre. J'avais ainsi résolu le problème pour les plus simples des fonctions non uniformes, c'est-à-dire pour les fonctions algébriques.

J'étais donc (73) naturellement porté à me demander si cette propriété est particulière aux fonctions algébriques, ou si l'on peut l'étendre à une fonction non uniforme quelconque. J'ai pu répondre à cette question et démontrer le théorème très général suivant:

Soit une fonction analytique quelconque de x, non uniforme. On peut toujours trouver une variable z, telle que x et y soient fonctions uniformes de z.

Mon point de départ a été la démonstration du principe de DIRICHLET donnée par M. SCHWARZ. Mais ce principe n'aurait pu à lui seul me permettre de triompher des difficultés qui provenaient de la grande généralité du théorème à démontrer. Il faut d'abord définir la surface de RIEMANN à une infinité de feuillets dont je cherche à faire, sur une partie du plan, la représentation conforme. Je choisis cette surface de façon qu'elle soit simplement connexe, que tous les points singuliers restent en dehors de la surface proprement dite et se trouvent pour ainsi dire sur sa frontière, et enfin de façon que la fonction y ne puisse prendre deux valeurs différentes en un même point de la surface.

Je découpe une portion finie R de cette surface, et j'en fais sur un cercle la représentation conforme, ce que le théorème de M. SCHWARZ me permet de faire. Cette représentation se fait à l'aide d'une certaine fonction analytique u. Faisons ensuite croître indéfiniment la région R; nous aurons la représentation conforme d'une portion de plus en plus étendue de notre surface de RIEMANN. Il me faut alors faire voir que la fonction analytique u dont je parlais plus haut tend vers une limite finie et déterminée. Quand cela est fait, les premières difficultés seules sont vaincues. En effet, il reste à démontrer que la limite de la fonction u est elle-même une fonction analytique. Pour cela, il faut que la fonction analytique u tende *uniformément* vers sa limite (*gleichmässig*), ce que je suis parvenu à démontrer.

Ainsi, l'étude des fonctions non uniformes est ramenée, dans tous les cas possibles, à l'étude bien plus facile des fonctions uniformes.

Je rattacherai à ces recherches, relatives aux fonctions d'une variable, les travaux que j'ai consacrés à l'étude des séries de polynômes (35, 73). Et en effet, il y a un fait qui joue un rôle très important dans la théorie des fonctions: c'est que les régions où une fonction quelconque peut être représentée par une série de puissances sont limitées par des cercles. On peut donc supposer qu'on pourra tirer un profit analogue de la connaissance des régions où conviennent des développements d'autre forme.

J'ai cherché, en particulier, les conditions de convergence des séries dont le $n^{\text{ième}}$ terme est un coefficient constant multiplié par un polynôme entier $P_n(x)$ de degré n, en supposant qu'il y ait entre un certain nombre de polynômes P_n consécutifs une relation de récurrence. Les séries ordonnées suivant les polynômes de LEGENDRE n'en sont évidemment que des cas particuliers. J'ai trouvé que les régions où ces séries convergent sont limitées par certaines *courbes de convergence* et j'ai déterminé ces courbes en remarquant que la série

$$\Sigma P_n z^n,$$

cosidérée comme fonction de z, satisfait à une équation différentielle linéaire dont les coefficients sont des polynômes entiers en z et en x.

VII. Théorie générale des fonctions de deux variables. (76, 77, 101).

Il semble d'abord que, pour étudier les fonctions de deux variables, il suffit d'appliquer, sans y rien changer, les principes qui ont servi à établir les propriétés des fonctions d'une seule variable. Il n'en est rien; il y a entre les deux théories des différences essentielles et l'on ne saurait passer de l'une à l'autre par une simple généralisation.

Cette différence apparaît dès que l'on considère les polynômes entiers qui sont décomposables en facteurs s'il n'y a qu'une variable et ne le sont plus dans le cas contraire. Je laisserai de côté pour le moment les difficultés que l'on a éprouvées en voulant généraliser la théorie des résidus de Cauchy, car j'y veux consacrer le paragraphe suivant. J'insisterai seulement sur un exemple qui met bien en évidence les différences dont je viens de parler: c'est l'étude des fonctions méromorphes dans tout le plan, c'est-à-dire des transcendantes qui ne présentent à distance finie d'autres singularités que des infinis.

On sait que M. Weierstrass a démontré que, si une fonction d'une seule variable est méromorphe dans tout le plan, elle peut être regardée comme le quotient de deux fonctions entières. Pour arriver à ce résultat, le célèbre géomètre de Berlin *construit* une fonction entière qui s'annule pour tous les infinis de la fonction méromorphe donnée. Le produit des deux fonctions, ne devenant plus infini, est une fonction entière. Pour construire la transcendante en question, il faut considérer *séparément* les différents infinis de la fonction méromorphe donnée.

La méthode de M. Weierstrass paraît donc, au premier abord, ne pas pouvoir s'étendre aux fonctions de deux variables, dont les infinis sont non plus des points isolés, mais des multiplicités continues, et ne peuvent par conséquent être envisagés *séparément*. Aussi les géomètres qui tentaient de généraliser le théorème de M. Weierstrass ont-ils été longtemps arrêtés (76, 77).

J'eus l'idée de tourner la difficulté en généralisant la notion de fonction de deux variables. Soit en effet $V + iW$ une fonction des variables imaginaires $x + iy$ et $z + it$. La partie réelle V satisfera à l'équation

$$(1) \qquad \Delta V = \frac{d^2 V}{dx^2} + \frac{d^2 V}{dy^2} + \frac{d^2 V}{dz^2} + \frac{d^2 V}{dt^2} = 0.$$

Mais cette condition n'est pas suffisante pour que V soit la partie réelle d'une fonction de nos deux variables. Il faut en outre que V satisfasse aux relations

$$(2) \qquad \frac{d^2 V}{dx^2} + \frac{d^2 V}{dy^2} = 0, \qquad \frac{d^2 V}{dx\,dz} + \frac{d^2 V}{dy\,dt} = 0.$$

Envisageons maintenant toutes les fonctions V qui satisfont à l'équation (1) sans être assujetties à satisfaire aux équations (2). On pourra alors construire une fonction qui remplira cette unique condition (1) et qui de plus admettra une partie seulement des infinis de la fonction méromorphe donnée *sans en admettre d'autres*. Cela était impossible au contraire quand cette fonction restait assujettie aux conditions (2).

Pouvant alors considérer séparement les infinis de notre fonction méromorphe, nous n'avons plus qu'à appliquer la méthode même de M. WEIERSTRASS pour construire une fonction e^V qui s'annule pour tous les infinis de la fonction méromorphe donnée et telle que V satisfasse à l'équation (1). On peut même en trouver une infinité. Soient en effet V_0 l'une d'elles et G une fonction entière, c'est-à-dire toujours finie, de x, y, z, t, satisfaisant à l'équation $\varDelta G = 0$; toutes les fonctions $V_0 + G$ rempliront, comme la fonction V_0 elle-même, les conditions énoncées plus haut. Il reste à faire voir que, parmi ces fonctions, $V_0 + G$, il y en a une qui peut être regardée comme la partie réelle d'une fonction de $x + iy$ et de $z + it$, ce qui veut dire que l'on peut disposer de la fonction entière G, de telle façon que

$$\frac{d^2(V_0 + G)}{dx^2} + \frac{d^2(V_0 + G)}{dy^2} = \frac{d^2(V_0 + G)}{dx\,dz} + \frac{d^2(V_0 + G)}{dy\,dt} = 0.$$

C'est ce que j'ai fait, démontrant ainsi le théorème suivant:

Si une fonction de deux variables imaginaires est partout méromorphe, elle sera le quotient de deux fonctions entières. Je suis revenu sur cette même question (101) et je suis parvenu à simplifier considérablement les démonstrations.

En ce qui concerne les fonctions non uniformes, j'ai contribué à l'étude de leurs propriétés dans le voisinage d'un point donné, par les lemmes que j'ai démontrés au début de ma thèse inaugurale. Supposons qu'une équation

$$F(z, x_1, x_2, \ldots, x_n) = 0,$$

définissant z comme fonction implicite de x_1, x_2, $\ldots$, x_n, soit satisfaite pour le système de valeurs

$$z = x_1 = x_2 = \cdots = x_n = 0$$

et que nous étudions la fonction dans un domaine voisin de ce système de valeurs. Je suppose de plus que dans ce domaine la fonction F soit holomorphe. On sait depuis longtemps que, si $\dfrac{dF}{dz}$ n'est pas nul, z est fonction holomorphe de x_1, x_2, $\ldots$, x_n. J'ai cherché ce qui se passe lorsque $\dfrac{dF}{dz}$ est nul en même temps que $\dfrac{d^2 F}{dz^2}$, $\dfrac{d^3 F}{dz^3}$, $\ldots$, $\dfrac{d^{m-1} F}{dz^{m-1}}$, mais que la $m^{\text{ième}}$ dérivée $\dfrac{d^m F}{dz^m}$ n'est pas nulle. J'ai démontré que dans ce cas z satisfait à une équation algébrique de la forme

$$z^m + B_{m-1} z^{m-1} + B_{m-2} z^{m-2} + \cdots + B_1 z + B_0 = o,$$

dont les coefficients A sont des fonctions holomorphes des x. J'ai obtenu ensuite un résultat analogue pour le cas où l'on a p fonctions implicites de n
variables définies par p équations simultanées.

VIII. Intégrales multiples. (84, 85, 89, 142, 100, 232, 92, 233, 234).

La théorie qui a le plus contribué à faciliter l'étude des fonctions d'une
variable est certainement celle des intégrales prises entre des limites imaginaires.
Elle conduit, comme on le sait, à envisager les périodes de ces intégrales et à
distinguer les périodes polaires (correspondant aux résidus) des périodes cycliques. Un des points les plus importants est d'ailleurs l'étude des intégrales
abéliennes, c'est-à-dire des intégrales de différentielles algébriques; cette théorie
est ordinairement présentée sous une forme géométrique, ce qui a amené à dire,
pour abréger, que ces intégrales «appartiennent à une courbe algébrique».

Quand on passe aux fonctions de deux variables, la notion de ces intégrales
et de leurs périodes peut se généraliser à deux points de vue différents: par les
intégrales de différentielles totales et par les intégrales doubles. Je ne m'étendrai pas beaucoup sur le premier de ces modes de généralisation. Il ne m'appartient pas, en effet: c'est M. PICARD qui en a tiré les premiers et les plus
beaux résultats. Je n'ai fait qu'appeler l'attention (84), à la suite de la Note
de M. PICARD, sur quelques points de détail. Ainsi ce géomètre avait démontré
qu'une surface algébrique ne possède d'intégrales abéliennes de différentielles
totales de première espèce que dans des cas particuliers.

Je veux dire que, si

$$f(x, y, z) = o$$

est l'équation d'une surface algébrique définissant z en fonction de x et de y, il
n'y aura pas, en général, de différentielle exacte

$$P\,dx + Q\,dy,$$

où P et Q soient rationnels en x, y, z, de telle façon que l'intégrale

$$\int P\,dx + Q\,dy$$

reste toujours finie.

Partant de là, j'ai trouvé les conditions pour qu'une surface du quatrième ordre possède de pareilles intégrales. Il faut et il suffit qu'elle soit une surface réglée ou qu'elle se ramène à une surface de révolution par une transformation linéaire. J'ai indiqué également un certain nombre de cas où il n'y a jamais, et d'autres où il y a toujours, des intégrales de première espèce.

J'ai reconnu que le théorème d'ABEL s'étendait immédiatement aux intégrales de différentielles totales de première espèce; mais il semblait au premier abord qu'il ne serait plus applicable aux surfaces qui ne possèdent pas de pareilles intégrales, c'est-à-dire la grande majorité des surfaces algébriques.

Il n'en était rien. J'ai démontré (85) le théorème suivant:

Si (x_1, y_1, z_1), (x_2, y_2, z_2), ..., (x_q, y_q, z_q) sont les q points d'intersection d'une surface algébrique S et d'une courbe algébrique C; si $(x_1 + dx_1, y_1 + dy_1, z_1 + dz_1)$, ... sont les q points d'intersection de cette même surface S avec une corbe C' infiniment peu différente de C, on aura un certain nombre de relations de la forme

$$X_1 dx_1 + X_2 dx_2 + \cdots + X_q dx_q = 0,$$

où X_i est une fonction rationnelle de x_i, y_i, z_i. Ces relations peuvent être regardées comme la généralisation du théorème d'ABEL.

Les difficultés qui s'attachent à l'étude des intégrales doubles et multiples étendues à un domaine imaginaire sont d'une nature différente. Il semble que la théorie des intégrales simples prises entre des limites imaginaires serait d'une exposition beaucoup plus laborieuse si l'on n'avait pour s'y guider une représentation géométrique. On perd ce guide quand on passe aux intégrales doubles; il faudrait alors recourir à la Géométrie à quatre dimensions, ce qui serait une complication plutôt qu'une simplification.

Cet obstacle ne paraît pas d'abord très sérieux; cependant il arrêta longtemps les géomètres. M. PICARD, à propos des fonctions hyperfuchsiennes, avait traité une question qui présente quelque analogie avec celle qui nous occupe, mais qui n'est pourtant pas la même; les quantités qu'il a ainsi introduites ne peuvent être en aucune façon regardées comme la généralisation des périodes des intégrales simples. Il importe de ne pas les confondre avec les périodes cycliques que ce même savant a étudiées peu de temps après la publication de ma première Note à ce sujet, et qui se rattache, au contraire, très directement à la théorie que j'ai cherché à fonder.

Je fus donc le premier à étudier méthodiquement cette importante question dans une Note (89) que j'ai eu l'honneur de présenter à l'Academie le 25 janvier 1886 et dont j'ai développé les résultats dans un Mémoire plus étendu (92).

Le premier point était d'imaginer un mode de représentation géométrique sans employer l'espace à quatre dimensions. On peut y arriver par diverses méthodes que je n'exposerai pas ici et dont j'ai fait tour à tour usage. Il faut ensuite donner une définition des intégrales doubles prises dans un domaine imaginaire. Grâce aux modes de représentation dont je viens de parler, on peut donner cette définition sans qu'il subsiste aucune équivoque. Il faut ensuite démóntrer le théorème fondamental, analogue à celui de CAUCHY, et d'après lequel une intégrale double prise le long d'un contour fermé est nulle en général. Cette démonstration ne présente aucune difficulté. On peut trouver sous une forme simple les conditions d'intégrabilité de différentielles doubles

$$A\,dy\,dz + B\,dx\,dz + C\,dx\,dy + \cdots,$$

qu'il faut d'abord définir sans ambiguïté. Ces conditions présentent presque la même forme que celles qui expriment l'intégrabilité d'une différentielle ordinaire. Seulement certains signes qui sont tous positifs pour les intégrales d'ordre pair, et, en particulier, pour les intégrales doubles, sont, au contraire, alternativement positifs et négatifs quand il s'agit d'intégrales d'ordre impair et en particulier d'intégrales simples. Ces conditions une fois trouvées, le théorème fondamental s'ensuit immédiatement.

Il admet cependant des exceptions, comme la proposition correspondante de la théorie de CAUCHY, et ce sont ces exceptions qui sont l'origine des périodes des intégrales doubles. Ces périodes se distinguent, comme dans le cas d'une seule variable, en périodes cycliques et en périodes polaires. Je me suis occupé, en particulier, des périodes polaires ou, si l'on veut, des résidus des intégrales doubles. M. PICARD a étudié ensuite les périodes cycliques.

J'ai envisagé l'intégrale d'une fonction rationnelle que j'ai écrite sous la forme suivante

$$\iint \frac{f(x,\,y)\,dx\,dy}{\varphi(x,\,y)\,\psi(x,\,y)},$$

en décomposant le dénominateur en facteurs irréductibles, et j'ai reconnu que cette intégrale présente trois sortes de périodes:

1º. Celles de la première sorte sont égales à $2i\pi$ multiplié par l'une des périodes de première espèce de l'intégrale abélienne

$$\int \frac{f\, dx}{\varphi \dfrac{d\psi}{dy}}$$

(rapportée à la courbe algébrique $\psi = 0$).

2°. Celles de la seconde sorte se rapportent aux divers points d'intersection des deux courbes $\varphi = \psi = 0$ et sont égales à

$$\pm\, 4\,\pi^2\, \frac{f(x_0,\, y_0)}{\varDelta(x_0,\, y_0)},$$

$\varDelta(x, y)$ étant le déterminant de φ et de ψ par rapport à x et à y; et x_0 et y_0 étant les coordonnées du point d'intersection.

3°. Enfin celles de la troisième sorte se rapportent aux divers points doubles de ces deux courbes et ont une expression analogue.

Mais la théorie serait incomplète si l'on se bornait à ces trois sortes de périodes. Il peut arriver que la fonction sous le signe intégral devienne infinie en divers points du contour d'intégration sans que l'intégrale elle-même cesse d'être finie. Cette circonstance ne pouvait pas se produire dans le cas des intégrales simples, lorsque la fonction à intégrer était rationnelle; il n'en est plus de même ici. D'un autre côté, on ne saurait exclure de parti pris les intégrales de cette sorte; car, autant qu'on en peut juger aujourd'hui, elles doivent jouer un rôle important dans les applications.

Or les intégrales de cette nouvelle sorte ont un caractère bien différent de celui des intégrales à périodes. Celles-ci, en effet, ou bien restent constantes quand on fait varier le chemin d'intégration d'une manière continue, ou bien s'accroissent par sauts brusques; celles-là, au contraire, varient d'une façon continue comme le chemin d'intégration lui-même. C'est là la principale différence entre la théorie nouvelle et celle de CAUCHY.

Ces résultats s'appliquent, *mutatis mutandis*, aux transcendantes et, en particulier, aux fonctions uniformes.

Cette théorie nouvelle sera-t-elle aussi féconde que l'ont été les découvertes de CAUCHY? Elle est encore trop jeune pour qu'on puisse se prononcer sur ce point. Certainement quelques-uns des résultats qu'on peut obtenir ainsi, et par une généralisation immédiate des méthodes de CAUCHY, auraient pu être atteints plus aisément par d'autres voies. Mais on peut espérer qu'il n'en sera pas toujours de même, et déjà je suis sur la voie de propositions réellement nouvelles sur la théorie des fonctions abéliennes.

Les périodes dont il a été question jusqu'ici sont analogues à celles qui se rapportent aux singularités polaires des intégrales simples. Mais lorsque la fonction sous le signe f n'est pas uniforme, les intégrales multiples peuvent outre ces périodes polaires présenter des périodes cycliques.

C'est une question de Mécanique Céleste, celle du développement de la fonction perturbatrice qui m'a amené à m'en occuper.

Si la fonction sous le signe f dépend d'un paramètre (comme par exemple les intégrales elliptiques du module) les périodes cycliques seront des fonctions de ce paramètre. De même que dans le cas des intégrales simples, ces fonctions seront définies par des équations linéaires à coefficients algébriques. J'ai étudié ces équations linéaires et leurs groupes (100). J'ai montré qu'il y a un lien intime entre l'équation linéaire qui définit les périodes des intégrales doubles dépendant du radical $\sqrt{F(x, y)}$ et celle qui définit les périodes des intégrales abéliennes simples engendrées par la courbe algébrique $F(x, y) = 0$. J'ai fait voir par quelle transformation on peut passer de l'une à l'autre.

Cette dernière recherche se rattache à mes travaux sur l'Analysis Situs.

D'un autre côté, étant donnée plusieurs intégrales multiples dépendant du radical $\sqrt{F(x, y)}$, on peut se proposer de faire une théorie de la réduction de ces intégrales, analogue à la théorie classique de la réduction des intégrales elliptiques (ou hyperelliptiques) à un petit nombre d'intégrales types (dites de $1^{\text{ère}}$, 2de et 3e espèces). J'ai résolu ce problème qui m'était utile au point de vue du développement de la fonction perturbatrice (232, 233, 234).

La réduction des intégrales doubles et celle des intégrales de différentielles totales se présentent d'ailleurs ici comme deux questions intimement liées.

Après cette revue des travaux que j'ai consacrés à la théorie générale des fonctions, je suis naturellement amené à passer l'étude de diverses fonctions particulières. J'ai déjà parlé plus haut des fonctions fuchsiennes. Il me reste à résumer mes recherches sur les fonctions elliptiques, sur les fonctions abéliennes et sur les fonctions hyperfuchsiennes.

IX. Fonctions elliptiques.

J'ai fait fort peu de chose sur les fonctions elliptiques. Cependant j'ai donné, dans un Mémoire d'Arithmétique (113, 119), une façon d'exprimer ces fonctions à l'aide d'une intégrale définie. On sait que les fonctions doublement périodiques peuvent se décomposer en éléments simples de la forme $\dfrac{\sigma'(u-\alpha)}{\sigma(u-\alpha)}$ ou de la forme

$$\frac{d^n}{du^n}\,\frac{\sigma'(u-\alpha)}{\sigma(u-\alpha)}.$$

Il suffit donc d'exprimer par une intégrale définie la fonction

$$\frac{\sigma'(u)}{\sigma(u)} = \frac{1}{u} + \sum\left(\frac{1}{u-w} + \frac{1}{w} + \frac{u}{w^2}\right),$$

où $w = 2\mu\omega + 2\mu'\omega'$ et où μ et μ' peuvent prendre tous les systèmes de valeurs entières positives et négatives, excepté $\mu = \mu' = 0$. On pourra évidemment décomposer la série du second membre en quatre autres: la première comprenant les termes où μ et μ' sont positifs, la seconde les termes où μ est positif et μ' négatif ou nul, la troisième ceux où μ est négatif ou nul et μ' positif, la quatrième enfin ceux où μ et μ' sont négatifs ou nuls. Cette décomposition est analogue à la décomposition de $\pi \cot x\pi$ en une somme de deux termes dépendant des fonctions eulériennes

$$\pi \cot x\pi = \frac{\Gamma'(x)}{\Gamma(x)} - \frac{\Gamma'(1-x)}{\Gamma(1-x)}.$$

Cette généralisation des fonctions eulériennes est analogue, mais non identique à celle qu'a donnée M. APPELL.

Il suffit alors d'exprimer, par une intégrale définie, la première de nos séries partielles, car les autres s'y ramènent aisément. On trouve que cette série partielle s'exprime par une intégrale prise par rapport à z entre les limites 0 et ∞, la fonction sous le signe $\int$ étant rationnelle par rapport à z et à diverses exponentielles de la forme $e^{\lambda z}$. Il est donc possible d'exprimer de la même manière toutes les fonctions doublement périodiques.

J'ai été conduit aussi d'une façon incidente à m'occuper des fonctions elliptiques en les considérant comme des cas particuliers des fonctions fuchsiennes.[1] J'ai retrouvé ainsi la plupart des formules connues et, en particulier, l'expression des fonctions à deux périodes par des séries trigonométriques. J'ai été conduit par la même voie à une formule que je crois nouvelle et qui permet d'exprimer les fonctions elliptiques par une série infinie d'une forme particulière. (Acta Mathematica, Tome 1, page 287.)

---

[1] *Sur les fonctions fuchsiennes (passim).*

X. Fonctions abéliennes. (70, 80, 82, 86, 88, 90, 87, 74, 82, 143, 97, 99, 101, 98).

La théorie des fonctions abéliennes est loin d'être aussi avancée que celle des fonctions elliptiques. Un grand nombre des propriétés de ces dernières transcendantes ne s'étendent pas ou ne s'étendent que difficilement au cas général. On ne doit pas s'en étonner si l'on se rappelle que beaucoup de propriétés des fonctions d'une variable ne sont plus applicables aux fonctions de plusieurs variables. C'est la même difficulté qui nous a occupés au paragraphe VII.

On a été conduit aux fonctions abéliennes par l'étude des courbes algébriques et des intégrales abéliennes. Je me suis occupé incidemment (143) des transformations birationnelles des courbes algébriques afin de démontrer qu'on peut toujours ramener ces courbes à des courbes gauches dépourvues de toute irrégularité. Un des premiers faits que l'on a remarqués est la possibilité de la réduction de ces intégrales abéliennes. JACOBI en a déjà rencontré quelques exemples; dans des cas assez nombreux, on voit des intégrales appartenant à une courbe de genre ϱ se réduire à des intégrales de genre inférieur à ϱ ou même à des intégrales elliptiques. Mais on est bientôt amené à se placer à un point de vue plus élevé; les fonctions Θ, qui doivent leur origine aux intégrales abéliennes de premère espèce, ne sont qu'un cas particulier des séries Θ les plus générales. Mais il est aisé de voir qu'à ces transcendantes plus générales appartiennent des intégrales, qui sont, il est vrai, des intégrales de différentielles totales, mais qui peuvent néanmoins être regardées comme la généralisation des intégrales de première espèce. Il est alors naturel d'appliquer à ces intégrales le procédé de la réduction; le problème primitif reçoit une importante extension: mais, par la suppression d'une restriction gênante, il est simplifié et non compliqué; car on peut désormais introduire dans ses raisonnements des fonctions Θ quelconques, sans avoir à s'inquiéter de leur origine.

Les géomètres se sont de longue date préoccupés de ce problème, qui doit nous fournir d'importantes données sur les fonctions algébriques, et qui est un des meilleurs chemins pour pénétrer dans le domaine mystérieux des fonctions abéliennes. Dans ces derniers temps, M. PICARD, par une série de brillants travaux, lui a fait faire plusieurs pas importants.

Mes premiers essais dans cet ordre d'idées ne portent que sur un cas particulier. Ainsi que je l'ai expliqué plus haut, dans le paragraphe intitulé: *Intégration des équations linéaires par les fonctions algébriques*, si l'intégrale générale d'une équation linéaire est algébrique et si, à l'aide de cette intégrale générale, on forme un système d'intégrales abéliennes de première espèce, il y a entre les périodes de ce système un grand nombre de relations intéressantes. J'avais là un moyen (40) de pénétrer plus profondément dans l'étude des fonctions abéliennes, et je résolus d'en profiter. Je choisis comme exemple particulier le système d'intégrales abéliennes que l'on peut former à l'aide de la résolvante de GALOIS de l'équation modulaire relative à la transformation du septième ordre. Je trouvai que les relations qui existent entre les périodes suffisent pour les déterminer complètement. Étant parvenu ainsi à calculer ces périodes, je m'aperçus que, parmi les intégrales abéliennes de ce système (qui est du genre 3), il y en a une infinité qui sont susceptibles d'être réduites aux intégrales elliptiques. C'était là un troisième exemple d'une circonstance remarquable, déjà signalée deux fois par M. PICARD.

Mon attention fut de nouveau attirée sur cette question par un Mémoire de Mme KOWALEVSKI, où se trouvaient cités deux théorèmes de M. WEIERSTRASS, sur la réduction des intégrales abéliennes aux intégrales elliptiques. Ces deux théorèmes avaient été communiqués à divers savants par des lettres du professeur de Berlin, mais la démonstration n'en avait pas été publiée. J'ai donné (82) deux démonstrations différentes de ces deux propositions; j'ignore encore si mes méthodes sont identiques à celles de M. WEIERSTRASS. Toutes deux sont empruntées à l'Arithmétique, et l'on ne doit pas s'en étonner, car le problème est en réalité purement arithmétique. La première démonstration se fonde sur la considération des formes bilinéaires. Dans la seconde, j'emploie un procédé particulier de réduction. Je suppose que dans un système d'intégrales de genre ϱ, il y en ait μ qui soient réductibles au genre μ. Leurs 2ϱ périodes, ou périodes anciennes, s'exprimeront alors à l'aide de 2μ quantités, qui seront les périodes nouvelles, par des polynômes linéaires à coefficients entiers. On peut donc dresser un Tableau de $4\varrho\mu$ nombres entiers qui caractérise la réduction. Mais ce Tableau peut être d'une infinité de manières; car on peut remplacer, soit le système des périodes anciennes, soit le système des périodes nouvelles par un système équivalent. Le problème est précisément de profiter de cette circonstance pour réduire le Tableau à sa plus simple expression. Dans ma seconde méthode, la réduction se fait par une série d'opérations toutes pareilles entre elles.

Je profitai des avantages de ces deux méthodes pour généraliser les deux théorèmes de M. WEIERSTRASS et les étendre au cas de la réduction des intégrales abéliennes à d'autres intégrales abéliennes.

Le théorème de M. WEIERSTRASS était plus général en un sens que le théorème de M. PICARD sur le même sujet; ce dernier ne s'appliquait en effet qu'à la réduction du genre 2 au genre 1; le géomètre allemand avait étudié la réduction d'un genre ϱ quelconque au genre 1. D'autre part, le théorème de M.

Picard contenait plus que celui de M. Weierstrass, car la réduction y était poussée plus loin. Était-il possible de trouver une proposition qui contînt à la fois celle de M. Weierstrass et celle de M. Picard, c'est-à-dire de pousser dans le cas général la réduction aussi loin que ce dernier analyste? L'application de ma seconde méthode m'a fait reconnaître (90, 87) que cela peut se faire sans difficulté.

Le même procédé me permit en même temps d'étudier le cas général (réduction d'un genre ϱ quelconque, non plus au genre 1, mais à un genre μ également quelconque) et de pousser la réduction beaucoup plus loin que je ne l'avais fait dans mon premier travail. Le théorème auquel je fus ainsi conduit contient, comme cas particulier, toutes les propositions antérieurement découvertes et résume ainsi toute la théorie.

Un cas particulier bien digne d'intérêt est celui où une infinité d'intégrales d'un même système se réduisent aux intégrales elliptiques. M. Picard en avait déjà rencontré deux exemples, et il semblait probable que, dans un système d'intégrales de genre ϱ, il ne pouvait y avoir plus de ϱ intégrales réductibles sans qu'il y en eût une infinité. J'ai démontré (82, 87) qu'il en était effectivement ainsi et j'ai trouvé en même temps les relations fort simples qui unissent entre elles les intégrales réductibles.

Les méthodes que je viens d'exposer permettent une classification rationnelle des cas de réduction. Mais cette classification, à côté d'incontestables avantages, présente un inconvénient grave: elle ne distingue pas du cas général les cas particuliers où les intégrales abéliennes à réduire appartiennent à une courbe algébrique. Ces derniers ne présentent pas d'intérêt spécial au point de vue de la théorie des fonctions abéliennes; mais ils en ont un fort grand, au contraire, si l'on se propose pour but l'étude des fonctions algébriques. Il importait donc de trouver une classification nouvelle, ne portant que sur ces cas particuliers et laissant de côté tous les autres. J'ai indiqué (88) un moyen d'arriver à ce résultat par l'étude de la transformation des fonctions fuchsiennes; mais je n'ai pas eu le temps d'approfondir cette théorie.

L'étude systématique des fonctions abéliennes devait naturellement commencer par l'examen des cas de réduction, la suite de cet exposé le fera suffisamment comprendre; mais ce n'était qu'un premier pas, et bien d'autres problèmes restaient à résoudre.

On vient de voir que les fonctions Θ, définies à l'aide des intégrales abéliennes de première espèce, ne sont que des cas très particuliers des séries Θ les plus générales.

Qu'est ce qui caractérise les fonctions Θ spéciales; c'est à dire celles qui doivent leur origine aux intégrales abéliennes? Qu'est ce qui permet de les distinguer parmi les fonctions Θ les plus générales? C'est la circonstance suivante; la variété

$$\Theta = 0$$

est pour employer le langage de Lie, une variété doublement de translation. C'est ce qu'on aurait pu déduire aisément des recherches antérieurs de Lie. Mais je suis parvenu au même résultat (97, 98) par une voie entièrement différente. Cette condition peut évidemment s'exprimer par une relation entre les périodes, mais cette relation est transcendante et ne peut s'exprimer que sous forme de série. Je me suis borné à indiquer les premiers termes de cette série, je veux dire ceux qui sont le plus sensibles quand la fonction Θ diffère peu d'un produit de fonctions Θ elliptiques. La forme en est curieuse, car il y entre des radicaux. On peut concevoir une infinité de fonctions de n variables, admettant $2n$ systèmes de périodes et ne rentrant pas dans la catégorie spécialement étudiée par Riemann. Ces fonctions peuvent-elles être toujours regardées comme le quotient de deux fonctions Θ? Riemann était parvenu à le démontrer; mais il n'a jamais publié sa démonstration. M. Weierstrass a retrouvé le même résultat, mais il n'a pas publié non plus de son vivant la méthode dont il s'est servi.

Abordant l'étude de ces fonctions, que je n'assujettissais qu'à la condition d'être périodiques (70), je reconnus qu'on pouvait toujours les tirer des fonctions abéliennes ordinaires, obtenues par la méthode d'inversion de Jacobi, en appliquant le procédé de la réduction des intégrales abéliennes.

Dans ces conditions, nous devions naturellement songer, M. Picard et moi, à unir nos efforts pour retrouver le résultat de Riemann. Nous reconnûmes (80) qu'il devait y avoir entre les périodes les mêmes relations que dans le cas particulier des fonctions nées de l'inversion des intégrales abéliennes. Il était aisé d'en conclure que toutes les fonctions à n variables et à $2n$ périodes s'expriment par le moyen des séries Θ.

La méthode que je viens d'exposer est celle même dont s'était servi M. Weierstrass et qu'il n'avait pas publiée; c'est ce que nous reconnûmes quand après la mort du savant géomètre, nous reçûmes les bonnes feuilles du troisième volume de ses oeuvres complètes.

Il n'y avait que quelques différences de détail; c'est ainsi que j'ai employé un moyen un peu différent pour démontrer un Lemme indispensable, d'après lequel il y a toujours une relation algébrique entre p fonctions à p variables et à $2p$ périodes. J'y suis revenu plus tard (99).

Revenons au théorème dont j'avais donné une démonstration, après M. Weierstrass et en commun avec M. Picard. Depuis M. Appell en a donné une nouvelle fondée sur de tout autres principes et j'en ai moi-même donné une troisième, entièrement différente des deux premières (99, 101). Je rappelle par quel artifice j'ai démontré qu'une fonction méromorphe de plusieurs variables est toujours le quotient de deux fonctions entières (vide supra § VII). Quelle est la nature de ces fonctions entières? Les perfectionnements apportés (101) à ma démonstration primitive m'ont permis de résoudre cette question, et de faire voir directement que si la fonction méromorphe est périodique, ces deux fonctions entières sont «des fonctions périodiques». Je me bornerai à dire que la démonstration présente quelques analogies avec celle par laquelle Weierstrass établit qu'il existe une fonction entière de genre 2 qui admet tous les zéros d'une fonction elliptique.

L'existence de ces fonctions périodiques, en face desquelles le procédé de l'inversion est impuissant, fait mieux ressortir la nécessité où nous nous trouvons d'établir la théorie des fonctions abéliennes en partant des séries Θ elles-mêmes. On sait qu'il est possible de fonder sur l'étude directe des fonctions Θ à une seule variable toute la théorie des fonctions elliptiques; le point de départ est ce fait que l'équation

$$\Theta(x) = 0$$

n'a qu'une seule racine à l'intérieur du parallélogramme des périodes. De là l'importance du problème suivant, dont la solution (86, 12) doit évidemment précéder toute étude directe des séries Θ à plusieurs variables: Combien les équations simultanées

$$(\mathrm{I})\begin{cases} \Theta(x_1 - a_1,\, x_2 - a_2,\, \ldots,\, x_n - a_n) \\ = \Theta(x_1 - b_1,\, x_2 - b_2,\, \ldots,\, x_n - b_n) = \cdots = \Theta(x_1 - l_1,\, x_2 - l_2,\, \ldots,\, x_n - l_n) = 0, \end{cases}$$

où les a, les b, ..., les l sont des constantes données, ont-elles de solutions distinctes? A l'aide d'une formule de M. Kronecker, j'ai pu démontrer que ce nombre est constant et indépendant des périodes, ainsi que des constantes a, b, ..., l. Il me fut facile ensuite, en envisageant le cas particulier où la fonction Θ se réduit à un produit de n fonctions Θ elliptiques, de démontrer que ce nombre est précisément $1 \cdot 2 \cdot 3 \ldots n$.

J'appliquai aussi la même méthode à des équations analogues aux équations (I), mais plus compliquées, et je trouvai le nombre des solutions distinctes qu'elles doivent avoir.

Mais il y a plus; dans le cas des fonctions elliptiques, on trouve aisément la valeur de la racine de l'équation

$$\Theta(x) = 0.$$

Si l'on a affaire à des équations analogues, mais plus compliquées, on peut encore trouver la somme des racines.

Revenant aux fonctions abéliennes et aux équations (1), on peut alors se demander (86, 87) s'il est possible de trouver la somme des valeurs de x_1, celle des valeurs de x_2, etc., qui satisfont à ces équations. Ce problème est plus compliqué que le précédent, dans lequel le nombre cherché était une constante; cette circonstance permettait de se restreindre à un cas particulier, et l'on était ainsi immédiatement ramené aux fonctions elliptiques. Il n'en est plus de même ici; les nombres cherchés ne sont plus des constantes, mais des fonctions des périodes.

Toutefois le problème est immédiatement résolu quand on est ramené aux fonctions elliptiques, c'est-à-dire quand on se trouve dans un des cas de réduction étudiés plus haut. Quand, dans le système d'intégrales abéliennes de genre n qui correspondent aux fonctions Θ envisagées, il y a n intégrales distinctes réductibles aux intégrales elliptiques, il est aisé de voir que les fonctions Θ abéliennes s'expriment très simplement à l'aide de fonctions Θ elliptiques. On peut alors, par l'application du théorème d'Abel généralisé (cf. § VIII) résoudre complètement le problème qui nous occupe.

Le système des périodes d'une fonction Θ quelconque diffère toujours infiniment peu d'un système de périodes correspondant à une cas de réduction. C'est là une circonstance qui donnera, je n'en doute pas, la clef de bien des problèmes. Elle nous donne en particulier la solution que nous cherchons.

Nous connaissons la somme cherché des valeurs de x toutes les fois que nous nous trouvons dans un cas de réduction. Or cette somme doit être une fonction continue des périodes; *nous la connaîtrons donc dans tous les cas possibles.* C'est ainsi que, si l'on connaît une fonction *continue* de x pour toutes les valeurs commensurables de la variable, on la connaîtra également pour toutes les valeurs incommensurables.

On peut encore se placer à un autre point de vue pour étudier les zéros des fonctions Θ. Considérons une fonction Θ de deux variables $\Theta(x, y)$. Soient (α, β), (γ, δ) deux périodes de cette fonction. Écrivons l'équation

$$\Theta(\alpha t + \gamma u, \beta t + \delta u) = 0,$$

où t et u sont des nombres assujettis à rester réels et compris ente 0 et 1.

Cette équation ainsi interprétée admettra un certain nombre de solutions. Je serai conduit, par des considérations qui ne sauraient trouver place ici, à les distinguer en deux espèces. Soient alors N_1 le nombre des solutions de la première espèce, T_1 la somme des valeurs correspondantes de t, U_1 celle des valeurs de u. Soient N_2, T_2 et U_2 les quantités analogues en ce qui concerne les solutions de la seconde espèce. On peut se proposer de déterminer les nombres $N_1 - N_2$, $T_1 - T_2$, $U_1 - U_2$.

Je suis parvenu par une méthode assez simple (86) à déterminer $N_1 - N_2$. On obtient ainsi divers renseignements importants au sujet du nombre total des solutions $N_1 + N_2$. Ce nombre est en effet toujours supérieur à $N_1 - N_2$ et il est de même parité.

On peut arriver au même résultat par l'emploi des intégrales doubles prises entre des limites imaginaires. J'ai lieu d'espérer de plus que la même considération pourra conduire à la valeur de $T_1 - T_2$ et de $U_1 - U_2$.

Enfin on peut se poser encore la question d'une autre manière. Soit une fonction Θ spéciale de p variables engendrée par une courbe algébrique de genre p

$$F(x, y) = 0.$$

Soient $u_1(x, y)$, $u_2(x, y)$, ..., $u_p(x, y)$ les p intégrales abéliennes de $1^{\text{ère}}$ espèce. J'écrirai pour abréger $\Theta(v_i)$ pour:

$$\Theta(v_1, v_2, \ldots, v_p) \quad [\text{et } u_i(x) \text{ pour } u_i(x, y)]$$

et je considérerai pq constantes e_{ik} où l'indice i varie de 1 à p et l'indice k de 1 à q. J'envisage alors les q équations:

$$\Theta[u_i(x_1) + u_i(x_2) + \cdots + u_i(x_q) - e_{i1}] = 0$$
$$\Theta[u_i(x_1) + u_i(x_2) + \cdots + u_i(x_q) - e_{i2}] = 0$$
$$\cdots \cdots \cdots \cdots \cdots \cdots \cdots \cdots \cdots$$
$$\Theta[u_i(x_1) + u_i(x_2) + \cdots + u_i(x_q) - e_{iq}] = 0$$

où x_1, x_2, ..., x_p sont les inconnues. Combien ces équations ont elles de solutions? C'est le problème que j'ai résolu par une formule simple (194), le cas de $q = p$ se réduisant à celui que j'ai traité plus haut, tandis que le cas de $q = 1$ n'est autre que celui de RIEMANN.

XI. Fonctions diverses. (42, 91, 94).

Les fonctions fuchsiennes sont des fonctions uniformes d'une variable, inaltérées par certaines substitutions linéaires. On est naturellement conduit à se poser le problème suivant: Former des fonctions uniformes de deux variables, qui demeurent inaltérées par certaines substitutions linéaires. C'est, comme on sait, ce que M. PICARD a fait avec un plein succès par l'invention des fonctions hyperfuchsiennes.

Le premier problème à résoudre était évidemment de trouver les groupes discontinus contenus dans le groupe linéaire à deux variables. M. PICARD est parvenu à en former un grand nombre par des considérations arithmétiques. J'ai moi-même (24) démontré l'existence de deux classes de ces groupes. La première classe comprend les substitutions semblables des formes quadratiques ternaires indéfinies quand les coefficients de ces formes et de ces substitutions sont des entiers complexes. La seconde classe ne diffère pas essentiellement des groupes_fuchsiens.

Si z désigne en effet une variable imaginaire

$$z = \xi + i\eta$$

et si l'on pose

$$x = \frac{2\xi}{1 + \xi^2 + \eta^2}, \qquad y = \frac{2\eta}{1 + \xi^2 + \eta^2},$$

à tout groupe fuchsien appliqué à z et admettant pour cercle fondamental

$$\xi^2 + \eta^2 = 1,$$

correspondra un groupe discontinu appliqué aux deux variables x et y. Ce groupe est discontinu lorsque x et y sont imaginaires, ou bien réels, mais de telle façon que

$$x^2 + y^2 < 1;$$

il n'est plus proprement discontinu si x et y sont réels et si

$$x^2 + y^2 > 1.$$

C'est cette circonstance qui explique ce fait remarquable: qu'il est impossible d'imposer à une forme quadratique binaire indéfinie des conditions de réduction, telles que chaque classe contienne une réduite unique.

Mais les groupes de cette nature sont beaucoup moins importants que les groupes hyperfuchsiens proprement dits. J'appelle ainsi ceux qui n'altèrent pas l'hypersphère

$$x x_0 + y y_0 = 1.$$

(Je désigne ici par x_0 et y_0 les quantités imaginaires conjuguées de x et de y.)

Cette hypersphère joue dans cette théorie tout à fait le même rôle que le cercle fondamental dans la théorie des fonctions fuchsiennes.

J'ai voulu contribuer à l'étude de ces groupes et j'ai commencé par m'occuper des substitutions elles-mêmes. J'ai reconnu (81) que la classification en substitutions elliptiques, paraboliques et hyperboliques s'étendait aux substitutions hyperfuchsiennes.

La classification des groupes fuchsiens en familles est également applicable aux groupes hyperfuchsiens. Si nous laissons de côté les familles mixtes, nous distinguerons les groupes de la première famille qui contiennent des substitutions elliptiques, ceux de la deuxième famille qui n'en contiennent pas d'elliptiques, mais en contiennent de paraboliques, ceux de la troisième famille qui n'en admettent que d'hyperboliques.

Tous les groupes antérieurement découverts par M. PICARD appartenant à la seconde famille, je signalai alors l'existence de toute une catégorie de groupes de la troisième famille et des fonctions hyperfuchsiennes correspondantes que plusieurs propriétés importantes distinguaient des fonctions déjà connues.

On se trouve ici en présence des mêmes difficultés que dans le problème de la formation des groupes fuchsiens. Il faut d'abord former un groupe tel que la fonction correspondante soit uniforme dans le voisinage de chaque point. Il faut ensuite reconnaître si ce groupe est effectivement discontinu. La première difficulté, bien que très grande, est d'ordre purement algébrique. La seconde exige, pour être résolue, l'emploi de considérations étrangères à l'Algèbre. On peut l'éviter tant que l'on se borne aux groupes de la deuxième et de la troisième famille; il est nécessaire de l'aborder au contraire si l'on veut étudier les groupes de la première famille.

Je l'avais résolue, dans le cas des groupes fuchsiens, par l'emploi de la pseudogéométrie de LOWATSCHEVSKI; j'avais reconnu en effet que certaines quantités (analogues à ce que LOWATSCHEVSKI aurait appelé *longueur* ou *surface*) étaient des invariants par rapport aux substitutions d'un groupe fuchsien quelconque. Je me suis donc demandé si les substitutions hyperfuchsiennes admettaient de semblables invariants (42). J'ai reconnu qu'il en était ainsi: par conséquent *tout groupe, tel que la fonction correspondante soit uniforme dans le voisinage de chaque point, sera discontinu.*

La recherche des groupes hyperfuchsiens est donc ramenée à un pur problème d'Algèbre; mais ce problème reste extrêmement difficile et il n'a été résolu par M. Picard que dans un cas particulier. J'ai cherché à généraliser d'une autre manière les transcendantes uniformes qui se reproduisent par des substitutions simples. J'ai cherché s'il n'existait pas des fonctions uniformes possédant un «théorème de multiplication«, c'est à dire subissant une transformation algébrique, quand la variable est multipliée par un facteur constant. J'ai trouvé (91, 94) qu'il existe une classe étendue de pareilles transcendantes. Ce qui est intéressant, c'est le mode de raisonnement dont je me suis servi et qui peut être appliqué avec avantage à quelques questions relatives aux fonctions abéliennes.

TROISIÈME PARTIE.

QUESTIONS DIVERSES DE MATHÉMATIQUES PURES.

XII. Algèbre. (125, 126, 81, 127, 121.)

C'est par un problème d'Arithmétique que j'ai été conduit à m'occuper d'Algèbre. La théorie des formes arithmétiques et des substitutions linéaires à coefficients entiers appliqués à ces formes est en effet intimement liée à l'étude algébrique de ces mêmes formes et des substitutions linéaires à coefficients quelconques qu'elles peuvent subir.

C'est ainsi que j'ai été amené, à deux reprises différentes, à rechercher quelles sont les formes algébriques qui ne sont pas altérées par une substitution linéaire donnée et quels sont les groupes continus formés par ces substitutions. Après avoir classé (115, 121) les substitutions linéaires en quatre catégories jouissant de propriétés différentes, j'ai cherché quelles étaient les formes cubiques ternaires et quaternaires qui sont reproduites par une substitution linéaire donnée et par un *faisceau* de substitutions, c'est-à-dire par un groupe de substitutions permutables deux à deux. J'ai résolu également le problème inverse, c'est-à-dire que j'ai déterminé les substitutions que reproduisent une forme cubique ternaire donnée, ce qui m'était nécessaire pour le but arithmétique que j'avais en vue.

Il restait à trouver les formes cubiques quaternaires qui ne sont pas altérées par diverses substitutions linéaires *non permutables entre elles*. J'y suis arrivé par une méthode qui est fondée sur l'emploi des «crochets de Jacobi» et dont M. Sophus Lie a fait usage dans des problèmes analogues. La méthode n'était d'ailleurs pas restreinte aux formes cubiques quaternaires et permettait de trouver quelles sont les surfaces qui ne sont pas altérées par deux transformations homologiques non permutables.

Depuis, j'ai étendu ces résultats (125) au cas général de la façon suivante. Ayant indiqué la manière de former les groupes continus contenus dans le groupe linéaire à n variables, j'ai étudié les formes homogènes par rapport à ces variables qui ne sont pas altérées par les substitutions d'un de ces groupes et j'ai reconnu que ces formes satisfont à un certain nombre d'équations aux dérivées partielles formant un «système complet». Les plus simples des groupes continus en question jouissent de quelques propriétés que je vais énoncer succinctement. Si l'on forme le déterminant des coefficients d'une substitution linéaire à n variables, qu'on ajoute $+ S$ à chacun des termes de la diagonale principale, et qu'on égale à o le déterminant ainsi obtenu, on a une certaine *équation en* S de degré n.

Un groupe continu contient toujours une infinité de faisceaux; on démontre que, s'il y a dans le groupe une substitution admettant une certaine équation en S, il y aura *dans tous les faisceaux du groupe* une substitution admettant cette même équation en S.

Parmi les groupes continus dont je viens de parler, les plus intéressants sont ceux qui donnent naissance à un système de nombres complexes à multiplication non commutative (comme sont, par exemple, les quaternions). J'ai démontré que toutes les équations en S des substitutions de ces groupes ont des racines multiples.

Je suis revenu depuis sur ces groupes particuliers (127). Les recherches de M. Sylvester sur les matrices avaient de nouveau attiré l'attention des savants sur les nombres complexes. On pouvait se demander s'il en existait d'autres que ces matrices et leurs combinaisons. J'ai montré qu'il y en avait encore d'autres classes parmi lesquelles j'ai signalé une classe de «ternions».

Je rattacherai à ces études algébriques une Note (126) où j'énonce un résultat analogue à un important théorème de M. Laguerre. Soit une équation algébrique ayant p racines positives; j'ai démontré qu'on pouvait toujours en multiplier le premier membre par un polynôme choisi de telle sorte que le produit n'ait que p variations. Parmi tous les polynômes qui satisfont à cette condition, il y en a évidemment un dont le degré est minimum; mais je n'ai pu le trouver que dans des cas particuliers.

XIII. Groupes Continus. (60, 61.)

On vient de voir comment mes recherches sur l'algèbre m'avaient amené à m'occuper des groupes continus. C'est ainsi que j'avais montré (127) le lien qui unit ces groupes aux nombres complexes: j'avais énoncé à ce sujet un théorème dont, détourné par d'autres travaux, je n'ai jamais publié la démonstration, mais qui a été depuis démontré par M. Study.

Je me suis servi également des groupes continus dans un travail relatif aux géométries non euclidiennes (402).

Mais ce n'est que beaucoup plus récemment que j'ai abordé la théorie générale de ces groupes.

Lie avait démontré au sujet de ces groupes trois théorèmes fondamentaux. D'après le troisième de ces théorèmes, il existe toujours un groupe qui admet des équations de structure données pourvu que ces équations satisfassent aux conditions jacobiennes.

Lie a donné de ce théorème deux démonstrations. La première s'applique seulement aux groupes qui ne contiennent pas de substitutions permutables à toutes les autres substitutions. Elle ne laisse rien à désirer au point de vue de la simplicité. La seconde s'applique à tous les groupes; elle est beaucoup plus indirecte et plus compliquée.

Je me suis proposé de donner de ce troisième théorème une démonstration directe et simple, applicable à tous les cas. J'y suis parvenu (60, 161) grâce à l'emploi d'une notation symbolique très abrégée.

Je dois dire quelques mots sur le caractère de cette démonstration. Etant données les équations de structure, c'est à dire les régles de la composition des substitutions infinitésimales, j'ai cherché à en déduire les règles de la composition des substitutions finies. Or ces règles s'expriment par des séries infinies; j'ai reconnu que ces séries pouvaient se sommer par des formules où n'entrent pas d'autre transcendantes que des exponentielles.

Je me plaçais ainsi au point de vue formel, en introduisant des formules qui, faisant complètement abstraction de la «matière» du groupe, sont également applicables à tous les groupes isomorphes. Mais ces formules elles-mêmes nous font connaître les transformations que subissent les paramètres qui définissent une substitution du groupe lorsque l'on compare cette substitution avec une autre substitution du groupe. Ces transformations forment un groupe isomorphe à celui que l'on se proposait de former; ce groupe porte le nom de groupe paramétrique.

Nos formules nous permettent donc de former effectivement ce groupe paramétrique. Ainsi non seulement elles démontrent l'existence d'un groupe de structure donnée, mais elles donnent le moyen de le former effectivement. Lie avait démontré que la formation d'un pareil groupe pouvait se ramener à l'intégration d'un système d'équations différentielles ordinares. J'ai fait voir que non seulement on pouvait sans intégration former les substitutions infinitésimales du groupe, mais que dans le cas le plus défavorable, la formation des substitutions finies pouvait se ramener à une simple quadrature.

Dans le cas particulier auquel s'appliquait la première démonstration de Lie, les formules auxquelles on parvient sont assez simples, moins simples toutefois que celles de Lie. En tout cas, elles sont différentes et on ne voit pas immédiatement comment on peut passer des unes aux autres. La comparaison des deux sortes de formules n'en est que plus instructive. Elle nous fait retrouver un certain nombre de théorèmes de Killing. L'étude des formules obtenues nous fait d'ailleurs, même dans le cas général, retomber sur ces mêmes théorèmes.

XIV. Algèbre de l'infini. (128, 132, 137.)

J'ai été conduit par diverses considérations à une généralisation de la théorie des déterminants et des procédés par lesquels on résout n équations linéaires à n inconnues.

Dans certaines questions d'Analyse on est conduit à envisager un système de relations que l'on peut regarder comme une infinité d'équations linéaires à une infinité d'inconnues.

Soit un système de nombres donnés formant un tableau infini à double entrée. Je désignerai le terme général de ce tableau par la notation

$$a_{np} \quad (n, p = 1, 2, \ldots, \infty).$$

Le problème à résoudre consiste à déterminer une infinité de nombres

$$x_1, x_2, \ldots, x_n, \ldots,$$

de telle façon que les séries

$$S_p = \sum_{n=1}^{n=\infty} a_{np} x_n \quad (p = 1, 2, \ldots, \infty)$$

soient absolument convergentes et aient pour somme 0.

Ces équations linéaires, que l'on peut écrire

$$\Sigma_n a_{np} x_n = 0,$$

se rencontrent en particulier dans les circonstances suivantes:

1° Quand on cherche le quotient de deux séries trigonométriques;

2° Quand, ayant à intégrer une équation différentielle linéaire dont les coefficients sont des séries trigonométriques, on cherche à y satisfaire par une autre série trigonométrique.

Ce dernier problème se rencontre souvent en Mécanique céleste.

Jusqu'à ces derniers temps, on ne s'était pas préoccupé de savoir à quelles conditions les règles ordinaires du calcul pouvaient être appliquées à de semblables équations. Cependant deux savants, ayant rencontré ce même problème dans deux ordres de recherches très différents, n'ont pas hésité à employer les règles de l'Algèbre ordinaire.

L'un d'eux est M. Appell, qui est arrivé à des équations de la forme que nous étudions en cherchant à développer les fonctions elliptiques en séries trigonométriques. Les traitant d'après les règles du fini, il est parvenu à des formules qui concordent avec les résultats bien connus ou conduisent les autres méthodes.

D'un autre côté, M. Hill, en voulant déterminer le mouvement du périgée de la Lune, a appliqué aussi au problème qui nous occupe les procédés ordinaires de l'Algèbre. Cependant, le nombre auquel il arrive diffère très peu du nombre observé, et la faible divergence qui subsiste provient simplement de l'inclinaison de l'orbite que M. Hill avait négligée.

La hardiesse de M. Appell et celle de M. Hill avaient donc été également heureuses; mais elles n'étaient justifiées que par le succès. Néanmoins ce succès lui-même devait faire désirer une étude rationnelle de la question.

C'est cette étude que j'ai entreprise dans deux courtes Notes insérées au *Bulletin de la Société mathématique de France* (128, 132). Je suis parvenu à démontrer rigoureusement que les équations considérées par MM. Appel et Hill admettent effectivement les solutions trouvées par ces auteurs. Mais elles en admettent en même temps une infinité d'autres; elles ne suffisent donc pas pour déterminer les inconnues. M. Appell, de même que M. Hill, cherchait à calculer les coefficients d'une série. Or ces coefficients ne devaient pas seulement satisfaire aux équations envisagées, ils devaient encore être tels que la série fût convergente. Or, parmi les solutions en nombre infini qui admettent ces équations, il se trouve qu'une seule remplit cette seconde condition, et c'est précisément celle des auteurs que je viens de citer.

C'est cette circonstance qui explique le succès obtenu par ces deux savants géomètres; leur méthode est maintenant à l'abri de toute objection; mais il est aisé de voir que les considérations qu'ils ont invoquées ne suffisaient pas pour la justifier.

Je vais maintenant parler des procédés qui m'ont fait parvenir à ces résultats. J'ai commencé par m'occuper du cas particulier où

$$a_{np} = a_n^p,$$

et j'ai reconnu que la solution du problème dépendait de la décomposition de la fonction méromorphe

$$\frac{1}{f(z)}$$

en fractions simples, en appelant $f(z)$ la fonction entière transcendante qui admet pour zéros les nombres a_n.

J'ai reconnu également qu'on peut faire usage de considérations analogues dans le cas général.

Enfin, j'ai rencontré un fait réellement inattendu et tout à fait particulier à cette théorie. Les égalités à traiter

$$\Sigma\, a_{np}\, x_n = 0,$$

qui sont en nombre infini, peuvent être remplacées par une infinité d'inégalités. Il suffit, en effet, pour que les nombres x_n satisfassent à ces équations, que certaines séries qui en dépendent soient absolument convergentes.

Dans l'étude de cette question, on est naturellement conduit à considérer des déterminants d'ordre infini. A cet effet, on écrira le tableau à double entrée des quantités a_{np}, on formera un déterminant avec les n premières lignes et les n premières colonnes de ce tableau, et l'on fera croître ainsi n indéfiniment. Il convient de supposer

$$a_{nn} = 1.$$

On doit alors se demander à quelle condition un pareil déterminant converge. J'ai trouvé pour ces déterminants une règle de convergence qui présente la plus grande analogie avec la règle relative aux produits infinis.

Mais en ce qui concerne l'application de la méthode de M. HILL à la Mécanique céleste toutes les difficultés n'étaient pas encore surmontées. Le déterminant de HILL dépend d'un certain paramètre. Il fallait démontrer d'abord que c'est une fonction entière de ce paramètre, puis que cette fonction entière se réduit à un cosinus.

J'y suis parvenu (190, Ch. XVII) par une application des mêmes principes; mais dans la première marche que j'ai suivie pour cela, il a été nécessaire de déterminer le genre de cette fonction entière et j'ai du pour cela me servir des théorèmes de M. HADAMARD cités plus haut (Ch. VI). Pour éviter ce détour, j'ai cru devoir revenir (137) sur la même question et j'ai simplifié considérablement ma première démonstration.

XV. Arithmétique.   (114, 115, 116, 118, 123, 129. 131, 117, 122, 133, 130, 119, 120, 29, 134, 135.)

Mes recherches arithmétiques ont presque exclusivement porté sur la théorie des formes. Je vais commencer par exposer les résultats que j'ai obtenus au sujet des formes quadratiques.

On sait (117) qu'on représente la forme quadratique définie

$$a x^2 + 2 b x y + c y^2, D = b^2 - a c < 0$$

par un réseau de parallélogrammes dont les sommets ont pour coordonnées

$$x \sqrt{a} + y \frac{b}{\sqrt{a}}, y \sqrt{\frac{-D}{a}}$$

ou bien encore

$$a x + b y, y \sqrt{-D}.$$

Ce mode de représentation ne peut pas s'étendre aux formes indéfinies. Je représente alors la forme quadratique par le réseau dont les sommets ont pour coordonnées

$$a x + b y, y,$$

mode de représentation qui s'applique à la fois aux formes définies et indéfinies. Je reconnus d'abord que les réseaux de parallélogrammes jouissent de propriétés analogues à celles des nombres, et j'ai esquissé une *arithmétique des réseaux* où l'on trouve des théories analogues à celles de la divisibilité, des plus grands communs diviseurs et des plus petits communs multiples et même des nombres premiers.

Ma manière de représenter les formes indéfinies me conduit à une définition nouvelle de la réduction de ces formes. L'unique condition de réduction, c'est que les coefficients extrêmes doivent être de signe contraire. Avec cette définition, la réduction continuelle d'une forme indéfinie est susceptible d'une interprétation géométrique très simple. Je représente une forme par un certain triangle T qui n'est autre, d'ailleurs, que le triangle fondamental de notre réseau de parallélogrammes. Si la forme est réduite, des deux droites $y = \pm x \sqrt{D}$, l'une traverse le triangle T, l'autre lui reste extérieure. Achevons le parallélogramme dont notre triangle est la moitié et partageons-le de nouveau en deux triangles en menant la seconde diagonale; de ces deux nouveaux triangles, un, et un seulement, sera traversé par l'une des droites $y = \pm x \sqrt{D}$. Ce triangle représentera la réduite contiguë à celle que représentait le triangle T. En poursuivant indéfiniment de la sorte, on trouve une série de triangles qui représentent la réduction continuelle de la forme envisagée.

On peut, au lieu des droites $y = \pm x\sqrt{D}$, considérer deux droites quelconques passant par l'origine. On trouve ainsi, appliquant les mêmes procédés à ces deux droites, une représentation géométrique des réduites successives d'une fraction continue. On est naturellement conduit à une généralisation immédiate. Passons, en effet, du plan à l'espace, remplaçons le réseau par un assemblage à la BRAVAIS et, au lieu de deux droites, faisons-en passer trois par l'origine. Les mêmes considérations seront applicables, et l'on sera ainsi amené à une généralisation des fractions continues, à laquelle j'ai consacré une Note (129), mais qui, malheureusement, ne donne pas une approximation très rapide.

Il me reste, pour terminer l'analyse de mon Mémoire sur les formes quadratiques (117), à signaler deux résultats:

Je retrouve, en poursuivant l'étude de cette représentation géométrique, les lois de la composition des formes démontrées par GAUSS.

Enfin, je termine ce Mémoire par l'étude des nombres idéaux, qui ont pour origine les formes quadratiques binaires.

On sait que, lorsqu'on fait subir à une forme algébrique des substitutions linéaires *quelconques*, certaines fonctions des coefficients demeurent inaltérées: ce sont les *invariants*. En dehors de ces invariants *algébriques*, dont l'étude a été poussée très loin, il y a, ainsi que je l'ai démontré (114, 119), d'autres fonctions des coefficients qui sont altérées quand on applique à la forme une substitution à coefficients fractionnaires ou incommensurables, mais qui se reproduisent au contraire quand on lui fait subir une substitution à coefficients entiers. Ce sont les invariants *arithmétiques*. Les formes linéaires binaires qui n'ont pas d'invariants algébriques ont, au contraire, des invariants arithmétiques dont l'étude se rattache à la théorie des fonctions elliptiques et à celle des fonctions modulaires et des fonctions fuchsiennes. Ces invariants peuvent être utilisés pour la solution des deux problèmes suivants:

1°. Trouver le plus petit nombre représenté par une forme quadratique binaire indéfinie;

2°. Reconnaître si deux formes quadratiques binaires indéfinies sont équivalentes.

A cet effet, on décompose chacune de ces formes en deux facteurs linéaires et l'on exprime en fonction des invariants de ces facteurs les coefficients de la substitution qui permet de passer d'une forme à l'autre, *à supposer qu'elles soient équivalentes*. Il est aisé ensuite de voir si les coefficients ainsi obtenus sont entiers et s'ils permettent effectivement de passer d'une forme à l'autre. Dans le cas où il n'en serait pas ainsi, on serait certain qu'il n'y aurait pas équivalence.

Les formes quadratiques binaires définies ou indéfinies possèdent également des invariants arithmétiques dont j'ai étudié les propriétés. Pour que deux formes soient équivalentes, il faut et il suffit que tous leurs invariants soient égaux. Toutefois, pour reconnaître rapidement l'équivalence, il est préférable de décomposer chaque forme en deux facteurs linéaires et d'envisager les invariants de ce système de formes linéaires.

Tous ces invariants sont susceptibles d'être exprimés: $1°$ par des intégrales définies; $2°$ par des séries.

L'un des problèmes les plus importants qui se posent au sujet des formes quadratiques ternaires indéfinies est l'étude des propriétés des groupes discontinus formés par les «substitutions semblables», c'est-à-dire par les substitutions linéaires qui n'altèrent pas ces formes (120, 131). Soit $F(x, y, z)$ une forme quadratique indéfinie. On peut choisir la constante K de telle façon que $F(x,y,z)=K$ représente un hyperboloïde à deux nappes. Les substitutions semblables changeront alors un point de cet hyperboloïde en un autre point de cette même nappe, de sorte que, le groupe étant discontinu, l'hyperboloïde se trouvera partagé en une infinité de polygones curvilignes, dont les côtés seront des sections diamétrales de la surface. Les substitutions semblables changeront ces polygones les uns dans les autres. Faisons maintenant une perspective en plaçant l'œil en un ombilic de la surface et prenant pour plan du tableau une section circulaire. Une nappe de l'hyperboloïde se projettera suivant un cercle, et les polygones que nous avons tracés sur cette nappe se projetteront suivant des polygones curvilignes, limités par des arcs de cercle reproduisant identiquement la figure dont nous avons parlé (p. 44 et suivantes), à propos de la théorie des groupes fuchsiens. Ainsi, l'étude des groupes de substitutions semblables des formes quadratiques est ramenée à celle des groupes fuchsiens, ce qui est un rapprochement inattendu entre deux théories très différentes et une application nouvelle de la Géométrie non euclidienne.

Après avoir signalé un certain nombre de propriétés de ces groupes fuchsiens particuliers, j'ai abordé une question un peu différente.

Les substitutions semblables sont celles qui reproduisent une forme quadratique et qui en même temps appartiennent au groupe G des substitutions à coefficients entiers. On peut rechercher alors les substitutions qui reproduisent la forme quadratique et qui en même temps appartiennent à un autre groupe, par exemple à un sous-groupe du groupe G. Cela nous permet en même temps de généraliser la théorie de l'équivalence des formes et de leur réduction.

On obtient aisément des groupes de ces substitutions semblables généralisées et l'on reconnaît que ce sont encore des groupes fuchsiens. En réfléchissant ensuite aux relations de ces divers groupes fuchsiens, j'ai démontré que les fonctions fuchsiennes correspondantes jouissent d'une propriété analogue au théorème d'addition des fonctions elliptiques, ce qui n'est pas vrai des fonctions fuchsiennes les plus générales.

Passons maintenant aux formes d'ordre supérieur au second (122). Le premier problème à résoudre est la réduction de ces formes et l'étude des conditions de leur équivalence. La solution a été trouvée par M. HERMITE; bien que le savant géomètre n'ait parlé que des formes binaires et des formes quadratiques, sa méthode s'applique, sans qu'on ait rien à y changer, à une forme tout à fait quelconque. C'est ainsi que M. JORDAN, étendant à un cas très général un théorème de M. HERMITE, a démontré que, toutes les fois que le discriminant n'est pas nul, toutes les formes qui ont mêmes invariants algébriques se répartissent en un nombre fini de classes. J'ai moi-même généralisé le théorème de M. JORDAN, en montrant qu'il subsiste, pourvu que certains invariants ne soient pas tous nuls à la fois.

J'ai cherché ensuite à appliquer la méthode générale aux formes cubiques ternaires que j'avais déjà étudiées au point de vue algébrique dans un Mémoire précédent. Je suis arrivé à trouver les limites supérieures des coefficients d'une réduite dont les invariants sont donnés, pourvu que le discriminant ne soit pas nul. Le nombre des classes est alors limité et, dans chaque classe, il n'y a qu'une réduite.

Lorsque la forme égalée à zéro représente une courbe de quatrième classe, le discriminant est nul et le nombre des classes est infini, mais chacune d'elles ne contient qu'une réduite. Si la courbe est de troisième classe, le nombre des classes est infini et chacune d'elles contient un nombre fini de réduites formant *une chaîne limitée à ses deux extrémités.* Si la courbe se décompose en une conique et une droite qui la coupe, le nombre des classes est tantôt fini et tantôt infini; de plus, la chaîne formée par les réduites d'une même classe est, tantôt limitée comme dans le cas précédent, tantôt illimitée de telle façon que les mêmes réduites s'y reproduisent périodiquement. Si enfin la droite est tangente à la conique, les réduites ne forment plus une chaîne, mais un réseau.

J'ai ensuite appliqué la même méthode, non plus à une forme unique, mais à un système de formes, et j'ai choisi comme exemple le système d'une forme quadratique ternaire et d'une forme (116, 133) linéaire dont j'ai étudié la *réduction simultanée.* La réduction continuelle d'un pareil système de formes est tout à fait analogue à celle d'une forme unique. Elle peut servir également à déterminer les substitutions semblables du système. Ces substitutions semblables existent toujours; mais, ayant voulu, dans un exemple particulier, calculer les coefficients de la plus simple d'entre elles, j'ai trouvé des nombres entiers de plus de huit chiffres.

Les lois de la réduction d'une forme quelconque étant connues, il est facile de reconnaître si deux formes sont équivalentes; mais ce n'est là qu'un premier pas. Le principal problème à résoudre, c'est de rechercher si un nombre donné peut être représenté par une forme donnée. Je me suis occupé spécialement de la représentation par une forme binaire (118, 130). Égalant la forme binaire à 0, on en tire pour le rapport $\dfrac{x}{y}$ une certaine valeur. Avec cette valeur, je forme un système de nombres complexes et d'idéaux. Le problème de la représentation des nombres par les formes se ramène à la recherche des idéaux de norme donnée. J'ai donné, en me fondant sur les mêmes principes que dans mon Mémoire intitulé *Sur un mode nouveau de représentation géométrique des formes quadratiques*, la manière de former tous les idéaux de norme N, de former tous les idéaux premiers et leurs puissances, de multiplier deux idéaux, de décomposer un idéal en facteurs premiers, etc. Pour cela j'envisage une certaine congruence, que je décompose en facteurs irréductibles. A chacun de ces facteurs irréductibles correspond un idéal.

On trouve toutes les représentations d'un nombre donné quand on connaît tous les idéaux dont la norme est le nombre donné, mais tous ces idéaux ne donnent pas naissance à une représentation du nombre. Il importerait donc de savoir distinguer *a priori* quels sont les idéaux qui conduiront à une pareille représentation. Tout ce que j'ai pu faire dans ce sens a été de montrer qu'ils devaient tous se trouver parmi les idéaux auxquels correspond un facteur irréductible *linéaire* de la congruence dont j'ai parlé plus haut (et par conséquent une racine *réelle* de cette congruence).

Dans deux Notes que j'ai eu l'honneur de présenter à l'Académie les 9 et 16 janvier 1882, j'ai cherché quelle était la véritable signification de la notion de genre définie par Gauss pour les formes quadratiques binaires et étendue par Eisenstein aux formes quadratiques ternaires, et je suis arrivé à en donner les définitions suivantes:

1°. Deux formes sont équivalentes suivant le module n, si l'on peut appliquer à la première de ces formes une substitution à coefficients entiers, telle que les coefficients de la transformée ainsi obtenue ne diffèrent de ceux de la seconde forme que par des multiples de n;

2º. Deux formes sont de même genre lorsqu'elles sont équivalentes suivant un module quelconque.

Il est clair que cette définition peut s'appliquer à des formes tout à fait quelconques auxquelles j'ai étendu également la définition de l'*ordre*. J'ai appliqué ces principes aux formes quadratiques quaternaires et cubiques binaires.

Dans un autre ordre d'idées, j'ai cherché à généraliser l'élégante méthode de TCHEBICHEFF pour l'étude de la distribution des nombres premiers. J'ai reconnu qu'elle pouvait s'appliquer presque sans changement aux nombres complexes de la forme $a + b\sqrt{-1}$ (134, 135). Au point de vue des nombres réels, cela permet de comparer la distribution des nombres premiers de la forme $4n + 1$ à celle des nombres premiers de la forme $4n + 3$.

XVI. Analysis Sitûs. (155, 156, 158, 154, 159, 160.)

L'Analysis Sitûs est la science qui nous fait connaître les propriétés *qualitatives* des figures géométriques non seulement dans l'espace ordinaire, mais dans l'espace à plus de trois dimensions.

L'Analysis Sitûs à 3 dimensions est pour nous une connaissance presque intuitive, L'Analysis Sitûs à plus de 3 dimensions présente au contraire des difficultés énormes; il faut pour tenter de les surmonter être bien persuadé de l'extrême importance de cette science.

Si cette importance n'est pas comprise de tout le monde, c'est que tout le monde n'y a pas suffisamment réfléchi. Mais que l'on pense aux avantages qu'ont tirés les analystes des représentations géométriques, même dans des questions d'Analyse Pure et d'Arithmétique; que l'on estime le soulagement que ces méthodes ont procuré à l'esprit des chercheurs. Combien il est regrettable que cet instrument merveilleux se trouve hors d'usage dès que le nombre des dimensions surpasse trois.

RIEMANN, qui avait fait de cet instrument l'usage que l'on sait, avait bien compris combien il serait important d'y suppléer et on a retrouvé dans ses papiers quelques notes, malheureusement un peu informes mais qui servent encore aujourd'hui de base à toutes nos connaissances sur l'Analysis Sitûs à plus de trois dimensions.

On a dit, écrivais-je (ou à peu près) dans une préface (154), que la géométrie est l'art de bien raisonner sur des figures mal faites. Oui, sans doute, mais à une condition. Les proportions de ces figures peuvent être grossièrement altérées, mais leurs éléments ne doivent pas être transposées et ils doivent conserver leur situation relative. En d'autres termes, on n'a pas à s'inquiéter des propriétés quantitatives, mais on doit respecter les propriétés qualitatives, c'est à dire précisément celles dont s'occupe l'Analysis Sitûs.

Cela doit nous faire comprendre qu'une méthode qui nous ferait connaître les relations qualitatives dans l'espace à plus de trois dimensions, pourrait, dans une certaine mesure, rendre des services analogues à ceux que rendent les figures. Cette méthode ne peut être que l'Analysis Sitûs à plus de trois dimensions.

Malgré tout, cette branche de la science a été jusqu'ici peu cultivée. Après RIEMANN est venu BETTI qui a introduit quelques notions fondamentales; mais BETTI n'a été suivi par personne.

Quant à moi, toutes les voies diverses où je m'étais engagé successivement me conduisaient à l'Analysis Sitûs. J'avais besoin des données de cette science pour poursuivre mes études sur les courbes définies par les équations différentielles (vide supra § V) et pour les étendre aux équations différentielles d'ordre supérieur et en particulier à celles du problème des trois corps. J'en avais besoin pour l'étude des fonctions non uniformes de 2 variables. J'en avais besoin pour l'étude des périodes des intégrales multiples et pour l'application de cette étude au développement de la fonction perturbatrice.

Enfin j'entrevoyais dans l'Analysis Sitûs un moyen d'aborder un problème important de la théorie des groupes, la recherche des groupes discrets ou des groupes finis contenus dans un groupe continu donné.

C'est pour toutes ces raisons que je consacrai à cette science un assez long travail (155, 156, 154). Je commence par donner plusieurs définitions des variétés de l'espace à plus de trois dimensions et par introduire la notion fondamentale de l'homéomorphisme qui est la relation de deux variétés qui ne sont pas distinctes au point de vue de leurs propriétés qualitatives.

Je suis amené ensuite à distinguer les variétés bilatères analogues aux surfaces ordinaires et les variétés unilatères analogues aux surfaces à un seul côté.

BETTI avait découvert certains nombres entiers relatifs aux variétés; analogues à ce qu'est pour une surface ordinaire ce qu'on appelle l'ordre de connexion. On sait que l'ordre de connexion d'une surface fermée dépend du nombre de trous qui y sont percés, de sorte que cet ordre est toujours impair, 1 pour une sphère, 3 pour un tore etc. On sait également quelle relation il y a entre le genre d'une courbe algébrique et l'ordre de connexion de la surface de RIEMANN correspondante.

J'ai fait voir que si l'on écrit la série des nombres de BETTI pour une surface fermée, les nombres également distants des extrêmes sont égaux.

M. HEEGAARD ayant attiré mon attention sur certains exemples où ce théorème paraissait en défaut, je revins sur la même question dans un autre travail

(159). La définition que j'avais donnée des nombres de BETTI ne concordait pas toujours avec celle qu'avait donnée BETTI lui-même. Le théorème, vrai pour les nombres de BETTI tels que je les avais définis, ne l'est pas toujours pour les nombres tels que BETTI les définissait lui-même.

On sait que l'ordre de connexion suffit pour déterminer une surface ordinaire au point de vue de l'Analysis Sitûs, c'est à dire que deux surfaces qui ont même ordre de connexion sont homéomorphes. On pouvait supposer que les nombres de BETTI suffisaient de même pour déterminer une variété. J'ai montré (154) qu'il n'en est rien, qu'à chaque variété correspond un groupe, nécessaire à sa détermination, et qu'à une même suite de nombres de BETTI ne correspond pas toujours un même groupe.

J'ai cru devoir multiplier les exemples, pensant que c'était le meilleur moyen de familiariser les esprits avec des idées aussi nouvelles.

On sait qu'EULER a démontré une relation entre le nombre des faces, des arêtes et des sommets d'un polyèdre convexe. Pour les polyèdres non convexes, il y a une relation analogue entre ces trois nombres et l'ordre de connexion. Existe-t-il des relations du même genre entre des éléments des polyèdres de l'espace à plus de trois dimensions? C'est la question que je me suis posée (154) et que j'ai résolue affirmativement, en faisant usage de plusieurs démonstrations distinctes. Il est à remarquer que si le nombre des dimensions de l'espace est pair, cette relation ne dépend pas des nombres de BETTI et qu'elle en dépend au contraire si le nombre des dimensions est impair.

Ces théorèmes sur les polyèdres ont une portée assez générale, car une variété fermée quelconque peut toujours être découpée en polyèdres; rectilignes ou curvilignes, cela n'importe pas au point de vue de l'Analysis Sitûs. Dans mes travaux ultérieurs (159, 160) j'ai généralement trouvé plus commode de supposer effectuée cette décomposition en polyèdres.

Il peut y avoir entre les figures tracées sur une variété, et en particulier entre les éléments d'un polyèdre, plusieurs sortes de relations qui sont susceptibles d'être représentées algébriquement par des équations symboliques et d'être combinées ensuite d'après les règles de l'algèbre ou d'après des règles analogues. J'ai appelé ces relations congruences, homologies, équivalences. Les congruences expriment tantôt que l'ensemble de tels éléments constitue une variété fermée, tantôt au contraire que cet ensemble constitue une variété ouverte dont la frontière complète est formé par l'ensemble de tels autres éléments.

Les homologies fondamentales expriment que l'ensemble de tels éléments constitue une variété fermée qui est la frontière complète d'une autre variété qui doit avoir une dimension de plus, mais qui reste indéterminée. Les homologies dérivées se déduisent des homologies fondamentales, mais il importe de distinguer celles qui s'en déduisent par addition et multiplication et celles qui s'en déduisent par division.

Les équivalences diffèrent des homologies parce qu'on ne se donne pas le droit d'y intervertir l'ordre des termes. C'est la considération de ces équivalences qui conduit au groupe dont j'ai parlé plus haut.

Toutes ces relations se présentent sous la forme d'équations linéaires à coefficients entiers. L'étude d'une variété se trouve ainsi ramenée à celle d'un certain nombre de tableaux formés de nombres entiers. Ces tableaux varient évidemment selon la manière dont la variété a été découpée en polyèdre; mais cependant tous les tableaux différents que l'on peut obtenir ainsi conservent certains caractères communs que l'on peut appeler iuvariants et qui restent les mêmes quelle que soit la manière dont la variété est découpée. Ces invariants sont les plus grands communs diviseurs de certains déterminants formés avec les éléments des tableaux.

Grâce à cette représentation arithmétique, les démonstrations deviennent plus faciles à suivre et j'ai pu ajouter divers résultats à ceux que j'avais déjà obtenus. Par exemple pour que les deux définitions des nombres de BETTI coïncident, il faut et il suffit que tous les invariants soient égaux à o ou à 1; on encore que le système des homologies obtenues par division n'en contienne pas que l'on ne puisse obtenir sans division; ou enfin que le polyèdre ne soit pas tordu, c'est à dire que toutes les variétés que l'on peut former avec ses éléments soient bilatères.

QUATRIÈME PARTIE.

MÉCANIQUE CÉLESTE.

XVII. Généralités sur les Equations de la Dynamique et de la Mécanique Céleste. (225, 226, 214, 229, 189, 191.)

Les équations de la Dynamique présentent des propriétés remarquables qui ont été mises en évidence par Jacobi dans ses Vorlesungen.

Quelles sont les conséquences plus ou moins immédiates de ces propriétés? Quel partie peut-on en tirer pour la mise en équation des problèmes de Dynamique et en particulier des problèmes de Mécanique Céleste? Telle est la première question dont je veux parler ici.

J'ai été amené à passer en revue les principales propriétés des équations canoniques (214, 189). Les propriétés sont classiques; et je n'ai eu qu'a perfectionner certains détails; en me servant surtout du caractère bien connu qui permet de reconnaître si un changement de variables conserve la forme canonique des équations.

Ce genre de transformations facilite la mise en équation du problème des trois corps; c'est ce que j'ai montré (225, 229). On sait que dans le procédé classique on rapporte toutes les planètes à des axes mobiles passant par le Soleil. L'inconvénient est que la fonction perturbatrice n'est pas la même pour toutes les planètes. Un autre procédé consiste à rapporter chaque planète au centre de gravité du système formé par le Soleil et toutes les planètes inférieures à celle que l'on considère. L'inconvénient est évité, mais la fonction perturbatrice est un peu plus compliquée. J'ai proposé un troisième procédé, dans lequel les coordonnées de chaque planète sont rapportées au Soleil, et sa vitesse à des axes fixes.

Malgré les travaux dont les équations canoniques ont été l'objet depuis Jacobi, toutes leurs propriétés ne sont pas connues, ou plutôt on n'a pas insisté sur toutes les formes que peuvent revêtir ces propriétés et qu'il peut être utile de connaître. Si par exemple on étudie les équations aux variations des équations de la Dynamique, c'est à dire les équations qui définissent une solution infiniment peu différente d'une solution donnée, on rencontre des propositions importantes sur lesquelles j'ai attiré l'attention (214, 189).

D'un autre côté, j'ai été amené à introduire une notion nouvelle, celle des invariants intégraux (214, 191). Ce sont certaines intégrales définies simples ou multiples qui demeurent constantes, quand le champ d'intégration varie conformément à une certaine loi définie par une équation différentielle. Si par exemple on envisage les équations différentielles au mouvement d'un fluide incompressible, le volume est un invariant intégral.

Les équations canoniques de la Dynamique possèdent des invariants intégraux remarquables et l'existence de ces invariants jette une grande lumière sur leurs propriétés.

Pour en finir avec ces généralités sur les équations de la Dynamique et le problème des 3 corps, je signalerai un dernier travail (226). On sait que BRUNS a démontré que le problème des 3 corps ne saurait admettre d'autre intégrale algébrique que les intégrales classiques. Malheureusement dans sa démonstration subsistait une lacune grave et particulièrement délicate à combler. J'ai été assez heureux pour mettre la belle et ingénieuse démonstration de M. BRUNS à l'abri de toute objection.

XVIII. Problème des Trois Corps; Propriétés qualitatives.
(200, 204, 225, 182, 214, 216, 189, 191.)

Ce qui va suivre est le développement naturel des méthodes dont il a été question plus haut au § V et leur application à la Mécanique Céleste.

J'ai montré de diverses manières (214, 189) qu'en dehors des intégrales classiques le problème des trois corps n'admet pas d'intégrale analytique et uniforme, et il en résultait que la plupart des séries proposées jusqu'ici pour l'intégration de ce problème, de même que celles dont il sera question dans le § suivant ne sont pas convergentes et ne peuvent être utitisées que dans un calcul approché.

D'après ce qui précède, il semble qu'il soit impossible en général d'exprimer les distances mutuelles des astres par des séries purement trigonométriques convergentes. Mais il est des cas particuliers où les séries auxquelles on est conduit ne contiennent qu'un seul argument et où leur convergence est évidente. En effet, j'ai démontré (200, 204) que, dans le problème des trois corps, on peut choisir les éléments initiaux du mouvement, de telle façon que les distances mutuelles des trois masses soient des fonctions périodiques du temps. On est ainsi amené à une solution particulière du problème, que l'on peut appeler *périodique*.

Ces solutions périodiques sont de trois sortes: dans les unes, les inclinaisons sont nulles et les excentricités très petites; dans d'autres, les inclinaisons sont nulles et les excentricités finies; dans d'autres, enfin, les inclinaisons sont finies et les excentricités très petites.

Je suis revenu (214, 189) sur ces solutions périodiques et je les ai étudiées en détail. Les procédés dont je me suis servi pour démontrer leur existence sont très simples et se ramènent au calcul des Limites.

Mais on peut arriver à cette démonstration par une voie toute différente, qu'il pourra être souvent utile d'adopter, mais dont je n'ai pas encore tiré tout le parti possible. Supposons par exemple que l'on recherche les géodésiques d'une surface indéfinie présentant la même forme générale qu'un hyperboloïde à une nappe. On sera certain alors qu'il doit y avoir une géodésique fermée (correspondant à une solution périodique) parce que parmi toutes les courbes fermées que l'on peut tracer sur la surface, et qui en font le tour il doit y en avoir une qui est plus courte que toutes les autres.

Les mêmes principes sont susceptibles d'être appliqués à divers problèmes de Mécanique, grâce au principe de moindre action que l'on peut employer soit sous la forme que lui a donnée HAMILTON, soit sous celle que lui a donnée MAU-PERTUIS. Je n'ai fait qu'esquisser cette méthode dont il y a sans doute encore beaucoup à tirer.

Outre les solutions périodiques, les équations du problème des 3 corps admettent aussi d'autres solutions remarquables que j'appelle asymptotiques (214, 189). Ce qui caractérise ces solutions c'est qu'elles se rapprochent indéfiniment d'une solution périodique, ou bien qu'elles s'éloignent sans cesse d'une solution périodique dont elles sont infiniment rapprochées pour $t = -\infty$.

D'autres solutions remarquables sont plus difficiles encore à apercevoir (191). Je citerai d'abord les solutions périodiques de 2e espèce, caractérisées par ce fait que deux des corps se rapprochent périodiquement de façon à presque se choquer.

Nous avons encore les solutions périodiques de 2e genre; si l'on fait varier d'une manière continue un des paramètres dont dépend le problème, par exemple l'une des masses, on voit une solution périodique du 1er genre se déformer l'une façon continue, sa période restant égale à T. A un certain moment, cette solution se dédouble pour ainsi dire, ou plutôt se détriple, je veux dire qu'à un certain moment on a trois solutions périodiques très peu différentes; l'une d'elles a encore pour période T, les deux autres ont pour période un multiple de T. Ce sont les solutions périodiques du 2e genre.

Je parlerai enfin des solutions doublement asymptotiques. Pour $t = -\infty$, elles sont infiniment voisines d'une solution périodique, puis elles s'en éloignent beaucoup, ensuite elles s'en rapprochent de nouveau de telle sorte que pour $t = +\infty$, elles en sont encore infiniment voisines.

Pour étudier les propriétés et les rapports de ces différentes solutions, je me suis servi des invariants intégraux. Ces rapports sont très compliqués de sorte que cette étude est éminemment propre à mettre en évidence la difficulté du problème des Trois Corps.

Je n'ai pu résoudre rigoureusement et complètement le problème de la stabilité du système Solaire, en entendant ce mot dans un sens strictement mathématique. L'emploi des invariants intégraux m'a cependant permis (214, 191) d'atteindre certains résultats partiels, s'appliquant surtout au problème dit restreint, où les deux corps principaux circulent dans des orbites sans excentricité, pendant que le corps troublé a une masse négligeable. Dans ce cas, si on laisse de côté certaines trajectoires exceptionnelles, dont la réalisation est infiniment peu probable, on peut démontrer que le système repassera une infinité de fois aussi près que l'on voudra de sa situation initiale. C'est ce que j'ai appelé la stabilité à la Poisson.

XIX. Problème des Trois Corps; Développements approchés et applications.
(203, 208, 211, 218, 221, 223, 224, 214, 230, 237, 241, 137, 242, 190.)

Tous les théorèmes dont il a été question dans le § précédent ont un caractère commun, ils sont rigoureux. Ceux dont je vais parler maintenant ne seront en général qu'approchés et auront par conséquent avant tout un intérêt pratique. Ce sont cependant les seuls dont les astronomes fassent et puissent faire effectivement usage.

Dans quelles conditions ces séries divergentes peuvent-elles être utilisées avec succès? C'est ce que j'ai cherché à éclaircir (190, Chapitre intitulé calcul formel). J'ai montré dans quelles limites, cet emploi est légitime, comme l'est, pour citer un exemple célèbre, celui de la série de Stirling et j'ai fait voir que les règles du calcul de ces séries sont les mêmes que celles de séries ordinaires.

J'ai justifié ainsi l'emploi que font les astronomes de ce genre de développements; et en particulier des développements trigonométriques. J'ai cherché en outre à perfectionner les méthodes qui permettent de les former.

Rappelons en quelques mots comment le problème se pose.

On ne peut résoudre le problème des n corps que par approximations suc-cessives, et la première idée qui s'est présentée a consisté à développer les coordonnées des astres en séries ordonnées suivant les puissances des masses. C'est sur cette idée qu'est fondée toute la Mécanique céleste ancienne. Mais quels que soient les services qu'aient rendus autrefois ces anciens procédés et qu'ils soient capables de rendre encore, on n'a pas tardé à s'apercevoir de leur insuffisance et de leur impuissance à donner une approximation indéfinie. Dans les développements auxquels ils conduisent, on voit en effet le temps entrer non seulement sous les signes sinus et cosinus, mais en dehors de tout signe trigonométrique. Ce fait suffit pour démontrer que le champ, où les anciennes méthodes conservent leur efficacité, quelque étendu qu'il puisse être, est certainement limité.

C'est ce qui explique les efforts qu'ont faits les géomètres pour remplacer les séries anciennes par des développements purement trigonométriques. Dans ces derniers temps, on a proposé deux méthodes remarquables qui paraissent atteindre complètement ce but. La première est celle de M. GYLDÉN, qui est fondée sur l'emploi des fonctions elliptiques; la seconde est celle de M. LINDSTEDT, dont je me suis surtout occupé.

Dans cette dernière méthode, un artifice ingénieux permet à chaque approximation de faire disparaître les termes séculaires qui peuvent s'être introduits. Il est aisé de voir que cet artifice réussira toujours s'il n'y a qu'un terme à faire disparaître; mais il n'en serait plus de même s'il s'était introduit à la fois deux termes séculaires. Il est facile de vérifier d'ailleurs que, dans les premières approximations, on n'a à se débarrasser que d'un seul terme; mais on peut se demander s'il doit en être toujours ainsi. Un examen superficiel pourrait faire croire le contraire, et même M. LINDSTEDT était disposé à penser que sa méthode ne réussirait que s'il n'y avait entre les arguments aucune relation linéaire.

Je suis parvenu à démontrer que le terme séculaire qui peut apparaître à chaque approximation est toujours unique et que, par conséquent, la méthode de M. LINDSTEDT est toujours applicable (208). Pour cela, j'ai eu recours à un théorème qui semblait ne se rapporter en aucune façon à la question, c'est-à-dire au théorème de GREEN.

J'aurais pu également, comme je l'ai fait ensuite, (211, 218) employer les invariants intégraux ou bien me servir des théorèmes de JACOBI sur les équations de la Dynamique.

La méthode de M. NEWCOMB est fondée sur les mêmes principes; mais elle s'applique plus naturellement aux cas les plus généraux du problème des Trois Corps. J'y ai apporté de notables perfectionnements (214, 190, 230). La même question se posait que pour la méthode de M. LINDSTEDT. On pouvait disposer de certaines arbitraires pour faire disparaître les termes séculaires; mais il y avait deux fois plus de termes à faire disparaître que d'arbitraires. Heureusement chaque fois que l'on fait disparaître l'un de ces termes, un autre disparaît spontanément. Je me suis servi de la méthode de JACOBI (190), pour démontrer cette disparition spontanée et la possibilité du développement. Une fois cette possibilité établie, j'ai donné (190, Chapitre XIV) le moyen de former effectivement et directement les séries.

Il y avait là toutefois un grave inconvénient, puisqu'il fallait dans deux analyses entièrement séparées, démontrer la possibilité du développement et en calculer les coëfficients. Ayant remarqué qu'une certaine expression devait être différentielle exacte, j'en ai profité pour modifier la méthode (190, Chapitre XV). Certaines intégrations peuvent être évitées et remplacées par des différentiations ou des opérations algébriques; il en résulte d'abord que la possibilité du développement devient presque évidente. De plus les calculs sont simplifiés. La partie la plus pénible du calcul, bien qu'elle ne présente aucune difficulté théorique, c'est en effet la substitution des valeurs approchées des variables dans les deux membres des équations différentielles. La méthode nouvelle permet de diminuer de près de moitié le nombre de ces substitutions.

Mais on peut aller plus loin encore dans cette voie. C'est ce que j'ai montré plus tard (230). Je me suis servi d'une expression différentielle exacte analogue à celle dont je viens de parler et j'ai pu diminuer encore le nombre des substitutions et des intégrations. On remarquera que le théorème de POISSON (invariabilité des grands axes en tenant compte du carré des masses) devient presque intuitif.

Tous ces procédés se trouvent en défaut quand les moyens mouvements sont près d'être commensurables. Il faut alors employer une méthode dérivée de celle de DELAUNAY. Dans l'invention première de cette méthode, j'ai été devancé de quelques jours par M. BOHLIN, mais j'y ai apporté divers perfectionnements en me servant toujours des mêmes principes. Je me bornerai à dire que cette méthode permet de discuter les cas singuliers dits «de libration» où il y a commensurabilité exacte entre certains mouvements moyens.

Quelle relation y a-t-il entre toutes ces méthodes et avec les anciens procédés? Telle est la question que je me suis posée (237) et j'ai montré qu'on pouvait envisager certaines séries qui embrassent pour ainsi dire toutes ces méthodes. En groupant les termes d'une certaine manière, on retombera sur les anciennes méthodes. Avec d'autres modes de groupement, on tombera sur la méthode de NEWCOMB, ou bien encore sur celle de BOHLIN.

Tous ces procédés s'appliquent naturellement à la Lune. Dans la théorie de cet astre, ils sont même plus nécessaires que partout ailleurs. C'est dans les théories lunaires les plus récentes, comme celles de HILL et de BROWN, que l'on voit le mieux l'importance des solutions périodiques dont j'ai parlé plus haut. J'ai donné (241) une méthode pour déterminer le mouvement du perigée de la Lune qui diffère de celle de HILL. Mais j'attirerai plutôt l'attention sur un autre mémoire (242) où j'applique à notre satellite les procédés du mémoire 230. Ces procédés seraient surtout utiles pour obtenir un développement *purement littéral* des coordonnées de la Lune. J'ai signalé en passant diverses circonstances curieuses et presque paradoxales.

Toutes les séries dont je viens parler ne peuvent être employées qu'au point de vue du calcul formel et par conséquent du calcul approché. J'ai insisté à diverses reprises sur ce point important (214, 189, 190, 223, 224).

Toutes ces méthodes si diverses peuvent être employées pour se servir mutuellement de vérification, sans trop de calculs supplémentaires. Je citerai surtout un procédé de vérification fondé sur l'emploi des invariants intégraux (221).

XX. Développement de la Fonction Perturbatrice.
(215, 232, 235, 233, 234, 231, 238, 189.)

Je me suis occupé de la fonction perturbatrice à plusieurs points de vue différents.

J'ai d'abord cherché la valeur approchée des coëfficients des termes de rang très élevé. M. FLAMME avait déjà employé pour cet objet la méthode de M. DARBOUX sur les fonctions de très grands nombres. Mais comme cette méthode sous sa forme primitive ne s'appliquait qu'aux fonctions d'une seule variable, M. FLAMME devait donc décomposer chaque terme en une somme de produits, où chacun des facteurs ne dépendait que d'une seule anomalie moyenne. J'ai préféré considérer directement la fonction comme dépendant des deux anomalies moyennes. Des procédés analogues, sont encore, comme je l'ai montré, applicables dans ce cas. Seulement ils exigent une discussion; j'ai donné les principes qui doivent diriger cette discussion et j'en ai fait l'application dans un cas simple (189).

On peut aussi considérer chaque coëfficient comme fonction des excentricités et des inclinaisons, étudier les divers modes de développement de ces fonctions et chercher les conditions de leur convergence. Les conditions auxquelles j'arrive (238) sont relativement simples.

On peut enfin chercher s'il y a des relations entre les divers coëfficients. J'en ai trouvé un certain nombre (233, 234, 231).

Dans toutes ces recherches je me suis servi des relations de ces coëfficients avec les périodes de certaines intégrales doubles.

XXI. Equilibre d'un fluide en rotation et figure des planètes.
(170, 171, 172, 176, 175, 173, 174, 206, 177, 178, 212. 213, 184.)

Je me suis occupé également d'une autre question de Mécanique céleste que l'on peut énoncer ainsi:

Une masse fluide homogène ou hétérogène est animée d'un mouvement de rotation autour d'un certain axe. De plus, ses molécules s'attirent d'après la loi de Newton. Quelles sont les formes d'équilibre qu'elle peut affecter?

C'est là un problème qui a beaucoup occupé les géomètres depuis plus d'un siècle et demi, et dont l'importance se comprend sans peine.

Dans le cas de l'homogénéité, auquel nous nous restreindrons, deux solutions étaient depuis longtemps connues: l'ellipsoïde de révolution et l'ellipsoïde à trois axes inégaux de Jacobi. Mais les conditions de stabilité de l'équilibre n'avaient pas été étudiées.

On ignorait s'il y avait d'autres formes possibles quand M. Mathiessen et, après lui, Sir W. Thomson, annoncèrent l'existence de figures annulaires d'équilibre. Mais la démonstration donnée par ces deux savants n'était pas parfaitement rigoureuse; d'ailleurs, M. Mathiessen supposait *a priori* que la section différait très peu d'une ellipse. J'ai montré (63) que cette hypothèse, légitime quand la section de l'anneau est très petite, est erronée dans le cas général, ce qui rend très douteuse l'existence de certains anneaux très aplatis que le savant de Rostock avait appelés *anneaux β*.

J'ai donc cru nécessaire de faire une étude plus approfondie de ces figures (170, 173, 174). J'ai mis à l'abri de toute objection la démonstration de leur existence et montré comment on peut en déterminer les principaux éléments avec une approximation indéfinie.

Pour la détermination de ces éléments, j'ai fait usage d'une méthode que Mme Kowalevski avait déjà employée dans son Mémoire sur l'anneau de Saturne et qui est fondée sur le développement des périodes d'une fonction elliptique en séries ordonnées suivant les puissances croissantes du module.

Il convient d'observer que ces anneaux sont des figures d'équilibre instable.

Dans un Mémoire plus étendu (175), j'ai repris la même question, en développant les résultats obtenus dans deux Notes antérieures (171, 172). Une première difficulté se présentait sur ma route. Quand il s'agit d'intégrer de simples équations différentielles, la méthode des approximations successives est parfaitement justifiée, parce que l'existence de l'intégrale a été tout d'abord rigoureusement démontrée. Il n'en est plus de même dans le problème actuel qui est beaucoup plus compliqué; il peut rester au sujet de la légitimité de cette méthode quelques doutes qu'il s'agissait d'abord de dissiper. Pour démontrer rigoureusement l'existence des diverses solutions du problème, j'ai employé un procédé tout à fait analogue à celui dont j'avais fait usage dans mes recherches sur les solutions périodiques du problème des trois corps et où je prends pour point de départ un théorème de M. Kronecker.

On reconnaît d'abord que les diverses figures d'équilibre d'une masse fluide forment des séries linéaires; dans une même série, ces figures dépendent d'un paramètre variable. Telles sont la série des ellipsoïdes de révolution et celle des ellipsoïdes de Jacobi. Mais il peut arriver qu'une même figure appartienne à la fois à deux séries différentes. C'est alors une figure d'*équilibre de bifurcation*.

A chaque figure est attachée une suite infinie de coefficients, que j'appelle *coefficients de stabilité*, parce que la condition de la stabilité, c'est qu'ils soient tous positifs. Quand un de ces coefficients s'annule, c'est que la figure correspondante est de bifurcation.

Ainsi, si en suivant une série de figures d'équilibre on voit s'annuler un des coefficients de stabilité, on saura qu'il existe une autre série de formes d'équilibre à laquelle appartient la figure de bifurcation.

Un autre résultat, c'est que les deux séries linéaires dont cette figure fait partie échangent leur stabilité. Si, en suivant l'une des séries, on ne rencontre que des équilibres stables jusqu'à la figure de bifurcation, on n'y trouvera plus ensuite que des figures instables. Les figures stables appartiendront à l'autre série.

Ces principes, appliqués à divers problèmes déjà traités par Laplace, m'ont permis d'en compléter la solution.

Pour trouver les formes d'équilibre d'une masse fluide en rotation qui diffèrent peu d'un ellipsoïde, il faut rechercher si, parmi les ellipsoïdes de révolution et ceux de Jacobi, il y a des figures de bifurcation. Pour cela il faut calculer les coefficients de stabilité de ces ellipsoïdes. On trouve que ces coefficients dépendent des fonctions de Lamé.

J'ai donc dû faire de ces fonctions une étude approfondie. J'ai démontré d'une manière nouvelle que ces polynômes ont toutes leurs racines réelles et j'ai étudié la manière dont ces racines se répartissent.

En égalant à o les divers coefficients de stabilité, on obtient des équations qui sont transcendantes, mais qui peuvent néanmoins se discuter d'une manière complète. Cette discussion montre que, parmi les ellipsoïdes de révolution, comme parmi les ellipsoïdes de JACOBI, il y a une infinité de figures de bifurcation.

Il résulte de là qu'il y a d'autres formes d'équilibre que les ellipsoïdes et les anneaux. Ces figures nouvelles sont en nombre infini; elles sont convexes et ont toutes un plan de symétrie. Quelques-unes n'en ont qu'un; d'autres sont de révolution; d'autres enfin ont plusieurs plans de symétrie passant par l'axe.

Il restait à étudier les conditions de stabilité de l'équilibre. J'ai distingué, à l'exemple de Sir W. THOMSON, la stabilité séculaire, qui subsiste lorsqu'on tient compte de la viscosité, et la stabilité ordinaire, qui n'a lieu que lorsqu'on néglige cette résistance. En ce qui concerne la première de ces stabilités, j'ai montré qu'elle appartient aux ellipsoïdes de révolution, moins aplatis que celui qui est en même temps un ellipsoïde de JACOBI, et que les ellipsoïdes de JACOBI, qui satisfont à une certaine condition, en jouissent également. Les autres ellipsoïdes sont instables et il en est de même des figures annulaires. Quant aux autres figures nouvelles que j'ai découvertes, elles sont toutes instables, à l'exception d'une d'entre elles qui est pour ainsi dire piriforme.

J'ai donné aussi une méthode pour déterminer les conditions de la stabilité ordinaire, mais je n'en ai fait qu'une application partielle qui permet, toutefois, de reconnaître que cette stabilité peut subsister quand la stabilité séculaire a cessé.

Je ne puis d'ailleurs mieux résumer tous ces résultats qu'en faisant l'hypothèse suivante:

Imaginons une masse fluide se contractant par refroidissement, mais assez lentement pour rester homogène et pour que la rotation soit la même dans toutes ses parties. D'abord très voisine d'une sphère, la figure de cette masse deviendra un ellipsoïde de révolution qui s'aplatira de plus en plus; puis, à un certain moment, se transformera en un ellipsoïde à trois axes inégaux. Plus tard, la figure cessera d'être ellipsoïdale et deviendra piriforme, jusqu'à ce qu'enfin la masse, se creusant de plus en plus dans sa partie médiane, se scinde en deux corps distincts et inégaux.

L'hypothèse qui précède ne peut certainement s'appliquer au système solaire. Quelques astronomes ont pensé qu'elle pourrait être vraie pour certaines étoiles doubles, et que des étoiles doubles du type de β de la Lyre présenteraient des formes de transition analogues à celles dont nous venons de parler.

Dans un de ces Mémoires (173), j'ai montré qu'aucune forme d'équilibre stable n'est possible si la vitesse de rotation dépasse une certaine limite.

On peut faire de ce principe une application aux anneaux de Saturne.

Clerk Maxwell a démontré que ces anneaux ne peuvent être solides et que, s'ils sont fluides, leur densité ne peut dépasser les $\frac{3}{100}$ de celle de la planète. D'autre part (206), je démontre que, si les anneaux sont fluides, ils ne peuvent être stables qui si leur densité est supérieure au seizième de celle de Saturne. L'analyse semble donc confirmer l'hypothèse de M. Trouvelot, qui considère les anneaux comme formés d'une multitude de satellites extrêmement petits et ne croit pas pouvoir expliquer autrement certaines apparences observées.

J'ai donné (177) une démonstration plus simple d'un théorème de M. Liapounoff, en vertu duquel la sphère correspond au maximum de la fonction potentielle, et je suis encore revenu à diverses reprises sur des questions analogues (178, 184) en donnant quelques égalités et inégalités curieuses.

M. Radau avait remarqué qu'aucune hypothèse ne peut rendre compte à la fois de l'aplatissement adopté et la valeur observée de la précession. Dans deux articles consacrés à la Figure de la Terre (212, 213) j'ai confirmé la conclusion de M. Radau.

XXII. Astronomie, Questions Diverses.
(199, 201, 205, 202, 210, 222, 227, 228, 207, 240, 244, 262.)

On a vu plus haut quel rôle jouent en Astronomie les séries trigonométriques.

J'ai donc au point de vue de ces applications été amené pour contribuer à la solution de cette question à étudier les conditions de convergence des séries trigonométriques (201, 202). J'ai reconnu ainsi deux faits principaux:

1º. Si une pareille série est absolument convergente pour certaines valeurs du temps, elle l'est éternellement; il n'en est plus de même quand la convergence n'est plus absolue.

2º. Une même fonction ne peut pas être représentée par deux séries différentes absolument convergentes.

Je n'ai pu résoudre, de façon à me mettre à l'abri de toute objection, la question de la convergence des séries particulières de M. Lindstedt; cependant, j'ai tout lieu de penser que ces séries ne convergent pas absolument, mais que, en ordonnant convenablement les termes, on peut les rendre semi-convergentes. La convergence pourrait alors ne subsister que pendant un intervalle de temps limité.

On croit d'ordinaire qu'une fonction représentée par une série trigonométrique absolument convergente ne peut croître au delà de toute limite. C'est même cette croyance qui sert de fondement aux démonstrations anciennes de la stabilité du système solaire et qui, depuis, a conduit les astronomes à faire tant d'efforts pour faire rentrer le temps sous les signes sinus et cosinus. Cette croyance est erronée; j'ai montré (199, 202) qu'une pareille fonction devient aussi grande que l'on veut si la convergence n'est pas uniforme. Mais il y a deux manières de croître au delà de toute limite: une fonction peut «tendre vers l'infini». Il arrive alors qu'elle finit par dépasser une quantité quelconque, si grande qu'elle soit, pour rester ensuite constamment supérieure à cette quantité. Une fonction peut encore subir une infinité d'oscillations successives, de façon que l'amplitude des oscillations croisse indéfiniment. J'ai montré (205) que les deux cas peuvent se présenter, en ce qui concerne la somme d'une série purement trigonométrique. En résumé, quand même on arriverait à représenter les coordonnées des astres par des séries trigonométriques convergentes, on n'aurait pas démontré la stabilité du système solaire.

J'ai ensuite (207) étudié des procédés destinés à augmenter la convergence de certaines séries trigonométriques dont divers astronomes et en particulier M. GYLDÉN avaient fait usage dans les quadratures mécaniques.

Je reviendrai plus loin sur le travail que j'ai consacré aux marées (227, 228) et qui intéresse également l'astronomie.

Le calcul des perturbations des comètes par les quadratures mécaniques devient particulièrement pénible quand la comète passe très près de l'astre troublant parce qu'à ce moment la distance des deux corps varie très rapidement. J'ai indiqué (240) comment un emploi judicieux des intégrales elliptiques pouvait faciliter ce calcul.

J'ai publié aussi un article (244) sur le rôle des observations du pendule en géodésie et j'ai montré comment les *seules* observations pendulaires, si elles étaient parfaites et complètes, pourraient suffire pour déterminer la forme de la Terre.

Enfin dans la Préface que j'ai écrite pour les leçons de TISSERAND sur la détermination des Orbites, j'ai comparé les diverses méthodes en usage et j'ai montré qu'une orbite parabolique pourrait se déterminer à l'aide de trois observations *quelconques*, par des formules où n'entrent que des fonctions rationnelles.

CINQUIÈME PARTIE.

PHYSIQUE MATHÉMATIQUE.

———

XXIII. Équations Différentielles de la Physique Mathématique.
(293, 294, 295, 297, 298, 300, 299, 220, 302, 58, 304, 101, 227, 228, 296, 301, 286, 192.)

Dans beaucoup de problèmes de Physique Mathématique on rencontre des équations aux dérivées partielles du 2^{d} ordre qui appartiennent toutes à peu près au même type et dont la plus simple est la célèbre équation de LAPLACE $\varDelta u = 0$.

Ces problèmes qui conduisent ainsi à des équations identiques ou presque identiques appartiennent cependant aux branches de la Physique les plus diverses. On rencontre ces équations en électrostatique, en magnétisme, en électrodynamique, dans la théorie de la propagation de la chaleur, dans celle de l'élasticité, en Optique, en Hydrodynamique. On les rencontre en Mécanique, dans la théorie du potentiel newtonien; et j'ajouterai enfin qu'elles jouent un rôle capital en Analyse Pure et servent de fondement à la théorie des fonctions analytiques.

La plus importante de ces équations est, comme je l'ai dit, l'équation de LAPLACE qui est l'équation fondamentale de la théorie de l'attraction newtonienne, de l'électrostatique, du magnétisme et de l'hydrodynamique. Dans d'autres questions on a à envisager des équations un peu plus compliquées telles que:

$$\varDelta u = ku, \ \varDelta u = k\frac{du}{dt}, \ \varDelta u = k\frac{d^2 u}{dt^2}.$$

Enfin en optique, en élasticité, on trouve des systèmes de trois équations à trois inconnues où figure encore le Laplacien $\varDelta$.

Nous parlerons plus loin dans une question d'Analyse pure, d'une équation analogue mais plus compliquée

$$\varDelta u = e^u.$$

338

Mais ce n'est pas seulement par la forme des équations que ces problèmes diffèrent entre eux, c'est surtout par les conditions aux limites. Tantôt la fonction inconnue u est assujettie à prendre des valeurs données sur une surface fermée; tantôt on se donne sur cette surface la dérivée normale $\dfrac{du}{dn}$, ou bien une relation entre u et $\dfrac{du}{dn}$.

Ces problèmes ont été envisagés à plusieurs points de vue différents. Tantôt on a cherché seulement à démontrer qu'ils sont possibles et à établir en ce qui les concerne des *«théorèmes d'existence»*.

Pour ces théorèmes d'existence eux-mêmes, on s'est quelquefois contenté de démonstrations par à peu près et à demi intuitives, dont le type est ce qu'on appelle le *principe de* DIRICHLET. Ces aperçus, dépourvus de véritable valeur mathématique, sont cependant de nature à satisfaire le physicien parce qu'ils laissent en évidence, pour ainsi dire, le mécanisme physique du phénomène. Ils sont généralement fondés sur la considération d'une intégrale définie simple ou multiple qui ne pouvant s'annuler doivent admettre un minimum.

D'autres fois, on s'est préoccupé de démontrer rigoureusement ces théorèmes d'existence et pour cela on a imaginé des procédés d'approximations successives dont on pouvait établir la convergence. Cette convergence est en général trop lente et les approximations trop compliquées pour que l'on puisse y voir autre chose qu'un moyen de démonstration et pour que le calcul effectif soit possible.

Enfin on a cherché à résoudre ces problèmes au moyen de séries procédant suivant certaines fonctions que l'on peut appeler *harmoniques,* parce que la représentation d'un phénomène quelconque par une pareille série est analogue à la décomposition d'un son complexe en ses harmoniques. Les types de ces séries sont la série de FOURIER, celle de LAPLACE qui· procède suivant les fonctions sphériques, et les diverses séries considérées par FOURIER dans l'étude du refroidissement des corps solides.

Dans l'application de cette méthode, on rencontre plusieurs difficultés successives. Il faut d'abord démontrer l'existence des fonctions harmoniques, ce qui se fait comme pour les autres théorèmes d'existence, par les procédés dont je viens de parler. On doit ensuite calculer les coëfficients des séries, ce qui généralement est facile. Il reste enfin à démontrer la convergence de la série et on a alors une difficulté grave à surmonter.

Tels sont les différents points de vue auxquels j'ai dû envisager successivement tous ces problèmes.

Je me suis occupé d'abord de l'équation de LAPLACE. La théorie de cette équation est intimement liée à celle du potentiel. Mais les propriétés du potentiel n'avaient pas toujours été démontrées ni avec assez de généralité, ni avec assez de rigueur. Je m'en suis aperçu quand j'ai voulu les enseigner et aussi quand j'ai voulu les appliquer à des questions d'Analyse. J'ai cherché (192) à perfectionner ces démonstrations et j'ai eu besoin également (101) de les étendre à l'espace à plus de trois dimensions.

En ce qui concerne l'équation de LAPLACE, bien des méthodes avaient déjà été proposées en vue de la démonstration rigoureuse du théorème d'existence. Les principales étaient celles de SCHWARZ et de NEUMANN. J'en ai proposé une troisième, entièrement nouvelle (293, 296) et qui est connue aujourd'hui sous le nom de méthode du *balayage*. Elle présente, comme méthode de démonstration, l'avantage de la généralité et elle permet de supprimer certains intermédiaires. En revanche elle semble, moins que celle de NEUMANN, susceptible de s'approprier au calcul numérique.

L'élégante méthode de NEUMANN ne paraissait applicable qu'aux surfaces convexes. J'ai reconnu (302, 304) qu'elle était beaucoup plus générale, j'ai montré rigoureusement qu'elle s'appliquait à toutes les surfaces simplement connexes et j'ai fait voir qu'elle était probablement applicable à une surface tout à fait quelconque; c'est ce qui a d'ailleurs été confirmé depuis. Je suis malheureusement obligé d'admettre le principe de DIRICHLET qu'il faut établir d'abord par exemple par la méthode du balayage, de sorte que la démonstration doit se faire en deux temps.

Les mêmes principes peuvent s'étendre au problème de l'équilibre d'un corps élastique qui est beaucoup plus compliqué puisqu'il y entre trois équations à 3 inconnues. J'ai montré (303) qu'on pouvait résoudre ce problème par une méthode tout à fait analogue à celle de NEUMANN. Je n'ai pas néanmoins développé cette idée et au lieu de démontrer rigoureusement la convergence du procédé, je me suis borné à un aperçu, analogue au «principe de DIRICHLET», qui, il est vrai, n'est pas suffisamment rigoureux, mais qui ne permet pas de douter sérieusement du résultat.

Je suis revenu à diverses reprises sur le problème de FOURIER relatif aux lois du refroidissement des corps solides. Dans mes premiers mémoires, je me suis placé plutôt au point de vue du physicien et j'ai montré l'existence des fonctions harmoniques par des aperçus analogues au principe de DIRICHLET (294, 295, 296). C'est à ce même point de vue que je me suis placé le plus souvent dans mon enseignement (286).

C'est seulement plus tard (303, 301) que j'ai donné une démonstration rigoureuse de ces théorèmes d'existence. L'équation est la même que celle des vibrations d'une membrane. M. SCHWARZ avait démontré l'existence de la première harmonique, M. PICARD celle de la seconde. J'ai démontré d'une façon plus simple l'existence de toutes les harmoniques. La méthode de démonstration, toujours fondée sur l'a considération de certaines intégrales si heureusement introduites par M. SCHWARZ, repose en outre sur l'étude d'une certaine fonction méromorphe d'un paramètre auxiliaire. En multipliant cette fonction par un polynôme convenable, on fait disparaître quelques-uns de ces pôles; le cercle de convergence s'étend et on peut aborder l'étude d'harmoniques d'ordre de plus en plus élevé.

Cette méthode ne s'applique pas seulement au problème du refroidissement et à celui des membranes. J'ai dit plus haut comment j'avais abordé l'étude de l'équation de LAPLACE par la méthode de NEUMANN et celle de l'équilibre des corps élastiques. Dans cette double étude je me suis encore servi des intégrales de SCHWARZ et de la 'fonction méromorphe auxiliaire dont je viens de parler.

Mais on peut encore employer les mêmes procédés dans des questions bien différentes. Je m'en suis servi par exemple pour étudier l'équilibre et le mouvement des mers (220, 227, 228). Dans le problème statique, par exemple, qui est sensiblement réalisé dans les marées à longue période, on rencontre, si l'on veut tenir compte de l'influence de la forme des continents, des fonctions harmoniques analogues aux fonctions sphériques et dont l'existence peut encore se démontrer par les mêmes méthodes. Le problème dynamique, tel qu'on doit le traiter pour les marées à courte période, présente évidemment plus de difficultés. Il se simplifierait si l'on pouvait négliger la force de Coriolis. L'équation où on serait ramené serait encore celle des vibrations d'une membrane dont la tension et la densité seraient variables. Si l'on tient compte de la force de Coriolis, tout devient plus complexe. Les intégrales de SCHWARZ doivent être remplacées par d'autres intégrales plus compliquées, quoique analogues; mais la marche générale du calcul reste la même.

Une équation de même forme $\varDelta u = e^u$ peut encore être traitée de la même manière. Ce n'est plus cette fois en vue d'applications physiques, mais en vue d'applications analytiques. J'ai dit plus haut quelle était l'importance de cette équation dans la théorie des fonctions fuchsiennes. M. PICARD l'a intégrée le premier. La méthode que j'ai proposée est entièrement différente et repose sur des principes analogues à ceux dont je viens de parler. Une difficulté nouvelle se présentait toutefois, c'est que notre équation n'est pas linéaire comme le sont d'ordinaire celles de la Physique Mathématique. Ce qui caractérise ma méthode et la distingue de celle de M. PICARD, c'est qu'elle embrasse tout de suite la totalité de la surface de RIEMANN envisagée, tandis que M. PICARD considère d'abord un domaine limité, et étend ensuite ses résultats de proche en proche jusqu'à ce qu'ils soient établis pour la surface entière (Vide supra § II).

Une autre équation linéaire aux dérivées partielles a été l'objet de mes recherches (297). C'est l'équation des télégraphistes, qui par sa forme tient le milieu entre l'équation des cordes vibrantes et celle de la chaleur. Les résultats aussi sont intermédiaires à ceux que donnent ces deux équations. Dans le cas des cordes vibrantes l'onde se propage avec une vitesse constante sans se déformer, sans s'étaler, sans envoyer d'avant-garde ni laisser d'arrière-garde. Dans le cas de la chaleur, l'onde s'étale immédiatement sur le fil conducteur tout entier, sans qu'on puisse lui assigner une commencement ou une fin. Dans le cas de l'électricité, la tête de l'onde s'avance régulièrement avec la vitesse de la lumière; mais l'onde laisse derrière elle un résidu, une sorte de queue.

Les conditions à remplir aux limites sont très différentes de celles que nous avons envisagées jusqu'ici. J'ai pu mettre en évidence les faits que je viens de signaler par une méthode où le rôle principal est joué par la théorie de CAUCHY sur les intégrales prises entre des limites imaginaires. Je suis revenu sur ce sujet à plusieurs reprises (286, 284).

Mais dans l'étude de ces équations aux dérivées partielles, je me suis également placé au point de vue du développement en séries harmoniques. L'existence des fonctions harmoniques était établie par les théorèmes d'existence dont j'ai parlé plus haut, mais la convergence des séries ne pouvait se démontrer sans difficulté. C'est CAUCHY qui a le premier imaginé une méthode générale applicable à un grand nombre de ces séries. J'ai présenté cette méthode de CAUCHY sous une forme particulière (286) qui en montre le véritable esprit et qui fait comprendre quels sont les obstacles qu'il reste à vaincre quand on veut l'appliquer à un problème particulier.

Dans le cas du refroidissement de la sphère ou du cylindre, les fonctions harmoniques sont faciles à former puisque ce sont les fonctions bien connues de BESSEL. J'ai pu dans ces cas particuliers pousser jusqu'au bout l'application de la méthode de CAUCHY (298, 286).

L'une des plus importantes de ces séries est celle de LAPLACE qui procède suivant les fonctions sphériques. La convergence a été démontrée par LEJEUNE DIRICHLET. J'ai introduit successivement (299, 286) de telles simplifications dans cette démonstration qu'elle devient presque méconnaissable. Je m'appuie sur la convergence de la série de FOURIER et sur certaines propriétés des fonctions de variables imaginaires. J'espère que la forme donnée au raisonnement en facilitera l'extension à d'autres problèmes analogues.

J'avais également à traiter la convergence des séries harmoniques que l'on rencontre dans la théorie de la propagation de la chaleur ou dans celle des vibrations d'une membrane. Je me suis contenté d'abord (294, 295, 296) de simples aperçus. Je montrais seulement par exemple que l'intégrale du carré de l'erreur commise tendait vers zéro quand on prenait dans la série un plus grand nombre de termes. Un pareil résultat, suffisant pour les applications physiques, ne pouvait satisfaire le mathématicien.

Depuis j'ai obtenu des démonstrations rigoureuses (301); mais à la conditions de faire des hypothèses restrictives sur la fonction à développer. Cela n'a pas d'ailleurs d'importance au point de vue des applications, car on peut toujours trouver une fonction qui satisfasse à ces hypothèses restrictives et qui diffère aussi peu que l'on veut de la fonction donnée.

XXIV. Critique des Théories Physiques.

(345, 346, 347, 348, 349, 350, 351, 356, 357, 360, 361, 362, 363, 366, 367, 368, 269, 475, 306, 304, 307-8-9-10, 364-365, 312, 313, 375, (352)-353-354, 357, 315, 277, 280, 278, 279, 285, 282, 283, 281, 286, 287, 289.)

Dans le paragraphe précédent, j'ai envisagé pour ainsi dire les problèmes de Physique mathématique; je vais les considérer maintenant par le côté physique.

J'ai publié dans une série de volumes mes cours de Physique Mathématique. Je reviendrai sur ces volumes au point de vue de l'enseignement (§ XXXI); j'en parlerai encore à propos des idées philosophiques exposées dans les Préfaces (§ XXIX). Pour le moment, je ne m'occuperai que de l'intérêt qu'ils présentent pour la physique proprement dite. J'ai été amené à passer en revue les différentes théories physiques et à les soumettre à la critique.

J'ai consacré d'ailleurs au même objet un certain nombre de notes et d'articles.

La chaire de Physique Mathématique a pour titre officiel: Calcul des Probabilités et Physique Mathématique. Ce rattachement peut se justifier par les applications que peut avoir ce calcul dans toutes les expériences de Physique; ou par celles qu'il a trouvées dans la théorie cinétique des gaz. Quoi qu'il en soit, je me suis occupé des probabilités pendant un semestre et mes leçons ont été publiées (287). La théorie des erreurs était naturellement mon principal but. J'ai dû faire d'expresses réserves sur la généralité de la «loi des erreurs»; mais j'ai cherché à la justifier, dans les cas où elle reste légitime, par des considérations nouvelles. J'ai montré que si l'erreur totale résulte de l'accumulation d'un grand nombre de petites erreurs partielles, si ces dernières suivent des lois quelconques, mais symétriques, l'erreur résultante totale sera soumise à la loi de GAUSS.

Je ne m'étendrai pas sur les leçons que j'ai consacrés aux Tourbillons (283) ou à la Capillarité (285); ce qu'elles peuvent contenir de nouveau intéresse plutôt l'enseignement.

J'ai consacré un volume à la Thermodynamique (281). Je signalerai seulement dans ce volume la discussion de l'inégalité de CLAUSIUS

$$\int \frac{dQ}{T} < 0.$$

La démonstration de cette inégalité avait donné lieu à de longues controverses; j'ai cherché à la mettre à l'abri de toute objection.

Dans la démonstration des théorèmes de GIBBS, on rencontre une difficulté qui avait échappé à plusieurs savants éminents. C'est celle qui se rapporte à l'évaluation de l'entropie d'un mélange gazeux. On en triomphe d'ordinaire aujourd'hui en faisant intervenir les propriétés de l'osmose. J'en suis venu à bout par un moyen tout différent, en faisant intervenir les propriétés de la dissociation du carbonate de chaux.

L'irréversibilité des phénomènes thermodynamiques a préoccupé de nombreux chercheurs. On en a voulu donner plusieurs explications mécaniques. La première est celle de HELMHOLTZ; j'ai montré (345) qu'elle ne suffit pas pour rendre compte des faits (vide infra § XXVIII). La seconde est celle qui se rattache à la théorie cinétique des gaz. J'y ai fait diverses objections qui me semblent la rendre peu vraisemblable, mais qui cependant ne sont pas décisives. J'ai discuté en particulier dans deux notes (356, 357) divers points de cette théorie, ce qui m'a donné l'occasion de rectifier une faute de calcul faite par MAXWELL à propos de la relation entre la conductibilité et la viscosité des gaz. J'ai d'autre part justifié le principe de BOLTZMANN et montré sa légitimité dans des cas où il avait été contesté (357).

J'ai exposé à deux reprises différentes la théorie générale de l'élasticité (277, in initio, 282). J'ai montré quelles relations il doit y avoir entre les 21 coëfficients d'élasticité. 1° dans l'hypothèse des forces centrales, 2° dans le cas où la pression est nulle à l'état d'équilibre. Ces résultats supposent qu'il n'y a que des molécules d'une seule sorte, et ne s'appliqueraient pas par conséquent, au moins immédiatement, à deux milieux différents se pénétrant mutuellement. Ils ont donné lieu à des objections que j'ai facilement réfutées (350).

Dans la suite de mon livre (282) j'ai signalé (page 134) une erreur commise par Lamé.

Je me suis occupé aussi d'un problème particulier d'élasticité (351) à propos d'une expérience de M. Cornu. J'ai montré que certaines relations, qui partout sont approchées, deviennent rigoureusement exactes sur les arêtes d'un prisme à base rectangulaire.

Les équations de l'élasticité s'appliquent immédiatement à l'Optique. J'ai cherché à réunir dans une exposition commune toutes les théories optiques des ondes. En particulier (277) j'ai montré comment dans le cas de la double réfraction, les théories de Fresnel, de Neumann et de Sarrau se déduisent des équations générales du milieu élastique. De même, grâce à l'emploi des «conches de passage», les diverses théories de la réflexion sont notablement simplifiées et on comprend mieux leurs rapports mutuels.

Que résulte-t-il de ce rapprochement entre la théorie de Fresnel où la vibration est supposée perpendiculaire au plan de polarisation, et celle de Neumann où elle est regardée comme parallèle à ce plan? On sait que ces deux théories ont jusqu'ici également bien rendu compte des faits; mais la comparaison que j'ai faite entre ces deux théories m'a montré la raison de ce fait. Cette raison est générale. Tout fait dont une des théories rendra compte, sera également bien expliqué par l'autre, de sorte qu'aucune expérience optique ne pourra décider entre elles.

On a voulu chercher un critère décisif dans les phénomènes de diffraction, dans la réflexion métallique et surtout dans l'expérience célèbre de Wiener. J'ai montré (347, 348) qu'on s'était fait là une illusion.

Le principe de Huyghens et ses applications à la diffraction m'a longtemps occupé. J'ignorais à cette époque les travaux de Kirchhoff; l'interprétation que j'ai proposée pour le principe de Huyghens présente les plus grandes analogies avec celle de Kirchhoff; elle repose également sur la considération d'une intégrale analogue au potentiel newtonien.

J'ai abordé aussi le problème d'une autre manière (277 page 300; 280, page 98) en déduisant directement des équations la propagation rectiligne de la lumière comme première approximation; et la diffraction comme seconde approximation. Je montre ainsi que la direction du rayon lumineux doit être conforme au principe de Huyghens; en ce qui concerne la seconde approximation, je cherche à illustrer la théorie générale par un exemple simple, en étudiant les cas de propagation anormale des ondes dans un faisceau très délié (330, 280).

Quoi qu'il en soit, la théorie de la diffraction restait très imparfaite. Non seulement la question de la polarisation était laissée de côté, mais certains phénomènes que M. Gouy appelle «diffraction éloignée» restaient complètement inexpliqués. J'ai cherché à donner une théorie de la diffraction éloignée et de la polarisation par diffraction (306, 311); malheureusement j'étais obligé de faire certaines hypothèses, assez voisines de la réalité pour donner une idée générale de la marche des phénomènes, pas assez toutefois pour en asseoir la théorie complète et définitive.

J'ai passé en revue dans mes leçons publiées les principales théories de l'électrodynamique (278, 279, 289). Mais j'ai consacré aussi plusieurs notes à certaines questions particulières. J'ai discuté (346) la loi électrodynamique de Weber et ses rapports avec le principe de la conservation de l'énergie. J'ai étudié (349) les conditions d'équilibre d'un liquide diélectrique placé dans un champ électrique, je voulais montrer qu'on pouvait faire cette théorie de la façon la plus élémentaire et sans faire intervenir les pressions de Maxwell. J'ai traité (314) des rapports du phénomène de Hall avec la théorie de Lorentz.

J'ai consacré (375) un article à l'induction unipolaire. Les lignes de force magnétiques sont-elles entraînées par les aimants en mouvement? Ma conclusion est que cette question si controversée est dénuée de sens.

Mais je me suis surtout occupé des rapports de l'Electrodynamique et de l'Optique. J'ai traité à plusieurs reprises du phénomène de Zeemann (312, 313, 289). L'un de ces articles a été critiqué par Lorentz. L'expérience décidera. Si l'on confirme définitivement les expériences de Zeemann sur la dissymétrie du triplet dans un champ faible, relatées par Lorentz dans son rapport au Congrès de Physique (Tome III, page 31), ce sera Lorentz qui aura raison.

J'ai passé en revue (278, 279, 289, 307, 308, 309, 310) les principales théories électromagnétiques de la Lumière, celles de Maxwell, de Helmholtz, de Hertz, de Lorentz et de Larmor. La plus satisfaisante m'a paru être celle de Lorentz. J'y ai fait cependant une objection; cette théorie n'est pas conforme au principe de l'égalité de l'action et de la réaction (supposé appliqué à la matière seule) (307, 308, 309, 310, 289). Je suis revenu (315) plus en détail sur les relations de la théorie de Lorentz avec ce principe.

J'ai été amené enfin à m'occuper des rayons cathodiques et des rayons Röntgen. J'ai donné (369) l'explication d'une expérience de M. Birkeland, en déterminant la trajectoire des rayons cathodiques dans un champ variable. J'ai réfuté (360, 361, 362, 363, 364, 365) les idées de M. Jaumann sur ces rayons.

J'ai fait (366, 367, 368) diverses observations à propos de certaines expériences relatives aux rayons Röntgen. C'est dans un article de vulgarisation (370) que j'ai émis au sujet des rayons Röntgen une hypothèse qui a exercé une certaine influence sur le développement de la Science. Je me suis demandé s'il n'y aurait pas quelque relation entre ces rayons et les phénomènes de phosphorescences. Plusieurs savants ont alors dirigé leurs recherches de ce côté. Quelques-uns n'ont obtenu que des succès partiels ou douteux. Mais M. Becquerel a réussi complètement et a découvert les rayons qui portent son nom. Les phénomènes qu'il a observés ne sont pas sans analogie avec ceux que j'avais prévus; mais ils sont beaucoup plus extraordinaires encore.

XXV. Oscillations hertziennes. (325, 329, 332, 333, 326, 327, 328, 279, 284.)

Je crois devoir mettre à part, à cause de leur nombre les articles que j'ai consacrés aux oscillations hertziennes. Le premier est une note (325) où j'ai rectifié une erreur de calcul commise par Hertz dans la détermination de la période. Cette rectification était facile, mais il importait de la faire promptement; car à ce moment, cette erreur, si elle était restée inaperçue, aurait pu arrêter les progrès de la Science.

J'ai donné dans d'autres notes et dans mes leçons, des procédés en vue du calcul plus ou moins approché de la période d'un excitateur, et j'ai discuté les différentes circonstances qui influent ou qui sembleraient devoir influer sur cette période (329, 326, 327, 284).

Le phénomène de la résonance multiple découvert par MM. Sarasin et de la Rive semblait assez paradoxal. J'en ai donné une explication simple (328, (284) fondée sur l'amortissement rapide des oscillations. Cette explication a été confirmée expérimentalement d'abord par M. Bjerknes, puis par d'autres physiciens.

J'ai étudié enfin (332, 333) la propagation d'une onde hertzienne le long d'un fil. J'ai montré que l'affaiblissement de cette onde est due à deux causes; le rayonnement et la chaleur de Joule. La première de ces causes croît avec le diamètre du fil, tandis que la seconde décroît. L'expérience semble indiquer que la seconde est prépondérante.

SIXIÈME PARTIE.

PHILOSOPHIE DES SCIENCES.

———

XXVI. Arithmétique et Analyse. (405, 408.)

Voulant approfondir la philosophie des sciences mathématiques, je devais d'abord étudier ces sciences dans les parties où elles sont le plus dégagées d'éléments sensibles et en général d'éléments étrangers. C'est là que j'avais le plus de chance de pénétrer la véritable nature du raisonnement mathématique. Or c'est en arithmétique que la notion fondamentale de nombre règne dans toute sa pureté. C'est donc à l'arithmétique que je devais emprunter mes exemples et non pas à la géométrie comme le font la plupart des philosophes.

J'ai reconnu pourtant (408) que même en arithmétique le raisonnement mathématique n'est pas un simple syllogisme, et que partout le syllogisme est stérile. Le raisonnement mathématique exige l'emploi d'une sorte d'induction qui diffère de l'induction ordinaire parce qu'elle entraîne la certitude absolue. C'est à cette induction que l'on a recours par exemple, quand on affirme qu'un théorème démontré pour le nombre 1, est vrai pour tous les nombres, si l'on établit qu'il est vrai de $n + 1$ s'il l'est de n.

Comment passe-t-on maintenant du nombre pur au continu et de l'Arithmétique à l'Analyse? J'ai étudié (405) les origines de la notion du continu mathématique; j'ai montré qu'elle ne saurait dériver de l'expérience et qu'elle diffère beaucoup de la notion du continu physique, qui nous vient des sens. Celle-ci est régie par la célèbre «loi de FECHNER» qui (si on voulait traduire littéralement les expériences qui lui servent de fondement) devrait s'exprimer par la formule contradictoire:

$$A > B, \ A = C, \ C = B.$$

C'est pour lever cette contradiction que l'esprit a *créé* le continu mathématique. Sa puissance créatrice n'est pas épuisée ainsi, et s'il ne l'exerce pas de nouveau c'est par ce que l'expérience ne lui en fournit pas l'occasion.

XXVII. Géométrie. (402, 409, 410, 412, 414, 411, 403, 404.)

Je me suis à plusieurs reprises, occupé d'éclaircir les origines de la géométrie et celles de la notion d'espace.

Je me suis demandé (402, 403) quel est le véritable caractère des vérités géométriques et en particulier du postulatum d'Euclide. Ce postulat est il un fait d'expérience, une nécessité logique, ou un jugement synthétique a priori? Rien de tout cela, c'est une convention et il n'est pas plus raisonnable de se demander si ce postulat est vrai et la géométrie de LOBATCHEFFSKI fausse que de rechercher si le système métrique est vrai et si le système de la toise et du pied est faux.

Comme LIE je crois que la notion plus ou moins inconsciente du groupe continu est la seule base logique de notre géométrie. Comme HELMHOLTZ je crois que l'observation des mouvements des corps solides en est l'origine psychologique. Mais je ne fais pas pour cela dériver la géométrie de l'expérience; loin de là. Les expériences sur les solides n'ont été que l'occasion qui, parmi tous les groupes continus dont nous aurions pu faire une géométrie, nous a fait choisir le groupe euclidien, non comme le seul vrai, mais comme le plus commode.

J'ai recherché également à analyser l'origine psychologique de la notion d'espace (409, 411). Comment parmi les changements que nos sens nous révèlent dans les objets extérieurs distinguons-nous les changements d'état des changements de position? C'est que ces derniers peuvent toujours être corrigés par des «changements internes» (mouvements du corps) qui eux-mêmes se distinguent des «changements externes» parce qu'ils sont volontaires et parce qu'ils sont accompagnés des sensations que l'on est convenu d'appeler musculaires. Il en résulte d'abord qu'un être immobile serait incapable de créer la géométrie.

La notion de l'espace ne peut donc faire partie intégrante d'aucune de nos sensations prise isolément. C'est seulement quand nous observons l'ordre dans lequel ces sensations se succèdent que cette notion peut prendre naissance. Or s'il est absurde de supposer que nous puissions imaginer des sensations différentes de nos sensations normales, nous pouvons au contraire avec quelque effort imaginer une succession de sensations, pareilles individuellement à nos sensations normales, mais se succédant dans un ordre anormal. Nous pouvons imaginer que ces sensations suivent d'autres lois et par exemple qu'elles s'ordonnent, non conformément à la structure du groupe euclidien, mais conformément à la structure d'un autre groupe. Des êtres qui éprouveraient nos sensations normales dans cet ordre anormal, créeraient une géométrie différente de la nôtre. C'est en ce sens que j'ai pu dire (403) sous une forme légèrement paradoxale: «quelqu'un qui voudrait y consacrer sa vie, arriverait à se figurer l'espace à quatre demensions». Dans un autre article (404), j'ai décrit un monde fictif dont les habitants seraient nécessairement conduits à créer la géométrie de LOBATCHEFFSKI.

Plus paradoxal encore est ce que je dis du point géométrique. Cette notion en apparence immédiate et primitive ne me semble pas résister à l'analyse et j'en cherche l'origine dans la structure du groupe euclidien et l'étude des sousgroupes qui y sont contenus. Je cherche à montrer que de quelque façon qu'on aborde la question, on sera toujours ramené à cette conséquence (409, 410, 411).

J'ai réfuté (412, 414) les idées de M. RUSSELL sur les fondements de la géométrie. Ce philosophe attribue à l'expérience un rôle important dans la genèse de la géométrie. Je lui ai objecté ce que j'avais dit ailleurs (256) qu'il est impossible d'expérimenter sur des droites ou des figures abstraites; que l'expérience ne peut porter que sur des corps matériels, et qu'alors si on opère sur les corps solides, on aura fait une expérience de mécanique, que si on opère sur les rayons lumineux, on aura fait une expérience d'optique, mais qu'on n'aura jamais fait une expérience de géométrie.

Bien que sur certans points je fusse d'accord avec M. RUSSELL et que je rendisse justice à son talent, j'ai été obligé de combattre également quelques autres idées de cet auteur.

XXVIII. Mécanique. (345, 406, 407, 373, 415, 183, 450, 281, préface.)

Mes recherches sur les principes de la Mécanique ont été inspirées par le même esprit.

La notion du temps nécessitait d'abord un examen critique analogue à celui que j'avais fait subir à la notion d'espace. On admet généralement, non seulement que le temps est relatif, mais que l'égalité de deux durées ne peut être perçue directement et ne peut même être définie que par une sorte de convention. J'ai cherché à montrer (373) que la notion même de la simultanéité de deux événements a aussi quelque chose d'un peu arbitraire, j'allais dire d'un peu conventionnel.

J'ai discuté (345, 406, 407) la portée philosophique du second principe de la Thermodynamique. Ce principe est ou semble être en contradiction avec l'hypothèse du mécanisme universel. J'ai parlé plus haut (§ XXIV) des diverses tentatives qui ont été faites pour lever cette contradiction et des discussions auxquelles elles ont donné lieu. J'ai cherché à mettre les philosophes au courant de ces controverses.

Les principes de la Mécanique sont-ils a priori ou bien leur origine est elle empirique? Aucune de ces deux solutions n'est entièrement satisfaisante. Quelle position convient-il d'adopter entre les deux? Je suis revenu sur ces problèmes à plusieurs reprises (415, 183, 452). Ce sont des faits expérimentaux qui nous ont conduits à adopter les principes fondamentaux de la Mécanique, celui de l'inertie, celui de l'égalité de l'action et de la réaction, celui du mouvement relatif. Devons-nous conclure que ces lois ne sont qu'approchées, que des expériences nouvelles nous conduiront à y ajouter de petits termes correctifs? Pour nous en rendre compte, il nous suffit d'essayer d'énoncer correctement ces principes, en les appliquant à la totalité de l'Univers. Nous reconnaissons alors que, quand on veut leur donner une forme absolue, ils deviennent invérifiables parce que nous ne pouvons connaître que les mouvements relatifs. Nous resterons donc toujours libres de les conserver, sans crainte d'être démentis, et nous les conserverons parce que l'expérience nous a appris qu'ils sont commodes. Voilà pourquoi l'expérience qui leur a donné naissance, ne pourra plus les détruire.

Les mêmes considérations s'appliquent à un principe qui touche de plus près encore à la Physique, celui de la conservation de l'énergie. C'est ce que je me suis efforcé de démontrer (281, préface; 415, 452). Quand on veut donner à ce principe une valeur absolue, on le voit pour ainsi dire s'évanouir en une tautologie. Il semble qu'il se soit mis ainsi hors des atteintes de l'expérience. Il n'est pas impossible toutefois que des expériences nouvelles nous empêchent de lui attribuer une portée universelle, en nous montrant qu'il ne pourrait recevoir une nouvelle extension sans perdre sa fécondité. C'est ce que j'ai cherché à expliquer (451).

Quoi qu'il en soit de ces considérations, je me suis efforcé de montrer pourquoi la Mécanique est et doit rester une science expérimentale (452).

XXIX. Physique. (413, 449, 451; préfaces, 287, 277, 278, 281, 289.)

En pénétrant dans le domaine de la Physique on doit renoncer à ce genre particulier de certitude qu'exigent les Mathématiciens. Nous devons nous contenter du probable. Le calcul des probabilités joue un rôle nécessaire et j'ai montré (413) que dans toute induction on faisait un calcul de probabilité inconscient. J'ai cherché (413, 287, préface) à pénétrer les principes de ce calcul, j'ai cru les découvrir dans la croyance à la continuité des phénomènes naturels.

La Physique ne peut se passer des Mathématiques qui lui fournissent la seule langue qu'elle puisse parler (449). De là les services mutuels et incessants que se rendent l'Analyse pure et la Physique. Chose remarquable, les travaux des analystes ont été d'autant plus féconds pour la physique qu'ils ont été plus exclusivement cultivée pour sa beauté propre. En revanche, la Physique, en posant de nouveaux problèmes, a été aussi utile au Mathématicien que le modèle l'est à l'artiste.

Le physicien (451) ne saurait se contenter de l'expérience toute nul. Son objet n'est pas le même que celui de l'historien et un fait isolé est pour lui sans valeur. De là l'utilité de la généralisation qui exige l'emploi des Mathématiques. Cette généralisation suppose une certaine croyance à l'unité et à la simplicité de la nature (451, 281, préface). Cette croyance, justifiée ou non, est nécessaire à la science.

D'ailleurs on ne saurait se passer de l'hypothèse et souvent une hypothèse fausse a rendu plus de services qu'une hypothèse vraie. Pour tirer parti de ces hypothèses, on a cherché à résoudre le phénomène complexe observable en un très grand nombre de phénomènes élémentaires obéissant tous aux mêmes lois. C'est ainsi que la Physique Mathématique est devenue possible.

Dans les théories physiques, il faut distinguer le fond et la forme. Le fond, c'est l'existence de certains rapports entre des objets inaccessibles. Ces rapports sont la seule réalité que nous puissions atteindre et tout ce que nous pouvons demander, c'est qu'il y ait les mêmes rapports entre ces objets réels inconnus et les images que nous mettons à leur place.

La forme n'est qu'une sorte de vêtement dont nous habillons ce squelette; ce vêtement, nous le changeons fréquemment, à l'étonnement des gens du monde, que cette instabilité fait sourire et qui proclament la faillite de la Science. Mais si la forme change souvent, le fond reste.

Les hypothèses relatives à ce que je viens d'appeler la forme ne peuvent pas être vraies ou fausses, elles ne peuvent être que commodes ou incommodes. Par exemple, l'existence de l'éther, celle même des objets extérieurs ne sont que des hypothèses commodes (451; 277, préface).

C'est pour cela que l'on voit renaître de leurs cendres en se transformant certaines théories que l'on croyait définitivement abandonnées. C'est pour cela aussi qu'il y a certaines catégories de faits qui s'expliquent également bien dans deux ou plusiers théories différentes, sans qu'aucune expérience puisse jamais décider. Cela est vrai en particulier pour les théories mécanistes. On peut en effet démontrer que, si un phénomène comporte une explication mécanique, il en comportera une infinité (451; 278, préface).

Le Mécanisme en tout cas n'est que l'un des vêtements dont la vérité peut s'habiller, et s'il satisfait notre esprit, il ne faut pas y attacher plus d'importance qu'il n'en mérite. Il oblige à introduire l'hypothèse de fluides auxiliaires tels que l'éther; j'expose quelques vues sur le plus ou moins de réalité de ce fluide.

Je termine (451) par l'exposé de l'état actuel de la Science et de ses progrès depuis cinquante ans. Ma conclusion est que l'on a marché vers l'unité; progrès important, car on ne doit pas oublier que le but véritable, ce n'est pas le Mécanisme, c'est l'unité.

XXX. Psychologie scientifique et Pédagogie. (522, 523, 450)

Dans la formation de la Science Mathématique et dans son enseignement, on distingue deux tendances opposées, la tendance analytique et la tendance intuitive. A mesure que la Scince a progressé, la première a tendu à prendre le pas sur la seconde. Toutes deux ont leur rôle nécessaire cependant. Dans l'enseignement, il est indispensable de faire appel à l'intuition pour développer certaines facultés de l'esprit, utiles au savant et surtout à l'ingénieur (523). Dans la Science même, l'intuition reste, sauf pour quelques esprits privilégiés, l'instrument principal de l'invention, tandis que l'analyse tend de plus en plus à devenir le seul instrument légitime de la démonstration (450).

Je me suis aussi préoccupé de l'influence que peut avoir dans l'enseignement, l'emploi de la notation différentielle (522). J'ai montré quels inconvénients peut entraîner l'usage prématuré de cette notation.

SEPTIÈME PARTIE.

ENSEIGNEMENT, VULGARISATION, DIVERS.

XXXI. Enseignement.

Voici le tableau des cours que j'ai professés à la Sorbonne depuis 1885.

Chaire de Mécanique Physique et Experimentale.

1885, 2. Frottement.

1886, 1. Cinématique et Mécanismes (167, 168).

1886, 2. Potentiel. Hydrostatique, Hydrodynamique (167, 168).

Chaire de Physique Mathématique.

1887, 1. Théorie du Potentiel.

1887, 2. Théorie du Potentiel et Propagation de la Chaleur.

1888, 1. Théorie mathématique de la Lumière (277).

1888, 2. Théorie de MAXWELL (278, 289).

1889, 1. Thermodynamique (281).

1889, 2. Capillarité (285).

1890, 1. Problème des Trois Corps.

1890, 2. Electrodynamique (189, 289).

1891, 1. Elasticité (282).

1891, 2. Electrostatique.

1892, 1. Optique (279).

1892, 2. Tourbillons (283).

1893, 1. Oscillations électriques (284).

1893, 2. Thermodynamique et théorie cinétique des gaz.

1894, 1. Propagation de la chaleur (286).

1894, 2. Calcul des Probabilités (287).

1895, 1. Potentiel newtonien (192).

1895, 2. Electrostatique.

1896, 1. Elasticité.

1896, 2. Optique.

Chaire de Mécanique Céleste.

1897, 1. Sur le Problème des Trois Corps et les Perturbations Planétaires.

1898, 1. Sur le Développement de la Fonction Perturbatrice et les méthodes de Gyldén et de Hansen.

1899, 1. Sur les nouvelles théories électrodynamiques (289).

1899, 2. Figure des Corps Célestes.

1900, 1. Théorie de la Lune.

1901, 1. Mouvement des Corps Célestes autour du leur centre de gravité.

La signification de ce tableau est aisée à comprendre; par exemple 1886, 1 signifie 1er semestre de l'année scolaire 1885—86. Les chiffres en caractères gros entre parenthèses après l'énoncé d'un cours sont des renvois à la bibliographie et se rapportent aux cours qui ont été publiés.

Ce serait faire double emploi que de revenir sur tous les cours publiés, dont j'ai déjà parlé au § XXIV. Je dirai seulement quelques mots, au point de vue de l'enseignement, des cours 1888, 2 et 1890, 2 (278, 279). A cette époque, on ne s'était pas encore familiarisé sur le continent avec les idées de Maxwell, et il était nécessaise de jeter un pont pour ainsi dire entre les anciennes manières de penser et les nouvelles. Maintenant que tout le monde est passé, ce pont peut paraître inutile et les précautions que j'ai cru devoir prendre étonneront quelques personnes. Néanmoins je crois qu'aujourd'hui encore les deux images artificielles dont je me suis servi pour faire comprendre les idées de Maxwell, celle du fluide inducteur, et celle des diélectriques cellulaires, peuvent faciliter pour certains esprits l'étude des théories électriques. Dans mes leçons sur l'électrodynamique (279) j'ai exposé les idées de Helmholtz, mais je crois les avoir rendues plus accessibles en employant les unités électromagnétiques au lieu des unités électrostatiques qui étaient aussi mal appropriées que possible au but que poursuivait le savant allemand.

On a vu que tous les cours n'avaient pas été publiés. Je parlerai seulement des cours d'électrostatique, des deux cours de 1896, et des leçons sur la théorie cinétique des gaz.

Je me suis efforcé d'exposer l'électrostatique sans prononcer le mot d'électricité et en parlant seulement de conducteurs électrisés. Je définissais d'abord l'égalité de potentiel de deux conducteurs par l'équilibre électrique comme on définit l'égalité de température par l'équilibre thermique, et ce n'est que tout à la fin du cours que j'obtenais l'expression habituelle du potentiel par une intégrale, en la déduisant d'un certain nombre de constatations expérimentales, et en particulier des expériences sur les écrans électriques. Je réduisais ainsi autant que possible le part de l'hypothèse.

Dans mon second cours sur l'élasticité (1896, 1), j'ai traité plusieurs questions nouvelles; j'ai exposé quelques-unes des idées de lord KELVIN et montré par exemple qu'en admettant plusieurs sortes de molécules (et pour ainsi dire plusieurs milieux se pénétrant mutuellement) on pouvait obtenir des équations contenant 21 coëfficients arbitraires et non pas 15 seulement, et cela sans renoncer à l'hypothèse des forces centrales. J'ai exposé ensuite d'une façon complète et systématique la théorie de l'éther gyrostatique, que les publications de lord KELVIN permettent de reconstituer, mais non sans quelque effort.

Dans mon troisième cours sur l'Optique (1896, 2), j'ai fait jouer un grand rôle aux idées de M. GOUY sur la façon dont se comporte une lumière polychromatique en traversant un appareil optique quelconque, à la représentation de cette lumière polychromatique par une intégrale de FOURIER, et aux conséquences qui s'en déduisent.

Dans mon cours sur la Thermodynamique et la Théorie Cinétique des Gaz (1893, 2), j'ai exposé la démonstation des deux principes de la Thermodynamique sous une forme un peu plus générale et plus abstraite que la forme ordinaire. J'ai ensuite expliqué la théorie du viriel, justifié le principe de BOLTZMANN— MAXWELL et analysé l'ouvrage de MAXWELL. J'ai discuté l'influence de la loi des grands nombres sur les équations de la Mécanique, mais sans arriver sur ce point à une conclusion définitive.

A la fin de mon cours sur les tourbillons (283), j'ai exposé les idées de HELMHOLTZ sur la formation des vagues; cette partie du cours n'a pas été publiée, parce que je ne suis pas parvenu à une solution qui m'ait entièrement satisfait.

On peut se faire une idée de ce qu'il y a eu de plus original dans mon enseignement de la Mécanique Céleste, en lisant les articles que je publiais en même temps dans le Bulletin Astronomique. Dans mon cours sur la Théorie de la Lune, j'ai exposé presque exclusivement les méthodes de HILL et de BROWN et n'ai consacré que quelques leçons à celles de DELAUNAY. Dans mon cours sur le mouvement des corps célestes autour de leur centre de gravité, j'ai fait usage des quaternions.

XXXII. Vulgarisation. (358, 372, 239, 217, 179, 357, 370, 290, 244.)

Dans ces articles qui portent sur la Mécanique Céleste, sur la géodésie, sur les oscillations hertziennes, sur les rayons RÖNTGEN, je me suis astreint à éviter complètement l'emploi des signes algébriques. Je rappallerai seulement que c'est dans l'un d'eux (370) que j'ai émis une hypothèse qui, historiquement, n'a pas été sans influence sur la découverte des rayons BECQUEREL.

XXXIII. Bibliographie. (460, 476, 461, 472, 471, 456, 457, 465, 467.)

Dans ces articles j'ai étudié la vie et les travaux de WEIERSTRASS, HERMITE, BERTRAND, LAGUERRE, HALPHEN, TISSERAND.

XXXIV. Rapports Divers. (463, 468, 470, 473, 474, 265, 263, 274, 275.)

Parmi ces rapports je citerai ceux que j'ai été chargé de faire sur le projet d'unification du jour civil et du jour astronomique et sur le projet de décimalisation du temps et de la circonférence. Ces deux projets de réforme n'ont pas réussi par suite de la difficulté d'une entente internationale. Je citerai également mon rapport sur la nouvelle mesure de l'arc de méridien de Quito.

Proceedings of Symposia in Pure Mathematics
Volume 39 (1983), Part 2

L'oeuvre Mathématique de Poincaré*

JACQUES HADAMARD

à Paris.

Poincaré lui-même a fourni aux lecteurs des *Acta* une analyse détaillée de son œuvre.[1]

On comprendra que, sur tous les points qui ont été portés à leur connaissance dans un des styles les plus lumineux, les plus définitifs que la langue scientifique — et la langue française — aient connus, nous nous croyions dispensés d'insister. Il nous arrivera donc très souvent de renvoyer à l'*Analyse* dont il s'agit.

Nous n'essaierons pas, d'autre part, de chercher dans tout l'ensemble de cette œuvre une unité, d'en dégager une personnalité intellectuelle. Cette tentative, qui s'imposerait pour tout autre, serait, à notre sens, chimérique en ce qui concerne Poincaré, et nous croirions diminuer en même temps que dénaturer son œuvre en nous y essayant. Ce serait méconnaître cette pensée «capable de faire tenir en elle toutes les autres pensées, de comprendre jusqu'au fond, et par une sorte de découverte renouvelée, tout ce que la science humaine peut aujourd'hui comprendre».[2]

Assurément, tout penseur tend à marquer de son sceau personnel ce que son cerveau façonne. Mais si cette tendance est une des forces de l'artiste, le savant, lui, bien loin de chercher à l'entretenir, la subirait plutôt. Elle est, chez lui, combattue par une nécessité toute contraire, celle de l'objectivité. »Nous sommes serviteurs plutôt que maîtres en mathématiques», aimait à dire Hermite, et l'adage tout analogue de Bacon est aussi vrai des mathématiques elles mêmes que des sciences expérimentales. Le savant — surtout le mathéma-

---

*Reprinted from Acta Math. 38 (1921), 203–287.

[1] Analyse de ses travaux scientifiques. Acta Math. tome 38.

[2] Painlevé, *Temps* du 18 Juillet 1912.

ticien — ne dispose guère, au fond, des moyens d'attaque. Tout au plus suit-il en général son tempérament dans le choix du terrain.

Poincaré ne fit même point ainsi. Il emprunta ses sujets d'étude non aux ressources de son esprit, mais aux besoins de la science. Il a été présent partout où il y avait une lacune grave à combler, un grand obstacle à surmonter. Lorsque nous aurons essayé d'énumérer — même aussi rapidement qu'il nous faudra le faire — les questions auxquelles il s'est attaqué, il nous semblera avoir touché à toutes celles auxquelles peuvent s'intéresser les mathématiciens et qui nécessitent encore leurs efforts. Son œuvre est devenue, dès lors, le patrimoine commun de tous. Si Poincaré a une «manière», si même on peut employer à son égard ce mot qui ressemble à «manie», nous en avons tous hérité, et elle est en chacun de nous.

De ses résultats se dégage souvent une unité; mais celle-ci n'est pas inhérente à l'auteur. Elle est, elle aussi, objective et réside dans les faits eux-mêmes. Nul mieux que Poincaré ne sut, en effet, découvrir, entre les diverses parties de la science, des relations imprévues, parce que personne ne sut mieux dominer cette science de tous les côtés à la fois.

Cette souplesse et cette universalité, cette adaptation rapide et parfaite à tous les problèmes posés par les mathématiques et leurs applications, se sont manifestées de manière d'autant plus éclatante qu'à notre époque, l'une des sciences qui dictent surtout ces problèmes, la Physique, évolue avec une plus déconcertante rapidité. On sait, — et d'autres le diront ici mieux que moi — à quel degré Poincaré, dès qu'il s'est mêlé à cette évolution, a su toujours la suivre et souvent la guider.

L'histoire de l'œuvre de Poincaré ne sera donc, au fond, autre chose que l'histoire même de la science mathématique et des problèmes qu'elle s'est posés à notre époque.

Le plus important d'entre eux est encore aujourd'hui le même qui est apparu à la suite de l'invention du Calcul infinitésimal.

Nous sommes loin d'avoir résolu les difficultés qu'il présente. Mais là même où nous y sommes arrivés, ce n'a été, le plus souvent, qu'en modifiant profondément nos idées sur ce qu'il faut entendre par «solution». Celles que nous avons acquises aujourd'hui se résument toutes dans la forte parole que Poincaré prononçait en 1908[1]:

«Il n'y a plus des problèmes résolus et d'autres qui ne le sont pas, il y a

---

[1] Conférence prononcée au Congrès international des Mathématiciens, Rome; t. i, p. 173 des *Actes du Congrès*.

seulement des problèmes *plus ou moins* résolus», — c'est-à-dire qu'il y a des solutions donnant lieu à des calculs plus ou moins simples, nous renseignant plus ou moins directement et aussi plus ou moins complètement sur l'objet de notre étude.

On peut dire, à ce point de vue, qu'une première solution est acquise dans la plupart des cas, — et cette conquête, ébauchée dès Newton, est surtout l'œuvre de Cauchy et de Weierstrass: — des relations entre états *infiniment* voisins, on sait déduire, ce qui est fort différent, la connaissance de tous les états *suffisamment voisins* d'un état donné. Si, par exemple, le phénomène à étudier dépend de la position d'un point dans un plan, on sait l'étudier dans toute une petite région entourant un point quelconque donné.

En un certain sens, il peut être ainsi considéré comme connu, puisque, avec de petites régions de cette espèce accolées les unes aux autres, on peut constituer des régions plus étendues et même aussi étendues qu'on le voudra.

Mais cette connaissance est souvent très insuffisante, beaucoup plus encore que ne le serait, pour un voyage d'un bout à l'autre d'un pays, la possession des feuilles partielles de la carte à quelqu'un qui ne disposerait d'aucune autre donnée géographique. Elle l'est à des degrés divers suivant la nature du problème posé; mais dans la plupart des cas, le résultat est connu, dans chaque domaine partiel, par des opérations d'une convergence médiocre, c'est-à-dire assez mal et assez péniblement; d'autant plus mal et d'autant plus péniblement même que le domaine en question est plus petit.

Quoi qu'il en soit, ces premiers résultats, même si l'on n'est pas réduit à s'en contenter, servent tout au moins d'intermédiaires obligés pour en obtenir de meilleurs, de sorte que, presque partout, la marche de la science mathématique actuelle comporte deux étapes:

La solution locale des problèmes;

Le passage de celle-ci à une solution d'ensemble, si cette sorte de synthèse est possible.

Le premier problème qui avait arrêté le Calcul infinitésimal, celui des quadratures, est, en somme, résolu, au sens précédent, d'une manière assez satisfaisante. Cette solution diffère assurément beaucoup de celle que cherchaient, — sans aucune chance de succès, nous le savons maintenant — les contemporains de Leibnitz. Elle contient cependant l'essentiel de ce qu'on peut savoir dans le cas général et des renseignements beaucoup plus importants dans tous les cas particuliers les plus usuels.

Mais le problème général des équations différentielles est autrement difficile. Les petites régions dont nous parlions ne peuvent même plus être considérées indépendamment les unes des autres. On doit les ranger dans un ordre déterminé, et les calculs relatifs à l'une d'elles ne peuvent être commencés sans qu'on ait exécuté jusqu'au bout ceux qui concernent les précédentes. En général, il arrive même qu'on ignore *a priori* jusqu'à l'amplitude des pas successifs que l'on peut ainsi effectuer, c'est-à-dire jusqu'aux dimensions des régions partielles successives: c'est ce que l'on ne connaît qu'au moment même où l'on atteint chacune d'elles.

Les difficultés dont nous venons de parler s'aggravent encore — et même d'autres toutes différentes apparaissent — si, au lieu d'équations différentielles ordinaires, on a à traiter des équations aux dérivées partielles.

L'intégration des équations différentielles et aux dérivées partielles est restée jusqu'ici le problème central de la mathématique moderne. Elle en restera vrai-semblablement encore l'un des problèmes capitaux, même si la Physique poursuit vers le discontinu l'évolution qui se dessine à l'heure actuelle.

La théorie des équations différentielles fut aussi la première à attirer l'atten-tion de POINCARÉ. Elle fait l'objet de sa Thèse (1879).

Notons cependant que, sous l'influence du maître qui gouverna la génération précédente, j'ai nommé HERMITE, le débutant ne craignait pas de suivre presque au même moment une voie pour ainsi dire opposée à la première, celle de l'Arithmétique.

La Thèse de POINCARÉ contient déjà sur les équations différentielles un résultat d'une forme remarquable, destiné à être plus tard pour lui un puissant levier dans ses recherches de mécanique céleste. Dès ce premier travail, il était, d'autre part, conduit à perfectionner le principal outil dont se fût servi jusque là, la théorie des équations différentielles, outil qu'il allait utiliser mieux que qui que ce soit, en même temps que, le premier, il allait enseigner à s'en passer: la théorie des fonctions analytiques.

Celle-ci allait, presque immédiatement après, lui devoir une de ses plus belles conquêtes: c'est en 1880 que les *fonctions fuchsiennes* vinrent désigner POINCARÉ à l'attention et à l'admiration de tous les géomètres.

I. La théorie des fonctions.

§ 1. Les fonctions fuchsiennes.

Auxiliaire puissant pour tout le Calcul infinitésimal, la Théorie des fonctions analytiques a fait ses preuves de façon particulièrement éclatante dans la résolu-tion du problème des quadratures, mais surtout lorsqu'il s'est agi de celles qui portent sur des fonctions algébriques, c'est-à-dire des intégrales elliptiques et abéliennes.[1]

On sait — et, avant de parler des fonctions fuchsiennes, nous rappellerons — les circonstances grâce auxquelles ce degré de perfection a pu être atteint.

La première d'entre elles n'est autre que la polydromie de l'intégrale cherchée, c'est-à-dire, par un phénomène qui n'est pas isolé en Mathématiques — n'a t-on pas dit de CAUCHY que ses deux grandes forces furent ce qui avait été l'effroi de ses prédécesseurs, l'infini et l'imaginaire! — le fait même qui paraissait devoir constituer la principale difficulté de son étude. C'est à cette polydromie que la fonction inverse, obtenue en prenant l'intégrale considérée u comme variable indépendante, doit sa double périodicité.

Cette fonction inverse doit, par contre, être uniforme et, pour qu'il en soit ainsi, on doit choisir de manière convenable l'intégrale elliptique qui sert de point de départ. Ici, ce choix — celui de l'intégrale de première espèce — est aisé à faire.

La double périodicité, à son tour, donne la clef de toutes les autres propriétés. Il y a plus. L'Analyse moderne laisse complètement de côté, au premier abord, le problème d'intégration posé et prend pour premier objet l'étude générale des fonctions doublement périodiques d'une variable. Parmi celles là, on découvre ensuite les solutions du problème en question.

On obtient ainsi tout l'ensemble de résultats qui font de l'intégration des différentielles elliptiques l'un des problèmes les mieux résolus de l'Analyse: celui même que POINCARÉ, dans la conférence à laquelle nous faisions allusion plus haut, prenait comme exemple typique à cet égard.

Les propriétés des fonctions abéliennes sont assurément moins simples et surtout moins commodes pour le calcul numérique que celles des fonctions elliptiques: elles sont toutes parallèles, néanmoins, et ne contentent pas moins complètement ce sens de la beauté dans lequel POINCARÉ nous a appris à discerner le véritable mobile du savant.

La notion de périodicité suffit à elle seule pour constituer ces deux théories, modèles d'harmonie et d'élégance.

Mais par cela même, on peut dire qu'elle avait rendu tous les services qu'on en pouvait attendre, et nulle autre notion fonctionnelle analogue ne paraissait offrir la même fécondité.

Les deux exemples qui ont inspiré POINCARÉ étaient cependant déjà con-

---

[1] POINCARÉ a eu ici même à retracer l'histoire de ces théories: voir sa Notice sur WEIERSTRASS, Tome 22 des Acta.

nus: je veux parler de la fonction modulaire et de l'inversion de la série hyper-géométrique, objets l'une des admirables travaux d'HERMITE, l'autre du Mémoire fondamental que l'on doit à M. SCHWARZ. Ils n'avaient pas fait apercevoir aux géomètres la généralisation hardie qui devait conduire aux *fonctions fuchsiennes*.

Cette généralisation était si audacieuse que le premier mouvement de POINCARÉ fut de la regarder comme impossible. Il nous apprend lui-même [1] qu'il s'efforça tout d'abord de montrer l'*inexistence* des fonctions dont il s'agit. C'est par une de ces intuitions d'apparence spontanée dont on verra l'histoire dans *Science et Méthode*, qu'il s'engagea dans la voie opposée.

POINCARÉ va se placer dans des conditions incomparablement plus générales et plus variées que les fondateurs de la théorie des fonctions elliptiques; mais la marche suivie sera cependant toute semblable de part et d'autre.

A la place du problème de quadrature, il considère une équation différentielle linéaire à coefficients algébriques. Ce problème dépasse le premier de toute la distance qui sépare l'intégration des équations différentielles de la simple recherche des primitives; parmi les équations différentielles, toutefois, les équations linéaires se présentaient à lui comme les plus simples de toutes.

La polydromie des fonctions obtenues par quadratures se retrouve chez celles qui sont définies par des équations linéaires; quoique plus complexe que le premier, ce nouveau mode de polydromie était bien connu par les recherches de FUCHS, auquel POINCARÉ sera ainsi conduit à dédier la nouvelle conception.

On aperçoit dès lors immédiatement ce qui devra correspondre à la notion de périodicité: ce rôle appartient à un certain *groupe* de substitutions linéaires.

Tout ceci apparaissait sur les deux exemples que nous citions tout à l'heure, de la fonction modulaire et de la série hypergéométrique. Dans ces deux cas, c'est bien un groupe de substitutions linéaires qui intervient, et ce groupe satisfait à la condition indispensable — nous renvoyons sur ce point à l'exposé de POINCARÉ [2] — d'être *discontinu*, c'est-à-dire tel que les transformés d'un même point n'aillent pas en s'accumulant en nombre infini dans le voisinage immédiat de l'un quelconque d'entre eux (sauf dans certaines régions particulières ou, plus exactement, le long de certaines lignes du plan).

---

[1] L'exemple des fonctions fuchsiennes est précisément, on le sait, celui que POINCARÉ a choisi pour décrire au point de vue psychologique, l'invention mathématique et montrer le rôle essentiel qu'y joue l'inconscient.

Ajoutons que, chez POINCARÉ, l'idée première d'une recherche est toujours mise en évidence avec une merveilleuse netteté qu'on est loin de trouver toujours au même degré chez les plus grands maîtres. C'est dire que l'accusation d'obscurité lancée parfois contre lui nous paraît, du moins au point de vue du lecteur qui va au fond des choses, exprimer le contraire de la vérité.

[2] Voir son *Analyse*, p. 43

Il possède même la seconde propriété par laquelle, entre les groupes linéaires discontinus, POINCARÉ distingue les groupes fuchsiens, à savoir celle de
laisser invariant un certain cercle (dit cercle *fondamental*).

Mais il restait à s'inspirer plus étroitement encore de l'exemple des fonctions elliptiques: je veux dire, conformément à ce qui précède, à partir *a priori*
du groupe en question, en laissant de côté d'abord l'équation différentielle.

Il fallait même faire un pas de plus, et cette première transformation de la
question, suffisante dans les cas traités antérieurement, ne l'était plus cette fois;
c'est sans doute pour cette raison que les découvertes mentionnées plus haut
d'HERMITE et de SCHWARZ étaient jusque là restées isolées et n'avaient pas mis
sur la voie de l'infinie multiplicité d'autres groupes analogues qui allait s'offrir à
POINCARÉ. Dans le problème actuel, on ne remonte pas assez loin en s'adres-
sant à la notion du groupe, trop complexe elle même pour nous servir de fondement premier. Il faut, si nous osons nous exprimer ainsi, placer plus
bas encore les fondations et appuyer à son tour la notion du groupe sur un
autre substratum.

Ce substratum est essentiellement géométrique: POINCARÉ le trouve dans
le *polygone générateur*, c'est-à-dire[1] dans la figure qui est au groupe ce que le
parallélogramme des périodes est à la double périodicité.

Cette notion intervenait forcément dans les exemples d'HERMITE et de SCHWARZ.
Mais POINCARÉ montre qu'elle caractérise tout groupe discontinu. Pure intuition
dans le premier Mémoire sur les groupes fuchsiens, ce fait est établi en toute
rigueur dans un des Mémoires suivants[2], et une règle générale est énoncée d'après
laquelle, à chaque point, on peut faire correspondre un de ses transformés et un
seul (à des cas limites près) de manière que, le premier prenant toutes les positions possibles, le second décrive le polygone générateur demandé.

Mais, nous l'avons dit, loin de chercher systématiquement ce dernier en partant du groupe, POINCARÉ suit bien plutôt la marche inverse et part de la notion
du polygone, aussi intuitive pour nous que celle du groupe nous est, au fond,
peu familière encore. L'expression de «polygone générateur» exprime d'une
manière parfaite comment les choses se passent dans l'Analyse de POINCARÉ:
c'est lui qui engendre véritablement le groupe. Non seulement il suffit entièrement à le définir, mais on lit immédiatement et sans la moindre difficulté, sur
cette figure, toutes les propriétés essentielles que l'on veut en connaître: substitutions fondamentales, relations entre ces substitutions, etc.

En particulier, il convient de noter, au point de vue des applications des

---

[1] Voir *Analyse,* p. 44.
[2] Acta, t. IV.

C'est une courbe jordanienne qui, comme le montre POINCARÉ, tient la place du cercle lorsqu'on passe de l'étude des groupes fuchsiens à celle des groupes kleinéens, et une courbe jordanienne dépourvue soit de tangente, soit de courbure en tous ses points.

Certes, les exemples de courbes sans tangentes sont classiques depuis RIEMANN et WEIERSTRASS; mais tout le monde comprendra la différence profonde qui existe entre un fait obtenu dans des circonstances rassemblées à plaisir, sans autre but et sans autre intérêt que d'en montrer la possibilité, sorte de pièce de musée tératologique, et le même fait rencontré au cours d'une théorie qui a toutes ses racines dans les problèmes les plus usuels et les plus essentiels de l'Analyse générale.

La théorie des groupes kleinéens offre le premier exemple de cette espèce, — le seul même que l'on connaisse, si nous ne nous trompons — en ce qui concerne la notion de courbe jordanienne. Le fait qu'on conduise ainsi nécessairement à cette notion nous fait sentir, dès cet exemple, combien les résultats de POINCARÉ pénètrent profondément dans la nature intime des choses.

*   *   *

La notion des groupes fuchsiens et kleinéens étant ainsi fondée, une *fonction fuchsienne* (ou kleinéenne) est celle qui reste invariante par toutes les substitutions d'un de ces groupes. Pour trouver un tel invariant, POINCARÉ forme d'abord un *invariant relatif*, c'est-à-dire une fonction qui, au lieu de rester inaltérée par une quelconque des substitutions en question, est multipliée par un facteur de forme connue et simple. C'est un intermédiaire classique dans beaucoup de recherches de cette nature non seulement dans la théorie des fonctions elliptiques (lesquelles se présentent comme quotients de fonctions thêta) mais dans celle des invariants de l'algèbre, ou même de certains invariants différentiels.[1] Pour les fonctions thêta fuchsiennes, comme pour les invariants algébriques, le facteur dont il s'agit est une puissance du déterminant de la substitution. Par là, et par leur mode même de formation, les fonctions thêta fuchsiennes diffèrent des fonctions thêta ordinaires et tendent bien plutôt à se rapprocher de la fonction elliptique pu elle même, telle que la forme WEIERSTRASS. Mais celle ci possède la propriété d'invariance absolue et évite, par conséquent, le détour employé dans la théorie actuelle, détour dont la nécessité, comme un peu de réflexion suffit à le faire apercevoir, est indissolublement liée à la présence des points singuliers du groupe.

---

[1] Voir, par exemple, plusieurs travaux de M. TRESSE.

Par contre, POINCARÉ a montré que la méthode ainsi modifiée s'applique à un groupe discontinu quelconque.

Une fois obtenues les fonctions thêta fuchsiennes, le quotient de deux d'entre elles, c'est-à-dire de deux invariants relatifs, donne, comme dans la théorie classique des formes algébriques, une des fonctions invariantes cherchées.
Les *fonctions fuchsiennes* sont formées.

La nouvelle notion ainsi créée, si supérieure en généralité, en extension, à celles sur le modèle desquelles elle avait été édifiée, ne leurs cède en rien sous le rapport de la compréhension. Si l'on ne dispose pas, cette fois, de séries rapidement convergentes à la façon des séries Θ, on peut dire que toutes les autres propriétés dont l'imposant ensemble forme la théorie des fonctions elliptiques trouvent encore leurs analogues. Une seule, le théorème d'addition, n'a été étendue par POINCARÉ qu'à certaines catégories de fonctions fuchsiennes, en relation, comme nous le dirons plus loin, avec les applications arithmétiques.

Mais l'une d'elles domine toutes les autres: les fonctions fuchsiennes présentent, comme les fonctions elliptiques, ce caractère que deux quelconques d'entre elles appartenant au même groupe, sont liées par une relation algébrique.

Dans le cas des fonctions elliptiques, cette relation est forcément très particulière: elle est de genre zéro ou un. Au contraire, ce qui fait l'importance des fonctions fuchsiennes, c'est que toute équation algébrique à deux variables donnée peut être obtenue par leur moyen.

Dans la démonstration de cette proposition résidait une autre, la plus profonde peut-être des grandes difficultés du problème.

L'opération qu'il s'agit d'effectuer est déjà de celles auxquelles s'appliquent les réflexions émises par POINCARÉ dans son Analyse,[1] c'est-à-dire qu'elle correspond, dans la nouvelle théorie, à ce qu'est le choix de l'intégrale de première espèce dans celle des fonctions elliptiques.[2] Mais autant sont simples les règles qui président à ce dernier choix, autant celui de la variable à l'aide de laquelle les coordonnées d'une courbe algébrique s'expriment par des fonctions fuchsiennes est un résultat caché.

POINCARÉ y parvint par une audacieuse méthode de continuité. M. KLEIN, qui avait été immédiatement frappé par la puissance de la nouvelle conception

---

[1] Page 48.

[2] C'est ainsi que la représentation d'une cubique par les fonctions elliptiques repose sur l'introduction de l'intégrale *de première espèce* attachée à cette courbe.

Ce sont également, pour une grande partie, les fonctions fuchsiennes qui lui permirent de traiter les cas singuliers des fonctions abéliennes: il s'agit des cas de *réduction*, dans lesquels, parmi les intégrales abéliennes attachées à une courbe algébrique, en figurent une ou plusieurs susceptibles de dériver d'une courbe plus simple, c'est-à-dire de genre inférieur.

On verra par son *Analyse*[1] comment, conduit une première fois à cette question par la précédente, il y fut un ramené peu plus tard par deux théorèmes de WEIERSTRASS. Lorsqu'il eut fourni et généralisé la démonstration de ces théorèmes (que WEIERSTRASS n'avait pas publiée), d'autres conséquences lui apparurent.

Ici encore, ce fut la théorie des fonctions fuchsiennes qui lui fit apercevoir quelques unes des plus lointaines, et cela non seulement parce qu'elle domine la question au point de vue analytique, mais aussi parce qu'elle apporta l'aide efficace de sa figuration géométrique, si lumineuse, nous l'avons dit, en ce qui regarde les relations d'un groupe fuchsien avec ses sous-groupes.

POINCARÉ considère en particulier les cas où la réduction entraine, entre deux courbes algébriques, une correspondance simplement rationnelle. De cette propriété ressortent, lorsqu'on lui applique les principes de la théorie des fonctions fuchsiennes, une série de conséquences aussi simples et aussi élégantes qu'elles sont cachées au premier abord.

Les cas de dégénérescence dont nous venons de parler ne furent pas simplement pour POINCARÉ des difficultés à résoudre. Ce furent, au contraire, les propriétés de ces fonctions dégénérées qui l'aidérent par la suite à pénétrer celles des autres fonctions abéliennes.

Mais cette deuxième catégorie de recherches découle d'une tout autre source, et, avant de les aborder, il nous faut avoir laissé de côté les transcendantes particulières pour nous occuper de la théorie générale des fonctions analytiques.

§ 2. Relations avec l'Arithmétique. Ensembles. Groupes continus.

Toutefois, avant d'abandonner les groupes et les fonctions fuchsiennes, nous parlerons de travaux qui, dans l'œuvre de POINCARÉ, s'y rattachent plus ou moins étroitement.

Tel est d'abord le cas pour la partie de cette œuvre qui touche à l'Arithmétique.

A côté des perspectives largement ouvertes de l'Analyse pure et de ses applications géométriques et physiques, la théorie des nombres, isolée, au moins en apparence, du reste de la Science, n'a pas cessé cependant d'être cultivée par

---

[1] Deuxième Partie, X.

les mathématiciens de race. Avec MM. JORDAN et PICARD, c'est surtout POIN
CARÉ qui contribua à perpétuer à cet égard, dans notre pays, la tradition d'HER
MITE. Nous avons dit que de cette tradition procèdent des notes presque contemporaines de la Thèse dont nous avons parlé en commençant. POINCARÉ
transporte dès cette époque les méthodes d'HERMITE au cas le plus général des
formes de degré quelconque à un nombre quelconque de variables.

Nul domaine où ces généralisations soient plus cachées que celui de l'Arithmétique qui nous occupe en ce moment. La discontinuité qui en fait le caractère essentiel s'y révèle en quelque sorte, au point de vue logique, par celle qui
sépare souvent les notions destinées à se montrer analogues entre elles, en ne
les laissant se rattacher les unes aux autres que par un fil ténu. En lisant les
Notes dans lesquelles POINCARÉ traite ainsi les notions de *genre* et d'*ordre* d'une
forme, on se convaincra à quel point de telles analogies sont difficiles à saisir.
POINCARÉ sut les rendre claires et évidentes et par conséquent, là comme
ailleurs, introduire la simplicité et la cohésion là où semblait devoir régner l'artifice. C'est ce qui apparaît encore à un haut degré dans ses recherches sur la
réduction des formes.

M. JORDAN venait de montrer que la méthode même d'HERMITE permet
d'établir pour les formes quelconques, le théorème d'après lequel le nombre
des classes algébriquement équivalentes est fini. POINCARÉ, poursuivant la même
voie, put ainsi généraliser la notion de réduction, généralisation que (comme la
précédente, d'ailleurs) HERMITE n'avait donnée, du moins pour des variables en
nombre supérieur à 2, que relativement aux formes décomposables en facteurs
linéaires.

Avec POINCARÉ, on peut dire que toute question disparait, en ce sens qu'une
idée d'une rare simplicité fournit d'un seul coup la règle applicable à tous les
problèmes de cette catégorie. La réduction demandée est décomposée en deux
opérations dont l'une ne dépend que de l'Algèbre: c'est la réduction de la forme
donnée, au sens purement algébrique du mot. L'autre opération est entièrement
indépendante de la forme considérée et ne dépend que des propriétés du groupe
arithmétique: c'est une sorte de réduction par rapport à ce groupe, des substitutions du groupe linéaire de l'Algèbre, réduction que l'on sait effectuer par cela
même que l'on sait réduire les formes quadratiques définies.

La solution de cette seconde partie du problème élimine, en somme, toutes
les difficultés de nature arithmétique.

La réduction des formes cubiques ternaires se présente comme application
immédiate de ce principe.

Ces recherches, ainsi que celles que POINCARÉ consacra à l'étude des points de coordonnées rationnelles sur une courbe du troisième degré, sont fondamentales dans la théorie, si peu explorée encore, des formes de degré supérieur.

La théorie des formes quadratiques dut, elle aussi, à POINCARÉ des progrès essentiels; et ceci nous ramène aux fonctions fuchsiennes.

C'est, on le sait, un titre de gloire de quelques uns des plus grands mathématiciens du XIXe siècle — de DIRICHLET, de RIEMANN, d'HERMITE entre autres — que d'avoir su éclairer l'Arithmétique à l'aide de l'analyse du continu qui semblait, au premier abord, ne devoir jamais y pénétrer.

Ce résultat remarquable a même été obtenu de deux manières entièrement différentes. Le point de départ de RIEMANN est le même que celui de DIRICHLET (et aussi, au fond, que celui qui a servi à JACOBI dans les *Fundamenta*). Mais celui d'HERMITE est sans rapport avec le premier.

Grâce aux fonctions fuchsiennes, POINCARÉ réussit à son tour à établir une alliance analogue, et cela sous deux formes, elles mêmes profondément distinctes, respectivement en relation avec les deux grands principes qui viennent d'être mentionnés. Aux idées d'HERMITE se rattachent les recherches que POINCARÉ entreprend sur les formes quadratiques, dans le cas qui appelle le plus de recherches, celui des formes indéfinies. La particularité qui fait la difficulté et l'intérêt de cette catégorie de formes est, on le sait, que chacune d'elles se reproduit sans altération par une infinité de substitutions linéaires formant un groupe discontinu. Or on est ainsi ramené aux groupes fuchsiens.

Non seulement ceux ci se trouvent ainsi — et cela, comme le montre POINCARÉ[1], par l'intermédiaire de la Géométrie non euclidienne — éclairer la théorie des nombres, mais il est remarquable qu'en l'espèce l'inverse a également lieu: c'est par cette voie qu'on étend à certaines fonctions fuchsiennes la seule propriété remarquable des fonctions elliptiques dont l'extension, dans ce domaine, ne paraît pas pouvoir se faire d'une manière entièrement générale, le théorème d'addition. POINCARÉ montre, en invoquant une fois de plus les propriétés géométriques des polygones générateurs, que cette extension depend d'une sorte de commensurabilité entre le groupe de la fonction fuchsienne et une substitution déterminée, non comprise dans ce groupe; et cette commensurabilité, qui n'existe pas dans le cas général, se présente au contraire lorsque le groupe fuchsien considéré a une origine arithmétique.

Mais ce rapprochement n'est pas, nous l'avons dit, le seul que POINCARÉ ait établi entre les fonctions fuchsiennes et l'Arithmétique. Dès le début de ses recherches, en effet, il a donné du problème de l'équivalence une solution géné-

---

[1] *Analyse,* page 97.

rale toute nouvelle, fondée sur l'extension au domaine arithmétique de la notion d'invariants. Grâce à la discontinuité des groupes auxquels conduit la théorie des nombres, les invariants arithmétiques existent là même où il n'y a point d'invariants algébriques et, indépendamment d'expressions par intégrales définies, ils en possèdent d'autres sous forme de séries sur lesquelles leur propriété d'invariance est mise immédiatement en évidence. Les séries auxquelles on aboutit ainsi sont très voisines des séries connues de DIRICHLET, mais leur formation, qui par conséquent se rattache aux recherches de ce géomètre, s'inspire cependant, comme on le voit, d'un principe d'une bien plus grande généralité et dont la relation avec les méthodes suivies en Algèbre apparaît immédiatement.

Elles sont, d'autre part, étroitement liées aux séries thêta d'une part, aux fonctions fuchsiennes de l'autre et montrent la relation qui existe entre ces deux sortes de fonctions par l'intermédiaire de la fonction modulaire. C'est ce point de vue qui a donné à POINCARÉ de nouveaux développements des fonctions elliptiques.

L'étude de la catégorie de fonctions fuchsiennes à laquelle appartient ainsi la fonction modulaire devait attirer à nouveau son attention; elle fait l'objet du dernier travail qu'il nous ait laissé.[1]

Outre la théorie des formes, les deux principaux chapitres de l'Arithmétique moderne sont la théorie des nombres premiers et celle des idéaux. POINCARÉ les a abordés tous deux ensemble dans le Mémoire intitulé *Extension aux nombres premiers complexes des théorèmes de* M. TCHEBICHEFF. La méthode du géomètre russe, conservée dans son principe général, a dû subir d'importantes modifications pour s'adapter à ces nouvelles circonstances. Il est remarquable que le résultat obtenu soit indépendant de deux éléments qui s'introduisent d'une manière nécessaire dans les calculs, le nombre des unités indépendantes du corps considéré et celui des classes d'idéaux qu'il renferme, et que d'autre part, ce résultat relatif aux nombre imaginaires puisse servir à étudier la distribution des nombres premiers réels entre les formes $4n + 1$, $4n + 3$.

Enfin, rappelons qu'une des premières publications de POINCARÉ avait enrichi, en même temps que rassemblé dans une synthèse particulièrement lumineuse, les propriétés des corps quadratiques et des idéaux correspondants, en les rattachant à une nouvelle théorie géométrique des réseaux (au sens de BRAVAIS). On sait avec quel succès une synthèse analogue fut reprise plus tard par M. KLEIN.

*   *   *

---

[1] Ce Mémoire (Ann. fac. sciences Toulouse, 1913) n'a paru qu'après la mort de POINCARÉ et ne figure pas dans la liste qui précède son *Analyse.*

D'autre part, vers le même temps où Poincaré se révélait, deux théories générales nouvelles sont venues modifier la marche de la science: la théorie des *groupes continus* de S. Lie et celle des *ensembles* de Cantor.

L'une et l'autre ne pouvaient manquer de recevoir de Poincaré d'importantes contributions.

La première lui doit une étude nouvelle de ses notions générales, qu'il éclaire[1] grâce à un remarquable emploi de l'intégrale de Cauchy, en montrant, en particulier, que les problèmes que Lie avait réussi à ramener à des équations différentielles peuvent se résoudre par quadratures, sinon par des opérations entièrement algébriques.

Mais la théorie des groupes continus vaut surtout par ses applications. On doit à Poincaré l'une des plus remarquables et des plus inattendues, celle qui est relative aux *quantités complexes* en général, c'est-à-dire aux diverses généralisations que, après Hamilton, Grassmann et d'autres, on peut essayer de donner à la théorie des imaginaires. Poincaré montre que ce problème se ramène entièrement à l'étude et à la discussion de certains groupes continus linéaires.

La théorie de Lie intervient d'ailleurs dans plusieurs autres travaux de Poincaré. Elle joue par exemple, un rôle essentiel dans les recherches dont nous parlerons plus loin sur la représentation conforme et les fonctions de deux variables et c'est par elle, ne l'oublions pas, qu'il guida la théorie naissante de la relativité.

Ailleurs, il montre que son emploi s'impose dans un sujet qui semblait épuisé, la mise en équations des problèmes de Mécanique rationnelle. Dans la méthode classique suivie à cet égard, les déplacements virtuels sur lesquels on opère sont obtenus en faisant varier isolément chaque paramètre; si au contraire, comme Poincaré est obligé de le faire en vue de certaines applications à la Mecanique céleste, ces déplacements virtuels sont choisis d'une manière quelconque, il montre qu'on doit les traiter comme des transformations infinitésimales et introduire la structure du groupe ainsi défini.

L'histoire des relations de Poincaré avec la théorie des ensembles est plus curieuse. Nous avons dit, en effet, qu'il la devança (Voir plus haut, page 211) en l'appliquant avant même qu'elle fût née, et cela dans un de ses résultats les plus saillants et les plus justement célèbres.

---

[1] Voir p. 93 de son *Analyse*.

Cette théorie s'est depuis constamment remontrée vers la plume. Qu'il s'agisse de théorie des fonctions, d'équations différentielles, on la verra toujours se présenter à lui, comme elle s'imposera désormais à tout géomètre qui, dans un domaine quelconque, tentera d'aller véritablement au fond des choses.

Les hauts problèmes qu'elle soulève en elle même ne pouvaient, eux non plus, laisser Poincaré indifférent. Il les a traités ici même[1] et repris souvent dans ses livres. Le terme de «définition non prédicative» qu'il a introduit, suffit à réfuter plusieurs des sophismes dont les notions fondamentales relatives aux ensembles étaient l'objet.[2]

§ 3. La théorie générale.

Avec Gauss, Cauchy, Riemann, Weierstrass, la notion précise de ce qu'on doit entendre par *fonction analytique* était acquise. La théorie en était faite, au fond, sur le modèle qu'offrait naturellement l'Algèbre. Toute fonction analytique peut être représentée, dans tout domaine suffisamment restreint, sauf au voisinage de certains points particuliers, par un développement en série entière.

Certaines d'entre elles peuvent être ainsi représentées, par un développement unique, pour toutes les valeurs de la ou les variables: ce sont les *fonctions entières*. Dans le cas d'une seule variable indépendante, Weierstrass avait réussi à étendre à ces fonctions le théorème de la décomposition en facteurs, sous une forme identique à celle des polynômes, à ceci près qu'aux facteurs binômes classiques de l'Algèbre venaient s'adjoindre, et cela à deux titres différents, des facteurs exponentiels.

Après ces fonctions entières viennent les fonctions *méromorphes*, analogues aux fonctions rationnelles et qui se comportent comme elles au voisinage d'un point quelconque (à distance finie). Grâce au théorème qui lui a donné la décomposition en facteurs, Weierstrass montre qu'une fonction méromorphe d'une seule variable est le quotient de deux fonctions entières, pendant que le théorème de M. Mittag-Leffler étend à ce domaine la décomposition en éléments simples.

Ces deux cas sont les plus élémentaires. D'autres beaucoup plus compliqués peuvent se présenter, même si l'on se borne aux fonctions uniformes. Mais celles-ci sont loin d'être la règle. La grande difficulté de la théorie est précisément l'existence des fonctions non uniformes, qui, en un certain sens, mettent en défaut la définition même de la notion de fonction.

---

[1] *Acta*, t. 32, 1909.

[2] Le seul raisonnement que nous défendrions contre les critiques de Poincaré à ce point de vue est celui de M. Zermelo sur la possibilité d'ordonner un ensemble quelconque.

De ces fonctions non uniformes, on n'avait qu'une notion purement négative, du moins dans le cas général. Quelques catégories particulières avaient seules été étudiées. A la plus classique d'entre elles, celle des fonctions algébriques, POINCARÉ avait, dès la Thèse dont nous avons parlé plus haut, adjoint sa généralisation la plus naturelle et la plus importante, celle des fonctions *algébroïdes*, que ses recherches de Mécanique analytique devaient ramener souvent sous sa plume.

Dès que le nombre des variables devenait supérieur à 1, il ne restait de tout cela que le point de départ: le développement en série entière, applicable à une fonction analytique quelconque dans le voisinage d'un point non singulier, et à une fonction entière dans tout l'espace. En particulier, la décomposition en facteurs de ces fonctions entières n'ayant plus lieu, la démonstration donnée par WEIERSTRASS de l'expression d'une fonction méromorphe par le quotient de deux fonctions entières disparaissait du même coup.

De l'outil qui permet de manier si sûrement les fonctions d'une variable, la théorie des fonctions de deux variables ne possédait que le manche.

Tel était l'état de cette branche de la science à la venue de POINCARÉ. Voyons comment, grâce à lui, l'évolution ultérieure fut possible.

Tout paraissait dit, en un sens, en ce qui regarde les *fonctions entières d'une variable*. Cependant, LAGUERRE avait montré, à l'aide de la formule de décomposition en facteurs, que, comme les polynômes, les fonctions entières ne devaient pas être placées toutes sur le même plan et présentaient des degrés de complication inégaux tout au moins sous ce point de vue. Avec une pénétration qui a été justement admirée, il avait appris à mesurer cette complication par un nombre, *le genre*, qui fait intervenir à la fois les deux espèces de facteurs exponentiels mentionnées plus haut.

Le problème se posa alors, pour POINCARÉ, de savoir si cette complication plus ou moins grande de la décomposition en facteurs de WEIERSTRASS avait ou non son retentissement sur les autres propriétés de la fonction. Il put montrer qu'en effet toute limitation supposée connue pour le genre en entraînait une correspondante pour l'ordre de grandeur du module de la fonction elle-même et aussi pour celui des coefficients de son développement, c'est-à-dire pour ses propriétés les plus simples et, en général, les plus aisément constatables.

Ainsi fut fondée la nouvelle théorie des fonctions entières. C'est, en effet, de ce résultat et aussi, ajoutons-le, d'un célèbre théorème dû à M. PICARD, qu'est sortie toute cette théorie, telle qu'elle s'est développée dans le cours de ces dernières années.

C'est d'ailleurs de la même source que découlent encore les progrès apportés, principalement par MM. Borel et Boutroux, à l'étude des fonctions méromorphes: car la méthode employée en cette circonstance dérive manifestement de celle qui est appliquée aux fonctions entières.

On peut même en dire autant pour le point essentiel: c'est en effet dans le même ordre d'idées que les fonctions présentant cette singularité, et elle seule, autrement dit les fonctions *quasi entières*, ont été traitées par M. Maillet.

Les *singularités non isolées* des fonctions uniformes sont un des sujets qui ont le plus particulièrement préoccupé Weierstrass et les géomètres qui, avec lui, ont exploré la théorie des fonctions, et la réalisation des diverses possibilités qui peuvent se présenter à cet égard a été l'un des buts principaux de leurs efforts. Or, parmi les dispositions les plus étranges qui peuvent se rencontrer, il n'en est pas une dont Poincaré n'ait formé, comme eux, des exemples, mais avec une signification nouvelle.

L'existence de coupures essentielles pour les fonctions dont il s'agit était connue depuis Weierstrass. Mais ce sont les fonctions fuchsiennes, — après la fonction modulaire, il est vrai — qui sont venues nons montrer combien il s'en fallait qu'on dût voir là un simple objet de curiosité.

En même temps, nous avons vu ces mêmes fonctions fuchsiennes imposer à Poincaré une nouvelle catégorie de singularités que l'imagination de ses prédécesseurs n'avait pu concevoir: les points singuliers formant un ensemble parfait non continu.

Reste enfin, à côté de la notion de ligne singulière, la notion toute voisine *d'espace lacunaire*. C'est à Poincaré que l'on doit, à cet égard, l'exemple peut être le plus général et en tout cas, le plus fécond, car la méthode qui y conduit, fondée sur l'introduction d'une série de fractions rationnelles, est celle qui, ultérieurement, a permis à M. Borel d'étendre nos connaissances sur ce sujet.

Mais ici encore, ce n'est pas uniquement pour elle même et pour mettre en évidence ses singularités que Poincaré forme la série dont il s'agit. Il y est amené nécessairement par les recherches sur les équations différentielles qui font l'objet de sa thèse. Les intégrales qu'il forme n'existent, comme nous le rappellerons plus loin, que moyennant des conditions d'inégalité convenables entre certains coefficients qui figurent dans l'équation et, dès lors, considérées comme fonctions d'un de ces coefficients, elles présentent précisément la singularité qui nous occupe.

Le rôle de POINCARÉ, à propos des fonctions à espaces lacunaires, a donc été le même que nous lui avions vu jouer vis-à-vis des lignes singulières, des ensembles parfaits discontinus, des courbes sans courbure.

Certes, même si l'une ou l'autre de ces circonstances avait été destinée à ne jamais se rencontrer dans les applications, leur découverte n'en aurait pas moins été importante pour nous. POINCARÉ nous a montré, dans un de ses discours,[1] combien il faut rendre grâces à l'Astronomie d'avoir élargi notre esprit par la seule notion de ses distances énormes et d'avoir permis ainsi que «notre imagination, comme l'oeil de l'aigle que le Soleil n'éblouit pas, puisse regarder la vérité face à face». Les singularités dont nous venons de parler tiennent une place analogue, à ceci près qu'elles ont soumis notre imagination à des épreuves autrement rudes encore, et jeté un désarroi passager, non seulement dans les habitudes que nous tenons de nos sens, mais dans celles que nous pouvions croire issues de notre logique elle même.

POINCARÉ n'a pas laissé à l'avenir le soin d'utiliser la leçon qui s'en dégageait. Au lieu de signaler de loin d'étranges régions que la science pouvait être exposée à rencontrer sur sa route, il les a traversées pour trouver, au delà, le but qu'elle avait à poursuivre. Ses découvertes semblent ainsi aller d'un coup aux limites, non seulement de ce que l'humanité d'aujourd'hui peut découvrir, mais de ce qu'elle peut comprendre.

C'est ce que la théorie des fonctions vient à nouveau de nous montrer. La même impression s'imposera à nous, et plus fortement même, lorsqu'il s'agira des équations différentielles. Plus complexe encore qu'en Théorie des fonctions, la vérité que nous verrons alors se dégager des travaux de POINCARÉ dépasse probablement la capacité actuelle de nos cerveaux.

La théorie des *fonctions non uniformes* fut tirée du néant grâce à un théorème d'une démonstration des plus délicates.

Une fonction analytique quelconque (par conséquent, non uniforme en général):

$$z = f(x)$$

étant donnée, on peut exprimer x en fonction *uniforme* d'une variable auxiliaire t, de manière que z soit aussi une fonction *uniforme* de t. La conclusion s'étend même à un nombre quelconque de fonctions d'une même variable.

La théorie des fonctions non uniformes est ainsi ramenée à celle des fonctions uniformes.

---

[1] Voir *La valeur de la science*, Chap. VI.

Un tel fait ne pouvait manquer de s'imposer à un POINCARÉ après la découverte des fonctions fuchsiennes. Celles-ci, nous l'avons vu, le mettaient en évidence, et fournissaient la variable auxiliaire cherchée, en ce qui regarde les fonctions algébriques. Il y a plus, elles permettent de le démontrer, sinon dans le cas général, du moins dans un cas très étendu, à savoir, toutes les fois que les points singuliers sont en nombre fini et tous réels.

Mais si l'on veut ne faire aucune restriction relativement aux points singuliers, d'autres moyens d'action sont nécessaires.

Ici (comme déjà d'ailleurs pour les fonctions fuchsiennes) ce sont les principes mêmes sur lesquels RIEMANN avait fondé la théorie des fonctions abéliennes qui s'élargissent entre les mains de POINCARÉ, et acquièrent l'ampleur nouvelle que la question comporte. D'une part, tout le calcul va reposer sur la formation d'un domaine géométrique, la surface de RIEMANN, par lequel on peut se représenter la variation simultanée de z et de x. En second lieu, un élément physico-mathématique, la théorie du potentiel, joue dans ce calcul le rôle principal. Mais son maniement exige une puissance d'analyse nouvelle, en raison de la complication de la surface de RIEMANN qui est ici à une infinité de feuillets. La notion d'une telle surface de RIEMANN, et surtout des fonctions harmoniques correspondantes est très délicate et ne peut être atteinte que par des passages à la limite appropriés.

On aboutit ainsi à la formation d'une certaine fonction analytique t. La propriété essentielle de cette quantité consiste en ce qu'elle ne prend jamais deux fois la même valeur sur la surface. C'est ce que l'on établit aisément à l'aide de l'intégrale classique de CAUCHY, étant donné que t est la limite de fonctions t_n qui possèdent la propriété en question.

Cette grandiose découverte de *l'uniformisation des fonctions analytiques* ne pouvait manquer de provoquer les recherches des géomètres, du moins de ceux qui, capables de ressentir son importance, avaient aussi les forces nécessaires pour aborder ce sujet.[1] A la suite de ces travaux, POINCARÉ revint lui-même sur sa découverte pour la compléter.

Dans l'intervalle, il avait donné, pour la résolution des problèmes fondamentaux de la théorie du potentiel, une méthode nouvelle, celle du *balayage*. Créée, semble-t-il, en dehors de la préoccupation du problème qui nous occupe, cette méthode se trouvait cependant s'y adapter d'une manière remarquablement parfaite. L'une des difficultés de la question est, nous l'avons dit, la présence d'une infinité de feuillets de la surface de RIEMANN, dont, par suite de cette

---

[1] Outre les auteurs dont nous parlerons dans un instant, nous nous contenterons de citer ici M. KOEBE.

circonstance, la forme *totale* et en particulier la frontière ne peuvent, au moins au premier abord, être définies sans de sérieuses difficultés. Or il se trouve que celles-ci ne gênent en aucune façon l'application de la méthode du balayage, pour laquelle il suffit de suivre la marche même, classique depuis WEIERSTRASS, de la définition d'une fonction analytique par une suite indéfinie d'«'éléments».

POINCARÉ avait, d'autre part, à l'occasion de son enseignement à la faculté des sciences de Paris, perfectionné toute la technique de la théorie des fonctions harmoniques: il avait, par exemple, reconnu tout le parti qu'elle peut tirer d'un remarquable théorème de HARNACK.

Fort de ces nouveaux moyens d'action, il put, répondant à un desideratum de M. HILBERT, écarter en toute certitude pour les fonctions cherchées, les trois points singuliers dont le raisonnement primitif n'excluait pas la possibilité.

D'autre part, les fonctions obtenues ont, en général, un domaine d'existence limité par une ligne sigulière essentielle (un théorème de M. PICARD est venu montrer que la solution n'est pas possible sans l'introduction de fonctions présentant ce caractère). MM. OSGOOD et BRODÉN s'étaient préoccupés de déterminer la forme exacte du domaine qui, dans le plan de la variable t, correspond ainsi à la surface de RIEMANN donnée. La nouvelle méthode de démonstration permet de préciser davantage les résultats de ces deux auteurs.

Enfin, chose plus précieuse encore, elle fournit la solution la plus simple, — et non plus une solution quelconque —, du problème posé. Dès lors, cette solution est parfaitement déterminée, du moins à une substitution linéaire près. De là ressort encore, comme conséquence, une propriété fonctionnelle qui place les fonctions obtenues à côté des fonctions fuchsiennes.

Ce n'est pas la seule contribution que POINCARÉ ait apportée à la théorie des fonctions non uniformes. Tout d'abord, c'est à lui qu'on doit la limitation, — au sens de la théorie des ensembles — de la multiplicité des valeurs que peut prendre une telle fonction pour une valeur unique de la variable et aussi des «éléments» (au sens de WEIERSTRASS) qui suffisent à la représenter: limitation essentielle d'ailleurs au second raisonnement par lequel il a établi le théorème d'uniformisation.

De plus, il a indiqué une méthode permettant d'établir que toute fonction analytique z — en général, non uniforme, — de la variable x peut être définie par une équation de la forme $G(z, x) = 0$, où G est une fonction entière: progrès moins essentiel peut-être que le théorème d'uniformisation, mais, néanmoins, extension importante aux fonctions transcendantes de la propriété fondamentale des fonctions algébriques.

Mais cette méthode est en relation avec les travaux dont nous avons à parler en second lieu, et qui concernent l'étude des fonctions de plusieurs variables.

* * *

Pour celle-ci plus encore que pour la précédente, on peut dire que les impulsions décisives viennent de POINCARÉ.

Dans cet ordre d'idées, un seul théorème, le «Vorbereitungssatz», avait été antérieurement obtenu. Mais, par deux fois, il était resté ignoré du public scientifique. WEIERSTRASS l'a réservé, comme il le faisait souvent, au cercle restreint de ses auditeurs, jusqu'en 1886, et M. LINDELÖF a été le premier à découvrir[1] que le véritable auteur en est CAUCHY.

Il peut n'être pas inutile, dans ces conditions, de noter que les résultats relatifs aux fonctions algébroïdes, obtenus par POINCARÉ dans sa Thèse, équivalent au théorème en question.[2]

Celui-ci d'ailleurs, pour POINCARÉ comme pour WEIERSTRASS, n'était que préparatoire. L'étude des fonctions de plusieurs variables ne fut véritablement inaugurée que lorsque, peu d'années après, POINCARÉ réussit à leur étendre le théorème de WEIERSTRASS sur les fonctions *méromorphes*.

Quel que soit le nombre des variables, une telle fonction est caractérisée par la propriété de se comporter au voisinage d'un point quelconque — autrement dit, *localement* — comme une fonction rationnelle. Localement donc, elle s'exprime par le quotient de deux séries entières convergentes dans un rayon suffisamment petit. C'est ce résultat qu'ils s'agit d'étendre à tout l'espace en exprimant la fonction considérée par le quotient de deux séries entières *toujours* convergentes.

Nous avons dit qu'on ne pouvait songer à employer, à cet effet, la méthode qui réussit dans le cas d'une variable. POINCARÉ recourt encore une fois à la théorie du potentiel ou plutôt à la théorie analogue dans l'espace à quatre dimensions, celle des fonctions V qui satisfont à l'équation

$$\frac{\partial^2 V}{\partial x^2} + \frac{\partial^2 V}{\partial y^2} + \frac{\partial^2 V}{\partial z^2} + \frac{\partial^2 V}{\partial t^2} = 0.$$

Cette méthode semble cependant, au premier abord, inapplicable au problème actuel. L'équation aux dérivées partielles précédente ne suffit plus, en effet, à caractériser la partie réelle d'une fonction analytique: cette partie réelle doit

---

[1] Voir ses *Leçons sur la théorie des résidus* (Paris, Gauthier-Villars, 1905, note de la page 27).

[2] Ils en fournissent même une extension, dans laquelle, au lieu de l'équation obtenue en égalant une fonction unique à zéro, on considère un système d'équations à plusieurs inconnues.

satisfaire à quatre équations aux dérivées partielles et non à une seule. Il semble donc que la formation d'un potentiel vérifiant l'unique équation ci-dessus écrite soit sans valeur au point de vue du résultat final.

Il n'en est rien: si l'on a obtenu ce potentiel, POINCARÉ montre qu'il suffit d'y ajouter une fonction régulière convenable pour intégrer l'ensemble des quatre équations mentionnées tout à l'heure et en déduire la fonction qu'il a en vue, à savoir le logarithme du dénominateur cherché.

Quant à la formation du potentiel en question, elle consiste en une sorte de raccordement analytique entre plusieurs fonctions (les logarithmes des dénominateurs des diverses fractions qui représentent localement la fonction donnée) définies chacune dans une portion seulement de l'espace, mais dont les différences mutuelle dans les régions où deux d'entre elles existent à la fois, sont régulières. Ce raccordement, l'emploi de potentiels, tout analogues aux potentiels de simples et de doubles couches, permet de l'opérer. Lorsqu'il s'agit enfin de passer à la limite pour le cas de l'espace indéfini en tous sens, la méthode à appliquer est connue: c'est celle par laquelle on démontre le théorème de M. MITTAG-LEFFLER sur le développement des fonctions méromorphes d'une variable en série d'éléments simples.

POINCARÉ a eu à revenir sur la démonstration de ce théorème, pour l'adapter à la théorie des fonctions abéliennes. Dans ce cas, en effet, la fonction donnée étant périodique, il importe de diriger le calcul de manière à ce que le numérateur et le dénominateur obtenus possèdent eux mêmes, non la périodicité proprement dite, mais, à la façon des fonction thêta (auxquelles, dès lors, ils se ramènent) — la périodicité de troisième espèce au sens d'HERMITE, qui est à la périodicité ordinaire ce que l'invariance relative est à l'invariance absolue. POINCARÉ reprend, à cet effet, la démonstration de son théorème général, tant celle qu'il avait donnée que celle qui avait été fournie depuis par M. COUSIN. Il en indique même, dans le même but, une autre, toute différente de celles dont il vient d'être question par la nature des potentiels employés. Au lieu d'être, dans l'hyperespace, les analogues d'un potentiel newtonien de surface, — comme dans la démonstration primitive — ceux ci peuvent, en effet, être analogues à des potentiels (newtoniens) de lignes attirantes, de sorte que nous devons à cette circonstance la connaissance des singularités (en général logarithmiques) des potentiels de cette espèce.

Un autre point important de la théorie des fonctions d'une variable attirait l'attention au point de vue de son extension au cas actuel: la notion de *résidu*, base des plus belles découvertes de CAUCHY. En général, c'est-à-dire dans toute région ne comprenant pas de points singuliers, l'intégrale d'une fonction analytique le long d'un contour fermé est nulle. Au contraire si ce contour contient à son intérieur un point singulier, l'intégrale est proportionnelle à un certain nombre déterminé, caractéristique et en quelque sorte, mesure de la singularité, qui est le résidu.

Cette pierre angulaire de la théorie de CAUCHY devait être transportée à la théorie des fonctions de deux variables, si l'on voulait fonder utilement celle-ci. Il fallait, à cet effet, considérer les intégrales doubles prises le long de multiplicités fermées de l'espace à quatre dimensions, et montrer tout d'abord que ces intégrales étaient indépendantes de la forme de la surface d'intégration (tant que celle-ci varie continûment sans rencontrer de singularités), une condition d'intégrabilité analogue à celle qui intervient pour les différentielles totales ordinaires étant vérifiée.

Mais ceci fait, le calcul de la valeur de cette intégrale autour d'une singularité donnée, présentait des difficultés inattendues. STIELTJES qui l'avait effectué dans un cas particulier, n'avait pu le publier, le résultat donnant lieu à une objection qui semblait sans réplique. Dans l'intégrale qu'il avait traitée, la quantité sous le signe $\iint$ était de la forme

$$\frac{P}{Q\,R},$$

$P,\ Q,\ R$ étant trois polynomes entiers dont les deux derniers s'annulent ensemble sur la singularité considérée. Or STIELTJES trouvait, pour le résidu, une valeur qui change de signe quand on permute entre eux les deux facteurs du dénominateur.

Pour faire cesser cette contradiction apparente, il fallait arriver à une vue exacte et pénétrante des propriétés géométriques d'une figure tracée dans l'hyperespace. POINCARÉ montra ainsi comment l'ordre des deux facteurs en question influe, dans cet exemple, sur le sens de l'intégration.

Ces deux séries de recherches de POINCARÉ restèrent longtemps la seule base des travaux entrepris sur les fonctions de deux variables.[1] Les plus importants, tels que celui de M. COUSIN, dérivent du théorème sur les fonctions méromorphes et fournissent une seconde démonstration de ce théorème.

---

[1] C'est seulement dans ces toutes dernières années que d'autres voies se sont ouvertes avec des travaux parmi lesquels nous nous contenterons de citer ceux de MM. FABER, HAHN, HARTOGS, E. E. LEVI, etc.

Ce vaste domaine des fonctions de plusieurs variables devait, plus tard, offrir encore à POINCARÉ un autre objet de méditations. La *représentation conforme* offre, dès le cas d'une variable, un remarquable exemple de la différence profonde qui existe entre les propriétés *locales* des fonctions et celles qui interviennent lorsqu'on les considère non plus au voisinage immédiat d'un point, mais dans tout leur domaine d'existence.

Le problème (problème *local*) qui consiste à représenter, par l'intermédiaire d'une fonction analytique, un *arc* (suffisamment petit) d'une courbe donnée *c* sur un arc d'une autre courbe donnée C a, en effet, une infinité de solutions dépendant d'une infinité d'arbitraires, tandis que le problème *étendu* qui consiste à représenter, dans les mêmes conditions, la courbe fermée *c tout entière*, sur la courbe fermée C (et l'aire *s* limitée par *c* sur l'aire *S* limitée par C) est, au contraire, déterminé à une substitution homographique près.

A cette différence, on aperçoit immédiatement deux raisons: la première résidant dans ce fait que les courbes *c* et C sont fermées et que dès lors le prolongement de la fonction cherchée tout le long de ces courbes doit présenter par rapport à l'arc de l'une d'elles, par exemple, une périodicité qui n'apparaissait point lorsqu'on se bornait à considérer des parties très petites des courbes en question; la seconde, dans celui que la fonction cherchée ne doit seulement être définie au voisinage de *c*, mais dans tout l'intérieur de *s*.

C'est cette étude que POINCARÉ transporte au cas de deux variables, en séparant même, par l'introduction d'un problème intermédiaire,[1] (dans lequel on demande que les fonctions cherchées soient régulières sur toute la frontière, mais non dans tout le domaine qu'elle limite) ces deux caractères qui différencient l'un de l'autre le problème local et le problème étendu. Les résultats changent d'ailleurs notablement de forme dans cette extension. Le problème local cesse lui même d'être possible en général. Une infinité de conditions de possibilité apparaissent et ces conditions de possibilité introduisent une série d'invariants différentiels, obtenus en écrivant que la transformation de l'une des frontières données en l'autre est possible, dans la région infiniment voisine d'un point donné, aux infiniment petits du $n^{\text{ème}}$ ordre près.

Ce Mémoire de POINCARÉ ouvre d'ailleurs la voie à toute une série de recherches où, comme il l'a montré, intervient d'une manière nécessaire l'étude approfondie de certains groupes continus.

*   *   *

---

[1] En vertu d'un théorème de M. HARTOGS, il se trouve que la solution de ce problème intermédiaire peut être utilisée pour celle du problème étendu.

C'est à ces propositions fondamentales sur les fonctions de plusieurs variables qu'il faut rattacher les résultats obtenus par POINCARÉ sur les fonctions abéliennes, ceux qui dérivent de l'application des fonctions fuchsiennes exceptés. Leur point de départ est la distinction qu'il établit entre la théorie des *fonctions* abéliennes et celle des *intégrales* abéliennes, théories que, depuis RIEMANN, on était habitué à confondre l'une avec l'autre.

Si, comme on le sait depuis RIEMANN, les intégrales des fonctions algébriques s'expriment par le moyen des séries Θ, la solution ainsi obtenue dépasse en quelque sorte le but. Certaines fonctions Θ correspondent à des problèmes de quadrature de l'espèce indiquée, mais elles sont *spéciales;* il en existe une foule d'autres qui n'ont point une origine de cette espèce.

Quelles sont les relations qui caractérisent ainsi les fonctions thêta spéciales? — et, d'autre part, que peut on dire des autres fonctions thêta?

Mais auparavant, une autre question analogue se présentait, qui s'était déjà posée à RIEMANN même (lequel l'avait signalée à HERMITE), puis à WEIERSTRASS et qui, en même temps que POINCARÉ, préoccupa MM. PICARD et APPELL,[1] celle de savoir si les fonctions périodiques obtenues comme quotients de fonctions thêta sont les plus générales parmi celles qui présentent le même nombre de périodes.

Cela est d'autant moins évident que les fonctions thêta ne peuvent pas être formées avec des périodes entièrement arbitraires. Au contraire il ne semble nullement a priori, que la définition des fonctions périodiques doive impliquer, entre ces périodes, une condition quelconque. C'est cependant ce qui a lieu et toute fonction méromorphe $2\,p$ fois périodique de p variables complexes peut être représentée par un quotient de séries thêta. POINCARÉ publia sur ce point, en collaboration avec M. PICARD, une démonstration qui, un peu plus tard fut reconnue identique à celle qu'avait obtenue WEIERSTRASS. Nous avons dit que le même fait se présenta plus tard à lui comme une simple conséquence du théorème fondamental sur les fonctions méromorphes, moyennant une étude plus approfondie des opérations par lesquelles on établit ce théorème.

Ceci élucidé, il fallait entreprendre l'examen des fonctions thêta indépendamment de toute origine algébrique, pour apprendre à distinguer entre les fonctions thêta appelées plus haut *spéciales* (c'est-à-dire celles qui ont une telle origine) et les autres.

Dans cet ordre d'idées, POINCARÉ, dès 1883, considère une système de p fonctions thêta à p variables, toutes aux mêmes multiplicateurs, et détermine le

---

[1] Voir *Analyse*, p. 82.

nombre des solutions (essentiellement distinctes, c'est-à-dire telles que la différence de deux quelconques d'entre elles ne soit pas une période) communes aux équations obtenues en égalant ces fonctions simultanément à zéro.

C'est à cette occasion que POINCARÉ utilise, pour la première fois, le théorème par lequel KRONECKER venait d'exprimer le nombre des solutions d'un système donné, admirable instrument qui semblait avoir été crée en vue d'un tel ouvrier, et que nous retrouverons à tant de reprises dans l'étude de son œuvre. Grâce à lui, il put montrer que le nombre en question ne dépend que du nombre p des variables et du degré des fonctions thêta.

La différence entre le point de vue de POINCARÉ et celui de ses prédécesseurs apparaît par la comparaison entre cette question et celle que s'était posée RIE-MANN relativement au nombre des zéros d'une fonction thêta du 1er degré. Si, dans une telle fonction, on substitue aux p variables les valeurs des p intégrales abéliennes de première espèce attachées à un même point M de la courbe, on a une équation (à une inconnue, cette fois) qui admet p solutions.

Cet énoncé diffère, on le voit, du précédent, non pas seulement en ce que la fonction thêta considérée doit être spéciale, mais en ce que les variables ne peuvent prendre que des valeurs très particulières, ne dépendant que d'un seul paramètre et non de p. POINCARÉ étend d'ailleurs le résultat de RIEMANN aux fonctions thêta de degré quelconque (le nombre des solutions devenant alors égal à np) et à une série de questions qu'on peut considérer comme intermédiaires entre les deux précédentes.

La relation entre ces différents points de vue est également mise en lumière dans la représentation géométrique qu'il emploie.

Il y a n^p fonctions thêta de degré n ayant de multiplicateurs donnés. Si l'on considère les valeurs de ces n^p fonctions thêta comme des coordonnées homogènes dans l'espace à $n^p - 1$ dimensions, le point qui a ces coordonnées — point qui reste inaltéré par l'addition aux variables d'une période quelconque, puisque toutes ses coordonnées sont multipliées par la même quantité, — décrit, dans cet espace, une variété p fois étendue V. Lorsque les fonctions Θ sont spéciales et dérivent d'une courbe algébrique C, si on remplace les variables par les intégrales abéliennes attachées à cette courbe, on a une courbe B située sur V. Le théorème de RIEMANN étendu par POINCARÉ montre que cette courbe est algébrique et fait connaitre son degré. POINCARÉ constate d'ailleurs qu'elle est plane, ou, plus exactement, qu'elle est située sur une variété plane à $(n-1)p$ dimensions. Quant au théorème mentionné plus haut sur les zéros communs à p fonctions Θ, il fait connaitre le degré de la variété V, laquelle est algébrique, et cela, cette fois, même si les fonctions Θ considérées ne sont pas spéciales.

Un seul résultat appartenant à cette catégorie porte, comme celui de Riemann, sur une seule équation à une seule inconnue, tout en n'exigeant pas que la fonction thêta qui y intervient soit spéciale. Il généralise la relation de Legendre entre les périodes des intégrales elliptiques de première et de deuxième espèce, relation que la théorie des fonctions nous a montrée dépendant du nombre des zéros de la fonction thêta dans un parallélogramme des périodes. Pour arriver à un résultat présentant ces caractères, il fallait vaincre une difficulté de nature géométrique par laquelle cette recherche se rapproche de celles que Poincaré avait développées sur la théorie générale des fonctions de plusieurs variables.

Après avoir, par son théorème de 1883, étendu une proposition classique sur le nombre des zéros d'une fonction elliptique, Poincaré va plus loin et donne une extension analogue à celle qui fait connaître la somme de ces zéros.

Dès le premier de ces deux théorèmes, on voit intervenir occasionnellement comme auxiliaires les cas de réduction dont nous avons parlé précédemment. Cette première intervention n'est toutefois qu'accessoire, pour ainsi dire: l'emploi du théorème de Kronecker ayant montré, comme nous l'avons dit, que le nombre des zéros communs est constant, l'examen d'un cas de réduction où tout se ramène aux fonctions elliptiques fournit simplement la valeur de cette constante.

Mais c'est seulement avec le théorème sur la somme des zéros que ces fonctions abéliennes réductibles jouent un rôle essentiel et, ainsi que les trajectoires périodiques de la Mécanique céleste auxquelles on pourrait à la rigueur les comparer, sont pour nous, toutes particulières qu'elles soient, le moyen d'atteindre toutes les autres fonctions abéliennes.

Poincaré remarque, en effet, qu'on peut trouver, d'une infinité de manières, des fonctions abéliennes réductibles aussi peu différentes qu'on le veut d'une fonctions abélienne quelconque donnée, de même qu'au voisinage d'une incommensurable donnée, on peut trouver une infinité de nombres commensurables. Il suffit dès lors de résoudre le problème pour les fonctions réductibles, la solution s'étendant immédiatement, par voie de continuité aux fonctions abéliennes quelconques.

Un nouvel emploi des cas de réduction va également permettre d'aborder le problème principal dont nous avons donné tout à l'heure l'énoncé: la recherche des conditions moyennnant lesquelles les fonctions thêta sont spéciales.

Ainsi est exploré tout d'abord l'aspect géométrique du problème. Comme on pouvait au fond, l'inférer des recherches de Lie, et comme Poincaré le démontre d'une façon nouvelle et particulièrement intuitive, la condition néces siare et suffisanté pour qu'un système de fonctions thêta soit spécial est que la variété S dont l'equation s'obtient en égalant l'une d'entre elles à zéro soit de translation et cela de deux manières différentes. L'étude de la courbe dont la translation produit ainsi la variété S n'est autre que celle de la courbe spéciale B dont il a été question plus haut (voir p. 232). On est ainsi conduit à définir cette courbe en adjoignant à l'équation de S un système d'autres équations analogues. Mais, quoique les variétés représentées par ces équations soient en nombre suffisant pour définir par leur intersection une courbe, elles ne fournissent pas uniquement celle que l'on cherche: l'intersection se décompose, et la courbe cherchée n'est qu'une des composantes.

Pour rendre possible l'intelligence complète du mécanisme de cette décomposition, POINCARÉ est obligé de faire intervenir à nouveau les cas de réduction. La marche suivie en cette circonstance est celle même qui est classique en Calcul infinitésimal: l'étude (au moins l'étude approchée) des cas infiniment voisins d'un premier cas donné dans lequel la solution est connue ou peut être obtenue. Ce cas initial est ici un cas de réduction. Toutefois l'emploi de la méthode est ici particulièrement difficile. Si, en effet, la discussion d'un cas singulier tel que le cas de réduction est ici la seule prise que nous ayons sur le cas général, les mêmes raisons qui nous la rendent accessible font — et nous retrouverons ce fait à propos des équations différentielles — qu'elle nous offre de ce cas général une image plus ou moins fortement déformée, où toutes les propriétés ont en quelque sorte dégénéré. Aussi ne faut il point s'étonner de ne la voir élucidée que par une dissection d'une finesse extrême et d'y trouver les interprétations aussi délicates que l'est pour le naturaliste celle d'organes dont les formes atrophiées ou régressives sont seules accessibles à l'observation.

Mais la condition qui caractérise une fonction abélienne spéciale doit s'exprimer, en dernière analyse, par une relation entre les périodes. C'est la partie la plus difficile du problème, celle pour laquelle POINCARÉ ne peut fournir qu'un commencement de solution. La méthode précédente donne cependant, sinon la forme complète des premiers membres des relations cherchées entre les périodes, du moins les premiers termes de leurs développements.

$$* \qquad * \qquad *$$

Peut-être convient-il de s'arrêter un instant pour jeter sur ce qui précède un coup d'œil chronologique. La théorie des fonctions fuchsiennes aurait à elle seule suffi pour fonder la gloire de POINCARÉ. Mais si elle fut d'abord la plus remarquée, d'autres, parmi les découvertes qui remontent à la même époque, ne lui cèdent nullement en importance et on ne peut enregistrer sans stupéfaction la rapidité avec laquelle elles se succédèrent à partir de 1879, date de la Thèse de doctorat de POINCARÉ. Parmi celles qui apparurent depuis cette date jusqu'en 1883, nous avons déjà signalé:

les fonctions fuchsiennes;

le théorème fondamental sur le genre, duquel découle toute la théorie des fonctions entières;

l'uniformisation des fonctions analytiques;

la représentation des fonctions méromorphes de deux variables par quotients de fonctions entières;

le théorème sur les zéros des fonctions thêta qui devait donner naissance à la nouvelle théorie des fonctions abéliennes;

l'extension des notions de genre et d'ordre aux formes de degré supérieur, et la notion d'invariants arithmétiques.

Nous avons essayé de donner une idée de l'importance fondamentale de ces différentes découvertes. Mais la plus essentielle peut-être nous reste à mentionner. Nous savons, en effet, que la théorie des fonctions, si grande que soit la place prise par elle dans les mathématiques contemporaines, n'est en somme qu'un moyen. On trouvera naturel, dès lors, que la *théorie des courbes définies par les équations différentielles*, dont nous aurons à parler tout à l'heure, ait eu sur toute l'œuvre de POINCARÉ et sur toute la marche de la science une influence plus décisive encore que les recherches même dont il a été question jusqu'ici. Or, dans ses deux premières parties, elle remonte à la même époque; et de cette période encore date une courte Note, grosse de toute une révolution dans nos conceptions astronomiques.

En quatre années, dans les domaines les plus divers, dans les directions les plus opposées, quelle armée de découvertes primordiales dont chacune aurait suffi à consacrer une réputation. Encore n'avons-nous cité que celles — et non peut-être toutes — qui marquent comme un tournant pour une branche de la science.

Il n'est pas vrai que le temps ne fasse rien à l'affaire, dans la vie d'un grand savant. N'oublions pas que celle de POINCARÉ, sans avoir la tragique brièveté de la carrière d'un GALOIS ou d'un ABEL devait être arrêtée en pleine fécondité.

L'accumulation de ces œuvres mémorables — un seul tome du *Bulletin de la Société mathématique de France* renferme trois de celles que nous venons de citer — n'en est d'ailleurs pas la seule caractéristique. Le dieu qui les inspirait manifeste son impatience dans leur style même. Dans nombre d'entre elles, —

particulièrement dans ces trois articles du *Bulletin de la Société mathématique de France* — deux ou trois pages lumineuses autant que concises, suffisent au «veni, vidi, vici» d'un triomphe de l'esprit humain.

II. Les équations différentielles.

1. Les voies classiques.

Le centre de la mathématique moderne est, nous l'avons dit, dans la théorie des équations différentielles et aux dérivées partielles.

Il nous faut maintenant montrer POINCARÉ aux prises avec ce double problème et tout d'abord, avec les équations différentielles.

La place n'est point de celles que l'on puisse emporter de haute lutte; il faut l'attaquer successivement sur toute sorte de points et se contenter d'avantages partiels. Essayons d'énumérer les directions à suivre.

I. On peut se préoccuper de perfectionner (spécialement autour des points singuliers) l'étude que nous avons appelée *locale* des solutions.

II. Il faut, d'autre part, savoir découvrir les cas où celles-ci s'expriment à l'aide de fonctions connues. C'était à eux que l'on réduisait le problème aux débuts du calcul infinitésimal. Tout déchus qu'ils soient de cette ancienne impor tance, il importe de ne pas les laisser échapper lorsque, exceptionnellement, ils existent.

III. A défaut des fonctions déjà existantes, il peut arriver que certaines transcendantes nouvelles, douées de propriétés qui en permettent l'étude et le calcul, gouvernent, d'autre part, une catégorie étendue d'équations différentielles dont elles permettent d'exprimer les intégrales.

IV. On peut étudier les solutions, supposées analytiques, au point de vue de la Théorie des fonctions et chercher les cas où elles se comportent à ce point de vue d'une manière remarquable.

V. On peut essayer de substituer, dans le cas général, aux développements en séries qui conviennent localement, des développements de forme différente valables pour toutes les valeurs de la variable, etc.

POINCARÉ suivit avec succès toutes ces voies, en même temps que nous le verrons en frayer d'autres sinon entièrement nouvelles, du moins presque inexplorées avant lui, et plus fécondes que les premières.

Sa Thèse marque surtout un progrès essentiel au premier point de vue qui, nous l'avons dit, dominait depuis CAUCHY, celui de l'étude locale des solutions.

Elle n'est, en un sens, qu'une généralisation des recherches de BRIOT et BOUQUET sur les points singuliers en lesquels la valeur de $\dfrac{d\,y}{d\,x}$ se présente sous la forme $\dfrac{o}{o}$: généralisation à un système d'équations du premier ordre, [1] au lieu que BRIOT et BOUQUET n'avaient traité qu'une équation unique. Mais ici cette généralisation fait apparaître des résultats de forme toute nouvelle. Dans l'exemple de BRIOT et BOUQUET, un seul coefficient influait sur la forme des résultats, et la discussion ne reposait que sur le signe de ce coefficient. Celle de POINCARÉ introduit au contraire plusieurs nombres (dépendant, comme le coefficient unique de BRIOT et BOUQUET, des termes du premier degré de l'équation) et les conditions d'inégalité que l'on doit former ne s'expriment aisément que sous forme géométrique, en circonscrivant un polygone convexe au système des points qui ont pour affixes les nombres en question. Le résultat obtenu entraîne dès lors que, considérées comme fonctions de l'un d'eux, les solutions présentent un espace lacunaire (en l'espèce, un polygone rectiligne); c'est l'exemple dont il a déjà été parlé plus haut et qui, comme on le voit, ne pouvait être soupçonné tant qu'on s'en tenait au cas de BRIOT et BOUQUET.

Une autre question qui, bien qu'appartenant à cette première catégorie des études locales, soulève de sérieuses difficultés, non encore complètement surmontées, est le calcul des intégrales irrégulières des équations linéaires, les seules que la méthode de FUCHS ne permette pas d'obtenir. Deux sortes de développements, très semblables au premier abord, complètement différents en réalité, peuvent être proposés pour représenter les solutions: les uns sont convergents, mais on ne sait pas en trouver les termes;[2] les autres peuvent être formés effectivement à l'aide des données de la question, mais ils sont divergents en général.

POINCARÉ, utilisant une transformation classique dûe à LAPLACE, montre, comme il le fera bientôt en Mécanique céleste, que ces développements divergents ont une signification: ils font connaître, jusqu'à tel ordre de petitesse qu'on le veut, l'allure de la fonction. De plus, il obtient par la même voie une condition nécessaire et suffisante pour qu'il y ait convergence.

Sur un point, — la recherche de la limite vers laquelle tend la dérivée logarithmique de la solution — la méthode employée se rapproche beaucoup de celles

---

[1] POINCARÉ considère plus spécialement, dans ce travail, l'équation aux dérivées partielles du premier ordre équivalente au système.

[2] On connait aujourd'hui, théoriquement parlant, grâce aux Mémoires de M. HELGE VON KOCH, un moyen de combler cette lacune en calculant les termes dont il s'agit: nous dirons plus loin comment ce résultat dérive des travaux de POINCARÉ lui même.

que nous retrouverons plus loin à propos de l'étude, non plus locale, mais générale du problème des équations différentielles; et dans le fait que la question dont nous parlons en ce moment n'est «locale» qu'en apparence réside sans doute la véritable raison des grandes difficultés de cette question qui mériterait encore tant de nouvelles recherches.

La connaissance des cas où l'intégration se fait par les fonctions classiques a également été notablement étendue par POINCARÉ. Il en a été ainsi en particulier en ce qui concerne l'intégration des équation linéaires par les fonctions abéliennes. Mais surtout, il s'est attaqué à la question si simple d'énoncé, si difficile en réalité, qui se posait après les recherches de M. DARBOUX et qui consiste à reconnaître si l'intégrale générale est algébrique. Il a pu, dans plusieurs catégories de cas nouveaux, obtenir le résultat essentiel, la limitation du degré. Ici encore, une partie de ses résultats est due à l'intervention des fonctions fuchsiennes.

Si grandes que soient les difficultés de cette question, on ne doit aujourd'hui, nous l'avons dit, voir là que le petit côté du calcul intégral. Au lieu de rechercher — non sans peine, nous venons de le dire, — si un extraordinaire hasard ne nous a pas mis en face d'une équation intégrable élémentairement, il est autrement important de disposer des transcendantes nécessaires pour intégrer les équations différentielles telles qu'elles se présentent en fait.

A ce point de vue, nul géomètre n'a remporté de victoire plus glorieuse que l'inventeur des fonctions fuchsiennes, qui permettent d'atteindre toutes les équations différentielles linéaires à coefficients algébriques.

L'étude des solutions analytiques au point de vue de la théorie générale des fonctions doit à POINCARÉ un travail qui a joué dans les recherches contemporaines un rôle primordial, quoique la conclusion en ait été essentiellement négative. L'hypothèse la plus simple que l'on puisse imaginer en ce qui regarde la disposition, inconnue en général, des points singuliers des intégrales d'une équation différentielle, celle des équations a points critiques fixes, avait été pour la première fois considérée par FUCHS. Ce savant était parvenu à écrire un système de conditions moyennant lesquelles les points critiques sont les mêmes pour toutes les solutions d'une même équation du premier ordre. Mais il n'y avait là que l'amorce d'une réponse à la question ainsi posée; il restait à savoir quelles étaient les équations différentielles remplissant ces conditions et si, par leurs intégrales, on pouvait être conduit à des transcendantes nouvelles. POINCARÉ, pour qui les équations de cette nature se présentaient nécessairement comme généralisation naturelle des équations linéaires qu'il venait d'intégrer, montra que toutes se ramènent à des cas déjà étudiés.

Ceci semblait uniquement terminer, sans laisser apercevoir d'issue nouvelle, les recherches de Fuchs.

Il n'en était rien: ce Mémoire, et particulièrement la méthode employée par Poincaré, — méthode sur laquelle nous reviendrons un instant plus loin — devaient servir de base à toute la théorie analytique des équations différentielles que l'on doit à M. Painlevé.

Enfin, dans le cas général, il importe tout d'abord, nous l'avons dit, de former des développements valables pour toutes les valeurs (au moins réelles) de la variable. Aucun résultat de cet ordre n'avait pu être atteint, et on voit quelle transformation essentielle un tel résultat devait opérer dans la question, puisque, jusque là, c'était uniquement à propos d'équations très particulières qu'on avait pu aboutir à autre chose qu'à une étude locale.

Il s'agissait donc déjà de faire faire un pas à la théorie dans une voie toute nouvelle.

Poincaré lui fit franchir ce pas important; il montra qu'il suffit à cet effet d'opérer sur la variable indépendante un changement convenable, après quoi le développement de Taylor lui même répond à la question.

Appliquée au problème des trois corps, cette méthode permet d'obtenir des développements valables pour toutes les valeurs du temps, sauf dans un seul cas d'exception, celui où, au cours du mouvement, deux corps viennent à se choquer.

C'est cette dernière lacune, — laquelle restait assez importante, car on ne sait pas, a priori, avec des circonstances initiales données, si le choc en question peut se produire, et encore moins quand il se produira — que les récents travaux de M. Sundman sont venus combler. L'idée première de sa belle analyse — à savoir, un prolongement analytique de la solution au delà de l'instant du choc, — a elle même, ajoutons le, ses racines dans *les Méthodes nouvelles de la Mécanique céleste*.

Mais Poincaré n'a entendu donner cette application au problème des trois corps qu'à titre d'exemple. Si utiles que puissent être les développements dont nous venons de parler, il ne les considère nullement comme résolvant le problème général. Tout en élucidant celui ci sous les différents aspects qui précèdent, il va montrer, en effet, qu'on en avait oublié d'autres plus difficiles encore, mais assurément non moins importants.

2. La théorie qualitative.

Le point de vue nouveau que nous allons voir apparaître est, en réalité, commun à toute sorte de questions mathématiques.

Dans les cas élémentaires, l'expression des inconnues par les symboles usuels fournit, en général, aisément à leur égard tous les renseignements que l'on se propose d'obtenir.

C'est ce qui a lieu pour tous les problèmes mathématiques suffisamment simples. Pour peu que la question se complique, il en est autrement. Dans la lecture, si j'ose m'exprimer ainsi, faite par le mathématicien des documents qu'il possède, POINCARÉ met en évidence deux grandes étapes, l'une que l'on peut appeler qualitative, l'autre quantitative.

Ici nous citerons les réflexions même qu'il enveloppe à cet égard: «...Pour »étudier une équation algébrique, on commence par rechercher, à l'aide du théorème »de STURM, quel est le nombre des racines réelles: c'est la partie qualitative; puis »on calcule la valeur numérique de ces racines, ce qui constitue l'étude quantita- »tive de l'équation. De même, pour étudier une courbe algébrique, on commence »par *construire* cette courbe, comme on dit dans les cours de Mathématiques spé- »ciales, c'est-à-dire qu'on cherche qu'elles sont les branches de courbes fermées, »les branches infinies, etc. Après cette étude qualitative de la courbe, on peut »en déterminer exactement un certain nombre de points».

»C'est naturellement par la partie qualitative qu'on doit aborder la théorie »de toute fonction et c'est pourquoi le problème qui se présente en premier lieu »est le suivant: *Construire les courbes définies par des équations différentielles.*

»Cette étude qualitative, quand elle sera faite complètement, sera de la plus »grande utilité pour le calcul numérique de la fonction.

. . . »D'ailleurs cette étude qualitative aura par elle-même un intérêt de »premier ordre. Diverses questions fort importantes d'analyse et de mécanique »peuvent en effet s'y ramener».

La plus importante d'entre elles est bien connue, et son exemple se présente de lui même à tout esprit que préoccupent les progrès de l'Astronomie: c'est la *stabilité du système solaire.* Le fait seul que cette question soit essentiellement qualitative suffit à montrer la nécessité du point de vue dont il s'agit.

Ainsi l'étude qualitative de la variation d'une grandeur ou du déplacement d'un point est indispensable à la fois en elle-même et comme précédant presque nécessairement l'étude quantitative.

Cependant ce point de vue avait été presque complètement délaissé et comme ignoré par les prédécesseurs de POINCARÉ. Quelques remarquables exceptions sont à citer: la démonstration du théorème de LAGRANGE sur la stabilité de l'équilibre par DIRICHLET; les travaux de STURM; ceux de LIOUVILLE. Mais même ceux d'entre eux qui avaient frappé les géomètres, — ce n'est pas le cas pour tous, nous le verrons plus loin — étaient resté isolés; l'exemple significatif qu'ils donnaient n'avait pas été suivi.

La faute en est, pour une part, au grand développement de la théorie des fonctions analytiques, aux services mêmes qu'elle avait rendus, et qui détournaient complètement les esprits du domaine réel.

En abandonnant cet auxiliaire, POINCARÉ eut à rompre avec une tradition vieille d'un quart de siècle et à laquelle l'Analyse devait tous ses progrès durant cette période.

D'autre part, la Science se trouvait du coup complètement désarmée en face des hautes difficultés des questions ainsi posées, les premières pour lesquelles cette théorie des fonctions analytiques n'apportait aucune solution.

Comment ces difficultés — ou plutôt certaines d'entre elles, car il reste beaucoup à explorer dans cet immense domaine qui n'était hier encore que mystère pour nous — furent-elles surmontées par POINCARÉ?

Ici se retrouve une circonstance qui était déjà apparue dans d'autres chapitres de l'histoire des mathématiques.

C'est ainsi que, dans la résolution algébrique des équations, il y eut une première période où l'on porta son attention sur la recherche d'une racine déterminée de l'équation proposée. Mais cette théorie ne passa d'un état en quelque sorte empirique à l'état de perfection logique où l'amenèrent LAGRANGE, RUFINI, ABEL, CAUCHY, GALOIS que lorsque l'on se décida, au contraire, à envisager simultanément toutes les racines cherchées. C'est en examinant les relations qui existent entre elles que furent conquis les principes modernes par lesquelles, dans cette question, tout s'éclaire, s'explique et se prévoit.

Dans les premières recherches sur les équations différentielles et exception faite précisément pour certains des travaux que nous citions il y a un instant, on avait généralement étudié une à une les intégrales d'une équation différentielle donnée quelconque: en examinant chacune d'elles, on avait fait abstraction de toutes les autres.

Les mémoires *sur les courbes définies par les équations différentielles* vinrent montrer que ce point de vue était insuffisant et que les solutions d'un système d'équations différentielles, comme les racines d'une équation algébrique, devaient, même en vue de l'intelligence de chacune d'elles, être envisagées dans leurs rapports mutuels.

Ceci fait comprendre tout d'abord l'importance que prend, dans l'œuvre de POINCARÉ, le théorème démontré dans le Mémoire de 1889 *sur le problème des trois corps* et dans *les Méthodes nouvelles de la Mécanique céleste*, relativement à la possibilité de développer les solutions d'un système différentiel suivant les puissances des paramètres qu'il renferme ou qui interviennent dans les données initiales.

L'un des Mémoires mentionnés précédemment relève déjà du principe dont nous parlons, de la considération simultanée de toutes les intégrales d'une même équation différentielle: c'est celui qui traite des équations du premier ordre à points critiques fixes. Si, dans cette question, POINCARÉ put dégager le résultat décisif qui vidait le débat et qui avait échappé à FUCHS, c'est en considérant les valeurs de l'inconnue y comme fonctions, non plus de la variable indépendante x qui figure avec elle dans l'équation différentielle, mais bien de leurs déterminations initiales y_0 pour une valeur fixe x_0 donnée à cette variable. La solution du problème est précisément due à ce que, entre y et y_0, existe une correspondance birationnelle.

Nous verrons plus loin une autre série de découvertes de POINCARÉ partir du même principe, je veux parler des recherches relatives à la figure d'équilibre du fluide en rotation. Tous les progrès qu'il réalise sur cette question sont dus à ce qu'il n'envisage pas une des figures d'équilibre cherchées en elle-même, mais bien dans ses relations avec les figures d'équilibre voisines.

POINCARÉ procède dans le même esprit, pour l'étude des équations différentielles réelles, dès le premier cas auquel il s'attaque. Ce cas est le plus simple de tous, celui d'une équation unique du premier ordre et du premier degré, donnant $\dfrac{dy}{dx}$ en fonction rationnelle de x et de y.

Quelles données possède-t-on sur les relations qui existent entre les différentes courbes intégrales de la même équation? Une seule apparaît au premier abord: le fait que deux quelconques de ces courbes, si elles ne coïncident pas, ne peuvent se couper, sauf en certains point singuliers.

Ceci à défaut de toute autre considération, montrait la nécessité de discuter à part les points dont il s'agit. C'est encore une question locale, qui, en un sens, n'est pas nouvelle (c'est elle que BRIOT et BOUQUET avaient traitée dans le cas des équations différentielles à coefficients analytiques) mais qu'il fallait reprendre, avec quelque difficultés nouvelles, du moment que la distinction entre le réel et l'imaginaire s'imposait.

Dès cette première étude, on aperçoit combien le nouveau point de vue est nécessaire et combien vaines étaient les anciennes recherches, celles qui avaient en vue l'intégration formelle.

Les points singuliers qu'elle fait apparaître sont, en effet, de quatre espèces:

1° les *noeuds*, où viennent se croiser une infinité de courbes définies par l'équation différentielle;

2° les *cols*, autour desquels les courbes cherchées ont une disposition analogue à celle des hyperboles $xy = $ const.;

3° les *foyers*, autour desquels ces courbes tournent en s'en rapprochant sans cesse à la façon d'une spirale logarithmique;

4° les *centres*, autour desquels ces courbes sont fermées et s'enveloppent mutuellement en enveloppant le centre à la façon d'ellipses homothétiques et concentriques.

Parmi toutes ces dispositions, quelles sont celles que l'on peut rencontrer lorsqu'on peut écrire l'intégrale générale de l'équation?

Il suffit, pour s'en rendre compte, de considérer l'exemple le plus familier que l'on puisse prendre à cet égard, celui des *lignes de niveau* sur une surface topographique quelconque. Il est clair que de telles lignes peuvent être considérées comme définies par une équation différentielle du premier ordre, dont l'intégrale générale est connue et s'obtient en égalant l'altitude à une constante arbitraire.

Quant aux points singuliers de cette équation, ils ne peuvent être ici que de deux espèces:

des *cols*, à savoir les points mêmes que la topographie désigne sous ce nom;

des *centres*, à savoir les fonds et les sommets du terrain.

Non seulement ces deux sortes de points singuliers sont les seules qui se présentent dans le problème des lignes de niveau, mais il en est de même toutes les fois que l'équation a une intégrale générale telle que

$$F = \text{const}^{\text{te}},$$

F étant une fonction holomorphe, ou plus généralement une fonction bien déterminée et partout finie.[1] Les points singuliers sont ceux où les deux dérivées partielles de F s'annulent à la fois: on a ainsi un centre lorsque F est maximum ou minimum, un col dans le cas contraire.

---

[1] Un noeud peut exister même si l'intégrale générale est univoque $\left(\text{exemple}; \dfrac{y}{x} = \text{const.}\right)$; mais alors cette intégrale F s'y présente sous la forme $\dfrac{\text{o}}{\text{o}}$ et peut y prendre des valeurs aussi grandes qu'on le veut.

Or si maintenant nous revenons à l'étude directe d'une équation différentielle quelconque, nous constatons que, des quatre espèces de points singuliers énumérés plus haut, trois se rencontrent dans le cas général (elles sont caractérisées par certaines conditions d'inégalité entre les termes de plus bas degré de l'équation au voisinage du point singulier): ce sont les nœuds, les cols, et les foyers.

Mais il en est tout autrement des centres, c'est-à-dire des seuls points singuliers qui, avec les cols, puissent se rencontrer, comme nous l'avons vu, dans le cas d'une intégrale générale uniforme et finie. Ces centres sont des points singuliers tout exceptionnels. Pour que l'un d'eux se présente, il faut qu'une infinité *d'égalités* (auxquelles conduit le calcul du développement en série de la fonction F) soient vérifiées.

C'est ce qui ne saurait avoir lieu pour une équation écrite au hasard, et même si cela était, il serait impossible de s'en assurer par un nombre fini d'opérations (du moins en l'absence de données particulières sur les propriétés de l'équation).

En dehors des points singuliers, on peut utiliser sans restriction la propriété fondamentale rappelée tout d'abord et d'après laquelle deux courbes intégrales dinstinctes ne se croisent pas.

Ce point de départ, si ténu qu'il soit, donne à lui tout seul la solution du problème difficile qui nous occupe. Il suffit, à cet effet, de l'appliquer, non seulement à des courbes, complètement différentes, mais à des arcs convenablement choisis d'une même courbe intégrale.

Mais si la méthode employée est, au fond, très simple, les résultats sont tout à fait imprévus et montrent encore que la solution n'était aucunement préparée par toutes nos connaissances antérieures sur ce sujet.

Dans le cas des lignes de niveau, toutes les courbes cherchées sont fermées.

C'est ainsi que l'on serait presque fatalement amené à se figurer les choses si l'on voulait s'en faire une idée d'après les cas où l'on sait écrire l'intégrale générale. C'est ainsi, en effet, qu'elles se passent toutes les fois que cette intégrale F est uniforme (ou même uniforme au point de vue réel, c'est-à-dire bien déterminée en tout point réel) et partout finie. Tout au plus, en considérant des formes fractionnaires de F, peut-on, comme nous l'avons vu, obtenir des courbes intégrales aboutissant à des nœuds.

Que cette vue elle même soit trop simpliste, à moins de compliquer encore notablement l'expression de F, c'est ce que l'on reconnaît dès l'exemple des lignes de pente. Ici on ne peut déjà plus, en général, obtenir l'intégrale élémentairement, mais il est évident que les lignes en question partent des sommets et aboutissent aux fonds (exception étant faite, toutefois, pour certaines d'entre elles, les lignes de faîte, qui aboutissent à un col).

Seulement, il y a, en général, plusieurs fonds et plusieurs sommets, et c'est l'un ou l'autre des fonds qui sert de point d'arrivée, suivant celle des courbes intégrales que l'on envisage: le passage des courbes qui aboutissent à un fond déterminé à celles qui aboutissent à un fond voisin se fait par l'intermédiaire d'une ligne de faîte.

Des dispositions de cette espèce sont déjà peu usuelles pour les équations différentielles dont l'intégrale générale a pu être écrite élémentairement.

Mais les résultats obtenus par POINCARÉ dans le cas général présentent un degré de complication de plus. Il existe alors un certain nombre de courbes intégrales qui sont des courbes fermées (des *cycles*, suivant la terminologie qu'il emploie). Toutes les autres, sauf celles qui aboutissent à des points singuliers,[1] s'enroulent asymptotiquement autour de certains de ces cycles, dits *cycles limites*. L'enroulement a d'ailleurs lieu autour de l'un ou de l'autre des cycles limites, suivant que la courbe intégrale considérée est située dans l'une ou l'autre de certaines régions déterminées.

Rien de tout cela ne pouvait être prévu à l'aide des exemples traités antérieurement. Non seulement ceux-ci donnaient une idée fausse des choses; mais, on le voit, il était inévitable qu'il en fût ainsi.

Nos résultats sont, en effet, plus encore que tout à l'heure, contradictoires avec l'existence d'une intégrale générale que l'on puisse écrire avec les procédés élémentaires. Ils ne pouvaient, par conséquent, se rencontrer dans les problèmes que l'on avait résolus avant POINCARÉ. L'opinion s'était faite, jusque-là sur des figures exceptionnelles, dégénérées en quelque sorte, parce que c'étaient les seules que l'on avait su tracer.

Ces résultats si extraordinaires demandaient à être complétés par la recherche *effective* des cycles limites lorsque l'équation est donnée. C'est une question d'une extrême difficulté, même si l'on entend se borner à une détermination approximative.

POINCARÉ triomphe plus ou moins complètement de cette difficulté, suivant les cas. Pour des équations de forme convenable[2], il détermine exactement le

---

[1] Dans le cas des lignes de pente, ces dernières existaient seules. Cet exemple et autres analogues (tels que les lignes de force du spectre magnétique) étaient donc, eux aussi, incapables de faire prévoir la solution générale.

[2] Voir *Analyse*, p. 59.

nombre des cycles limites et obtient certaines régions dans lesquelles chacun d'eux doit nécessairement se trouver.

Il emploie, à cet effet, un second principe qui était déjà intervenu dans son étude des points singuliers et qui sert de fondement à toutes les autres recherches entreprises sur ce genre de questions.

Analytiquement, il consiste à chercher le sens dans lequel varie une fonction convenablement choisie des coordonnées, lorsqu'on se déplace le long d'une courbe intégrale. On sait avec quel succès un principe de cette nature fut appliqué, peu d'années après, par M. LIAPOUNOF, dans son célèbre Mémoire sur la stabilité du mouvement.

POINCARÉ l'applique, non seulement à une trajectoire déterminée, mais à toutes celles qui traversent une courbe donnée. Géométriquement parlant, cela revient à considérer en chaque point d'une courbe arbitrairement donnée, le sens dans lequel elle est traversée par la courbe intégrale qui passe en ce point. Ce sens, qui peut être déterminé par des opérations élémentaires du moment que l'équation différentielle est donnée, ne change qu'en un point où les deux courbes sont tangentes. On comprend dès lors l'importance que prennent, dans la discussion, les courbes fermées ou cycles «sans contact» c'est-à-dire qui ne sont tangents en aucun de leurs points à une courbe intégrale et le long desquels, par conséquent, ce sens ne peut changer.

La manière dont varie, le long d'une courbe fermée quelconque, le sens dont il s'agit, est d'ailleurs liée à la disposition et à la nature des points singuliers de l'équation par une relation simple qui est d'un grand secours dans les discussions dont nous venons de parler, et que POINCARÉ retrouvera lorsqu'il passera aux équations d'ordre supérieur. Les considérations qui la fournissent équivalent, au fond, au théorème de KRONECKER mentionné plus haut et que plus tard POINCARÉ introduira explicitement.

Les résultats précédents ne subsistent pas pour toutes les équations du premier ordre et de degré supérieur au premier en $\dfrac{dy}{dx}$; mais ils s'étendent cependant d'eux-mêmes à un grand nombre d'entre elles.

Ce n'est pas, en effet, le degré qui joue ici un rôle essentiel: POINCARÉ rencontre une notion qui était apparue une première fois dans la science avec RIEMANN, mais dont les recherches que nous résumons en ce moment devaient montrer la véritable signification. C'est la *géométrie de situation*, la science des propriétés géométriques qui ne changent pas quelles que soient les déformations subies par une figure, pourvu qu'il n'y intervienne ni déchirure, ni soudure.

Tant que l'on se borne au point de vue local, rien ne fait prévoir la nécessité d'une pareille étude. Sinon toutes les figures que les géomètres ont pu imaginer, du moins toutes celles dont ils se sont servis effectivement soit pour les étudier en elles-mêmes soit pour représenter des relations analytiques, sont identiques entre elles au point de vue de la géométrie de situation lorsqu'on se borne à les considérer dans leurs portions suffisamment petites, pourvu qu'elles aient simplement le même nombre de dimensions: par exemple, toute portion suffisamment restreinte de surface quelconque peut être remplacée à ce point de vue par un petit disque circulaire.

Aussi cette découverte est-elle de celles qui se firent le plus attendre. La théorie des fonctions algébriques, à laquelle elle est indispensable, avait été inlassablement étudiée et perfectionnée avant que la nécessité en fût aperçue: cette nécessité avait échappé à CAUCHY lui-même.

Puis, lorsqu'à cette occasion, RIEMANN l'eut mise en évidence d'une manière éclatante, ses successeurs ne virent point que la portée de ce principe n'était pas limitée à la circonstance particulière qui l'avait fait apparaître.

Mais, après le second exemple fourni par POINCARÉ, cette portée est clairement établie. Elle est indissolublement liée à ce passage du local au général qui est la grande préoccupation du Calcul infinitésimal. Dans tout passage de cette nature, on peut s'attendre à voir la géométrie de situation jouer son rôle.

Pour l'appliquer au problème qui nous occupe, on doit regarder x, y, et $\dfrac{dy}{dx}$ comme trois coordonnées cartésiennes et considérer la surface définie, dans ces conditions, par l'équation différentielle. Quel que soit le degré de celle-ci, si cette surface est de genre zéro, c'est-à-dire a une forme analogue à celle d'une sphère, on aura, pour les courbes intégrales, la même disposition générale que dans l'équation du premier degré.

Pour d'autres formes de surfaces les conclusions peuvent être totalement différentes. Lorsque, après l'étude de la sphère, POINCARÉ entreprend, au même point de vue, celle du tore, il constate que ce second cas peut offrir une foule de circonstances nouvelles que le premier ne permettait nullement de prévoir. Encore s'en faut-il qu'il arrive toujours à déterminer exactement ce qui se passe. Les difficultés, elles aussi, sont nouvelles, et telles qu'il est obligé de se poser un certain nombre de questions sans les résoudre.

Ces questions, qui soulèvent des problèmes ardus d'arithmétique, sont, depuis, restées sans réponse.

Avec le cas du second ordre, qui fait l'objet du quatrième et dernier Mémoire de cette série, ce sont déjà les difficultés du cas général qui sont abordées. Les remarques faites dans le cas précédent subsistent, mais ne suffisent plus, à elles seules, à résoudre le problème.

Celui-ci étant mis sous la forme de la recherche de courbes tracées dans l'espace ordinaire et vérifiant un système de deux équations du premier ordre, POINCARÉ généralise sans difficulté la classification des points singuliers obtenue pour une équation du premier ordre unique.

Il existe encore une relation entre leur distribution et les *surfaces fermées sans contact*, qui sont ici les analogues des cycles sans contact, c'est-à-dire les surfaces qui ne sont tangentes, en aucun de leurs points, à une courbe intégrale. Seulement, cette fois, la relation en question ne pourrait être démontrée si POINCARÉ ne partait de la formule de KRONECKER.

C'est surtout dans la théorie actuelle, en effet, que cette formule se présente comme l'auxiliaire indiqué et même indispensable dont l'apparition, à l'heure même où l'œuvre de POINCARÉ allait naître, semble répondre à une sorte d'harmonie préétablie. Deux caractères: la manière dont il dépasse d'emblée le domaine local et, d'autre part, le peu d'hypothèses qu'il implique, font que nul autre n'a pu, jusqu'ici, lui être substitué à ce point de vue.

POINCARÉ en a notablement augmenté la puissance par une remarquable proposition qui, dans beaucoup de cas, dispense même du calcul de la formule en question. Celle-ci, — si, pour fixer les idées, nous la considérons dans l'espace ordinaire — fait, comme on le sait, intervenir un système de trois fonctions F, G, H et exprime le nombre des zéros communs à ces trois fonctions dans un volume déterminé V (ces zéros étant comptés avec des signes convenables) à l'aide des valeurs que les fonctions en question prennent sur la frontière S de ce volume.

Or, POINCARÉ trouve une condition très simple et très générale moyennant laquelle on est certain que le nombre ainsi obtenu ne change pas lorsqu'on remplace le système des fonctions F, G, H par un autre analogue quelconque f, g, h. Ou bien, en effet, on peut affirmer que le nombre en question restera inaltéré dans cette substitution, ou bien il existera sur S, au moins un point où F, G, H seront proportionnels à f, g, h et cela même avec un facteur de proportionnalité de signe connu à l'avance. Cette proposition a été obtenue à nouveau, un peu plus tard, sous une autre forme et avec une autre démonstration, par M. BOHL, à qui elle a fourni toute un nouvelle série de résultats dynamiques.

Elle s'applique immédiatement aux surfaces fermées sans contact, en prenant f, g, h proportionnels aux connus directeurs de la normale à une telle surface S. Le nombre trouvé par la formule de KRONECKER dépend alors de la *courbure totale* de S.

Mais cette première conclusion se simplifie encore, et tout se ramène à une question de géométrie de situation, la courbure totale ainsi introduite dépend uniquement du genre de S. Les résultats de ce type devaient donner lieu, on le sait, à d'importantes recherches de M. W. Dyck.

Avec le cas du second ordre apparaissent également les deux notions qui ont eu sur l'œuvre de Poincaré, dans le domaine de la Mécanique et, particulièrement, de la Mécanique céleste, la plus grande influence.

L'honneur d'avoir recherché spécialement, entre toutes les solutions des équations différentielles du mouvement des planètes, une *solution périodique*, telle, autrement dit, 'que les différents corps mobiles décrivent des courbes fermées (tout au moins par rapport à un système d'axes convenablement choisi) — revient à l'astronome Hill, qui a donné un premier exemple remarquable à cet égard, en ce qui concerne le problème de trois corps.

Mais c'est à Poincaré qu'il appartient d'avoir montré dans les solutions périodiques un instrument, l'un des plus puissants dont on dispose, pour la recherche et l'étude des autres solutions.

Que les solutions périodiques soient capables de jouer ce rôle capital, c'est ce que, après les réflexions qui précèdent, nous pouvons faire comprendre d'un mot. Une courbe intégrale fermée déterminée étant supposée connue, Poincaré considère toutes les courbes intégrales voisines de celle-là.

On voit immédiatement qu'une telle question est à cheval sur les deux points de vue entre lesquels pivote toute la théorie des équations différentielles; et cela, en combinant les avantages de tous deux. Accessible aux mêmes procédés qui s'appliquent au domaine local, elle est d'emblée cependant en dehors de ce domaine, puisque les nouvelles trajectoires obtenues n'évoluent nullement au voisinage d'un point unique et sont étudiées sur des parcours aussi étendus que la solution périodique primitive elle-même.

Ainsi s'explique comment les solutions périodiques «se sont montrées la seule brèche par où nous puissions essayer de pénétrer dans une place jusqu'ici réputée inabordable».[1]

En faisant pour le voisinage d'une solution périodique ce que nous avons fait pour le voisinage d'un point unique, c'est la même marche ascensionnelle que nous entreprendrons, mais avec un point de départ plus élevé.

Cette identité de méthode se vérifie bien lorsqu'on examine le détail des opérations. De même que tout le calcul infinitésimal repose sur la comparaison

---

[1] Poincaré, *Les Méthodes nouvelles de la Mécanique céleste.*

approchée des valeurs d'une fonction en un point et aux points infiniment voisins, on commencera par étudier, en vue du nouveau problème, les solutions infiniment voisines d'une solution donnée.

En prenant l'écart entre les deux solutions comme un infiniment petit principal et en en négligeant les puissances supérieures à la première, on est conduit ainsi, avec POINCARÉ, à introduire systématiquement les équations linéaires qu'il a appellées *équations aux variations* pendant que, de son côté, M. DARBOUX qui en a, lui aussi, découvert l'importance, leur donnait le nom d'équations *auxiliaires*.

Si la solution prise comme point de départ est périodique, il en est de même des coefficients des équations aux variations. POINCARÉ se trouvera ainsi ramené quel que soit l'ordre, à des systèmes dont les propriétés sont connues et dépendent essentiellement de certaines constantes qui vont jouer un rôle essentiel dans ses recherches dynamiques, les *exposants caractéristiques*. A chacun de ceux-ci correspond, pour le système, une solution possédant, non pas la périodicité proprement dite, mais une périodicité relative (périodicité de seconde espèce, au sens d'HERMITE) caractérisée par le fait que toutes les valeurs des inconnues sont multipliées par un même facteur constant lorsque la variable angmente d'une quantité égale à la période des coefficients.

Par ces exposants caractéristiques se trouveront ainsi définies les principales relations entre une solution périodique et les solutions infiniment voisines. En particulier, toutes les propriétés analytiques de l'équation auront leur répercussion sur celles de ces exposants.

Cette étude prépare celle des courbes intégrales *suffisamment* (et non plus infiniment) voisines de la courbe fermée donnée. POINCARÉ entreprend cette dernière, en ce qui regarde le second ordre, dès le Mémoire dont nous parlons. L'analogie que nous avons essayé de faire ressortir tout à l'heure se manifeste d'une manière tout à fait imprévue dans les résultats. La disposition des courbes nouvelles L' autour de la courbe primitive L rappelle d'une manière frappante les formes rencontrées précédemment dans l'étude des équations *du premier ordre* au voisinage immédiat d'un point singulier.

Imaginons, en effet, en un point quelconque P de L, un petit élément de surface normal a L. Toute courbe intégrale L' suffisamment voisine de L percera cet élément de surface en un nombre infini ou, en tout cas, très grand de points successifs P'.

La figure formée par ces points suffit à nous faire connaitre la disposition des arcs successifs de la seconde courbe intégrale L'. Chacun d'eux nous renseigne en effet, sur l'arc qui le contient, puisque tous ces arcs, de part et d'autre de notre surface, cheminent plus ou moins parallèlement les uns aux autres et à la courbe primitive.

Si maintenant on joint chacun des points P' au suivant, on aura une ligne, variable d'ailleurs avec celle des courbes L' que l'on considère: c'est la disposition de ces lignes qui est tout analogue à celle des courbes intégrales d'une équation du premier ordre autour d'un point singulier.

Poincaré met d'ailleurs en évidence la raison de ce parallélisme. Elle doit être cherchée dans l'étroite parenté qui existe entre l'étude des équations différentielles et celles, beaucoup moins avancée, des équations aux différences finies. Ce n'est pas la première fois que Poincaré éclairait, par le même rapprochement, cette dernière question. Les intégrales irrégulières des équations différentielles linéaires (voir p. 237) lui avaient fourni une illustration du même principe, dont les travaux ultérieurs devaient montrer la fécondité.

Conformément à l'analogie dont il vient d'être parlé, il y a quatre dispositions principales possibles, correspondant aux quatre espèces de points singuliers de l'équation du premier ordre. Les exposants caractéristiques permettent (ainsi que le faisaient précédemment les coefficients des termes de plus bas degré autour du point singulier) de reconnaître trois d'entre elles, celles qui correspondent aux noeuds, aux foyers, et aux cols.

Dans chacune de celles ci, la même analogie nous montre que les points P' peuvent aller en se rapprochant indéfiniment de P (puisque, dans chacune des trois hypothèses correspondantes relatives à l'équation du premier ordre, tout ou partie des courbes intégrales aboutissent au point singulier). On voit alors que toute la nouvelle courbe L' va en se rapprochant indéfiniment de L, du moins si on la suit dans un sens convenable, ainsi que le faisaient tout à l'heure les courbes intégrales de l'équation du premier ordre vis à vis des cycles limites: c'est une *solution asymptotique*. Il peut en être ainsi quel que soit le choix de la courbe L' dans le voisinage de L (ou, ce qui revient au même, celui du point P' initial dans le voisinage de P): c'est le cas correspondant à celui d'un nœud ou à celui d'un foyer.

Dans le cas correspondant à celui des cols, au contraire, le point P' doit être choisi d'une façon convenable, à savoir sur l'une ou l'autre de deux courbes qui se croisent en P (de sorte que les courbes intégrales asymptotiques à L se distribuent sur l'une ou l'autre de deux surfaces passant par L). Une page du quatrième Mémoire *sur les courbes définies par les équations différentielles* résout, par une remarquable application du Calcul des limites de Cauchy, la question, en réalité difficile, du calcul de ces courbes et transforme ainsi la théorie des équations aux différences finies en intégrant une des catégories les plus étendues d'équations de cette espèce qu'il ait été possible de traiter jusqu'ici.

Plus tard, lorsqu'il eut à passer au problème des trois corps, cette même recherche se présenta à lui pour des systèmes d'ordre supérieur au second. La généralisation, remarquons-le, n'était pas évidente ou, plus exactement, ne l'aurait pas été sans le complément que la Thèse de POINCARÉ avait préalablement apporté à l'étude des systèmes différentiels au voisinage des points singuliers. Nous savons, en effet, que, dans ce cas, l'introduction de plusieurs inconnues crée une difficulté d'un genre nouveau dont on ne savait pas triompher avant le travail en question. C'est donc grâce à lui qu'il peut démontrer l'existence de ces solutions asymptotiques qui sont une importante conquête de la Mécanique analytique.

Jusqu'au moment dont nous parlons, d'ailleurs, celle-ci n'a pas été envisagée d'une manière spéciale. Les résultats précédents concernent un système quelconque d'équations différentielles.

3. Les cas des équations de la Dynamique.

Les propriétés particulières des équations de la Dynamique apparaissent une première fois dès le quatrième Mémoire *sur les courbes définies par une équation différentielle*, et cela, à propos de la dernière hypothèse qui reste à examiner relativement aux courbes L', d'après l'analogie même qui nous a guidés jusqu'ici: c'est celle qui correspondrait, pour l'équation du premier ordre, au cas d'un centre.

La disposition correspondante, pour le problème actuel, est celle où les points P' sont disposés, autour de P, le long d'une ligne fermée, la même pour chaque courbe L', les diverses lignes formées ainsi obtenues s'emboîtant les unes les autres autour de P.

Notre courbe primitive L sera alors entourée d'une famille de surfaces fermées tubulaires (analogues aux tores contenant à leur intérieur une circonférence de l'espace) telles que chacune d'elles soit un lieu de courbes intégrales.

Absolument comme lorsqu'il s'agissait d'un centre, une telle disposition implique, comme condition nécessaire, l'évanouissement d'une infinité d'expressions constantes C.

C'est seulement si toutes ces constantes C sont nulles que les développements trigonométriques figurant dans l'équation polaire des courbes lieux du point P' — et, par conséquent, dans celle des surfaces tubulaires — pourront être écrits.

Or c'est ce que, en l'absence d'autres renseignements, les calculs ne permettent jamais d'affirmer, si loin qu'on les pousse.

Pour les équations de la Dynamique il en est autrement, et l'on sait *a priori* que toutes les constantes C sont nulles.

Pour le démontrer, un nouveau principe intervient, la notion d'*invariant intégral*. Cette fois encore il s'agit, mais sous une nouvelle forme, de la considération simultanée des différentes courbes intégrales et des relations qu'elles ont entre elles.

Représentons nous notre système d'équations différentielles comme définissant le mouvement d'une molécule fluide. Au lieu de considérer une seule trajectoire, c'est-à-dire le mouvement d'une molécule unique et déterminée, on considérera toutes les molécules qui, à un instant déterminé t, remplissent un volume déterminé V de l'espace (plus exactement de l'espace à $2n$ dimensions, s'il s'agit d'un problème de dynamique dans lequel l'état du système à étudier dépende de n paramètres).

Si maintenant on considère les nouvelles positions de ces mêmes molécules à un instant ultérieur T, celles-ci rempliront un nouveau volume V'.

Or, dans le cas des équations de la Dynamique, quel que soit T, ce nouveau volume est équivalent à l'ancien. Autrement dit V reste constant lorsque le temps varie: c'est, dans la terminologie de Poincaré, un *invariant intégral*.

Ainsi qu'il a été reconnu ensuite, cette belle découverte, qui est, au fond, une, propriété de la notion de multiplicateur, est déjà ancienne: on doit la faire remonter à Liouville.[1]

Mais, lors de sa première apparition, elle était passée inaperçue. Un autre inventeur génial l'avait même — tant elle joue un rôle essentiel dans toute recherche profonde de Dynamique — rencontrée à son tour sur son chemin: Boltzmann l'avait énoncée (1871), ignorant le résultat de Liouville comme Poincaré a ignoré l'un et l'autre; elle est, depuis cette date, à la base de toutes les théories cinétiques.[2]

Mais à ce premier invariant intégral, Poincaré en joindra toute une série d'autres dont il indiquera les relations avec le premier. Le «volume», considéré tout à l'heure, s'exprime par une intégrale d'ordre $2n$ étendue à une portion de l'espace. Poincaré constate que toute une série d'intégrales de tous les ordres,

---

[1] Journal de Mathématiques t. 3, 1838; p. 348.

[2] Le théorème de la stabilité à la Poisson, l'une des applications les plus importantes des invariants intégraux, a été également énoncé et démontré par Gibbs, mais en 1898 seulement.

Il ne se trouve pas, à notre connaissance, dans les travaux de Boltzmann.

c'est-à-dire simples, doubles, etc., le volume n'étant que la dernière d'entre elles, possèdent la même propriété d'invariance.

Quoiqu'il se soit jusqu'ici montré le plus fécond, les autres invariants que POINCARÉ a formés et dont il établit qu'ils se déduisent tous les uns des autres (en particulier de l'invariant intégral simple) constituent autant de propriétés importantes des équations de la Dynamique.

Dans le Mémoire qui nous occupe actuellement, le volume suffit à trancher la question relative aux constantes C ci-dessus mentionnées, c'est-à-dire à montrer que toutes ces expressions sont nulles. La liaison entre ces deux faits est encore dûe à la notion de surface sans contact: elle résulte de ce que, en présence d'un invariant intégral, une surface fermée sans contact ne peut exister. Or, comme le prouve POINCARÉ, on en pourrait tracer autour de la courbe donnée si l'une quelconque des constantes C était différente de zéro.

Avec l'analyse précédente, POINCARÉ entre de plain-pied dans le domaine de la Mécanique céleste.

Les développements en séries qui peuvent être écrits grâce aux conditions $C = o$ sont, pour ce problème particulier, ceux par lesquels LINDSTEDT s'est proposé de représenter les éléments des orbites planétaires et les conditions dont il s'agit ne sont autres que celles qui, dans cette méthode, permettent de faire disparaître les termes séculaires.

C'est, au fond, dans l'existence des invariants intégraux que réside, par conséquent, la véritable raison de la validité (au point de vue formel) de la méthode de LINDSTEDT, validité qui est d'ailleurs établie sans les hypothèses restrictives que LINDSTEDT lui-même était obligé de faire.

Les questions qualitatives liées aux calculs précédents sont des questions de stabilité tout analogues à celles qui préoccupent les astronomes.

POINCARÉ nous a appris à distinguer plusieurs sens du mot «stabilité» et nous a montré la fécondité de celui que POISSON avait substitué à l'acception primitive de LAGRANGE. Toutes les fois qu'il existe, dans le voisinage de L, un système de surfaces fermées sans contact, les courbes L' ne pourront jouir de la stabilité à la POISSON, c'est-à-dire qu'elles ne repasseront pas dans le voisinage immédiat de leur point de départ. C'est, nous l'avons vu, ce qui arriverait si l'une des constantes C était différente de zéro.

L'instabilité (toujours au sens de POISSON) est également la règle pour les courbes L' asymptotiques à L, telles qu'elles se présentent dans les trois premières hypothèses examinées précédemment.

Au contraire, dans l'hypothèse actuelle — et du moment que toutes les constantes C sont nulles — la stabilité devient possible.

Des conclusions analogues s'appliquent à la stabilité de la trajectoire primitive L elle même. Mais le sens que l'on doit adopter alors (et que Poincaré adoptera également en Mécanique céleste, lorsqu'il étudiera, au point de vue de la stabilité, les solutions périodiques) est encore différent des deux premiers.[1] C'est celui qui avait déjà été considéré dans plusieurs cas importants par les auteurs anglais, mais qui n'a été précisé d'une manière complète et générale que quelques années après, par M. Liapounof, dans le Mémoire déjà cité *sur la stabilité du mouvement* où le géomètre russe a repris, pour les systèmes dépendant d'un nombre quelconque de variables, les questions mêmes dont nous parlons en ce moment. Au lieu que la stabilité à la Lagrange ou à la Poisson est une propriété intrinsèque d'une solution déterminée, la stabilité à la Liapounof, seule analogue d'ailleurs à la notion d'équilibre stable, concerne l'écart entre cette solution et les solutions voisines.

Mais, en raison même de la signification astronomique de ses résultats, Poincaré se trouve du même coup aux prises avec les difficultés fondamentales de la Mécanique céleste, et particulièrement avec la plus classique d'entre elles, celle des «petits diviseurs». Dans le cas du premier ordre, le fait, supposé établi, de l'évanouissement des constantes C aurait suffi pour mettre en évidence d'une manière certaine l'existence d'un centre: car Poincaré démontre la convergence du développement en série que l'on peut écrire dans ces conditions. Il n'en est plus de même cette fois. Nos calculs nous permettent d'écrire le développement; mais les petits diviseurs interviennent: ce développement peut n'être et n'est en général, que formel, de sorte que l'existence des surfaces tubulaires n'est nullement démontrée.

Par l'examen de ces difficultés, les Mémoires *sur les courbes définies par les équations différentielles* inaugurent l'immense œuvre dynamique et astronomique de Poincaré.

*　*　*

Cette œuvre se poursuit dans l'ouvrage qui devait être pour la jeune gloire de son auteur une consécration mondiale. C'est avec le Mémoire *Sur le problème des trois corps et les équations de la Dynamique* que Poincaré remporta le prix dans le grand concours international ouvert à Stockholm en 1889, entre les Mathématiciens du monde entier.

---

[1] Toute solution périodique est, par définition, stable au sens de Lagrange ou de Poisson.

Le grand traité intitulé: *les Méthodes nouvelles de la Mécanique céleste* prolonge à son tour les deux Mémoires précédents: c'est dans ces trois ouvrages, et aussi dans une série d'articles insérés au *Bulletin Astronomique*, que se développent les idées de POINCARÉ sur le problème des n corps.

Il sera parlé ici même de ces problèmes au point de vue astronomique avec plus de compétence que nous ne pourrions le faire. Au point de vue analytique, — que nous ne saurions même épuiser tant il offre d'aspects divers dans *les Méthodes nouvelles de la Mécanique céleste*, — l'œuvre dont il s'agit est double: elle présente un côté positif et un côté négatif. Ce dernier, comme il résulte de ce qui vient d'être dit en dernier lieu, se dessina, lui aussi, dès les mémoires *Sur les courbes définies par les équations différentielles.* Il était même apparu auparavant, car les résultats dont nous allons avoir à parler sur ce point ne sont que l'application de la note à laquelle nous avons fait allusion plus haut (p. 235).

Examinons donc comment, tant dans ces deux travaux que dans ceux qui les suivirent, POINCARÉ limite la portée des méthodes qui avaient été appliquées avant lui.

L'intégration, au sens élémentaire du mot, avait, depuis longtemps, été abandonnée. Pourait-on songer à faire des progrès dans ce sens, c'est-à-dire à chercher de nouvelles intégrales? Pour les équations de la Mécanique céleste, le nombre des intégrales connues est de dix. En peut-il, en général, exister d'autres exprimables par les moyens classiques de l'Analyse? Il était vraisemblable que non.

La preuve rigoureuse d'impossibilités de cette nature est une catégorie de questions dont la difficulté a, de tout temps, éveillé l'intérêt des géomètres vraiment supérieurs. La démonstration de l'incommensurabilité entre le côté d'un carré et sa diagonale, dans l'antiquité, celles de l'impossibilité de la quadrature du cercle et de la non-résolubilité des équations algébriques au delà du quatrième degré, dans les temps modernes, comptent à juste titre, parmi les plus belles conquêtes des mathématiques.

En ce qui concerne les intégrales des équations de la Mécanique céleste, une démonstration de l'espèce en question avait été partiellement fournie par BRUNS; mais c'est à POINCARÉ qu'il fut donné de la compléter et d'établir en toute rigueur l'inexistence non seulement d'intégrales algébriques, mais plus généralement, d'intégrales *uniformes* autres que les intégrales classiques.

Le résultat ainsi obtenu n'intéresse pas moins l'analyste pur que l'astronome. Sa portée n'est pas limitée au système différentiel particulier qui fait l'objet de la mécanique céleste. La même méthode qui l'a fourni, permet de discuter le nombre des intégrales uniformes des problèmes de la mécanique classique, et, lorsque ce nombre est insuffisant pour l'intégration, de trouver les seuls cas où il puisse s'accroître. Cette méthode est donc nécessairement à la base de toutes les recherches ultérieures sur ces sujets.

Elle ne doit pas moins attirer l'attention par les principes qu'elle fait intervenir. Elle a conduit POINCARÉ à étudier le développement de la fonction perturbatrice sous un jour nouveau, en en considérant, non plus seulement les premiers termes qu'ils ont pu former explicitement, mais au contraire les termes d'ordres très élevés. Dans cette étude, POINCARÉ utilise non seulement les résultats de la théorie des fonctions dus à ses prédécesseurs et particulièrement à M. DARBOUX, mais leur généralisation aux fonctions de plusieurs variables, telle que la lui ont fournie ses recherches sur les résidus et les périodes des intégrales doubles. La Théorie des fonctions est ainsi appliquée d'une façon toute nouvelle à celle des équations différentielles.

Ces recherches fournissent, entre les coefficients successifs du développement, une infinité de relations qui montrent que, considérés comme fonctions des éléments des orbites, ils se réduisent à un nombre fini de transcendantes.

Un nouveau chapitre de la Mécanique céleste a été ainsi ouvert et a donné lieu, depuis, aux travaux de plusieurs de nos jeunes géomètres et astronomes.

Mais l'impossibilité d'intégrer sous forme élémentaire se dégage également, à un autre point de vue, des résultats qualitatifs.

Dès l'équation du premier ordre, et à propos du cas le plus simple, celui de la sphère, nous avons vu que, par leur aspect même, les formes des courbes intégrales ne sont pas de celles qu'on aurait pu obtenir à l'aide des moyens classiques.

Des faits du même ordre se passent dans le cas général de la Mécanique céleste, dès que le nombre des corps en présence est supérieur à 2. L'existence même des solutions asymptotiques est déjà du nombre. Mais plus topique encore est l'exemple des solutions *doublement asymptotiques*, dont la mise en évidence a été l'une des grandes difficultés qu'ait surmontées POINCARÉ sur ce sujet.

Soit une solution périodique L instable: elle admettra des solutions L' asymptotiques pour $t = \infty$, et aussi des solutions L'' asymptotiques pour $t = -\infty$. Les premières engendreront une surface S' passant par L, les secondes, une surface analogue S''.

Peut-il exister des solutions qui soient à la fois des courbes L' et des courbes L'', c'est-à-dire qui après avoir été, pour $t = -\infty$, infiniment voisines de L, s'en écartent d'une quantité quelconque pour s'en rapprocher ensuite indéfiniment pour $t = +\infty$?

Cela revient à se demander si les surfaces S' et S'' se coupent ailleurs que suivant L. Cette question est une des plus difficiles que POINCARÉ ait abordées. Ce sont les invariants intégraux qui, dans une hypothèse particulière (telle que les surfaces en question passent très près l'une de l'autre) lui ont permis d'y répondre. Eux seuls pouvaient évidemment remplir ce rôle, puisque (en l'absence d'intégrales connues) eux seuls renseignent sur ce que deviennent les trajectoires au bout de très longs intervalles de temps. Non seulement leur considération montre que les surfaces S' et S'' se coupent, de sorte qu'il existe des solutions doublement asymptotiques, mais ces surfaces se coupent une infinité de fois, et la disposition des courbes d'intersection est extrêmement compliquée. En effet, sur une surface asymptotique quelconque, entre deux solutions doublement asymptotiques quelconques, il y en a une infinité d'autres.

On comprendra mieux encore ce que ce résultat a de singulier si l'on réfléchit que, au contraire, une surface S' ou une surface S'' ne peut jamais se couper elle-même.

Avec POINCARÉ, substituons aux deux surfaces en question les courbes obtenues en coupant par un plan. «Que l'on cherche à se représenter la figure formée par ces deux courbes et leurs intersections en nombre infini dont chacune correspond à une solution doublement asymptotique, ces intersections forment une sorte de treillis, de tissu, de réseau à mailles infiniment serrées; chacune de ces courbes ne doit jamais se recouper elle-même, mais elle doit se replier sur elle-même d'une manière très complexe pour venir recouper une infinité de fois toutes les mailles du réseau».

«On sera frappé de la complexité de cette figure, que je ne cherche même pas à tracer. Rien n'est plus propre à nous donner une idée de la complexité du problème des trois corps et en général de tous les problèmes de Dynamique où il n'y a pas d'intégrale uniforme[1] . . . . .»

D'autres conséquences du même ordre découlent des mêmes prémisses.

Au lieu d'une seule solution doublement asymptotique à L, considérons en plusieurs, $L_1, L_2, \ldots$: toutes ces courbes seront, pour $t = -\infty$ situées sur S'' et, pour $t = +\infty$, sur S'.

Mais il résulte des faits établis par POINCARÉ que l'ordre dans lequel elles se succèdent sur S' est sans rapport avec celui dans lequel elles se succédaient sur S''. Si de deux solutions la première est plus voisine que la seconde de la solution périodique pour $t = -\infty$, il pourra arriver que pour $t = +\infty$, la première soit plus éloignée que la seconde de la solution périodique, mais il pourra arriver aussi que ce soit le contraire.

---

[1] *Les Méthodes nouvelles de la Mécanique céleste*, t. III, p. 389.

«Cette remarque est encore de nature à nous faire comprendre toute la complication du problème des trois corps et combien les transcendantes qu'il faudrait imaginer pour le résoudre diffèrent de toutes celles que nous connaissons.»[1]

La voie de l'intégration proprement dite étant ainsi fermée, la Mécanique céleste procède par approximations successives. La tâche qui s'offre à POINCARÉ est de discuter la valeur des méthodes imaginées dans ce but.[2]

On savait que, grâce surtout aux petits diviseurs, la convergence de toutes ces méthodes est très douteuse. Il se trouve cependant, — POINCARÉ montrera par quel mécanisme — qu'elles suffisent, en général, aux calculs numériques usuels.

Mais ceux ci ne sont pas seuls en jeu. «Il ne s'agit pas seulement de calculer les éphémérides des astres, quelques années d'avance, pour les besoins de la navigation ou pour que les astronomes puissent retrouver les petites planètes déjà connues. Le but final de la Mécanique céleste est plus élevé: il s'agit de résoudre cette importante question: la loi de NEWTON peut-elle expliquer à elle toute seule tous les phénomènes astronomiques? Le seul moyen d'y parvenir est de faire des observations, aussi précises que possible, de les prolonger pendant de longues années ou même de longs siècles et de les comparer ensuite aux résultats du calcul. Il est donc inutile de demander au calcul plus de précision qu'aux observations, mais on ne doit point non plus lui en demander moins. Aussi l'approximation dont nous pouvons nous contenter aujourd'hui deviendra-t-elle un jour insuffisante.»[3]

Or, non seulement les séries classiques ne pouvaient nous assurer cette exactitude de plus en plus grande; mais, en raison de leur forme même, on ne pouvait leur demander de conduire à coup sûr à de bons résultats pour une période par trop longue.

A plus forte raison ne pouvaient-elle nous renseigner sur la question de *la stabilité*, laquelle fait intervenir l'indéfinie durée des siècles.

Aussi, au XIXe siècle, des développements en séries de forme nouvelle ont-ils été proposés pour exprimer les éléments des orbites planétaires.

---

[1] Loc. cit. p. 391.

[2] Leur nombre et la variété (au moins apparente) de leurs principes vient en quelque sorte, dans l'état actuel de la Science, ajouter un obstacle nouveau à toutes les difficultés qui entourent l'étude de la Mécanique céleste.

On doit à POINCARÉ d'avoir montré (voir en particulier t. 14, 15 du Bulletin Astronomique) comment on peut passer des unes aux autres en changeant le groupement des termes.

[3] POINCARÉ, *Revue Générale des Sciences*, tome II, 1891, p. 1—2.

Ils ont pour but de diriger le calcul de manière à ne jamais introduire que des termes périodiques.

Une première difficulté de la question (celle qui provient des termes «séculaires»), est ainsi évitée. Mais celle des petits diviseurs subsiste; et, par conséquent une question préjudicielle se pose: les séries ainsi obtenues — celles de LINDSTEDT, par exemple, — convergent-elles?

Jusqu'à POINCARÉ, il paraissait de toute évidence qu'une réponse à cette question, dans le sens de l'affirmative, démontrait la stabilité. On était même tenté de présumer celle-ci par l'existence seule de séries telles que celles de LINDSTEDT.

En d'autres termes, si, grâce aux «petits diviseurs», les développements en séries formés pour rendre compte des mouvements des corps célestes sont divergents, on était porté à admettre qu'ils peuvent cependant fournir sur certaines propriétés des solutions — particulièrement sur les propriétés qualitatives — les indications qu'on se serait cru autorisé à en déduire en toute rigueur en cas de convergence.

POINCARÉ va décider ces questions en sens tout contraire. Non seulement les séries de LINDSTEDT sont, en général, divergentes; mais il y a plus — et cette paradoxale découverte, qui a bouleversé les conceptions des astronomes, remonte aux premières années de son labeur —, la convergence même de séries de cette nature ne permettrait pas, à elle seule, d'affirmer la conclusion demandée, celle à laquelle on serait conduit en se fiant au calcul formel.

POINCARÉ montrera par des exemples que cette conclusion peut être fausse. Cette démonstration est donnée sur le cas du second ordre, où les représentations géométriques sont plus simples. Ici, toutefois, ce ne sont pas elles qui jouent le rôle important, et le point de vue purement analytique reprend ses droits.

Une Note, contemporaine, nous l'avons dit, des premiers travaux de POINCARÉ, contient les principes essentiels de la solution. Les développements habituellement considérés en Mécanique céleste sont, on le sait, des séries trigonométriques

$$\Sigma \left[A_n \cos (\alpha_n t) + A'_n \sin (\alpha_n t) \right]$$

mais bien différentes des séries de FOURIER en ce que les arguments des sinus et cosinus s'obtiennent en multipliant la variable indépendante (autrement dit, le temps) par des coefficients α_n qui ne croissent pas nécessairement à l'infini et qui peuvent même tendre vers zéro.

C'est la théorie mathématique de ces séries qui a été fondée par POINCARÉ en quelques pages des *Comptes rendus de l'Académie des Sciences*, puis du *Bulletin*

Astronomique. Les résultats en offrent, pour le moins, autant de singularités que ceux qui sont relatifs aux séries de FOURIER; mais certaines propriétés essentielles de ces dernières trouvent, moyennant modification convenable, leur généralisation. La plus importante est l'expression des coefficients A_n, A'_n par des intégrales définies: seulement celles-ci, étant donnée que les séries dont il s'agit ne sont plus périodiques, doivent être étendues, non plus à un intervalle fixe, mais à un intervalle indéfiniment croissant.

C'est cette expression qui permet à POINCARÉ de mettre en évidence d'une manière irréfutable l'erreur que l'on commet en voyant dans la convergence d'une série trigonométrique de cette espèce pour toutes les valeurs de la variables une preuve du fait que la somme de cette série reste finie même lorsque cette variable augmente indéfiniment. L'expression en question montre en effet que la somme de la série ne peut rester finie si les coefficients A, A' eux-mêmes ne sont pas tous inférieurs en valeur absolue à une même limite fixe.

Or l'hypothèse que les coefficients A, ou certains d'entre.eux, aillent en augmentant indéfiniment n'est nullement incompatible avec la convergence de la série, du moment que les coefficients α correspondants peuvent tendre vers zéro. Il en est ainsi même dans le cas de la convergence absolue, celui où surtout on pouvait être porté à croire le contraire, par analogie avec les autres types de séries connus.

A cet égard, les deux séries partielles formées, l'une par les termes cosinus, l'autre par les termes sinus, se comportent très différemment. La première

$$\Sigma A_n \cos (\alpha_n t)$$

ne saurait évidemment converger absolument pour $t = 0$ sans converger *uniformément* pour toutes les valeurs réelles de t (la série des coefficients A étant absolument convergente) et représenter une fonction bornée.

Il en est autrement pour la série partielle des sinus

$$\Sigma A'_n \sin (\alpha_n t).$$

Tout ce qu'on en peut dire, c'est que, si elle converge absolument dans un intervalle, si petit qu'il soit, autour de l'origine elle est absolument convergente pour toute valeur réelle de t: — théorème qui s'étend dès lors aisément à la série totale et au cas où l'intervalle où la convergence est donnée ne comprend pas l'origine —; mais cette convergence, si elle est absolue, n'est pas nécessairement uniforme et des exemples tels que celui de la série $\Sigma 2^n \sin \left(\dfrac{t}{3^n}\right)$ montrent qu'elle peut avoir lieu avec des coefficients indéfiniment croissants.

Les principes ainsi établis ne servent pas seulement à discuter les questions de stabilité dont nous parlons en ce moment. Combinés avec ceux que POINCARÉ indique d'un mot à une autre occasion,[1] ils ont donné naissance à toute la théorie des fonctions quasi-périodiques que l'on doit à MM. BOHL et ESCLANGON et qui est destinée à prendre une place importante en Mécanique céleste.

Si maintenant on applique ces principes aux trajectoires L, L' considérées plus haut, et aux séries correspondantes qui, nous l'avons dit, ne sont autre chose que des séries de LINDSTEDT, on voit que non seulement ces développements en séries ne suffisent pas à démontrer l'existence des surfaces tubulaires, mais qu'en fait ces surfaces en question n'existent pas toujours et que plusieurs dispositions très différentes, tant stables qu'instables, sont possibles.

On voit alors «à quel point les difficultés que l'on rencontre en mécanique céleste, par suite des petits diviseurs et de la quasi-commensurabilité des moyens mouvements, tiennent à la nature même des choses et ne peuvent être tournées. Il est extrêmement probable qu'on les retrouvera, quelle que soit la méthode que l'on emploie».

Ajoutons que si l'on passe au problème des n corps lui-même, la divergence des séries de LINDSTEDT, du moins en général, — car la convergence reste encore, à la rigueur, possible quoique très improbable, pour des valeurs particulières des constantes d'intégration — ressortira, elle aussi, des propriétés des solutions et, en particulier, de celles des exposants caractéristiques.

Sur cette question, d'ailleurs, les conclusions de POINCARÉ ne furent pas purement négatives. S'il constate la divergence des séries en question, c'est lui qui a montré — à l'aide de principes déjà acquis par ses recherches sur les intégrales irrégulières des équations linéaires et qui ont reçu une portée nouvelle par les travaux ultérieurs de M. BOREL — pourquoi elles peuvent être néanmoins utiles et dans quelles conditions on pouvait en faire un usage légitime: pourquoi, autrement dit, tout en étant incapables de fournir une approximation indéfinie, même si on les poursuivait indéfiniment, elles permettent néanmoins, les masses perturbatrices étant petites, de pousser cette approximation jusqu'à un certain point, heureusement suffisant en pratique.

En même temps que POINCARÉ est amené à faire les réserves que nous venons d'indiquer sur la puissance des principaux moyens d'action employés avant lui, nous avons déjà vu qu'il en apporte, à son tour, de nouveaux.

---

[1] Bull. Astr. t. XIV.

Les invariants intégraux viennent rendre des services sinon égaux, du moins analogues à ceux qu'auraient pu fournir ces intégrales uniformes à la poursuite desquelles la mécanique céleste doit renoncer. Comme elles, ils représentent des quantités qui restent constantes pendant tout le cours du mouvement, seule propriété qui permette d'établir des relations directes entre des phases éloignées de celui-ci.

Quant aux solutions périodiques et aux solutions asymptotiques qui en dérivent, nous avons dit qu'elles servent, non seulement en elles-mêmes, mais comme intermédiaires permettant d'arriver aux autres solutions.

C'est sous ce jour que les solutions périodiques apparaissent déjà dans les recherches dont nous avons précédemment parlé. Mais leur puissance, à cet égard, va surtout se manifester avec les méthodes même par lesquelles POINCARÉ démontre leur existence.

Nul sujet n'a retenu davantage son attention. On peut dire qu'il s'en est préoccupé toute sa vie. Le premier travail qu'il y consacra date, en effet, de 1883; et l'ombre de la mort planait déjà sur lui lorsqu'il écrivit le dernier,[1] en l'ouvrant par les nobles et mélancoliques paroles, véritable testament scientifique, que nul d'entre nous n'a oubliées.

Pour la formation des solutions périodiques, le Mémoire de 1883 emploie le théorème de KRONECKER. Celui-ci se présente, en l'espèce, comme la généralisation naturelle au cas des systèmes d'équations à plusieurs inconnues (problème auquel peut se ramener en dernière analyse la détermination des solutions périodiques dont il s'agit) de la méthode la plus élémentaire qui existe pour déceler les racines d'une équation unique, celle qui est fondée sur les changements de signe du premier membre.

Une autre méthode classique qui permet évidemment, elle aussi, de montrer l'existence des solutions des systèmes d'équations peut être considérée comme une généralisation du théorème de ROLLE: elle consiste à utiliser l'existence du maximum ou du minimum d'une fonction convenablement choisie des inconnues. On aura ainsi assurément une solution des équations obtenues en égalant à zéro les dérivées partielles de cette fonction. POINCARÉ ne l'emploie pas seulement sous cette forme, mais sous celle, sur laquelle nous reviendrons plus loin, du calcul des Variations.

Ces différents procédés sont combinés entre eux, et surtout, comme nous allons le voir, avec les résultats que donne la théorie des fonctions implicites, en

---

[1] Rendic. del Circ. Mat. di Palermo, t. XXXIII (1er semestre 1912) pp. 375—407.

vue de l'étude plus particulière du problème des trois corps et des équations de la Mécanique céleste.

Au point de vue analytique, le système planétaire se présente comme un système dynamique dépendant d'un paramètre μ (masse perturbatrice ou facteur proportionnel aux masses pertubatrices) auquel on ne donne que de très petites valeurs. Pour $\mu = 0$, l'intégrale générale est connue: tous les points matériels qui composent le système décrivent des ellipses suivant la loi de KEPLER.

Lorsqu'un système d'équations (en termes finis) à un nombre égal d'inconnues dépend d'un paramètre et que son jacobien n'est pas nul, le théorème classique relatif aux fonctions implicites montre l'existence d'une solution pour les petites valeurs de ce paramètre dès que la solution existe pour la valeur zéro.

POINCARÉ a parfois l'occasion d'appliquer ce principe sous la forme que nous venons de rappeler; — et le théorème précédemment cité (page 242) sur la dépendance des intégrales des équations différentielles par rapport aux données initiales et aux paramètres lui permet même d'affirmer l'analyticité des solutions. Mais, en général, dans le type de problèmes qu'il traite, les choses se passent de manière un peu plus compliquée. Les équations relatives à $\mu = 0$, c'est-à-dire celles qu'on obtient quand on ne tient pas compte des perturbations, admettent une infinité de solutions périodiques, à savoir toutes celles dans lesquelles les moyens mouvements sont tous commensurables entre eux. Mais c'est précisément cette infinité, — d'une manière plus précise, l'infinité continue de solutions qui correspondent à un seul et même système de valeurs des moyens mouvements — qui fait ici la difficulté: car elle entraîne cette conséquence que le jacobien est nul.

Le théorème classique ne suffit donc plus, et une étude plus approfondie des fonctions implicites dont il s'agit doit être entreprise. Géométriquement parlant, si, aux coordonnées initiales qui définissent la solution cherchée, on joint la valeur de μ pour définir ainsi un point de l'hyperespace, les équations qui expriment que la solution est périodique définissent, dans cet hyperespace, une variété dont certaines parties continues sont situées sur le domaine $\mu = 0$. On ne pourra avoir une série continue de solutions de ces équations correspondant à μ variable et dépendant analytiquement de μ que lorsque l'on aura une courbe de l'hyperespace appartenant à la variété en question et coupant le domaine $\mu = 0$, d'où résultera un *point multiple* de cette variété.

POINCARÉ, en usant des deux moyens d'actions indiqués plus haut et en reliant entre eux par des lemmes remarquables qui permettent d'établir l'existence de fonctions implicites dans des cas étendus où le jacobien est nul, établit l'existence de tels points multiples: à partir de l'un d'entre eux, la méthode classique devient applicable moyennant des modifications convenables et fournit une série de mouvements périodiques dont les éléments représentatifs sont développables suivant les puissances entières ou fractionnaires de μ.

Mais la méthode devait devenir plus souple et plus générale, grâce aux recherches que POINCARÉ développait vers le même temps (1889) sur la figure des planètes et dont il sera question plus loin. Là, on a à résoudre, et dans des conditions beaucoup plus difficiles encore, puisqu'il s'agit d'une infinité de variables, des questions de même nature. Les principales notions qu'il va introduire à cette occasion, celles de *forme de bifurcation* et de *coefficients de stabilité*, trouvent ici leurs analogues. Les formes de bifurcation correspondent aux points doubles de notre variété et, constituent par conséquent les éléments essentiels qui permettent de la construire; les coefficients de stabilité ne sont ici autres que les carrés des exposants carectéristiques, effectivement liés à la stabilité (à la LIAPOUNOF) d'une solution périodique quelconque.

Comme dans la théorie de la figure des planètes, il y a une sorte d'échange des stabilités chaque fois qu'on passe par une forme de bifurcation. Un fait du même ordre se produit d'ailleurs dans le cas qui s'oppose à un certain point de vue à celui de la bifurcation, celui où, au cours de la variation de μ, il y a disparition de solutions périodiques. Cette disparition se fait par couples comme celle des racines réelles des équations algébriques et les solutions qui disparaissent ensemble sont de stabilités différentes.

Mais, sur un arc de courbe tracé dans notre hyperespace et le long duquel μ varie constamment dans le même sens, le théorème de l'échange des stabilités admet au contraire une réciproque: il ne peut y avoir changement dans les stabilités autrement qu'en passant par les bifurcations. On a ainsi un nouveau moyen efficace de mettre en évidence celles-ci et un nouvel exemple des services que peut rendre l'introduction des exposants caractéristiques.

Ainsi élargie, la méthode se généralise d'elle même et suffit à faire apparaître des résultats d'une complication inattendue, lorsqu'on passe à ce que POINCARÉ appelle les solutions périodiques du *second genre*.

Il donne ce nom à celles qui sont voisines d'une solution périodique déterminée de période T et qui sont également périodiques, mais dont la périodicité ne se retrouve qu'après k révolutions, de sorte que leur période est voisine non de T, mais de kT. Leurs points représentatifs, dans notre hyperespace, engendreront une variété analogue à la précédente, laquelle en fera d'ailleurs évidemment partie. Mais il y aura en outre, des branches nouvelles et, par conséquent, des points multiples nouveaux, intersections de ces branches nouvelles avec les anciennes; et nous trouverons ainsi de nouvelles séries de solutions périodiques, greffées, en quelque sorte, sur les premières.

L'emploi des exposants caractéristiques montre bien, en effet, la condition qui caractérise les nouveaux points doubles comme plus large que celle qui caractérisait les anciens.

Reste, il est vrai, à s'assurer, même lorsque cette condition est remplie, si les nouvelles branches de courbes dont elle fait prévoir l'existence sont réelles. Ce sont les invariants intégraux qui permettent de triompher de cette difficulté en la ramenant à l'étude des maxima et minima d'une certaine fonction, étroitement liée d'ailleurs au principe de la moindre action. Un lemme (analogue à ceux dont nous avons parlé à la fin de la page 264 ainsi qu'à ceux dont il sera question à propos de la figure des planètes), qui constitue en lui même un progrès essentiel pour l'étude des fonctions implicites, fournit le moyen de constater que la condition précédemment écrite est bien suffisante.

Or, cette condition est que l'un des exposants caractéristiques soit un multiple de $\dfrac{2\,i\,\pi}{kT}$.

Comme k est un entier quelconque, on peut le prendre assez grand pour que les multiples de $\dfrac{2\,i\,\pi}{kT}$ soient aussi rapprochés les uns des autres que l'on veut. Comme ces exposants caractéristiques varient continûment avec μ, les bifurcations dont il s'agit se produiront dès lors à intervalles aussi petits qu'on le voudra, au cours de la variation de ce paramètre. Ce ne sont donc plus un certain nombre de familles de solutions périodiques qui sont ainsi mises en évidence, mais un réseau extrêmement compliqué de familles de cette espèce, distribuées comme le sont les nombres commensurables dans la suite totale des nombres. Les périodes correspondantes seront, par contre, indéfiniment croissantes, puisque ce seront des multiples plus ou moins éloignés de la période primitive T.

Il est aisé de comprendre qu'un tel résultat éclaire d'un jour nouveau les précédents et ouvre de nouvelles perspectives.

Nous avons vu POINCARÉ rattacher aux solutions périodiques toutes celles qui en sont suffisamment voisines. Etant donnée la manière dont les solutions périodiques dépendent des nombres commensurables, ne peut-on atteindre, par cette voie, toutes les solutions possibles (du moins toutes les solutions stables), de même que, à l'aide des nombres commensurables, on peut représenter par approximation tous les nombres réels?

On aurait ainsi une voie conduisant en un sens à l'intégration complète du problème. Les choses se passent d'ailleurs effectivement de cette façon pour certains problèmes de Dynamique.[1]

Un récent travail, auquel la question ainsi soulevée a conduit M. BIRKHOFF, est venu modifier nos idées sur ce point. Mais, en nous amenant à élargir le principe précédent, il ne tend pas, loin de là, à en affaiblir la portée. M. BIRKHOFF, en effet, arrive à établir la possibilité d'une approximation indéfinie, analogue à celle qu'avait en vue POINCARÉ, en remplacant, toutefois, les solutions périodiques par une autre catégorie de solutions un peu plus générale.

Il semblerait, à un examen superficiel, que nous ayons ainsi épuisé toutes les solutions périodiques du problème de la Mécanique céleste correspondant aux valeurs suffisamment petites de μ, ou du moins toutes celles qui forment des séries continues. Nous savons, en effet, que toute série de cette espèce doit, à la limite, pour $\mu = o$, donner une solution périodique du problème primitif, qui est celui où on ne tient pas compte des perturbations. Or il semble que nous ayons passé en revue toutes les solutions périodiques de ce problème primitif, et qu'il suffise, par conséquent, de chercher celles qui sont voisines de celles là pour μ voisin de zéro.

Mais nous avons déjà vu, avec POINCARÉ, les difficultés d'un genre tout particulier que l'on rencontre lorsque, dans les questions vraiment ardues et vraiment mystérieuses comme celles auxquelles il s'attaque, on cherche à préjuger de la solution par l'étude des cas particuliers que l'on sait traiter La simplification s'achète par une déformation où il peut arriver que tous les phénomènes deviennent méconnaissables. Nous sommes bien obligés d'accepter le marché (le cas des courbes définies par une équation différentielle du premier ordre est le seul où POINCARÉ ait pu opérer autrement) du moment que, en dehors de lui, nous serions condamnés à l'impuissance absolue; mais nous devons compter avec les pièges auxquels il nous expose. Nulle lecture n'est plus instructive à cet égard que celle des derniers paragraphes du *Mémoire sur le problème des trois corps*, ou du Chapitre correspondant des *Méthodes nouvelles de la Mécanique céleste*.[2]

Une chose rend suspecte, ici, la conclusion provisoire à laquelle nous songions tout à l'heure. Parmi les paramètres dont dépend l'état du système, un certain nombre (les anomalies des planètes sur leurs orbites osculatrices, ou les longitudes des périhélies ou des nœuds de ces orbites) sont angulaires: il ne serait dès lors pas nécessaire, pour la périodicité, que ces paramètres reviennent à leurs valeurs primitives au bout de la période T; il suffit que chacun d'eux (ou plutôt

---

[1] Par exemples pour les géodésiques des surfaces à courbure négative.
[2] Tome III, Ch. XXXII.

chacune de leurs différences mutuelles) ait alors augmenté de $2\,p\pi$, p étant un entier quelconque. Or, en ce qui concerne certains d'entre eux (les longitudes mentionnées en dernier lieu), cet entier p a toujours la valeur zéro lorsqu'il s'agit du mouvement Keplérien sans perturbation. Il en est forcément de même sur toutes les solution périodiques dont l'existence a été jusque là établie pour μ voisin de zéro, puisque p ne peut, sans discontinuité, passer de zéro à une valeur entière non nulle.

Cependant, l'absence, pour μ différent de zéro, de solutions périodiques dans lesquelles les entiers p soient quelconques, nous apparaît, non seulement comme très peu probable, mais même comme tout à fait absurde lorsqu'on tient compte de ce que l'annulation des entiers p est une conséquence des propriétés toutes particulières du problème envisagé et n'aurait plus lieu si on le remplaçait par un autre problème de Dynamique infiniment voisin.

Il faut donc qu'il existe d'autres systèmes de solutions périodiques dégénérant, pour $\mu = 0$, en courbes limites autres que celles dont nous avons parlé jusqu'ici. C'est en effet ce qui a lieu. POINCARÉ en indique la raison, pour la première fois, dans la conclusion du *Mémoire sur le problème des trois corps*.

»Si $\mu = 0$, c'est que les masses des deux planètes sont infiniment petites et «qu'elles ne peuvent agir l'une sur l'autre d'une manière sensible, *à moins d'être à* «*une distance infiniment petite l'une de l'autre*. Mais si ces planètes passent infini- «ment près l'une de l'autre, leurs orbites vont être brusquement modifiées comme «si elles s'étaient choquées. On peut disposer des conditions initiales de telle «façon que ces chocs se produisent périodiquement et on obtient ainsi des solutions «discontinues qui sont de véritables solutions périodiques du problème du mouve- «ment Képlerien et que nous n'avons pas le droit de laisser de côté.»

Autour de ces courbes, composées chacune de plusieurs ellipses Képleriennes et présentant des points anguleux, se groupent les nouvelles solutions périodiques, dites *de deuxième espèce*, que POINCARÉ examine d'ailleurs sommairement, dans *les Méthodes nouvelles*, en raison de leur peu d'analogie avec les orbites observées, mais qui, comme on le voit, n'en sont pas moins d'un haut intérêt analytique.

POINCARÉ reprend la recherche des solutions périodiques sous une autre forme, dix ans plus tard, dans un Mémoire des *Transactions* de la Société mathématique américaine. Nous dirons plus loin comment il lui applique les données du Calcul des Variations.

C'est à ce même problème enfin, et sous sa forme la plus difficile, qu'est allée l'une des dernières méditations de sa vie, ce Mémoire des *Rendiconti del*

Circolo Mat. di Palermo qui a douloureusement ému tous ses admirateurs par le triste pressentiment qui s'y trouve exprimé.

Poincaré y cherche à ne plus se borner, comme il l'avait fait dans *les Méthodes nouvelles*, aux petites valeurs de μ, c'est-à-dire à obtenir des solutions périodiques même si l'on n'est pas au voisinage d'un cas d'intégration connu.

Par une méthode de forme toute nouvelle, il montre que tout se ramène à un théorème de géométrie relatif aux transformations des figures planes (existence d'un point invariant sous des conditions très-générales imposée à la transformation) et que, par conséquent, la démonstration de ce théorème équivaudrait à la résolution de la question posée, au moins dans le premier cas que l'on soit conduit à aborder.

Cette démonstration, que Poincaré s'excusait de ne pouvoir fournir, fut donnée, peu de mois après sa mort, par M. Birkhoff, de sorte que les résultats qu'il énonçait à titre hypothétique sont définitivement acquis aujourd'hui.

Invariants intégraux, solutions périodiques, solutions asymptotiques, sont les matériaux dont sont tissées *les Méthodes nouvelles de la Mécanique céleste*. C'est par leur réaction mutuelle que sont obtenues les conquêtes qui ont fait l'admiration des géomètres et des astronomes.

Non seulement les notions ainsi créées sont grosses pour la Mécanique céleste de résultats nouveaux, mais elles constituent, pour les résultats obtenus par ailleurs, un important moyen de contrôle. Les invariants intégraux, par exemple, donnent une série de vérifications pour tous les calculs entrepris par les méthodes connues.[1]

Les propriétés des solutions périodiques ont, à cet égard, fait leurs preuves d'une manière remarquable à l'occasion des mémorables travaux de G. Darwin.[2] Les calculs du grand astronome anglais ont, on le sait, dans un exemple numérique déterminé, abouti à la formation d'une série d'orbites périodiques de formes entièrement nouvelles et souvent inattendues. Ces orbites sont de plusieurs catégories différentes; elles sont tantôt stables et tantôt instables. Certaines des transformations qu'elles subissent, lorsque la constante de Jacobi varie continûment, obéissaient bien aux lois établies par Poincaré. En particulier, on voyait à un certain moment apparaître simultanément deux d'entre elles, l'une stable et l'autre instable, et cependant très peu différentes l'une de l'autre lors de leur apparition.

---

[1] Les invariants intégraux possèdent d'ailleurs d'importantes propriétés formelles; S-Lie et depuis, MM. Koenigs et Goursat leur ont consacré leurs efforts. On sait qu'on doit à M. de Donder plusieurs exposés de leurs propriétés.

[2] Acta, tome XXI.

Au contraire, une des familles d'orbites périodiques trouvées passait de la stabilité à l'instabilité dans des conditions où ce passage n'aurait pu se faire que moyennant échange de stabilités et, par conséquent, bifurcation. Celle ci n'apparaissant pas en l'espèce, POINCARÉ fut conduit à présumer que les orbites instables n'étaient pas la continuation des orbites stables.

C'est ce qu'a confirmé ultérieurement, dans ce journal même,[1] M. HOUGH, en reprenant l'étude des transformations mutuelles des orbites précédentes. On retrouve, dans cet exemple, les phénomènes généraux décrits dans les *Méthodes nouvelles de la Mécanique céleste*.

En particulier, serrant les calculs de plus près au voisinage du passage mis en doute par POINCARÉ, M. HOUGH constate que, effectivement, les apparences constatées par DARWIN sont dues à ce que, en vertu des données numériques adoptées, l'apparition d'une famille de satellites coïncide approximativement avec la disparition de l'autre.

Ce sont les invariants intégraux qui ont permis à POINCARÉ de s'attaquer au problème de la stabilité des trajectoires, qui correspond, pour un système dynamique quelconque, a celui de la stabilité du système solaire.[2]

LAPLACE a, on le sait, démontré cette dernière stabilité en première approximation et POISSON a passé à l'approximation du second ordre. Mais nous savons maintenant que les méthodes d'approximation ne peuvent donner ici de réponse valable: on peut seulement en inférer une certaine stabilité temporaire, nous renseignant pour une très longue période.

C'est la stabilité au sens de POISSON (moins précis que celui de LAPLACE) que dans une catégorie étendue de mouvements (laquelle toutefois n'embrasse pas notre système solaire) POINCARÉ a pu démontrer d'une manière rigoureuse et non plus approximative.

Par contre, son résultat a une signification toute différente de ceux qui avaient été obtenus antérieurement. Il ne gouverne pas toutes les trajectoires sans exception, mais seulement à *des trajectoires exceptionnelles près*.

Les mots «trajectoires exceptionnelles» doivent s'interpréter, ici, à l'aide du Calcul des probabilités: ils veulent dire que, une trajectoire étant prise au hasard,

---

[1] Acta, tome XXIV, p. 257—288.

[2] Encore ne s'agit-il ici que de la question prise au point de vue théorique. POINCARÉ a soin de rappeler (Annuaire du Bureau des des Longitudes, 1898) que le problème analytique ainsi posé est tout différent du problème physique, l'influence des éléments négligés (les marées, entre autres, et le frottement qu'elles produisent) ne pouvant manquer de devenir, en fin de compte, prépondérante.

la probabilité pour qu'elle soit une de celles qui mettent en défaut le théorème est *infiniment petite* (et non pas seulement très petite).

Autrement dit, il n'est pas absolument certain qu'une trajectoire arbitraire possède la stabilité à la Poisson, mais il y a infiniment peu de chances qu'il en soit autrement.

*   *   *

Le *Calcul des Probabilités*, auquel Poincaré était une première fois amené par la Dynamique, devait, par la suite, tenir une place importante dans son œuvre.

En même temps qu'il s'occupait d'en élucider les principes, il est un de ceux qui en ont poussé le plus loin l'application. Nous aurions à insister sur ces deux aspects de son oeuvre si, à quelques exceptions près, le premier d'entre eux ne concernait le philosophe et le second le physicien.

C'est le développement des théories moléculaires qui a imprimé au génie de Poincaré cette dernière orientation. Au point de vue du mathématicien, les théories en question ont eu pour effet: 1° de faire passer au second plan les équations aux dérivées partielles, au profit des équations différentielles ordinaires; 2° de faire reposer toutes les déductions sur le Calcul des probabilités.

De là, et du rôle directeur que Poincaré sut prendre dans ce grand mouvement, découlent, par une conséquence nécessaire, les recherches qu'il eut à entreprendre dans cette dernière direction, recherches qu'il ne nous appartient pas de retracer dans leur ensemble.

Contentons nous d'en rappeler l'aspect proprement mathématique, tel qu'il est traité dans la deuxième édition des *Leçons sur le Calcul des Probabilités* professées à la Faculté des Sciences de Paris.

Tout cet ouvrage renferme des aperçus qu'il conviendrait de signaler: — telle est, par exemple, l'application nouvelle de la méthode des moindres carrés à l'interpolation, si adéquate, comme l'a montré M. Quiquet,[1] aux besoins de la pratique par la manière dont les calculs faits en vue de l'approximation par un polynôme de degré déterminé peuvent être utilisés pour former le polynôme d'approximation de degré supérieur. Mais la question fondamentale au point de vue de l'application du calcul des probabilités aux phénomènes moléculaires fait l'objet du dernier chapitre. C'est déjà elle qui est abordée lorsque Poincaré étudie le battage des cartes.

Pourquoi, lorsque le jeu a été battu assez longtemps, admettons-nous que toutes les permutations des cartes, c'est-à-dire tous les ordres dans lesquels ces cartes peuvent être rangées, doivent être également probables? Le joueur qui

---

[1] Congrès de Cambridge, 1912, t. II, p. 385.

bat les cartes a cependant des habitudes instinctives et, grâce à elles, si l'ordre primitif des cartes est donné, on doit supposer que, pour l'ordre obtenu après un seul battage, certaines permutations sont plus probables que d'autres. POINCARÉ va montrer en toute rigueur que, si le nombre des battages est grand, le résultat obtenu sera totalement indépendant de ces habitudes inconnues du joueur, toutes les permutations ayant finalement la même probabilité.

La question qui se pose à la base des théories cinétiques est tout analogue, mais avec des difficultés nouvelles; car l'énoncé comporte, cette fois, des cas d'exception. Il doit, tout d'abord, subir une modification chaque fois que les équations différentielles du problème admettent des intégrales. Mais cette réserve n'est pas la seule à laquelle soit conduit POINCARÉ, au moins théoriquement; et quoique, physiquement parlant, la conclusion visée (à savoir que, une fois connues toutes les intégrales univoques, la probabilité de chaque état final du système peut être connue *a priori*, indépendamment du mécanisme de mélange) reste vraisemblable, on voit que les conditions de sa validité mathématique sont à préciser.

Les nouveaux aspects que prenait ainsi la théorie physique ont mis une fois de plus en évidence, en en faisant sentir tout le prix, cette universalité, cette maîtrise simultanée des domaines les plus divers, qui est une des caractéristiques du génie de POINCARÉ.

La substitution des équations différentielles ordinaires aux équations aux dérivées partielles tendait évidemment à rapprocher les méthodes de la Physique mathématique des précédentes, de celles de la Mécanique céleste. Mais, grâce aux recherches ci-dessus mentionnées de POINCARÉ, on voit que l'introduction du Calcul des probabilités se trouvait agir dans le même sens. C'est, notons-le, sous la même forme que le Calcul des probabilités intervenait de part et d'autre: nous avons vu précédemment que le principe fondamental, à savoir l'existence de l'invariant intégral le plus usuel, est commun aux théories moléculaires et à la Dynamique de POINCARÉ.

Ce rapprochement entre les méthodes se retrouve d'une manière remarquable dans les résultats. Un astrologue aurait sans doute trouvé une preuve de l'identité du microcosme et du mégacosme dans cette similitude constatée entre l'étude de molécules dont il entre des millions de millions dans un millimètre cube et celle d'astres séparés par des distances que la lumière met des milliers d'années à franchir, celles-là étant considérées pendant quelques milliardièmes de seconde et ceux-ci pendant des millions de siècles.

Ce sont, tout d'abord, nos connaissances sur le mouvement des planètes qui nous ont aidés à comprendre la vie des molécules.

Mais l'inverse s'est produit lorsque, d'un unique système planétaire tel que le nôtre, on a voulu passer à la foule de ceux qui composent le monde stellaire, même limité à notre voie lactée. C'est LORD KELVIN qui émit pour la première fois une idée de ce genre; mais c'est POINCARÉ qui montra tout ce qu'elle est capable de donner. Il suffit de parcourir son livre sur les Hypothèses Cosmogoniques pour voir combien de relations nous commençons à pénétrer, qui nous resteraient encore incompréhensibles si nous n'avions à notre disposition les études statistiques entreprises par les physiciens sur le perpétuel et inextricable grouillement des molécules.

Ce livre fut un des derniers de son existence. Il était digne d'en marquer le couronnement.

Nul ouvrage pour lequel il fût plus nécessaire. Pour éclairer les propriétés des molécules par celles des nébuleuses et inversement, il fallait dominer à la fois les unes et les autres. Il fallait un successeur de LAPLACE, qui fût en même temps un successeur des CLAUSIUS et des BOLTZMANN, pour écrire les *Leçons sur les hypothèses cosmogoniques*.

4. Analysis situs. Calcul des Variations. Déterminants infinis.

D'autres parties de l'oeuvre de POINCARÉ se rattachent encore à ses travaux sur les équations différentielles.

Ceux ci devaient, tout d'abord, l'amener logiquement à perfectionner la *Géométrie de situation.* Nous l'avons vu montrer qu'on ne saurait faire des progrès importants dans la théorie qualitative des équations différentielles sans rencontrer sur son chemin cette doctrine.

Pour les lignes et les surfaces de l'espace ordinaire, l'Analysis situs, du moins dans les conditions où les applications l'introduisent, tient tout entière dans les données utilisées par RIEMANN. Mais dès que l'on augmente le nombre des dimensions, les résultats se compliquent énormément, pendant qu'il devient impossible de les atteindre par l'intuition directe.

POINCARÉ se trouvait donc amené à traiter la géométrie de situation dans les espaces à plusieurs dimensions.

Il en est, en un sens, le premier fondateur, non qu'il ait été le premier à l'avoir abordée; mais seul, il a indiqué exactement les éléments qu'on doit se donner pour définir, à cet égard, une figure: ces éléments avaient été énumérés incomplètement avant lui.

Une première solution avait été fournie que l'on pouvait croire, au premier abord suffisante.

Betti avait généralisé à un nombre quelconque n de dimensions la notion d'ordre de connexion: il avait défini $n-1$ nombres jouant visiblement un rôle tout analogue à celui de l'ordre en question. De même qu'une surface est complètement définie, au point de vue de l'Analysis situs, par son ordre de connexion, (lorsqu'on donne, en outre, le nombre de ses frontières), on admettait implicitement qu'une multiplicité quelconque était suffisamment caractérisée, au même point de vue, par ses nombres de Betti.

Pour fonder véritablement l'Analysis situs à plusieurs dimensions, Poincaré eut à corriger cette erreur. Il montra que, au point de vue dont il s'agit, la définition précise et complète d'une multiplicité à un nombre quelconque de dimensions exige la connaissance d'un certain groupe de substitutions que l'on peut en déduire. Ce groupe, — et, par conséquent, les propriétés topologiques de la multiplicité — peuvent être altérées dans certains cas où, cependant, les nombres de Betti conservent tous leurs valeurs.

Une loi remarquable fut énoncée par Poincaré sur ces nombres de Betti. A la suite d'une pénétrante critique de M. Heegaard, elle l'amena à constater que la définition de ces nombres peut être précisée de plusieurs façons différentes et que l'une de ces modifications s'impose, au point de vue de l'exactitude de la loi en question.

D'autres nombres que ceux de Betti furent découverts dans la suite de ces recherches et jouent également un rôle important en l'espèce: ce sont les *coefficients de torsion*, liés aux variétés à un seul côté que l'on peut tracer sur la variété donnée. Mais un exemple montre que, pour caractériser celle-ci, la connaissance simultanée des nombres de Betti et des coefficients de torsion est encore insuffisante.

La nouvelle Analysis situs ainsi fondée devait également, comme celle de Riemann, conduire à des applications algébriques.

La théorie des fonctions algébriques de deux variables venait, en effet, d'être fondée par les travaux de M. Picard. L'analogue de la surface de Riemann est, dans cette théorie, un domaine à quatre dimensions dont il est nécessaire d'étudier la connexion. C'est une étude que M. Picard avait déjà commencée. Poincaré y appliqua les données nouvelles dont il disposait.

Il faut dès lors définir et étudier, sur les surfaces algébriques, des *périodes d'intégrales doubles*, notion qui n'est pas sans relation avec celle des résidus des intégrales doubles considérées plus haut, mais qui est à celle ci ce que les périodes cycliques des intégrales abéliennes classiques sont aux périodes polaires, et qui devaient jouer un rôle important dans les résultats qu'il obtint ensuite sur le développement de la fonction perturbatrice en Mécanique céleste.

Nous nous bornerons à ces indications en ce qui concerne les fonctions algébriques de deux variables. Il faudrait, si nous voulions insister et montrer quelle aide POINCARÉ a pu apporter à l'effort des PICARD, des CASTELNUOVO, des ENRIQUES, des SEVERI, retracer, plus longuement que nous ne saurions le faire ici, les principes de cette théorie et l'important développement qu'elle a pris dans ces dernières années.

Nous avons aussi constaté les relations de l'oeuvre dynamique et astronomique de POINCARÉ avec le *Calcul des Variations.*

Tous les travaux de POINCARÉ sur les équations differentielles, toutes les *Méthodes nouvelles de la Mécanique céleste* — et aussi toutes les recherches de POINCARÉ sur la figure des planètes — sont comme le remarque M. HILBERT,[1] du Calcul des variations, si l'on prend ce mot au sens le plus large: l'étude des relations d'une fonction avec les fonctions voisines.

Quant au Calcul des variations proprement dit, il a été également, nous l'avons dit, essentiel à la recherche des solutions periodiques. POINCARÉ a même indiqué brièvement à cet égard des voies qu'il importerait de poursuivre[2] et par lesquelles on pourra démontrer immédiatement l'existence de solutions périodiques toutes les fois que, par des conventions convenables, on pourra considérer les lignes cherchées comme tracées sur des variétés à connexion (linéaire) multiple, en particulier chaque fois que certains des entiers désignés plus haut (page 268) par p (nombre total de circonferences dont a augmenté pendant un période un paramètre angulaire) seront différents de zéro.

La question est beaucoup plus difficile lorsqu'on ne peut introduire de connexion multiple, ainsi qu'il arrive pour les géodésiques des surfaces convexes, auxquelles est consacré le mémoire des *Transactions of the Amer. Math. Soc.* mentionné plus haut. La propriété de minimum habituellement employée pour caractériser les géodésiques suffit encore à établir l'existence de géodésiques fermées si la surface est très peu différente d'une sphère, — c'est-à-dire, une fois de plus, dans une hypothèse infiniment voisine d'un cas d'intégration classique. Mais c'est en posant le problème d'une manière toute nouvelle, en en faisant un problème d'extremum lié, que POINCARÉ arrive au même résultat sur une surface fermée convexe quelconque.

Dans le Mémoire dont nous venons de parler, il utilise le Calcul des Variations sans se préoccuper de le compléter. Au contraire, dans *les Méthodes nouvelles de la*

---

[1] Congrès Int^{al} des Mathématiciens. Paris 1900, page 107.
[2] Voir *Analyse,* page 106.

Mécanique Céleste, il avait examiné la question de savoir si une courbe solution des équations différentielles fournit ou non un extremum de l'action.

On sait que la méthode obtenue par WEIERSTRASS pour décider de l'extremum entre deux extrémités données, comme plusieurs des découvertes dont nous avons eu l'occasion de parler précédemment, avait été une de celles que le géomètre allemand s'était contenté d'enseigner oralement (les premiers ouvrages portant trace de cet enseignement, la thèse de M. ZERMELO et, — pour les intégrales doubles, le Mémoire de M. KOBB, — paraissaient vers le même moment que l'ouvrage de POINCARÉ et l'exposé complet de M. KNESER, quelques années plus tard). Par contre, M. DARBOUX avait fait connaître, sur l'exemple particulier des lignes géodésiques, une autre méthode conduisant à la solution: M. KNESER devait (dans l'ouvrage auquel nous venons de faire allusion) l'étendre au cas général.

POINCARÉ a-t-il retrouvé, de son côté, les résultats de WEIERSTRASS? Dans la notice qu'il lui a consacrée,[1] le Calcul des Variations n'est pas mentionné. D'autre part, dans *les Méthodes nouvelles*,[2] la condition de WEIERSTRASS est énoncée, mais non sous sa forme connue, quoique celle-ci et celle de POINCARÉ soient équivalentes.

Quoi qu'il en soit à cet égard, le résultat que POINCARÉ a en vue, en l'espèce, est nouveau et lui appartient en propre: c'est l'obtention des conditions nécessaires et suffisantes pour l'extremum lorsque la ligne arbitraire est assujettie non plus, comme dans le problème classique qui est celui auquel s'était attaqué WEIERSTRASS, à joindre deux points donnés, mais à être fermée.

A cet effet, POINCARÉ emploie d'avance, mais sans en faire la théorie générale, quelques uns des moyens dont s'est servi M. KNESER pour généraliser la méthode de M. DARBOUX. En particulier, si la notion d'orthogonalité suffit à l'étude de l'action correspondant au mouvement absolu, POINCARÉ est amené à construire des »transversales», lorsqu'il passe au mouvement relatif. Le «champ» qu'il fait intervenir est d'ailleurs remarquable: il n'a pas, à notre connaissance, eu d'autres applications et n'a sans doute pas rendu tous les services qu'on est en droit d'en attendre: il et formé par une des familles de solutions asymptotiques à la courbe fermée envisagée.

Le problème d'extremum tel qu'il se pose sous la forme classique, c'est-à-dire entre deux points donnés, figure d'ailleurs également dans les recherches de POINCARÉ sur les solutions périodiques, et la manière dont la solution dépend

---

[1] Acta, tome 22.

[2] Tome III, p. 261.

de la nature des foyers (suivant que ceux-ci sont des points ordinaires ou des points de rebroussement de l'enveloppe de la famille d'extrémales correspondantes), y est indiquée. POINCARÉ applique surtout sa discussion au cas où les deux extrémités données coïncident en un même point A. Avec une autre remarque géométrique également importante, l'influence du sens de l'angle ainsi formé au point anguleux A, cette étude est, pour lui, le moyen d'arriver à une conception simple des solutions périodiques du deuxième genre et de période k^{uple} qui naissent, comme nous l'avons dit, au voisinage de certaines solutions du premier genre convenablement choisies. Un exposant caractéristique étant, sur celles-ci, commensurable avec $\dfrac{2\,i\,\pi}{T}$ il en résulte que chaque point sera à lui-même son $2\,p^{ième}$ foyer, p étant un entier convenable; tous ces foyers seront d'ailleurs points de rebroussement, et de même sorte (c'est-à-dire que le rebroussement y aura lieu dans le même sens).

Pour arriver à une solution périodique du deuxième genre, POINCARÉ fait varier d'une petite quantité, dans un sens convenable, le paramètre μ et constate que chaque point M de la courbe fermée primitive C peut être joint à lui-même par une ligne satisfaisant aux équations différentielles avec la nouvelle valeur du paramètre. Si enfin on choisit M de manière que l'action le long de cette ligne soit le plus grande ou le plus petite possible, le point anguleux en M disparaît, et on a une solution périodique du deuxième genre, coupant C en $2\,p$ points.

C'est également, du moins pour une partie, à propos des trajectoires de la Dynamique, — en s'occupant de légitimer la méthode qui avait donné à HILL ses solutions périodique du problème des trois corps et, en même temps, celle que M. APPELL avait appliquée à un nouveau développement des fonctions elliptiques — que POINCARÉ a été amené à doter l'Analyse d'un nouveau mode de passage à la limite, voisin de celui qu'elle allait devoir à M. FREDHOLM, les *déterminants infinis* et la résolution des équations linéaires à une infinité d'inconnues.

Nous ne redirons pas après lui[1] les circonstances remarquables qu'il a rencontrées dans cette recherche. On sait que le nouvel algorithme ainsi fondé a dû d'importantes applications à M. HELGE VON KOCH. Nous avons déjà rappelé que ce savant a pu ainsi calculer (sous leur forme convergente) les intégrales irrégulières des équations linéaires, question liée d'ailleurs directement à l'intégration des équations à coefficients périodiques.

---

[1] *Analyse*, page 92—94.

III. Les équations aux dérivées partielles et les problèmes de la Physique mathématique.

Même après l'évolution qui a augmenté l'importance des équations différentielles ordinaires pour la Physique mathématique, celle-ci continue — et continuera — à s'appuyer sur les *équations aux dérivées partielles*.

Pour ces dernières également, et plus nettement même que pour les précédentes, la solution telle qu'on la concevait primitivement, — ce qu'on a appelé l'intégration *formelle* — est hors de cause. Non seulement l'*intégrale générale*, — par le moyen des symboles élémentaires connus, est le plus souvent introuvable; mais même une fois obtenue, elle ne rend pas les mêmes services que dans le cas des équations différentielles et ne dispense pas de recherches aussi difficiles ou plus difficiles que celles qui ont conduit à l'écrire, lorsqu'on veut l'appliquer aux véritables problèmes qui se posent le plus généralement.

Les difficultés que ceux ci présentent peuvent être, suivant les cas, de nature très différente.

Il peut arriver qu'elles ressemblent, avec des différences de degré, à ce qu'elles sont pour les équations différentielles, de sorte que la solution puisse être considérée, au point de vue théorique, comme fournie localement par les méthodes de Cauchy, quitte, dans une seconde partie du travail, à faire la synthèse des différents éléments de solution ainsi obtenus.

C'est ce qui se passe — l'équation étant supposée introduite par l'étude d'un phénomène physique — lorsque celui-ci se déroule librement dans l'espace illimité, et où, par conséquent, pour définir son évolution, il suffit de se donner les *conditions initiales*, c'est-à-dire son état à un instant déterminé.

Mais si le phénomène a pour théâtre une enceinte limitée par des parois, — de sorte que pour achever de le définir, il faut écrire un système de *conditions aux limites*, exprimant le rôle joué par les parois en question, — une difficulté d'un tout autre ordre apparaît.

Il est encore vrai que, au voisinage d'un point quelconque, la solution est le plus souvent représentable par des développements en séries du même type que dans les problèmes précédents. Mais, cette fois, aucun de ces *éléments* de solution, — non pas même le premier, [1] comme il arrivait pour les équations différentielles ordinaires — ne peut être déterminé isolément: la connaissance de chacun d'eux est inséparable de celle de *tous* les autres.

---

[1] Rien ne conduit d'ailleurs à établir entre les éléments en question un ordre déterminé ; à considérer spécialement l'un d'entre eux plutôt qu'un autre comme le premier.

C'est le renversement du principe même qui, en toutes les autres circonstances, guide la marche du calcul intégral: la division de la difficulté en une difficulté locale et une difficulté de synthèse. Une telle division est ici radicalement impossible.

Aussi l'apparition de ces sortes de problèmes — et surtout du premier de tous, celui qui leur a servi de type, le problème de DIRICHLET — a-t-elle changé profondément toute l'allure de la mathématique moderne.

Cet exemple est précisément celui que POINCARÉ a choisi pour montrer comment la physique impose aux mathématiques des problèmes auxquels elle n'aurait pas songé à elle seule. On voit qu'il n'en pouvait exister de plus typique.

Un tel problème ne pouvait manquer d'attirer l'attention de POINCARÉ comme il avait attiré celle de plusieurs de ses prédécesseurs. La nouvelle solution qu'il y apporta, la méthode du *balayage*, s'inspire très directement de la nature même de la question, de cette interdépendance mutuelle de toutes les parties de la solution telle que nous venons de la signaler.

Mais, alors que la méthode du balayage elle-même se rattache aux autres travaux antérieurement consacrés à la théorie du problème de DIRICHLET,[1] cette théorie devait peu après entrer dans une phase toute nouvelle et subir une révolution profonde dont l'utilité ressort, elle aussi, des remarques précédentes.

Son principe consiste à remplacer l'équation *aux dérivées partielles*, ainsi que les autres conditions auxquelles doit satisfaire la fonction inconnue, par une équation *intégrale*. Au lieu de faire figurer l'inconnue sous des signes de dérivation, on la fait apparaître sous un signe d'intégration.

Les premiers sont évidemment une sorte de microscope par laquelle on représente des relations dans l'infiniment petit. Le second, au contraire, est essentiellement synthèse et non analyse. Point n'est besoin dès lors de longues explications pour comprendre comment son emploi est autrement bien adapté aux circonstances dont nous avons parlé que celui de la différentiation.

Ce changement complet d'orientation dans l'étude du problème de DIRICHLET et de tous les problèmes analogues de la physique mathématique évoque, tout d'abord, le nom de M. FREDHOLM.

On se tromperait cependant du tout au tout en n'y rattachant pas également, et d'une manière très étroite, celui de POINCARÉ. Ce serait méconnaître

---

[1] Elle se distingue des méthodes d'approximations successives proposées jusqu'à lui surtout en ce que celles ci, dans le choix des expressions destinées à servir de points de départ, se préoccupaient tout d'abord de satisfaire dès l'abord à l'équation aux dérivées partielles, les autres conditions du problème devant être vérifiées par le jeu des retouches successives. POINCARÉ, le premier, guidé par l'interprétation physique de ses calculs, songea à faire l'inverse.

cette vérité aujourd'hui banale que les manifestations les plus importantes, les plus inattendues de l'esprit humain sont le produit non seulement du cerveau de leur auteur, mais de toute l'époque qui les a vu naître.

Or notre époque, au point de vue mathématique, c'est avant tout, POINCARÉ.

Voyons comment son œuvre a été une condition indispensable, la naissance de la nouvelle méthode.

La première étape qui devait conduire à celle-ci peut être cherchée dans le célèbre travail de M. SCHWARZ inséré, à l'occasion du jubilé de WEIERSTRASS, dans les *Acta Societatis Fennicae* (1885).

Le point de départ de M. SCHWARZ est une question de pure analyse empruntée au Calcul des Variations. Mais le résultat obtenu admet une interprétation physique immédiate. L'équation aux dérivées partielles considérée par M. SCHWARZ est immédiatement liée à celle qui gouverne les vibrations d'une membrane tendue et ce qu'il obtient, c'est le *son fondamental* lequel se présente comme correspondant à la valeur qu'il faut donner à un certain paramètre λ qui figure dans l'équation aux dérivées partielles.

Dans l'étude de tout phénomène vibratoire dans un milieu limité l'expérimentateur constate, on le sait, l'existence d'un tel son fondamental, ou, s'il s'agit d'autre chose que d'acoustique, d'une telle *fréquence fondamentale*. Mais, de plus, cette fréquence fondamentale n'est pas la seule *fréquence propre:* en acoustique, par exemple, le son fondamental s'accompagne d'une série indéfinie d'*harmoniques* dont les propriétés, sous les rapports les plus essentiels, sont analogues à celles du premier.

Expérimentalement, l'existence de toutes ces fréquences propres est manifeste. Mathématiquement, M. SCHWARZ était le premier à démontrer par sa savante méthode celle de la plus simple d'entre elles, la fréquence fondamentale. Il est clair qu'un tel résultat demandait à être complété par son extension aux sons harmoniques. Dix ans après, en effet, M. PICARD parvenait à établir l'existence du premier d'entre eux, c'est-à-dire du second son propre.

C'est à POINCARÉ qu'est due la solution générale, c'est-à-dire la démonstration de l'existence de tous les harmoniques successifs.

Par l'emploi de profonds lemmes géométriques, il démontre que, multipliant la solution par un polynôme en λ à coefficients indéterminés, on peut toujours choisir ces coefficients de manière à ce que le développement du produit suivant les puissances de λ converge dans un rayon plus grand qu'avant la multiplication, et même aussi grand qu'on le veut si le degré du polynôme a été pris suffisamment élevé. Ceci équivaut à dire que cette solution est une fonction méromorphe de λ. Son numérateur seul est fonction de la position d'un point dans le domaine que remplit le milieu considéré: son dénominateur et, par conséquent, ses pôles en sont indépendants.

Ce sont eux qui fournissent les fréquences propres cherchées. Les résidus correspondants ou *fonctions fondamentales* qui donnent la forme des vibrations propres, représentent une seconde partie importante de la découverte ainsi réalisée.

Ce résultat capital, véritable fondement de toute cette partie de la physique mathématique, ne suffisait cependant pas à préparer l'évolution dont nous avons parlé tout à l'heure. En particulier, il n'aurait pas à lui seul rendu possible l'application de la méthode des équations intégrales au problème de DIRICHLET. Il a fallu d'abord que POINCARÉ reprît au même point de vue la plus connue et la plus importante des méthodes indiquées (indépendamment de celle du balayage) pour la résolution de ce problème, la méthode de NEUMANN.

Ce qui fait peut-être du mémoire sur la *Méthode de Neumann et le principe de Dirichlet* un des plus beaux triomphes du génie de POINCARÉ, c'est que rien ne faisait prévoir l'analogie qu'il allait établir entre ce problème et le précédent.

Nous avons rappelé que les constatations expérimentales indiquaient *a priori* l'existence, dans le problème considéré par SCHWARZ une série d'harmoniques, sinai que de fonctions fondamentales correspondantes.

Rien de pareil ne se présentait à propos de la méthode de NEUMANN; et même, rien ne conduisait à introduire dans cette nouvelle question le paramètre indéterminé λ qui s'introduit de lui-même dans celle des harmoniques.

L'analogie analytique était à peine plus utilisable que l'analogie physique. Il est vrai que la solution fournie par POINCARÉ fait apparaître dans les deux cas les mêmes résultats essentiels, mais non pour les mêmes raisons.

En un mot, les fonctions fondamentales, au lieu d'être suggérées par une interprétation physique simple, devaient ici sortir tout armées du cerveau de l'analyste.

POINCARÉ montra cependant, ici encore, que la vraie signification de la méthode de NEUMANN n'était autre que le développement de la solution par rapports aux puissances d'un certain paramètre qu'il introduit dans les données du problème, et que toutes les autres circonstances principales rencontrées à propos de l'étude des sons harmoniques se retrouvent ici.

Ces résultats étaient d'ailleurs essentiels pour la méthode de NEUMANN elle-même: car ils permettaient d'en établir la légitimité sans les restrictions qu'avait apportées son auteur.

Avec eux, — et aussi, ajoutons-le, après la méthode de ROBIN, d'une part, à côté de laquelle il faut citer, de l'autre, les travaux bien connus de M. VOLTERRA, — tout était prêt pour l'entrée en scène de la méthode de M. FREDHOLM.

Celle-ci, en effet, suit pas à pas la marche que nous venons de retracer. Elle repose essentiellement sur l'introduction du paramètre λ de POINCARÉ et sur la manière dont il figure dans l'expression de l'inconnue. Seulement, grâce à sa belle méthode de résolution des équations intégrales, M. FREDHOLM peut écrire, sous forme de développements en séries immédiatement connus, le numérateur et le dénominateur que POINCARÉ n'obtenait que par de délicates approximations successives.

Ainsi les solutions de tous ces problèmes fondamentaux de la physique mathématique, — et en particulier, la détermination des sons propres, où la forme des domaines intervient d'une manière si mystérieuse — sont acquises dès POINCARÉ.

Seulement, pour reprendre la parole même que nous citions en commençant, ces mêmes problèmes sont «plus» résolus par la méthode de FREDHOLM.

Les recherches précédentes ne s'appliquent pas uniquement à la Physique mathématique; elles intéressent également la Mécanique céleste par le problème des marées. POINCARÉ montrait effectivement, dans deux Mémoires *Sur l'équilibre et le mouvement des mers*, comment l'emploi des fonctions fondamentales qu'il venait de découvrir permet, quoiqu'avec des difficultés nouvelles,[1] de tenir compte de l'élément le plus compliqué du problème, l'influence des continents. Nous nous avancerions encore ici sur un domaine qui n'est pas le nôtre en analysant les conséquences auxquelles il est ici parvenu; et nous ne saurions, pour la même raison, insister sur celles qu'il a obtenues lorsque, après l'apparation de la méthode de FREDHOLM, il est revenu sur ce sujet dans sa *théorie des Marées*, une des premières et des plus importantes applications qui (après celles en vue desquelles elle avait été imaginée) aient été données de la méthode en question.

$$*\quad*\quad*$$

Mais nous avons à rappeler, quoique sommairement et au strict point de vue des principes analytiques, les recherches qui ont eu pour objet la figure des corps célestes, c'est-à-dire la figure d'équilibre d'une masse fluide en rotation. Ce problème occupe une place à part, la plus haute en un sens, dans la Philo-

---

[1] Voir *Analyse*, p. 119.

sophie naturelle. Si difficiles que soient les problèmes de Physique mathématique étudiés tout à l'heure, un caractère leur est commun, qui est une notable simplification: ils sont tous linéaires. Si l'on a obtenu la solution du problème de DIRICHLET, pour une surface donnée, avec les données à la frontière V_1 d'une part, et avec les données V_2 de l'autre, cette solution sera connue par cela même, si les données ont les valeurs $V_1 + V_2$. Il est aisé de se convaincre que toutes les théories imaginées pour la résolution de ce problème et de tous ceux qui s'y rattachent reposent essentiellement sur ce fait.

Le problème de l'équilibre d'une masse fluide en rotation est, parmi toutes les applications physiques ou mécaniques des équations aux derivées partielles, la seule pour lequel la simplification précédente ne se produise pas; et, par cela même, il se montre d'un ordre de difficulté supérieur à tous les autres. C'est aussi le seul[1] pour lequel, en même temps que la fonction qui doit vérifier une équation aux dérivées partielles, le domaine même dans lequel cette fonction est définie soit inconnu.

Aussi les théorèmes d'existence les plus simples manquaient-ils eux mêmes dans cette théorie.

Ces hautes difficultés ne pouvaient tarder à tenter POINCARÉ. Voyons par quelles méthodes, dès 1885, il travailla, ici même,[2] à les résoudre.

Nous avons dit que ces méthodes relèvent toutes du Calcul des Variations si l'on prend ce mot dans son acception la plus large. Les considérations qui font l'objet propre du Calcul des Variations classique, celles de maximum et de minimum, y interviennent également et, outre l'usage qui en est fait pour la démonstration des théorèmes d'existence, POINCARÉ a repris et complété les résultats de M. LIAPOUNOF sur la sphère considérée comme donnant le potentiel d'attraction maximum.

Mais l'essence de son analyse est dans l'extension aux problèmes à une infinité d'inconnues, des méthodes de discussion que fournissent, pour le cas d'une inconnue unique, le Calcul différentiel et la géométrie analytique.

Considérons une équation à une seule inconnue x, mais contenant un paramètre μ, soit

$$f(x, \mu) = 0.$$

Si l'on tient compte des deux variables qui y entrent, on sera conduit à la représenter par une courbe plane où μ sera l'abscisse et x l'ordonnée.

Si en un point (x_0, μ_0) de cette courbe, la dérivée $\dfrac{\partial f}{\partial x}$ s'annule, on aura, en général, une tangente parallèle à l'axe des x et, lorsque μ passera par la valeur μ_0, l'équation précédente, considérée comme l'équation en x, perdra deux racines (celles-ci venant se confondre entre elles en x_0 pour devenir ensuite imaginaires) ou, au contraire, en acquerra de nouvelles: μ_0 est ce que l'on peut appeler une valeur *limite* pour μ. Mais si μ traverse la valeur μ_0 sans que l'équation en x cesse d'avoir, tant avant qu'après cette valeur, des racines voisines de x_0, le point (x_0, μ_0) est, en général, point multiple. Les discussions qui apprennent à décider s'il en est bien ainsi sont élémentaires, mais POINCARÉ leur emprunte un énoncé d'une forme nouvelle. Pour qu'il y ait *bifurcation*, autrement dit point multiple, il suffit (sur un arc de courbe réel où μ est supposé pouvoir prendre des valeurs tant immédiatement supérieures qu'immédiatement inférieures à μ_0) que $\dfrac{\partial f}{\partial x}$, en s'annulant, change de signe.

Si maintenant on remplace l'équation unique qui précède par un système d'équations à un nombre égal d'inconnues, dépendant également du paramètre μ, on sait que le rôle de la dérivée considérée tout à l'heure est rempli par un déterminant fonctionnel. Grâce au théorème de KRONECKER, POINCARÉ étend à ces nouvelles conditions la conclusion précédente: en d'autres termes, si, au cours d'une variation continue dans laquelle μ est constamment croissant ou constamment décroissant, le déterminant fonctionnel en question change de signe, il y a bifurcation.

Ceci suffirait théoriquement, dans un grand nombre de cas, en ce qui concerne les équations ordinaires. Mais POINCARÉ se propose d'introduire ces notions dans un domaine nouveau. Un problème tel que celui de l'équilibre de la masse fluide en rotation peut être considéré comme conduisant à un système d'équations, mais en nombre infini et à une infinité d'inconnues.

Rien ne semblait alors devoir subsister de toute la discussion précédente, car la notion qui en formait le pivot, celle du jacobien, faisait défaut. Du moins il en était ainsi au moment ou POINCARÉ poursuivait les recherches dont nous parlons. La méthode de FREDHOLM seule devait, quelques années plus tard, fournir le moyen de combler directement cette lacune; et c'est d'elle en effet que s'est servi M. LIAPOUNOF lorsqu'il a continué les recherches de POINCARÉ et, là où celles-ci avaient simplement abouti à la démonstration de théorèmes d'existence, formé, pour représenter les solutions, des séries convergentes.

En 1889, POINCARÉ n'avait pas le déterminant de FREDHOLM à sa disposition. Mais il y a plus: les quantités qu'il va introduire pour parer à cet inconvénient seront, par le fait même de leur nombre, appelées à rendre des services qu'on ne pourrait obtenir de la considération du seul jacobien. C'est ce que nous a déjà montré l'exemple analogue des solutions périodiques où, cependant, la définition du jacobien ne souffrait aucune difficulté.

Les quantités en question ne sont autres, dans les questions d'équilibre ainsi abordées par POINCARÉ, que les *coefficients de stabilité* c'est-à-dire ceux par l'examen desquels on reconnait (conformément au théorème de LAGRANGE-DIRICHLET[1]), le minimum du potentiel. Ce calcul consiste, comme on sait, dans la décomposition en carrés d'une certaine forme quadratique: les coefficients de stabilité seront les coefficients des carrés ainsi obtenues.

Pour un système dont la position dépend d'un nombre fini de paramètres, ces coefficients de stabilité ont un produit précisément égal au jacobien: par conséquent, leur liaison avec les considérations qui précèdent est évidente et entraîne les conséquences suivantes:

Une solution des équations d'équilibre étant supposée connue pour une certaine valeur μ_0 de μ, si tous les coefficients de stabilité correspondants sont différents de zéro, les équations admettront encore une solution pour μ voisin de μ_0 (puisqu'alors le jacobien sera aussi différent de zéro). En second lieu, dans une série continue de figures d'équilibre telles que μ varie constamment dans le même sens, tout changement de signe de l'un des coefficients de stabilité correspond à une figure de bifurcation.

Toutefois, si l'on ne se servait que du jacobien, il faudrait, si plusieurs coefficients changent de signe à la fois, supposer que leur nombre total est impair. En réalité, cette restriction est inutile, et on voit déjà ici un cas où il y a avantage à employer les coefficients de stabilité.

Par leur moyen d'autre part, on va triompher de la difficulté capitale du problème. Une fois mis sous la forme précédente, les énoncés conserveront un sens, même pour un système dépendant d'une infinité de paramètres, dès que les coefficients de stabilité auront pu être définis. Moyennant une hypothèse toujours vérifiée dans les applications qui se sont présentées, POINCARÉ établit (par des considérations d'extremum) qu'ils restent exacts.

Ainsi les quantités qui, d'après leur définition, n'intéressaient que la stabilité de l'équilibre, se trouvent gouverner l'existence même de cet équilibre.

---

[1] Ce théorème, d'ailleurs, n'est plus seul en jeu dans la discussion proprement dite de la stabilité et, ici encore, POINCARÉ est conduit, avec LORD KELVIN et TAIT à une nouvelle distinction entre plusieurs espèces de stabilité possibles.

En même temps, dans ces mêmes formes d'équilibre de bifurcation que POINCARÉ enseignait à reconnaître les coefficients dont nous venons de parler obéissent à une loi remarquable, non moins importante au point de vue de l'application concrète qu'au point de vue purement analytique, celle de *l'échange des stabilités,* d'après laquelle le nombre des coefficients positifs s'échange entre deux séries de formes d'équilibre qui se rencontrent suivant une forme de bifurcation. Si donc l'une des séries était stable jusque à la valeur μ_0 du paramètre qui correspond à la bifurcation, c'est l'autre série qui possède cette propriété lorsque μ varie au delà de μ_0.

A l'aide du premier des principes cités tout à l'heure, POINCARÉ démontre aisément l'existence des figures annulaires d'équilibre, simplement affirmée par LORD KELVIN et TAIT dans leur traité de Philosophie naturelle. Il lui suffit, à cet effet, de partir d'un premier équilibre (obtenu, il est vrai, en assujettissant d'abord le système à une liaison supplémentaire) et de constater que, dans ce premier état, aucun coefficient de stabilité n'est nul.

C'est le second principe qui a permis la découverte des nouvelles figures d'équilibre dérivées des ellipsoïdes de JACOBI. Le lecteur verra dans *l'Analyse* de POINCARÉ,[1] comment, en effet, la série des ellipsoïdes de JACOBI (comme celle des ellipsoïdes de MACLAURIN, du reste) comprend une infinité de formes de bifurcation, servant de point de départ aux nouvelles formes dont il s'agit.

On verra également, au même endroit, comment un théorème sur l'impossibilité d'un équilibre stable au delà d'une certaine valeur de la vitesse de rotation a fourni à POINCARÉ la réponse à la question que pose l'explication des anneaux de Saturne.

Nous arréterons ici cette revue déjà trop longue et cependant si incomplète.

Sans même parler des applications aux sciences de la Nature qui seront étudiées ici même, il resterait tout au moins à traiter le côté philosophique de l'œuvre de POINCARÉ, qui tient une si grande place dans sa pensée et dans toute la pensée contemporaine. Nous n'avons pas qualité pour le faire, et cependant, sur combien de points cette œuvre philosophique n'est elle pas indissolublement liée aux découvertes scientifiques elles-mêmes. Qu'il s'agisse de géométrie non euclidienne, de théorie des ensembles, de relativité, de calcul des probabilités surtout, — la seule science mathématique qui, des trois états d'Auguste Comte, n'ait pas entièrement dépassé le second — une même impulsion est commune deux domaines; et l'on s'explique déjà, dans une certaine mesure la puissante

---

[1] P. 113.

contribution apportée par POINCARÉ, à la mécanique statistique, lorsqu'on lit les réflexions sur le hasard qui figurent dans *la Science et l'Hypothèse,* ou dans *Science et Méthode.*

D'autre part, les œuvres aussi grandes et aussi géniales que celles de POINCARÉ, dont l'étendue se refuse à une analyse détaillée et ne peut être parcourue qu'à grands traits, sont cependant celles qu'on peut le moins abréger sans les trahir. Chez lui comme chez tous les créateurs vraiment grands, il serait essentiel, au contraire, de faire sentir, ainsi que nous l'avons tenté à une ou deux reprises, comment chaque détail est souvent fécond en conséquences, chaque ligne, en quelque sorte suggestive et grosse de travaux ultérieurs.

Ceux que l'inspiration Poincaréenne a déjà engendrés, et dont nous avons pu à peine signaler, chemin faisant, quelques-uns, remplissent, à eux seuls, plusieurs des Chapitres les plus importants des mathématiques contemporaines. Cependant, nul géomètre n'en doute, l'Analyse de POINCARÉ n'est pas près, tant s'en faut, d'avoir donné la mesure. Même, dans un grand nombre des voies qu'il a ouvertes, sa marche audacieuse nous a emportés sans que nous puissions encore songer à la poursuivre. Si grande qu'elle nous apparaisse, la pensée de POINCARÉ, comme celle d'un GAUSS ou d'un CAUCHY, ne laissera découvrir toute sa puissance qu'à nos successeurs, à la lumière des découvertes futures.

Proceedings of Symposia in Pure Mathematics
Volume **39** (1983), Part 2

Lettre de M. Pierre Boutroux à M. Mittag-Leffler[1]

Vous voudriez avoir, cher Monsieur, quelques détails sur la vie intime de mon oncle, sur la façon dont il travaillait, sur ses habitudes et son caractère? Je n'ai cependant rien d'extraordinaire à vous raconter. Les enquêtes sensationnelles, faites un peu bruyamment par certains psychologues modernes, tendraient à nous faire croire qu'un savant est un être anormal dont tous les actes doivent être étranges. Vous savez pourtant qu'on ne pourrait imaginer une existence plus simple, plus exempte d'évènements, plus uniforme en apparence, que celle d'Henri Poincaré. L'activité de sa pensée lui suffisait et se suffisait. Point ne lui était besoin de chercher des excitations au dehors, ou d'entretenir chez lui par des moyens artificiels cette exaltation spéciale, cette fièvre intellectuelle, sans laquelle certains inventeurs ne sauraient produire. Il ne fuyait pas, il recherchait même, les distractions, les voyages, les plaisirs artistiques; mais c'est qu'il y était porté par un intérêt véritable, par une curiosité naturelle très étendue, en même temps que par le besoin de se délasser. C'est chez lui, en famille, c'est dans le calme de son existence journalière, qu'il a accompli la plus grande partie de sa tâche.

Dans son paisible cabinet de travail, rue Claude Bernard, ou sous les ombrages de son jardin, à Lozère, Henri Poincaré s'asseyait quelques heures par jour devant une main de papier écolier réglé, et l'on voyait alors les feuillets se couvrir, avec une rapidité et une régularité surprenantes, de son écriture fine et anguleuse. Presque jamais une rature, très rarement une hésitation. En quelques jours un long mémoire se trouvait achevé, prêt à être imprimé, et mon oncle ne s'y intéressait plus désormais que comme à une chose du passé. A peine consentait-il — ses éditeurs en savent quelque chose — à jeter un rapide coup d'œil sur les épreuves.

Voilà à quoi se bornait le travail, je veux dire le travail apparent d'Henri Poincaré. A quel labeur sa pensée avait-elle dû se livrer au préalable, lui seul l'a jamais su. Il pensait dans la rue lorsqu'il se rendait à la Sorbonne, lorsqu'il

---

[1]Reprinted from Acta Math. **38** (1921), 197–201.

441

allait assister à quelque réunion scientifique, ou lorsqu'il faisait, après son déjeuner,
une de ces grandes marches à pied dont il était coutumier. Il pensait dans son
antichambre, ou dans la salle des séances de l'Institut, lorsqu'il déambulait à
petits pas, la physionomie tendue, en agitant son trousseau de clefs. Il pensait
à table, dans les réunions de famille, dans les salons même, s'interrompant souvent brusquement au milieu d'une conversation, et plantant là son interlocuteur,
pour saisir au passage une pensée qui lui traversait l'esprit. Tout le travail de
découverte se faisait mentalement chez mon oncle, sans qu'il eût besoin, le plus
souvent, de contrôler ses calculs par écrit ou de fixer ses démonstrations sur le
papier. Il attendait que la vérité fondît sur lui comme le tonnerre, et il comptait
sur son excellente mémoire pour la conserver.

On a souvent remarqué qu'HENRI POINCARÉ gardait jalousement pour lui
ses pensées. A l'inverse de certains savants, il ne croyait pas que les communi-
cations orales, l'échange verbal des idées, pussent favoriser la découverte. Cette
réserve de mon oncle me frappa spécialement lorsque, passant quelques mois à
Göttingen, je fus témoin d'habitudes toutes différentes. On sait quel admirable
foyer de pensée en commun et de travail collectif est la célèbre université alle-
mande. Là tout se passe au grand jour. A peine l'étranger est-il débarqué dans
la petite cité hanovrienne, qu'il sait déjà quels sont les travaux dont s'occupent
les illustrations du lieu, jusqu'où elles sont parvenues et quelles difficultés les arrê-
tent. Les idées, colportées, confrontées, discutées, au cours des promenades dans
la forêt et aux séances de la Société mathématique, mûrissent d'elles-mêmes dans
ce milieu fertile, où la curiosité toujours alerte et la néomaïeutique de M. KLEIN
contribuent à entretenir un ferment inépuisable. Le profit que peuvent retirer
les jeunes gens d'un contact aussi intime avec leurs maîtres est manifeste. Ce
n'est point, cependant, par accident, ou par besoin égoïste de solitude, que mon
oncle s'abstenait d'imiter sur ce point ses collègues allemands. Nul n'était plus
liant que lui, nul n'était plus porté à la sympathie, pour les jeunes en particulier. Mais mon oncle se faisait de la découverte mathématique une idée qui
excluait toute possibilité de collaboration. La recherche telle qu'il la comprenait
doit être une lutte à deux. C'est un corps-à-corps avec la réalité fuyante et
rebelle, qu'il s'agit de frapper au cœur. Dans un tel duel il n'y a pas de place
pour des témoins. L'intuition, par où s'opère la découverte, est une communion
directe, sans intermédiaires possibles, de l'esprit et de la vérité. Il ne convient
pas, il faut se garder, de troubler ce tête-à-tête.

Sans doute, une fois l'idée conquise, il peut être utile de se mettre à
plusieurs pour l'exploiter. Mais c'est là une besogne, en partie mécanique, qui
n'avait qu'un intérêt secondaire, il faut bien le dire, aux yeux d'HENRI POINCARÉ.

— Avez-vous l'idée, demandait-il? Si vous ne l'avez pas, je ne puis vous être d'aucun secours pour la découvrir. En revanche, je suis prêt à vous faire crédit. Quoi qu'il me semble de la voie où vous vous engagez, je ne vous adresse aucune critique, aucune objection de principe. Je sais trop bien que la vérité surgit souvent aux carrefours où l'on s'attendait le moins à la rencontrer.

Je m'explique ainsi que mon oncle ait été, à l'égard des débutants, l'un des juges les plus bienveillants, les plus larges d'esprit, que j'aie rencontrés, et, en même temps, l'un des plus sévères. Loin de prétendre entraîner ses élèves à sa suite et de leur dicter leur tâche, il voulait laisser à chacun une initiative complète; il était toujours disposé à s'intéresser aux recherches les plus inusitées, les plus paradoxales même; aucune nouveauté ne lui faisait peur. Mais, quand venait le moment d'apprécier les résultats, il se montrait extrêmement exigeant. Si vous ne lui apportiez que des propositions qu'il considérait comme acquises — et, dans sa tendance à aller de l'avant, il regardait comme virtuellement acquis tout ce dont nous n'étions plus séparés par des difficultés de principe — si vous ne lui ouvriez pas des aperçus nouveaux pour lui, on devinait qu'il avait aux lèvres l'éternel et décourageant «à quoi bon?»; non que vous eussiez, selon lui, perdu votre temps; mais vous lui aviez appris que votre méthode — sur laquelle il avait jusque-là réservé son jugement — n'offrait, en réalité, aucun avantage.

Ceux qui approchèrent mon oncle de près ont été surpris de le voir rarement se servir de livres. Il lisait peu, en effet — je ne parle ici, bien entendu, que de ses lectures scientifiques —, et il lisait d'une façon très particulière. Henri Poincaré ne pouvait s'astreindre à suivre la longue chaîne de déductions, la trame serrée de définitions et de théorèmes, que l'on trouve généralement dans les mémoires de mathématiques. Mais, allant tout droit au résultat qui lui paraissait le centre du mémoire, il l'interprétait et le repensait à sa manière; il le contrôlait par ses propres moyens; après quoi, seulement, reprenant le livre en mains, il jetait un rapide regard circulaire sur les propositions, lemmes et corollaires, qui constituaient la garniture du mémoire.

Il faut insister sur ces détails, car nous touchons ici peut-être à l'un des caractères distinctifs de la pensée de mon oncle. Au lieu de suivre une marche linéaire, son esprit rayonnait du centre de la question qu'il étudiait vers la périphérie. De là vient que dans l'enseignement et même dans la conversation ordinaire, il était souvent difficile à suivre et parfois semblait obscur. Qu'il exposât une théorie scientifique, ou qu'il contât une anecdote, il ne commençait presque jamais par le commencement. Mais, *ex abrupto*, il lançait en avant le fait saillant, l'évènement caractéristique, ou le personnage central, personnage qu'il n'avait point même pris le temps d'introduire et dont parfois son interlocuteur ignorait jusqu'au nom.

Cette tournure d'esprit explique comment la pensée d'HENRI POINCARÉ a pu être si agile et s'appliquer à tant d'objets différents, comment, par suite, il lui a été possible de satisfaire une curiosité presque universelle.

Habitué à négliger les détails et à ne regarder que les cimes, il passait de l'une à l'autre avec une promptitude surprenante; et les faits qu'il découvrait, se groupant d'eux-mêmes autour de leurs centres, étaient instantanément et automatiquement classés dans sa mémoire. D'ailleurs mon oncle n'était pas de ceux qui vivent sur les trésors acquis et qui se complaisent à faire chez eux le tour du propriétaire. Il se contentait de savoir qu'il possédait et, sans regarder en arrière, il travaillait sans relâche à remplir de nouvelles cases de son cerveau.

HENRI POINCARÉ avait un goût marqué pour la géographie et pour les voyages. Conformément à ses tendances ordinaires, il voulait voir dans chaque pays les sites et les monuments les plus caractéristiques, et il n'éprouvait point le désir de s'écarter des routes traditionnelles. Il était l'opposé de ces romantiques qui voyagent pour donner un cadre à leurs rêveries et qui, souhaitant ce cadre inédit, s'efforcent de s'isoler du flot des touristes. Ses jouissances à lui étaient d'un ordre tout intellectuel. Extrayant d'ailleurs du premier coup, et traduisant immédiatement en concepts, les traits essentiels des impressions qu'il recueillait, il n'avait que rarement besoin de voir deux fois les mêmes contrées. Sans doute, il est possible qu'à la fin de sa vie, mon oncle ait été sensible, lui aussi, à l'attrait qu'exercent sur presque tous les hommes l'évocation de leurs souvenirs et les lieux qui leur sont déjà familiers. Cependant le besoin incessant de voir du nouveau, a bien été, si je ne me trompe, un trait dominant de son caractère.

Dès sa jeunesse HENRI POINCARÉ lisait avec un intérêt passionné les récits de voyage du «Tour du Monde» et suivait au jour le jour les progrès de l'exploration du continent africain. C'est, je crois, un sentiment du même genre qui, en toutes circonstances et dans tous les domaines, le lançait vers la poursuite de l'inconnu, et lui faisait assigner à sa vie et à la science un but simple et précis: comme les grands voyageurs de l'Afrique, remplir les espaces blancs de la carte du monde.

Je me rappelle qu'un jour, parlant devant HENRI POINCARÉ d'un mathématicien qui quittait ses études pour d'autres occupations, quelqu'un laissa échapper cette remarque: «Tout se vaut, après tout; il sera sans doute aussi heureux que s'il avait continué à faire des mathématiques». Mon oncle eut un mouvement de protestation qui arrêta la conversation. Venant d'un spécialiste enfermé dans des études étroites, pareille intransigeance n'eût point étonné, et on l'eût mise sur le compte d'une foi un peu naïve. Mais Henri Poincaré n'avait point les défauts des spécialistes; il avait des goûts très variés et ne prétendait nullement placer ses propres occupations au-dessus de toutes les autres. Que signifiait donc sa protestation? Très catégoriquement, je crois, mon oncle estimait que si l'on s'est une fois mis au service de la science, on n'a plus le droit de déserter son poste. Tant qu'il reste des blancs sur la carte du monde, il ne nous est pas permis de nous reposer.

En effet, bien qu'il ait été sensible autant qu'aucun autre à la grandeur et la beauté de la science, mon oncle n'appartenait pas à cette école de dilettantes qui se livrent aux mathématiques parce qu'elles leur procurent des jouissances esthétiques. La recherche était pour lui un devoir, d'autant plus attachant qu'il lui coûtait plus de peine. Je n'ai jamais entendu mon oncle parler du travail scientifique — du sien ou de celui d'autrui — qu'avec le plus grand sérieux et le plus grand respect: lui, si gai à ses heures de délassement, lui qui aimait et pratiquait l'ironie, il n'en avait point lorsque la science était en cause.

Voilà, cher Monsieur, quelques-unes des réflexions qui me venaient à l'esprit, voilà ce que je sentais ou croyais deviner quand j'avais le bonheur de converser avec mon oncle. Henri Poincaré, je vous l'ai dit, ne parlait guère de ses travaux; encore moins se fût-il complu à décrire ses sentiments intimes et les ressorts de son intelligence; mais il aimait faire causer les autres, et, lorsqu'on se trouvait exprimer une idée qui lui était chère, lorsqu'on découvrait une pensée conforme à la sienne, son sourir et son regard révélaient le plaisir qu'il éprouvait. C'est par de tels signes à peine perceptibles qu'Henri Poincaré manifestait sa sympathie et sa bienveillance. Lui qui, par discrétion, n'a pas voulu se faire des disciples, lui que sa réserve naturelle faisait passer pour froid, il avait un cœur chaud, un grand désir de se sentir entouré, un profond besoin d'affection.

Paris, le 18 Juin 1913.

Proceedings of Symposia in Pure Mathematics
Volume 39 (1983), Part 2

Bibliography of Henri Poincaré

Arranged by broad topics following the general form of the eleven volume collected works and compiled from

E. Lebon. Henri Poincaré. 2 éd., Paris, 1912, 111p.

Analyse des travaux scientifiques de Henri Poincaré, faite par lui—même. — Acta mathematica, 1921, 38, 3—135.

Oeuvres de Henri Poincaré, t. I—XI. Paris, Gauthier-Villars, 1916—1956.

J. J. Mooij. La philosophie des mathématiques de Henri Poincaré. Paris, Gauthier-Villars, 1966, p. 159—171.

Catalogue of scientific papers. Completed by the Royal Society of London, v. XI. 1896, p. 39—41; v. XVII. 1921, p. 942—944.

J. C. Poggendorffs Biographisch-Literarisches Handwörterbuch der exacten Naturwissenschaften, Bd. IV. 1904, S. 1178—1180; Bd. V. 1925, S. 990.

Selected Works of H. Poincaré (in Russian) edited by N. Bogoliubov and V. I. Arnold, 3 volumes, Moscow, 1972. (Bibliography compiled by E. I. Pogrebiskoi in Vol. 3.) The items in the following bibliography were reprinted from this source; Russian items were translated into English.

The principal subdivisions of the bibliography are:
    I. Mathematics
        1. Differential equations
        2. Theory of functions
        3. Algebra and arithmetic
        4. Geometry
        5. Topology
    II. Mechanics
        1. Analytical mechanics. Hydrodynamics
        2. Celestial mechanics, astronomy, and geodesy
        3. Forewords and reports.
    III. Theoretical and mathematical physics
        1. Books
        2. Differential equations of mathematical physics
        3. Physical theory
        4. Electromagnetic theory
        5. Criticism and extension of physical theories
    IV. General problems of science

V. Diverse writings
   1. Addresses to International Congresses
   2. Analysis of mathematical works
   3. Pedagogy
   4. Other writings

I. Mathematics

1. *Differential equations.*

1. Note sur les propriétés des fonctions définies par les équations différentielles. — J. l'École Polytechnique, 1878, 45e Cahier. 13—26; Oeuvres, t. I, p. XXXVI—XLVIII.

2. Sur les propriétés des fonctions définies par les équations aux différences partielles. — Thèses presentées à la Faculté des Sciences de Paris, 1er août 1879. Paris, Gauthier-Villars, 1879, 93 p. Oeuvres, t. I, p. XLIX—CXXXI.

3. Perfectionner en quelque point important la théorie des équations différentielles linéaires à une seule variable indépendante. (Memoir written for the Concours of 1880 for the Grand Prix de Mathématiques. Only part II was published in (68).)

4. Sur les courbes définies par une équation différentielle. — C. r. Acad. sci., 1880, 90, 673—675; Oeuvres, t. I, p. 1—2.

5. Sur les fonctions fuchsiennes. — C. r. Acad. sci., 1881, 92, 333—335; Oeuvres, t. II, p. 1—4.

6. Sur les fonctions fuchsiennes. — C. r. Acad. sci., 1881, 92, 395—398; Oeuvres, t. II, p. 5—7.

7. Sur les équations différentielles linéaires à intégrales algébriques. — C. r. Acad. sci., 1881, 92, 698—701; Oeuvres, t. III, p. 95—97.

8. Sur une nouvelle application et quelques propriétés importantes des fonctions fuchsiennes. — C. r. Acad. sci., 1881, 92, 859—861; Oeuvres, t. II, p. 8—10.

9. Sur l'intégration des équations linéaires par le moyen des fonctions abéliennes. — C. r. Acad. sci., 1881, 92, 913—915; Oeuvres, t. III, p. 98—100.

10. Sur les fonctions fuchsiennes. — C. r. Acad. sci., 1881, 92, 957; Oeuvres, t. II, p. 11.

11. Sur les fonctions fuchsiennes. — C. r. Acad. sci., 1881, 92, 1198—2000; Oeuvres, t. II, p. 12—15.

12. Sur les fonctions fuchsiennes. — C. r. Acad. sci., 1881, 92, 1274—1276; Oeuvres, t. II, p. 16—18.

13. Sur les fonctions fuchsiennes. — C. r. Acad. sci., 1881, 92, 1484—1487; Oeuvres, t. II, p. 19—22.

14. Sur les groupes kleinéens. — C. r. Acad. sci., 1881, 93, 44—46; Oeuvres, t. II, p. 23—25.

15. Sur une fonction analogue aux fonctions modulaires. — C. r. Acad. sci., 1881, 93, 138—140; Oeuvres, t. II, p. 26—28.

16. Sur les fonctions fuchsiennes. — C. r. Acad. sci., 1881, 93, 301—303; Oeuvres, t. II, p. 29—31.

17. Sur les fonctions fuchsiennes. — C. r. Acad. sci., 1881, 93, 581—582; Oeuvres, t. II, p. 32—34.

18. Mémoire sur les courbes définies par une équation différentielle (I partie).— J. math. pures et appl., 3e sér., 1881, 7, 375—422; Oeuvres, t. I, p. 3—44.

19. Sur les courbes définies par les équations différentielles. — C. r. Acad. sci., 1881, 93, 951—952; Oeuvres, t. I, p. 85—86.

20. Mémoire sur les courbes définies par une équation différentielle (II partie [18]). — J. math. pures et appl., 3e sér., 1882, 8, 251—296; Oeuvres, t. I, p. 44—84.

21. Sur les points singuliers des équations différentielles. — C. r. Acad. sci., 1882, 94, 416—418; Oeuvres, t. XI, p. 3—5.

22. Sur l'intégration des équations différentielles par les séries. — C. r. Acad. sci., 1882, 94, 577—578; Oeuvres, t. I, p. 162—163.

23. Sur les fonctions fuchsiennes. — C. r. Acad. sci., 1882, 94, 163—166; Oeuvres, t. II, p. 35—37.

24. Sur les groupes discontinus. — C. r. Acad. sci., 1882, 94, 840—843; Oeuvres, t. II, p. 38—40.

25. Sur les fonctions fuchsiennes. — C. r. Acad. sci., 1882, 94, 1038—1040; Oeuvres, t. II, p. 41—43.

26. Sur les fonctions fuchsiennes. — C. r.
Acad. sci., 1882, 94, 1166—1167;
Oeuvres, t. II, p. 44—46.

27. Sur une classe d'invariants relatifs
aux équations linéaires. — C. r. Acad.
sci., 1882, 94, 1402—1405; Oeuvres,
t. II, p. 47—49.

28. Sur les fonctions fuchsiennes. — C. r.
Acad. sci., 1882, 95, 626—628; Oeuvres, t. II, p. 50—52.

29. Sur les fonctions uniformes qui se
reproduisent par des substitutions linéaires. — Math. Ann., 1882, 19,
553—564; Oeuvres, t. II, p. 92—104.

30. Sur les fonctions uniformes que se
reproduisent par des substitutions linéaires (extrait d'une lettre adressée
à F. Klein). — Math. Ann., 1882,
20, 52—53; Oeuvres, t. II, p. 106—107.

31. Sur la théorie des fonctions fuchsiennes. — Mémoires Acad. nationale des
Sciences, Arts et Belles—Lettres de
Caen, 1882, 3—29; Oeuvres, t. II,
p. 75—91.

32*. Théorie des groupes fuchsiennes. —
Acta math., 1882, 1, 1—62; Oeuvres,
t. II, p. 108—168.

33*. Sur les fonctions fuchsiennes. —
Acta math., 1882, 1, 193—294; Oeuvres, t. II, p. 169—257.

34. Mémoire sur les groupes kleinéens. —
Acta math., 1883, 3, 49—92; Oeuvres,
t. II, p. 258—299.

35. Sur les séries de polynômies. —
C. r. Acad. sci., 1883, 96, 637—639;
Oeuvres, t. I, p. 223—225.

36. Sur les groupes des équations linéaires. — C. r. Acad. sci., 1883, 96,
691—694; Oeuvres, t. II, p. 53—55.

37. Sur les groupes des équations linéaires. — C. r. Acad. sci., 1883, 96,
1302—1304; Oeuvres, t. II, p. 56—58.

38. Sur les fonctions fuchsiennes. — C. r.
Acad. sci., 1883, 96, 1485—1487;
Oeuvres, t. II, p. 59—61.

39. Sur l'intégration algébrique des équations linéaires. — C. r. Acad. sci.,
1883, 97, 984—985; Oeuvres, t. III,
p. 101—102.

40. Sur l'intégation algébrique des équations linéaires. — C. r. Acad. sci.,
1883, 97, 1189—1191; Oeuvres, t. III,
p. 103—105.

41. Sur les courbes définies par les équations
différentielles.-C. r. Acad. sci., 1884, 98,
287–289; Oeuvres, t. I, p. 87–89.

42. Sur les groupes hyperfuchsiennes.-C. r.
Acad. sci., 1884, 98, 503–504; Oeuvres, t.
II, p. 62–63.

43*. Sur les groupes des équations linéaires.-
Acta math., 1884, 4, 201–311; Oeuvres,
t. II, p. 300–401.

44. Mémoire sur les fonctions zétafuchsiennes.
Acta math., 1884, 5, 209–278; Oeuvres, t.
II, p. 402–462.

45. Sur un théorème de M. Fuchs.-C. r.
Acad. sci., 1884, 99, 75–77; Oeuvres, t.
III, p. 1–3.

46. Sur un théorème de M. Fuchs.-Acta
math., 1885, 7, 1, 1–32; Oeuvres, t. III,
p. 4–31.

47. Sur les équations linéaires aux différenti-
elles ordinaires et aux différences finies.-
Amer. J. Math., 1885, 7, N 3, 1–56;
Oeuvres, t. I, p. 226–289.

48. Sur les intégrales irrégulières des équations
linéaires.-C. r. Acad. sci., 1885, 101, 939–
941; Oeuvres, t. IV, p. 611–613.

49. Sur les intégrales irrégulières des équations
linéaires.-C. r. Acad. sci., 1885, 101, 990–
991; Oeuvres, t. IV, p. 614–615.

50. Sur les courbes définies par les équations
différentielles (III partie [18]).-J. math.
pures et appl., 4e sér., 1885, 167–244;
Oeuvres, t. I, p. 90–161.

51. Sur les courbes définies par les équations
différentielles (IV partie [18]).-J. math.
pures et appl., 4e sér., 1886, 2, 151–217;
Oeuvres, t. I, 167–222.

52. Sur les intégrales irrégulières des équations
linéaires.-Acta math., 1886, 8, 295–344;
Oeuvres, t. I, p. 290–332.

53. Les fonctions fuchsiennes et l'arithmétique. — J. math. pures et appl.,
4e sér., 1887, 3, 405—464; Oeuvres,
t. II, p. 463—511.

54. Remarques sur les intégrales irrégulières des équations linéaires (Réponse à M. Thomé). — Acta math.,
1887, 10, 310—312; Oeuvres, t. I,
p. 333—335.

55. Sur l'intégration algébrique des équations différentielles — C. r. Acad.
sci., 1891, 112, 761—764; Oeuvres,
t. III, p. 32—34.-

56. Sur l'intégration algébrique des équations
différentielles du premier ordre et du
premier degré.-Rendiconti Circolo mat.
Palermo, 1891, 5, 161–191; Oeuvres, t.
III, p. 35–58.

57. Sur l'intégration algébrique des équations
différentielles du premier ordre et du
premier degré.-Rendiconti Circolo mat.
Palermo, 1897, 11, 193–239; Oeuvres, t.
III, p. 59–94.

58. Les fonctions fuchsiennes et l'équation
$\Delta u = e^u$.-C. r. Acad. sci., 1898, 126,
627–630; Oeuvres, t. 11, p. 67–70.

59*. Les fonctions fuchsiennes et l'équation $\Delta u = e^u$. – J. math. pures et appl., 5e sér., 1898, 4, 137–230; Oeuvres, t. 11, p. 512 –591.

60. Sur les groupes continus.–C. r. Acad. sci., 1899, 128, 1065–1069; Oeuvres, t. 3. p. 169–172.

61. Sur les groupes continus.–Cambridge Philos. Trans., 1899, 18, 220–225; Oeuvres, t. III, p. 173–212.

62. Quelques remarques sur les groupes continus.–Rendiconti Circolo mat. Palermo, 1901, 15, 321–368; Oeuvres, t. III, p. 213–260.

63. Sur l'intégration algébrique des équations linéaires et les périodes des intégrales abéliennes.–J. math. pures et appl., 5e sér., 1903, 9, 139–212; Oeuvres, t. III, p. 106–166.

64. Nouvelles remarques sur les groupes continus.–Rendiconti Circolo mat. Palermo, 1908, 25. 81–130; Oeuvres, t. III, p. 261–321.

65. Fonctions modulaires et fonctions fuchsiennes.–Ann. Fac. sci. Toulouse, 3e sér., 1912, 3, 125–149; Oeuvres, t. II, p. 592–618.

66. Lettres à L. Fuchs (1880, 1881).–Acta math., 1921, 38, 175–184; Oeuvres, t. XI, p. 13–25.

67. Correspondance d'Henri Poincaré et de Felix Klein (1881, 1882). – Acta math., 1923, 39, 94–132; Oeuvres, t. XI, p. 26–65.

68. Sur les fonctions fuchsiennes. (Extrait d'un Mémoire inédit de Henri Poincaré). – Acta math., 1923, 39, 59–93; Oeuvres, t. I, p. 336–373.

2. *Theory of functions.*

69. Sechs Vorträge über ausgewählte Gegenstände aus der reinen Mathematik und mathematischen Physik (Göttingen, 22–28. IV 1909). Leipzig u. Berlin, 1910, 60 S. Includes [109], [110], [255], [340], [439], [440].

70. Sur les fonctions abéliennes. – C. r. Acad. sci., 1881, 92, 958–959; Oeuvres, t. IV, p. 299–301.

71. Sur une propriété des fonctions uniformes. – C. r. Acad. sci., 1881, 92, 1335–1336; Oeuvres, t. IV, p. 9–10.

72. Sur les transcendantes entières. – C. r. Acad. sci., 1882, 95, 23–26; Oeuvres, t. IV, p. 14–16.

73. Sur un théorème de la théorie générale des fonctions. – Bull. Soc. math. France, 1883, 11, 112–125; Oeuvres, t. IV, p. 57–69.

74. Sur les Θ fonctions. – Bull. Soc. math. France, 1883, 11, 129–134; Oeuvres, t. IV. p. 302–306.

75. Sur les fonctions entières. – Bull. Soc. math. France, 1883, 11, 136–144; Oeuvres, t. IV, p. 17–24.

76. Sur les fonctions de deux variables. – C. r. Acad. sci., 1883, 96, 238–240; Oeuvres, t. IV, p. 144–146.

77. Sur les fonctions de deux variables. – Acta math., 1883. 2, 97–113; Oeuvres, t. IV. p. 147–161.

78. Sur les fonctions à espaces lacunaires. – C. r. Acad. sci., 1883 96. 1134–1136; Oeuvres, t. IV, p. 25–27.

79. Sur les fonctions à espaces lacunaires. – Acta Societatis scientiarum Fennicæ, 1883, 12, 343–350; Oeuvres, t. IV, p. 28–35.

80. Sur un théorème de Riemann relatif aux fonctions de n variables indépendantes admettant $2n$ systémes de périodes (en collaboration avec E. Picard). – C. r. Acad. sci., 1883, 97, 1284–1287; Oeuvres, t. IV, p. 307–310.

81. Sur les substitutions linéaires. – C. r. Acad. sci., 1884, 98. 349–352; Oeuvres, t. IV, p. 531–533.

82. Sur la réduction des intégrales abéliennes. – Bull. Soc. math. France, 1884, 12, 124–143; Oeuvres, t. III, p. 333–351.

83. Sur la réduction des intégrales abéliennes. – C. r. Acad. sci., 1884, 99, 853–855; Oeuvres, t. III, p. 352–354.

84. Sur les intégrales de différentielles totales. – C. r. Acad sci., 1884, 99, 1145–1147; Oeuvres, t. III, p. 355–356.

85. Sur une généralisation du théorème d'Abel. – C. r. Acad. sci., 1885, 100, 40–42; Oeuvres, t. III, p. 357–359.

86. Sur les fonctions abéliennes. – C. r. Acad. sci., 1885, 100, 785–787; Oeuvres, t. IV, p. 311–313.

87. Sur les fonctions abéliennes. – Amer. J. Math., 1886, 8, 289–342; Oeuvres, t. IV, p. 318–378.

88. Sur la transformation des fonctions fuchsiennes et la réduction des intégrales abéliennes. – C. r. Acad. sci., 1886, 102, 41–44; Oeuvres, t. IV, p. 314–317.

89. Sur les résidus des intégrales doubles.– C. r. Acad. sci., 1886, 102, 202–204; Oeuvres, t. III, p. 437–439.

90. Sur la réduction des intégrales abéliennes. – C. r. Acad. sci., 1886, 102, 915–916; Oeuvres, t. III, p. 360–361.

91. Sur une classe étendue de transcendantes uniformes. — C. r. Acad. sci., 1886, 103. 862—864; Oeuvres, t. IV, p. 534—536.

92. Sur les résidus des intégrales doubles. — Acta math., 1887, 9, 321—380; Oeuvres, t. III, p. 440—489.

93. Sur une propriété des fonctions analytiques. — Rendiconti Circolo mat. Palermo, 1888, 2, 197—200; Oeuvres, t. IV, p. 11—13.

94. Sur une classe nouvelle de transcendantes uniformes. — J. math. pures et appl., 4e sér., 1890, 6, 313—365; Oeuvres, t. IV, p. 537—582.

95. Sur les fonctions à espaces lacunaires. — Amer. J. Math., 1892, 14, 201—221; Oeuvres, t. IV, p. 36—56.

96. Sur une propriété d'une fonction algébrique d'un arc. (Réponse à une question proposée par M. H. Dellac). — Intérmédiaire Mathématiciens, 1894, 1, 141—144.

97. Sur les fonctions abéliennes. — C. r. Acad. sci., 1895, 120, 239—243; Oeuvres, t. IV, p. 379—383.

98. Remarques diverses sur les fonctions abéliennes. — J. math., pures et appl., 5e sér., 1895, 1, 219—314; Oeuvres, t. IV, p. 384—468.

99. Sur les fonctions abéliennes.–C. r. Acad. sci., 1897, 124, 1407–1411; Oeuvres, t. IV, p. 469–472.

100. Sur les périodes des intégrales doubles.–C. r. Acad. sci., 1897, 125, 995–997; Oeuvres, t. III, p. 490–492.

101. Sur les propriétés du potentiel et sur les fonctions abéliennes.–Acta math., 1898, 22, 89–178; Oeuvres, t. IV, p. 162–243.

102. Sur les fonctions abéliennes.–Acta math., 1902, 26, 43–98; Oeuvres, t. IV, p. 473–526.

103. Sur les périodes des intégrales doubles.–J. math. pures et appl., 6e sér., 1906, 2, 135–189; Oeuvres, t. III, p. 493–539.

104. Les fonctions analytiques de deux variables et la représentation conforme.–Rendiconti Circolo mat. Palermo, 1907, 23, 185–220; Oeuvres, t. IV, p. 244–289.

105. Sur l'uniformisation des fonctions analytiques.–Acta math., 1908, 31, 1–63; Oeuvres, t. IV, p. 70–139.

106. Remarques sur l'équation de Fredholm.–C. r. Acad. sci., 1908, 147, 1367–1371; Oeuvres, t. III, p. 540–544.

107. Sur la réduction des intégrales abéliennes et les fonctions fuchsiennes. — Rendiconti Circolo mat. Palermo, 1909, 27, 281—336; Oeuvres, t. III, p. 362—428.

108. Sur quelques applications de la méthode de M. Fredholm. — C. r. Acad. sci., 1909, 148, 125—126; Oeuvres, t. III, p. 545—546.

109. Ueber die Fredholmschen Gleichungen. – In: Sechs Vorträge über ausgewählte Gegenstände aus der reinen Mathematik und mathematischen Physik von H. Poincaré. Leipzig u. Berlin, 1910, S. 1—10; Oeuvres, t. III, p. 547—554 (на франц. яз.).

110. Ueber die Reduction des Abelschen Integrale und die Theorie der Fuchsschen Funktionen. – In: Sechs Vorträge über ausgewählte Gegenstände aus der reinen Mathematik und mathematischen Physik von H. Poincaré. Leipzig u. Berlin, 1910, S. 33–41; Oeuvres, t. III, p. 429–436 (на франц. яз.).

111. Remarques diverses sur l'équation de Fredholm.–Compt. rend. Sessions Assoc. franç. avancement sci., 38e Session, Lille 1909, p. 1–28; Acta math., 1910, 33, 57–86; Oeuvres, t. III, p. 555–582.

112. Lettres à M. Mittag-Leffler (1 juin 1881, 29 juin 1881, 26 juillet 1881).–Acta math., 1921, 38, 147–160.

3. Algebra and arithmetic.

113. Sur quelques propriétés des formes quadratiques.–C. r. Acad. sci., 1879, 89, 344–346; Oeuvres, t. V, p. 189–191.

114. Sur les formes quadratiques.–C. r. Acad. sci., 1879, 89, 897–899; Oeuvres, t. V, p. 192–194.

115. Sur les formes cubiques ternaires.–C. r. Acad. sci., 1880, 90, 1336–1339; Oeuvres, t. V, p. 25–27, 291–292.

116. Sur la réduction simultanée d'une forme quadratique et d'une forme linéaire.–C. r. Acad. sci., 1880, 91, 844–846; Oeuvres, t. V, p. 337–339.

117. Sur un mode nouveau de représentation géométrique des formes quadratiques définies ou indéfinies.–J. École Polytechn., 1880, Cahier 47, 177–245; Oeuvres, t. V, p. 117–180.

118. Sur la représentation des nombres par les formes. — C. r. Acad. sci., 1881, 92, 777—779; Oeuvres, t. V, p. 397—399.

119. Sur les invariants arithmétiques. — Assoc. franç. avancement sci., 10° Session. Alger, 1881, p. 109—117; Oeuvres, t. V, p. 195—202.

120. Sur les applications de la géometrie non-euclidienne à la théorie des formes quadratiques. — Assoc. franç. avancement sci., 10° Session. Alger, 1881, p. 132—138; Oeuvres, t. V, p. 267—274.

121*. Sur les formes cubiques ternaires et quaternaires. I partie. — J. Ecole Polytechn., 1881, Cahier 50, 199—253; Oeuvres, t. V, p. 28—72.

122*. Sur les formes cubiques ternaires et quaternaires. II partie. — J. École Polytechn., 1882, Cahier 51, 45—91; Oeuvres, t. V, p. 293—334.

123. Sur une extension de la notion arithmétique de genre. — C. r. Acad. sci., 1882, 94, 67—71; Oeuvres, t. V, p. 435—437.

124. Sur une extension de la notion arithmétique de genre. — C. r. Acad. sci., 1882, 94, 124—127; Oeuvres, t. V, p. 438—440.

125. Sur la reproduction des formes. — C. r. Acad. sci., 1883, 97, 949—951; Oeuvres, t. V, p. 73—75.

126. Sur les équations algébriques. — C. r. Acad. sci., 1883, 97, 1418—1419; Oeuvres, t. V, p. 81—82.

127. Sur les nombres complexes. — C. r. Acad. sci., 1884, 99, 740—742; Oeuvres, t. V, p. 77—79.

128. Remarques sur l'emploi d'une méthode proposée par M. P. Appell intitulée Méthode élémentaire pour obtenir le développement en série trigonométrique des fonctions elliptiques. — Bull. Soc. math. France, 1884, 13, 19—27; Oeuvres, t. V, p. 85—94.

129. Sur une généralisation des fractions continues. — C. r. Acad. sci., 1884, 99, 1014—1016; Oeuvres, t. V, p. 185—187.

130. Sur la représentation des nombres par les formes. — Bull. Soc. math. France, 1885, 13, 162—194; Oeuvres, t. V, p. 400—432.

131. Sur les fonctions fuchsiennes et les formes quadratiques ternaires indéfinies. — C. r. Acad. sci., 1886, 102, 735—737; Oeuvres, t. II, p. 64—66; t. V, p. 275—277.

132. Sur les déterminants d'ordre infini. — Bull. Soc. math. France, 1886, 14, 77—90; Oeuvres, t. V, p. 95—107.

133. Réduction d'une forme quadratique et d'une forme linéaire. — J. École Polytechn., 1886, Cahier 56, 79—142; Oeuvres, t. V, p. 340—393.

134. Sur la distribution des nombres premiers. — C. r. Acad. sci., 1891, 113, 819; Oeuvres, t. V, p. 441.

135. Extension aux nombres premiers complexes des théorèmes de M. Tchebicheff. — J. math. pures et appl., 4° sér., 1891, 8, 25—68; Oeuvres, t. V, p. 442—479.

136. Sur le théorème de Goldbach relatif aux nombres premiers. (Question proposée en commun avec E. Catalan). — Iutérmédiaire Mathématiciens, 1894, 1, 91.

137. Sur le déterminant de Hill. — Bull. astron., 1900, 17, 134—143; Oeuvres, t. V, p. 108—116; t. VIII, p. 383—391.

138*. Sur les propriétés arithmétiques des courbes algébriques. — J. math. pures et appl., 5° sér., 1901, 7, 161—233; Oeuvres, t. V, p. 483—548.

139. Sur les invariants arithmétiques. — J. reine und angew. Math., 1905, 129 : 2, 89—150; Oeuvres, t. V, p. 203—265.

140. Sur les invariants arithmétiques. Talk at London University, 10 May 1912.

4. *Geometry.*

141. Démonstration nouvelle des propriétés de l'indicatrice d'une surface. — Nouvelles Annales Math., 2° sér., oct. 1874, 13, 449—456. Первая печатная работа А. Пуанкаре.

142. Sur les transformations des surfaces en elles-mêmes. — C. r. Acad. sci., 1886, 103, 732—734; Oeuvres, t. VI, p. 1—5.

143. Sur les transformations birationnelles des courbes algébriques. — C. r. Acad. sci., 1893, 117, 18—23; Oeuvres, t. VI, p. 6—11.

144. Sur le faisceau de cubiques passant par huit points d'un plan. (Question proposée). — Intérmédiaire Mathématiciens, 1894, 1, 2.

145. Sur le réseau de quadriques passant par sept points donnés dans l'espace. (Question proposée). — Intérmédiaire Mathématiciens, 1894, 1, 3.

146. Sur les courbes gauches particulieres. (Question proposée en commun avec M. Leon Autonne). — Intérmédiaire Mathématiciens, 1894, 1, 90.

147. Sur certaines familles de courbes algébriques. (Question proposée). — Intérmédiaire Mathématiciens, 1894, 1, 145.

148. Sur certaines familles des courbes algébriques. (Question proposée). — Intérmédiaire Mathématiciens, 1900, 7, 114—115.

149. Sur les surfaces de translation et les fonctions abéliennes. — Bull. Soc. math. France, 1901, 29, 61—86; Oeuvres, t. VI, p. 13—36.

150. Sur la généralisation d'un théorème élémentaire de géométrie.–C. r. Acad. sci., 1905, 140, 113–117; Oeuvres, t. XI, p. 8–12.

151*. Sur les lignes géodésiques des surfaces convexes.–Trans. Amer. Math. Soc., 1905, 6, 237–274; Oeuvres, t. VI, p. 38–84.

152. Sur les courbes tracées sur les surfaces algébriques.–C. r. Acad. sci., 1909, 149, 1026–1027; Oeuvres, t. VI, p. 86–87.

153*. Sur les courbes tracées sur les surfaces algébriques.–Ann. scient. École Normale supér., 3e sér., 1910, 27, 55–108; Oeuvres, t. VI, 88–139.

154*. Sur les courbes tracées sur une surface algébrique.–Sitzungsberichte der Berlin. math. Ges., 1911, 10, 28–55; Completed in Arch. Math., 1911, 18; Oeuvres, t. VI, p. 140–178.

5. *Topology.*

155. Sur l'Analysis situs.–C. r. Acad. sci., 1892, 115, 633–636; Oeuvres, t. VI, p. 189–192.

156. Sur la généralisation d'un théorème d'Euler relatif aux polyèdres.–C. r. Acad. sci., 1893, 117, 144–145; Oeuvres, t. XI, p. 6–7.

157*. Analysis situs.–J. École Polytechniques, 2e sér., 1895, Cahier 1, 1–121; Oeuvres, t. VI, p. 193–288.

158. Sur les nombres de Betti.–C. r. Acad. sci., 1899, 128, 629–630; Oeuvres, t. VI, p. 289.

159*. Complément à "l'Analysis situs".–Rendiconti Circolo mat. Palermo, 1899, 13, 285–343; Oeuvres, t. VI, p. 290–337.

160*. Second complément à "l'Analysis situs".–Proc. London Math. Soc., 1900, 32, 277–308; Oeuvres, t. VI, p. 338–370.

161. Sur "l'Analysis situs".–C. r. Acad. sci., 1901, 133, 707–709; Oeuvres, t. VI, p. 371–372.

162. Sur la connexion des surfaces algébriques.–C. r. Acad. sci., 1901, 133, 969–973; Oeuvres, t. VI, p. 393–396.

163*. Sur certaines surfaces algébriques; troisième complément à "l'Analysis situs".–Bull. Soc. math. France, 1902, 30, 49–70; Oeuvres, t. VI, p. 373–392.

164*. Sur les cycles des surfaces algébriques; quatrième complément à "l'Analysis situs". J. math. pures et appl., 5e sér., 1902, 8, 169–214; Oeuvres, t. VI, p. 397–434.

165*. Cinquième complément à "l'Analysis situs".–Rendiconti Circolo mat. Palermo, 1904, 18, 45–110; Oeuvres, t. VI, p. 435–498.

166*. Sur un théorème de géométrie.–Rendiconti Circolo mat. Palermo, 1912, 33, 375–407; Oeuvres, t. VI, p. 499–538.

II. Mechanics

1. *Analytical mechanics. Hydrodynamics.*

167. I. Cinématique pure. Mécanismes. II. Potentiel et Mécanique des fluides. Paris, 1886, Autographié, I—140 p., II — 140 p.

168. Cinématique et Mécanismes. Potentiel et Dynamique des fluides, 2° éd. Paris, G. Carré et C. Naud, 1899, 385 p.

169. Figures d'équilibre d'une masse fluide. Paris, G. Carré et C. Naud, 1902, 211 p.

170. Sur l'équilibre d'une masse fluide animée d'un mouvement de rotation. — C. r. Acad. sci., 1885, 100, 346–348; Oeuvres, t. VII, p. 14–16.

171. Sur l'équilibre d'une masse fluide animée d'un mouvement de rotation. — C. r. Acad. sci., 1885, 100, 1068–1070; Oeuvres, t. VII, p. 34–36.

172. Sur l'équilibre d'une masse fluide animée d'un mouvement de rotation. — C. r. Acad. sci., 1885, 101, 307–309; Oeuvres, t. VII, p. 37–39.

173. Sur l'équilibre d'une masse fluide animée d'un mouvement de rotation. — Bull. astron., 1885, 2, 109–118; Oeuvres, t. VII, p. 17–25.

174. Sur l'équilibre d'une masse fluide animée d'un mouvement de rotation.—Bull. astron., 1885, 2, 405—413; Oeuvres, t. VII, p. 26—33.

175. Sur l'équilibre d'une masse fluide animée d'un mouvement de rotation. — Acta math., 1885, 7, 259—380; Oeuvres, t. VII, p. 40—140.

176. Sur l'équilibre d'une masse fluide en rotation. — C. r. Acad. sci., 1886, 102, 970—972; Oeuvres, t. VII, p. 141—142.

177. Sur un théorème de M. Liapounoff, relatif à l'équilibre d'une masse fluide. — C. r. Acad. sci., 1887, 104, 622—625; Oeuvres, t. VII, p. 143—146.

178. Sur l'équilibre d'une masse hétérogène en rotation. — C. r. Acad. sci., 1888, 106, 1571—1574; Oeuvres, t. VII, p. 147—150.

179. Les formes d'équilibre d'un masse fluide en rotation. — Rev. gén. sci. pures et appl., 1892, 3, 809—815; Oeuvres, t. VII, p. 203—217.

180. Sur le problème de la rotation d'un corps solide autour d'un point fixe. (Réponse à une question proposée par M. Appell). — Intérmédiaire Mathématiciens, 1894, 1, 41—42.

181. Sur les solutions périodiques et le principe de moindre action. — C. r. Acad. sci., 1896, 123, 915—198; Oeuvres, t. VII, p. 224—226.

182. Sur les solutions périodiques et le principe de moindre action. — C. r. Acad. sci., 1897, 124, 713—716; Oeuvres, t. VII, p. 227—230.

183. Les idées de Hertz sur la Mécanique. — Rev. gén. sci. pures et appl., 1897, 8, 734—743; Oeuvres, t. VII, p. 231—250.

184. Sur l'équilibre d'un fluide en rotation. — Bull. astron., 1899, 16, 161—169; Oeuvres, t. VII, p. 151—158.

185. Sur une forme nouvelle des équations de la mécanique. — C. r. Acad. sci., 1901, 132, 369—371; Oeuvres, t. VII, p. 218—219.

186. Sur la stabilité de l'équilibre des figures piriformes affectées par une masse fluide en rotation. (Résumé). — Proc. Roy. Soc. London, 1901, 69, 148—149; Oeuvres, t. VII, p. 159—160.

187. Sur la stabilité de l'équilibre des figures piriformes affectées par une masse fluide en rotation. — Philos. Trans., sér. A, 1902, 198, 333—373; Oeuvres, t. VII, p. 161—202.

188. Sur une généralisation de la méthode de Jacobi. — C. r. Acad. sci., 1909, 149, 1105—1108; Oeuvres, t. VII, p. 220—223.

2. Celestial mechanics, astronomy, and geodesy.

189*. Les méthodes nouvelles de la Mécanique céleste, t. 1. Paris, Gauthier-Villars, 1892, 385 p.

190*. Les méthodes nouvelle de la Mécanique céleste, t. 2. Paris, Gauthier-Villars, 1893, VIII+479 p.

191*. Les méthodes nouvelle de la Mécanique céleste, t. 3. Paris, Gauthier-Villars, 1899, 414 p.

192. Théorie du potentiel newtonien. Paris, G. Carré et Naud, 1899, 366 p.

193. Cours d'Astronomie générale, avec un Supplement intitulé Mécanique céleste. École Polytechnique, autographié, 208 p.

194. Leçons de Mécanique céleste, t. 1. Paris, Gauthier-Villars, 1905, VI+367 p.

195. Leçons de Mécanique céleste, t. 2, partie I. Paris, Gauthier-Villars, 1907, IV+167 p.

196. Leçons de Mécanique céleste, t. 2, partie II. Paris, Gauthier-Villars, 1909, IV+137 p.

197. Leçons de Mécanique céleste, t. 3. Paris, Gauthier-Villars, 1910, IV+472 p.

198. Leçons sur les hypothèses cosmogoniques. Paris, Hermann et Fils, 1911, XV+294 p.; 2 éd., 1913, XV+294 p.

199. Sur les séries trigonométriques. — C. r. Acad. sci., 1882, 95, 766—768; Oeuvres, t. IV, p. 585—587.

200. Sur certaines solutions particulières du problème des trois corps. — C. r. Acad. sci., 1883, 97, 251—252; Oeuvres, t. VII, p. 251—252.

201. Sur les séries trigonométriques. — C. r. Acad. sci., 1883, 97, 1471—1473; Oeuvres, t. IV, p. 588—590.

202. Sur la convergence des séries trigonométriques.–Bull. astron., 1884, 1, 319–327; Oeuvres, t. IV, p. 591–598.

203. Sur une équation différentielle.–C. r. Acad. sci., 1884, 98, 793–795; Oeuvres, t. VII, p. 543–545.

204. Sur certaines solutions particulières du problème des trois corps.–Bull. astron., 1884, 1, 65–74; Oeuvres, t. VII, p. 253–261.

205. Sur les séries trigonométriques.–C. r. Acad. sci., 1885, 101, 1131–1134; Oeuvres, t. I, p. 164–166.

206. Note sur la stabilité de l'anneau de Saturne.–Bull. astron., 1885, 2, 507–508; Oeuvres, t. VIII, p. 457–458.

207. Sur un moyen d'augmenter la convergence des séries trigonométriques. — Bull. astron., 1886, 3, 521—528; Oeuvres, t. IV, p. 599—606.

208. Sur une méthode de M. Lindstedt. — Bull. astron., 1886, 3, 57—61; Oeuvres, t. VII, p. 546—550.

209. Sur la figure de la Terre. — C. r. Acad. sci., 1888, 107, 67—71; Oeuvres, t. VIII, p. 120—124.

210. Sur les satellites de Mars. — C. r. Acad. sci., 1888, 107, 890—892; Oeuvres, t. VIII, p. 459—460.

211. Sur les séries de M. Lindstedt. — C. r. Acad. sci., 1889, 108, 21—24; Oeuvres, t. VII, p. 551—554.

212. Sur la figure de la Terre. — Bull. astron., 1889, 6, 5—11; Oeuvres, t. VIII, p. 125—131.

213. Sur la figure de la Terre. — Bull. astron., 1889, 6, 49—60; Oeuvres, t. VIII, p. 132—142.

214*. Sur le problème des trois corps et les équations de la Dynamique. — Acta math., 1890, 13, 1—270; Oeuvres, t. VII, p. 262—479.

215. Sur le développement approché de la fonction perturbatrice. — C. r. Acad. sci., 1891, 112, 269—273; Oeuvres, t. VIII, p. 5—9.

216*. Sur le problème des trois corps. — Bull. astron., 1891, 8, 12—24; Oeuvres, t. VII, p. 480—490.

217. Le problème des trois corps. — Rev. gén. sci. pures et appl., 1891, 2, 1—5; Oeuvres, t. VIII, p. 529—537.

218. Sur l'application de la méthode de M. Lindstedt au problème des trois corps. — C. r. Acad. sci., 1892, 114, 1305—1309; Oeuvres, t. VII, p. 491—495.

219. Note accompagnant la présentation d'un ouvrage relatif aux Méthodes nouvelles de la Mécanique céleste. — C. r. Acad. sci., 1892, 115, 905—907.

220. Sur l'équilibre des mers. — C. r. Acad. sci., 1894, 118, 948—952; Oeuvres, t. VIII, p. 193—197.

221. Sur un procédé de vérification, applicable au calcul des séries de la Mécanique céleste. — C. r. Acad. sci., 1895, 120, 57—59; Oeuvres, t. VII, p. 555—557.

222. Observations au sujet de la communication de M. Delandres (intitulée: "Recherches spectrales sur la rotation et les mouvements des planètes").—C. r. Acad. sci., 1895, 120, 420—421.

223. Sur la divergence des séries de la Mécanique céleste.—C. r. Acad. sci., 1896, 122, 497—499; Oeuvres, t. VII, p. 558—560.

224. Sur la divergence des séries trigonométriques. — C. r. Acad. sci., 1896, 122, 557—559; Oeuvres, t. VII, p. 561—563.

225. Sur une forme nouvelle des équations du problème des trois corps. — C. r. Acad. sci., 1896, 123, 1031—1035; Oeuvres, t. VII, p. 496—499.

226. Sur la méthode de Bruns. — C. r. Acad. sci., 1896, 123, 1224—1228; Oeuvres, t. VII, p. 512—516.

227. Sur l'équilibre et les mouvements des mers. — J. Math. pures et appl., 5e sér., 1896, 2, 57—102; Oeuvres, t. VIII, p. 198—236.

228. Sur l'équilibre et les mouvements des mers. – J. Math. pures et appl., 5e sér., 1896, 2, 217–262; Oeuvres, t. VIII, p. 237–274.

229. Sur une forme nouvelle des équations du problème des trois corps. – Bull. astron., 1897, 14, 53–67; Oeuvres, t. VII, p. 500–511.

230. Sur l'intégration des équations du problème des trois corps. – Bull. astron., 1897, 14, 241–270; Oeuvres, t. VII, p. 517–542.

231. Sur le développement de la fonction perturbatrice. – Bull. astron., 1897, 14, 449–466; Oeuvres, t. VIII, p. 10–26.

232. Sur les périodes des intégrales doubles et les développement de la fonction perturbatrice. — C. r. Acad. sci., 1897, 124, 199—200; Oeuvres, t. VIII, p. 48—49.

233. Sur les périodes des intégrales doubles et le développement de la fonction perturbatrice. — J. Math. pures et appl., 5e sér., 1897, 3, 203—270; Oeuvres, t. VIII, p. 50—109.

234. Sur les périodes des intégrales doubles et le développement de la fonction perturbatrice. — Bull. astron., 1897, 14, 353—354; Oeuvres, t. VIII, p. 110—111.

235. Sur le développement approché de la fonction perturbatrice. — C. r. Acad. sci., 1898, 126, 370—373; Oeuvres, t. VIII, p. 27—30.

236. Développement de la fonction perturbatrice. — Bull. astron., 1898, 15, 70—71; Oeuvres, t. VIII, p. 31—32.

237. Sur la façon de grouper les termes des séries trigonométriques qu'on rencontre en Mécanique céleste. — Bull. astron., 1898, 15, 289—310; Oeuvres, t. VII, p. 564—582.

238. Développement de la fonction perturbatrice. — Bull. astron., 1898, 15, 449—464; Oeuvres, t. VIII, p. 33—47.

239. Sur la stabilité du système solaire. —
Rev. scient., 4ᵉ sér., 1898, 9, 609—
613; Oeuvres, t. VIII, p. 538—547.

240. Sur les quadratures mécaniques. —
Bull. astron., 1899, 16, 382—387;
Oeuvres, t. VIII, p. 461—466.

241. Sur le mouvement du périgée de
la Lune. — Bull. astron., 1900, 17,
87—104; Oeuvres, t. VIII, p 367—
382.

242. Sur les équations du mouvement
de la Lune. — Bull. astron., 1900,
17, 167—204; Oeuvres, t. VIII,
p. 297—331.

243. Sur la théorie de la précession.— C. r.
Acad. sci., 1901, 132, 50—55; Oeuvres, t. VIII, p. 113—117.

244. Les mesures de gravité et la Géodésie. — Bull. astron., 1901, 18, 5—39;
Oeuvres, t. VIII, p. 143—174.

245. Sur les déviations de la verticale en
Géodésie. — Bull. astron., 1901, 18,
257—276; Oeuvres, t. VIII, p. 175—
192.

246. Observations au sujet de l'article
de F. H. Seares, intitulé: «Sur les
quadratures mécaniques». — Bull. astron, 1901, 18, 406—420; Oeuvres,
t. VIII, p. 467—479.

247. Les solutions périodiques et les
planètes du type d'Hécube. — Bull.
astron., 1902, 19, 177—198; Oeuvres,
t. VIII, p. 417—436.

248. Sur les planètes du type d'Hécube. —
Bull. astron., 1902, 19, 289—310;
Oeuvres, t. VIII, p. 437—456.

249. Sur un théorème général relatif aux
marées. — Bull. astron., 1903, 20,
215—229; Oeuvres, t. VIII, p. 275—
288.

250. Sur la mèthode horistique de Gyldén. — C. r. Acad. sci., 1904, 138,
933—936; Oeuvres, t. VII, p. 583—
586.

251. Sur la méthode horistique. Observa-
tions sur l'article de M. Backlund. —
Bull. astron., 1904, 21, 292—295;
Oeuvres, t. VII, p. 619—621.

252. Sur la méthode horistique de Gyldén. — Acta math., 1905, 29, 235—
271; Oeuvres, t. VII, p. 587—618.

253. Sur la détermination des orbites par
la méthode de Laplace. — Bull. astron., 1906, 23, 161—187; Oeuvres,
t. VIII, p. 393—416.

254. Sur les petits diviseurs dans la théorie de la Lune. — Bull. astron.,
1908, 25, 321—360; Oeuvres, t. VIII,
p. 332—366.

255. Anwendung der Theorie der Integralgleichungen auf die Flutbewegung
des Meeres. — In: Sechs Vorträge über
ausgewählte Gegenstände aus der reinen
Mathematik und mathematischen Physik
von H. Poincaré. Leipzig u. Berlin, 1910,
S. 12—19; Oeuvres, t. VIII, p. 289–296.

256. Présentation du tome III des Leçons
de Mécanique céleste professées a la
Sorbonne. — C. r. Acad. sci., 1910,
150, 667.

257. Sur la précession des corps déformab-
les. — Bull. astron., 1910, 27, 321—
356; Oeuvres, t. VIII, p. 481—514.

258. Présentation des Leçons sur les Hy-
pothèes cosmogoniques. — C. r. Acad.
sci., 1911, 153, 795.

259. Remarque sur l'hypothèse de Laplace. — Bull. astron., 1911, 28, 251—
266; Oeuvres, t. VIII, p. 515—
528.

260. Note sur la XVIᵉ Conférence de l'Associa-
tion géodésique internationale.–Annuaire
Bureau Longitudes, 1911, p. A.1–A.29;
Oeuvres, t. VIII, p. 548–563.

261. Le démon d'Arrhénius.–Hommage à Louis
Olivier. Paris, 1911, p. 281–287; Oeuvres,
t. VIII, p. 564–569.

3. *Forewords and reports.*

262. Préface. – In: F. Tissorand. Leçons sur
la determination des orbites. Paris,
Gauthier-Villars, p. V–XIV; Bull. Sci.
math., 2ᵉ sér., 1809, 23, 107–117.

263. Rapport sur la proposition d'unification
des jours astronomique et civil.–Annuaire
Bureau Longitudes, 1895, p. E. 1–E. 10;
Oeuvres, t. VIII, p. 642–647.

264. Rapport sur les résolutions de la Commission chargée de l'étude des projets
de Décimalisation du Temps et de la
Circonférence.–Arch. Bureau Longitudes,
1897, 12 p.; Oeuvres, t. VIII, p. 648–664.

265. Rapport sur le projet de révision de l'arc
méridien de Quito.–C. r. Acad. sci., 1900,
131, 215–236; Oeuvres, t. VIII, p. 571–
592.

266. Rapport présenté au nom de la Commission chargée du contrôle scientifique des
opérations géodésiques de l'Équateur.–
C. r. Acad. sci., 1902, 134, 965–972;
Oeuvres, t. VIII, p. 593–601.

267. Rapport sur les opérations géodésiques
de l'Équateur.–Compt. rend. Séances 14ᵉ
Conf. gén. Assoc. Géodésique internat.
(4–13. VIII 1903), 1905, p. 113–127;
Oeuvres, t. VIII, p. 602–620.

268. Rapport présenté au nom de la Commission chargée du contrôle scientifique des opérations géodésiques de l'Équateur. — C. r. Acad. sci., 1903, 136, 861—871.

269. Rapport présenté au nom de la Commission chargée du contrôle scientifique des opérations géodésiques de l'Équateur. — C. r. Acad. sci., 1904, 138, 1013—1019.

270. Rapport présenté au nom de la Commission chargée du contrôle scientifique des opérations géodésiques de l'Équateur. — C. r. Acad. sci., 1905, 140, 998—1006.

271. Rapport présenté au nom de la Commission chargée du contrôle scientifique des opérations géodésiques de l'Équateur. — C. r. Acad. sci., 1907, 145, 366—370.

272. Rapports sur les opérations géodésiques de l'Équateur en 1903, 1904 et 1905, presentés à l'Academie des Sciences au nom de la Commission chargée du contrôle scientifique des opérations géodésiques de l'Équateur.-Compt. rend. Séances 15e Conf. gén. Assoc. Géodésique internat. (20–28. IV 1906), 1908, p. 289–304; Oeuvres, t. VIII, p. 621–641. [268], [269], [270].

273. Lettres à M. Mittag-Leffler concornant le Mèmoire couronné du prix de S. M. le roi Oscar II (18. IV 1883, 16. VII 1887, 5.II 1889, 1. III 1889, 5.III 1889).-Acta math., 1921, 38, 161–173; Oeuvres, t. XI, p. 66–78.

274. A propos de la décimalisation de l'heure.-Éclairage électrique, 1897, 12, 40.

275. La décimalisation de l'heure et de la circonférence.-Éclairage électrique, 1897, 11, 529–531; Oeuvres, t. VIII, p. 676–679.

276. Conférence sur les comètes.-Bull. Soc. industrielle de Milhouse, 1910, 80, 311–323; Oeuvres, t. VIII, p. 665–675.

III. Theoretical and mathematical physics.

1. Books.

277. Théorie mathématique de la lumiére, t. 1. Paris, G. Carré et Naud, 1889, IV+408 p.

278. Électricité et optique, t. 1. Les théories de Maxwell et la théorie électromagnetique de la lumière. Paris, G. Carré et Naud, 1890, XIX+ 314 p.

279. Électricité et optique, t. 2. Les théories de Helmholtz et les expériences de Hertz. Paris, G. Carré et Naud, 1891, XI+262 p.

280. Théorie mathématique de la lumière, t. 2. Nouvelles études sur la diffraction. Théorie de la dispersion de Helmholtz. Paris, G. Carré et Naud, 1892, VI+310 p.

281. Thermodynamique. Paris, G. Carré et Naud, 1892, XIX+432 p.

282. Leçons sur la théorie de l'elasticité. Paris, G. Carré et Naud, 1892, 210 p.

283. Théorie des tourbillons. Paris, G. Carré et Naud, 1893, 212 p.

284. Les oscillations électriques. Paris, G. Carré et Naud, 1894, 343 p.

285. Capillarité. Paris, G. Carré et Naud, 1895, 189 p.

286. Théorie analytique de la propagation de la chaleur. Paris, G. Carré et Naud, 1895, 316 p.

287. Calcul des probabilités. Paris, G. Carré et Naud, 1896, 275 p.

288. La théorie de Maxwell et les oscillations hertziennes. La Télégraphie sans fil. Paris, G. Carré et Naud, 1899, 80 p.

289. Électricité et optique. Leçons en 1888, 1890 et 1899. 2e éd. Paris, Gauthier-Villars, 1901, 632 p.

290. La théorie de Maxwell et les oscillations hertziennes. 2e éd. Paris, G. Naud, 1904, 80 p.; 3e éd., Gauthier-Villars, 1907, 97 p.

291. Thermodynamique, 2e éd. Paris, Gauthier-Villars, 1908, XIX+458 p.

292. Calcul des probabilités. 2e éd., rev. et augm. Paris, Gauthier-Villars, 1912, IV+335 p.

2. Differential equations of mathematical physics.

293. Sur le problème de la distribution électrique. — C. r. Acad. sci., 1887, 104, 44—46; Oeuvres, t. IX, p. 15—17.

294. Sur la théorie analytique de la chaleur. — C. r. Acad. sci., 1887, 104, 1753—1759; Oeuvres, t. IX, p. 18—23.

295. Sur la théorie analytique de la chaleur. — C. r. Acad. sci., 1888, 107, 967—971; Oeuvres, t. IX, p. 24—27.

296. Sur les équations aux dérivées partielles de la physique mathématique. — Amer. J. Math., 1890, 12, p. 211—294; Oeuvres, t. IX, p. 28—113.

297. Sur la propagation de l'électricité. — C. r. Acad. Sci., 1893, 117, 1027—1032; Oeuvres, t. IX, p. 278—283.

298. Sur certains développements en séries que l'on rencontre dans la théorie de la propagation de la chaleur. — C. r. Acad. sci., 1894, 118, 383—387; Oouvres, t. IX, p. 114—118.

299. Sur la série de Laplace. — C. r. Acad. sci., 1894, 118, 497—501; Oeuvres, t. IV, p. 607—610.

300. Sur l'équation des vibrations d'une membrane. — C. r. Acad. sci., 1894, 118, 447—451; Oeuvres, t. IX, p. 119—122.

301. Sur les équations de la physique mathématique. — Rendiconti Circolo mat. Palermo, 1894, 8, 57—155; Oeuvres, t. IX, p. 123—196.

302. Sur la méthode de Neumann et le problème de Dirichlet. — C. r. Acad. sci., 1895, 120, 347—352; Oeuvres, t. IX, p. 197—201.

303. Sur l'équilibre d'un corps élastique. — C. r. Acad. sci., 1896, 122, 154—159; Oeuvres, t. IX, p. 273—277.

304. La méthode de Neumann et le problème de Dirichlet. — Acta math., 1896—1897, 20, 59—142; Oeuvres, t. IX, p. 202—272.

305. Fourier's séries (lettre à A. A. Michelson). — Nature, 1899, 60, May 18, 52.

3. *Physical theory.*

306. Sur la polarisation par diffraction. — Acta math., 1892—1893, 16, 297—339; Oeuvres, t. IX, p. 293—330.

307. A propos de la théorie de M. Larmor. — Éclairage électrique, 1895, 3, 5—13; Oeuvres, t. IX, 369—382.

308. A propos de la théorie de M. Larmor. — Éclairage électrique, 1895, 3, 289—295; Oeuvres, t. IX, p. 383—394.

309. A propos de la théorie de M. Larmor. — Éclairage électrique, 1895, 5, 5—14; Oeuvres, t. IX, p. 395—413.

310. A propos de la théorie de M. Larmor. — Éclairage électrique, 1895, 5, 385—392; Oeuvres, t. IX, p. 414—426.

311. Sur la polarisation par diffraction. II partie (cf. [306]). — Acta math., 1896—1897, 20, 313—355; Oeuvres, t. IX, p. 331—368.

312. La théorie de Lorentz et les expériences de Zeeman.-Éclairage électrique, 1897, 11. 481–489; Oeuvres, t. IX, p. 427–441.

313. La théorie de Lorentz et le phénomène de Zeeman.-Éclairage électrique, 1899, 19, 5–15; Oeuvres, t. IX, p. 442–460.

314. Le phénomène de Hall et la théorie de Lorentz.-C. r. Acad. sci., 1899, 128, 339–341; Oeuvres, t. IX, p. 461–463.

315. La théorie de Lorentz et le principe de réaction.-Arch. Neerl. sci. exactes et natur., 2e sér., 1900, 5, 252–278; Oeuvres, t. IX, p. 464–488.

316*. Sur la dynamique de l'électron.-C. r. Acad. sci., 1905, 140, 1504–1508; Oeuvres, t. IX, p. 489–493.

317*. Sur la dynamique de l'électron.-Rendiconti Circolo mat. Palermo, 1906, 21, 129–176; Oeuvres, t. IX, p. 494–550.

318*. Réflexions sur la théorie cinétique des gaz.-J. Phys. théoret. et appl., 4e sér., 1906, 5, 369–403; Bull. Soc. franç. phys., 1906, 150–184; Oeuvres, t. IX, p. 587–619.

319*. La dynamique de l'électron.-Revue gén. sci. pures et appl., 1908, 19, 386–402; Oeuvres, t. IX, p. 551–586.

320*. Sur la théorie des quanta.-C. r. Acad. sci., 1911, 153, 1103–1108; Oeuvres, t. IX, p. 620–625.

321*. Sur la théorie des quanta.-J. Phys. théoret. et appl., 5e sér., 1912, 2, 5–34; Oeuvres, t. IX, p. 626–653.

322*. L'hypothèse des quanta. — Revue scient., 4° sér., 1912, 17, 225—232; Oeuvres, t. IX, p. 654—668.

323. Les rapports de la matière et de l'éther. — J. Phys. théoret. et appl., 5° sér., 1912, 2, 347—360; Oeuvres, t. IX, p. 669—682.

324. La theorie du rayonnement. Lecture at London University, May 11, 1912, cf. [320], [321], [322].

4. *Electromagnetic theory.*

325. Contribution à la théorie des expériences de M. Hertz.-C. r. Acad. sci., 1890, 111, 322–326; Oeuvres, t. X, p. 1–5.

326. Contribution à la théorie des expériences de Hertz.-Arch. sci. phys. et natur. Genève, 3e période, 1890, 24, 285–290.

327. Sur le calcul de la période des excitateurs horizions.-Arch. sci. phys. et natur. Genève, 3e période, 1891, 25, 5–25; Oeuvres, t. X, p. 6–19.

328. Sur la résonance multiple des oscillations hertziennes.–Arch. sci. phys. et natur. Genève, 3e période, 1891, 25, 609–627; Oeuvres, t. X, p. 20–32.

329. Sur la théorie des oscillations hertziennes. C. r. Acad. sci., 1891, 113, 515–519; Oeuvres, t. X, p. 33–37.

330. Sur un mode anormal de propagation des ondes.–C. r. Acad. sci., 1892, 114, 16–18; Oeuvres, t. X, p. 38–40.

331. Sur la propagation des oscillations hertziennes.–C. r. Acad. sci., 1892, 114, 1046–1048; Oeuvres, t. X, p. 41–43.

332. Sur la propagation des oscillations électriques.–C. r. Acad. sci., 1892, 114, 1229–1233; Oeuvres, t. X, p. 44–47.

333. Observations sur la Communication précédente de M. M. Birkeland et Sarasin.–C. r. Acad. sci., 1893, 117, 622–624; Oeuvres, t. X, p. 48–52.

334. Sur la diffraction des ondes électriques; à propos d'un article de M. Macdonald.–Proc. Roy. Soc., London, 1903, 72, 42–52; Oeuvres, t. X, p. 53–64.

335. Les ondes hertziennes et l'équation de Fredholm. — C. r. Acad. sci., 1909, 148, 449–453; Oeuvres, t. X, p. 65–69.

336. Sur la diffraction des ondes hertziennes. — C. r. Acad. sci., 1909, 148, 812–817; Oeuvres, t. X, p. 70–75.

337. Sur la diffraction des ondes hertziennes. — C. r. Acad. sci., 1909, 148, 966–968; Oeuvres, t. X, p. 76–77.

338. Les ondes hertziennes et l'equation de Fredholm. — C. r. Acad. sci., 1909, 148, 1488–1490; Oeuvres, t. X, p. 89–91.

339. Sur la diffraction des ondes hertziennes. — C. r. Acad. sci., 1909, 149, 621–622; Oeuvres, t. X, p. 92–93.

340. Anwendung der Integralgleichungen auf Hertzsche Wellen. — In: Sechs Vorträge über ausgewahlte Gegenstände aus der reinen Mathematik und mathematischen Physik. Leipzig u. Berlin, 1910, S. 21–31; Oeuvres, t. X, p. 78–88.

341. Sur la diffraction des ondes hertziennes. — Lumière électrique, 2e sér., 1910, 10, 355–362, 387–394; 11, 7–12.

342. Sur la diffraction des ondes hertziennes. — Rendiconti Cricolo mat. Palermo, 1910, 29, 169–259; Oeuvres, t. X, p. 94–203.

343. Über einige Gleichungen in der Theorie der Hertzschen Wellen. — Math. Naturwiss. Blätter (Berlin), 1910, 8, N 4; Oeuvres, t. X, p. 204–213.

344. Sur la diffraction des ondes hertziennes. — C. r. Acad. sci., 1912, 154, 795–797; Oeuvres, t. X, p. 214–215.

5. Criticisms and extensions of physical theories.

345. Sur les tentatives d'explication mécanique des principes de la thermodynamique. — C. r. Acad. sci., 1889, 108, 550–553; Oeuvres, t. X, p. 231–233.

346. Sur la loi électrodynamique de Weber. — C. r. Acad. sci., 1890, 110, 825–829; Oeuvres, t. X, p. 292–296.

347. Sur l'expérience de M. Wiener. — C. r. Acad. sci., 1891, 112, 325–329; Oeuvres, t. X, p. 271–277.

348. Sur la réflexion métallique. — C. r. Acad. sci., 1891, 112, 456–459; Oeuvres, t. X, 278–286.

349. Sur l'équilibre des diélectriques fluides dans un champ électrique. — C. r. Acad. sci., 1891, 112, 555–557; Oeuvres, t. X, p. 297–298.

350. Sur la théorie de l'élasticité. — C. r. Acad. sci., 1891, 112, 914–915; Oeuvres, t. X, p. 221–227.

351. Sur la théorie de l'élasticité. — C. r. Acad. sci., 1892, 114, 385–389; Oeuvres, t. X, p. 228–230.

352. Réponse à l'article de P. G. Tait: Poincaré's Thermodynamics. — Nature, 1892, 45, 414–415; Oeuvres, t. X, p. 234–235.

353. Réponse à P. G. Tait. — Nature, 1892, 45, 485; Oeuvres, t. X, p. 236–237.

354. Réponse à P. G. Tait. — Nature, 1892, 46, 76; Oeuvres, t. X, p. 238–239.

355. Sur une objection à la théorie cinétique des gaz. — C. r. Acad. sci., 1893, 116, 1017–1021; Oeuvres, t. X, p. 240–243.

356. Sur la théorie cinétique des gaz. — C. r. Acad. sci., 1893, 116, 1165–1166; Oeuvres, t. X, p. 244–245.

357. Sur la théorie cinétique des gaz. — Rev. gén. sci. pures et appl., 1894, 5, 513–521; Oeuvres, t. X, p. 246–263.

358. La lumière et l'électricité d'après Maxwell et Hertz. — Annuaire Bureau Longitudes, 1894, A. 1—A. 22; Rev. scient., 4ᵉ sér., 1894, 1, 106—111; Oeuvres, t. X, p. 557—569.

359. Sur le spectre cannelé. — C. r. Acad. Sci., 1895, 120, 757—762; Oeuvres, t. X, p. 287—291.

360. Remarque sur un Mémoire de M. Jaumann intitulé: «Longitudinales Licht». — C. r. Acad. sci., 1895, 121, 792—793; Oeuvres, t. X, p. 299—306.

361. Observations au sujet de la communication précédente (de M. G. Jaumann). — C. r. Acad. sci., 1896, 122, p. 76.

362. Observations au sujet de la communication précédente (de M. Jaumann). — C. r. Acad. sci., 1896, 122, 520.

363. Observations au sujet de la communication de M. Jaumann. — C. r. Acad. sci., 1896, 122, 990.

364. Les rayons cathodiques et la théorie de Jaumann. — Éclairage éléctrique, 1896, 9, 241—251; Oeuvres, t. X, p. 314—332.

365. Les rayons cathodiques et la théorie de Jaumann. — Éclairage éléctrique, 1896, 9, 289—293; Oeuvres, t. X, p. 333—340.

366. Observations au sujet de la communication de M. J. Perrin "Quelques propriétés des rayons de Röntgens."–C. r. Acad. sci., 1896, 122, 188; Oeuvres, t. X, p. 307.

367. Observations au sujet de la communication de M. G. de Metz "Photographie à l'intérieur du tube de Crookes".–C. r. Acad. sci., 1896, 192, 881; Oeuvres, t. X, p. 308.

368. Observations au sujet de la communication de M. G. de Metz: "Photographie à l'intérieur du tube de Crookes".–C. r. Acad. sci., 1896, 123, 356; Oeuvres, t. X, p. 309.

369. Remarques sur une expérience de M. Birkeland.–C. r. Acad. sci., 1896, 123, 530–533; Oeuvres, t. X, p. 310–313.

370. Les rayons cathodiques et les rayons Röntgen.–Rev. gén. sci. pures et appl., 1896, 7, 52–59; Oeuvres, t. X, p. 570–583.

371. Observations au sujet de la note de M. J. J. Thomson (intitulée "On the cathode Rays").–Éclairage électrique, 1897, 12, 186.

372. Les rayons cathodiques et les rayons Röntgen.–Annuaire Bureau Longitudes, 1897, D.1–D.35; Rev. scient., 4ᵉ sér., 1897, 7, 72–81; Oeuvres, t. X, p. 584–603.

373*.La mesure du temps.–Rev. métaphys. et morale, 1898, 6, 1–13.

374. L'énergie magnétique d'après Maxwell et d'après Hertz.–Éclairage électrique, 1899, 19, 361–367; Oeuvres, t. X, p. 341–351.

375. Sur l'induction unipolaire.–Éclairage électrique, 1900, 23, 41–53; Oeuvres, t. X, p. 355–371.

376. Sur les excitateurs et résonateurs hertziens (à propos d'un article de M. Johnson).–Éclairage électrique, 1901, 29, 305–307; Oeuvres, t. X, p. 352–354.

377. A propos des expériences de M. Crémieu. — Rev. gén. sci. pures et appl., 1901, 12, 994—1007; Oeuvres, t. X, p. 391—420.

378. Sur les propriétés des anneaux à collecteurs. — Éclairage éléctrique, 1902, 30, 77—81; Oeuvres, t. X, p. 372—377.

379. Sur les propriétés des anneaux à collecteurs.–Éclairage électrique, 1902, 30, 301–310; Oeuvres, t. X, p. 378–390.

380. Sur les expériences de M. Grémieu et une objection de M. Wilson.–Éclairage électrique, 1902, 31, 83–93; Oeuvres, t. X, p. 421–437.

381. Notice sur la télégraphie sans fil.–Annuaire du Bureau des Longitudes, 1902, A.1–A.34; Rev. scient., 4ᵉ sér., 1902, 17, 65–73; Oeuvres, t. X, p. 604–622.

382. Éntropy.–Électrician, 1903, 50, 688–689; Oeuvres, t. X, p. 264–270.

383. Théorie de la balance azimutale quadrifilaire. — C. r. Acad. sci., 1904, 138, 869–874; Oeuvres, t. X. p, 438–444.

384. Étude de la propagation du courant en période variable sur une ligne munie de récepteur.–Éclairage électrique, 1904, 40, 121–128, 161–167, 201–212, 241–250; Oeuvres, t. X, p. 445–486.

385. Étude du récepteur téléphonique.–Éclairage électrique, 1907, 50, 221–234, 257–262, 329–338, 365–372, 401–404; Oeuvres, t. X, p. 487–539.

386. Sur quelques théorèmes generaux relatifs à l'Electrotechnique.–Éclairage électrique, 1907, 50, 293–301; Oeuvres, t. X, p. 540–551.

387. Sur la théorie de la commutation.–Lumière électrique, 2e sér., 1908, 2, 295–297; Oeuvres, t. X, p. 552–556.

388. Sur la télégraphie sans fil.–Lumière électrique, 2e sér., 1908, 4, 259–266, 291–297, 323–327, 355–359, 387–393; Conférences sur la Télégraphie sans fil. Paris, 1909, 86 p.

389. La télégraphie sans. fil. — J. Univ. annales, 1909, 1, 541–542.

390. Sur les signaux horaires destinés aux marins. — C.. r. Acad. sci., 1910, 150, 1471–1472.

391. Sur l'envoi de l'heure par la télégraphie sans fil. — C. r. Acad. sci., 1910, 151, 911.

392. Sur diverses questions relatives à la télégraphie sans fil. — Lumière électrique, 2e sér., 1911, 13, 7–12.

393. Sur diverses questions relatives à la télégraphie sans fil. — Lumière électrique, 2e sér., 1911, 13, 35–40.

394. Sur diverses questions relatives à la télégraphie sans fil. — Lumière électrique, 2e sér., 1911, 13, 67–72.

395. Sur diverses questions relatives à la télégraphie sans fil. — Lumière électrique, 2e sér., 1911, 13, 99–104.

396. Présentation de la 2e édition de l'ouvrage Calcul des Probabilités. — C. r. Acad. sci., 1911, 153, 795.

IV. General Problems of Science

397. La science et l'hypothèse. Paris, Flammarion, 1902, 284 p.

398. La valeur de la science. Paris, Flammarion, 1905, 278 p.

399. La science et l'hypothèse. Éd. revue et corrigée. Paris, Flammarion, 1906, 281 p.

400. Science et méthode. Paris, Flammarion, 1908, 314 p.

401. Dernières pensées. Paris, Flammarion, 1913, 258 p. Includes [322], [323], [435], [445], [447], [448].

402. Sur les hypothèses fondamentales de la géométrie.–Bull. Soc. math. France, 1887, 15, 203–216; Oeuvres, t. XI, p. 79–91.

403. Les géométries non euclidiennes. — Rev. gén. sci. pures et appl., 1891, 2, 769–774.

404. Lettre à M. Mouret sur les géométries non euclidiennes. — Rev. gén. sci. pures et appl., 1892, 3, 74–75.

405. La continu mathématique. — Rev. métaphys. et morale, 1893, 1, 26–34.

406. Mécanisme et expérience. — Rev. métaphys. et morale, 1893, 1, 534–537.

407. Mécanisme et expérience (réponse à M. Léchalas). — Rev. métaphys. et morale, 1894, 2, 197–198.

408. Sur la nature du raissonnement mathématique. — Rev. métaphys. et morale, 1894, 2, 371–384.

409. L'espace et la géométrie. — Rev. métaphys. et morale, 1895, 3, 631–646.

410. Réponse à quelques critiques (relatives aux [406], [409]). — Rev. métaphys. et morale, 1897, 5, 59–70.

411. On the foundations of geometry. — Monist, 1898–1899, 9, 1–43.

412. Sur les fondements de la géométrie, à propos d'un livre de M. Russell. — Rev. métaphys. et morale, 1899, 7, 251–279.

413. Réflexion sur le calcul des probabilités. — Rev. gén. sci. pures et appl., 1899, 10, 262–269.

414. Sur les principes de la géométrie. Réponse à M. Russel. — Rev. métaphys. et morale, 1900, 8, 73–86.

415. Comptes rendus des Séances du Congres de Philosophie, discussion. — Rev. métaphys. et morale, 1900, 8, 556–561.

416. Sur la valeur objective de la science.— Rev. métaphys. et morale, 1902, 10, 263–293.

417. Grandeur de l'astronomie. — Bull. Soc. astron. France, 1903, 17, 253–259.

418. L'espace et ses trois dimensions. — Rev. métaphys. et morale, 1903, 11, 281–301.

419. L'espace et ses trois dimensions. — Rev. métaphys. et morale, 1903, 11, 407–429.

420. La Terre tourne-t-elle? — Bull. Soc. astron. France, 1904, 18, 216–217.

421. Les mathématiques et la logique. — Rev. métaphys. et morale, 1905, 13, 815–835.

422. Une image de l'Univers. — Bull. Soc. astron. France, 1905, 19, 30–31.

423. Cournot et les principes du calcul infinitésimal. — Rev. métaphys. et morale, 1905, 13, 293–306.

424. Les mathématiques et la logique. —
Rev. metaphys. et morale, 1906, 14,
17—34.

425. Les mathématiques et la logique. —
Rev. métaphys. et morale, 1906,
14, 294—317.

426. A propos de la logistique. — Rev.
métaphys. et morale, 1906, 14,
866—868.

427. Lettre à M. G. F. Stout. — Mind,
1906, 15, 141—143.

428. La Voie Lactée et -la théorie de
gaz. — Bull. Soc. astron. France.
1906, 20, 153—165.

429. La fin de la matière.–Atheneum (London),
17 Feb. 1906, 201–202; cf. also [399].

430. La relativité de l'espace.–Année Psychol.,
1907, 13, 1–17.

431. Le Hasard.–Rev. du Mois, 1907, 3, 257–
276; [400], Chap. IV.

432. Comment se fait la Science.–Le Matin, 25
nov. 1908.

433. Comment on Invente. Le travail de
l'inconscient.–Le Matin, 24 déc. 1908.

434. La Mécanique nouvelle.–Compt. rend.
Sessions Assoc. franç. avancement sci.
Conférence. Paris, 1909, p. 38–48; Rev.
scient., 1909, 47, 170–177.

435. La logique de l'Infini.–Re. métaphys. et
morale, 1909, 17, 461–482; [401], chap.
IV.

436. Le choix des faits.–Monist, 1909, 19, 231–
239; [400], chap. I.

437. Réflexions sur deux notes de M. A. S.
Schönflies et de M. E. Zermelo.–Acta
math., 1909, 32, 195–200; Oeuvres, t. XI,
p. 144–119.

438. Sur la nécessité de la culture scientifique.–
Palmarès du Lycée Henri IV. Paris, 1909–
1910, p. 31–36; Rev. internat. Enseignement, 1909, 58, 342–345.

439. Über transfinite Zahlen.–In: Sechs Vorträge über ausgewählte Gegenstände aus
der reinen Mathematik und mathematischen Physik. Leipzig u. Berlin, 1910.
S. 43–48; Oeuvres, t. XI, p. 120–124.

440. La Mécanique nouvelle. – In: Sechs Vorträge über ausgewählte Gegenstände aus
der reinen Mathematik und mathematischen Physik. Leipzig u. Berlin, 1910,
S. 49–58.

441. La Mécanique nouvelle. In: Hummel und
Erde (Leipzig), 1910, 23, 97–116; H.
Poincaré. Die neue Mechanik. Leipzig u.
Berlin, 1911, 22 S.

442. La morale et la science.–Foi et Vie (Paris),
1910, 13, 323–329; Questions du temps
présent. Paris, 1910, p. 49–69; Revue de
Jean Finot, 1910, 86, 289–302.

443. Le libre examen en matière scientifique.
Bruxelles, M. Weissenbruch, 1910, p. 97–
106.

444. Vue d'ensemble sur les hypothèses cosmogoniques. – Rev. du Mois, 1911, 12,
385–403.

445. L'espace et le temps. — Scientia
(Rivista di Scienza) 1912, 12,
159—171; [401], Chap. II.

446. Les conceptions nouvelles de la matière.
– Foi et Vie (Paris), 1912, 15, 185–191;
In: Le matérialisme actuel. Paris, E.
Flammarion, 1916, p. 49–67.

447. Pourquoi l'espace a trois dimmen-
sions. — Rev. métaphys. et morale,
1912, 20, 483—504; [401], Chap. III.

448. La logique de l'infini. — Scientia
(Rivista scienza), 1912, 12, 1—11;
[401], Chap. V.

V. Diverse topics

1. *Addresses at International Congresses.*

449. Sur les rapports de l'analyse pure
et de la physique mathématique. —
Acta math., 1897, 21, 331—341;
Rev. gén. sci. pures et appl., 1897,
8, 857—861; Verhandl. I internat.
mathematiker Kongress in Zürich
(1897). Leipzig, 1898, S. 81—
90.

450. Du rôle de l'intuition et de la logique
en mathématiques. — Compt. rend. II
Congrès internat. Mathématiciens.
Paris, 1900, p. 115—130.

451. Relations entre la Physique expérimentale et de la Physique mathématique. — Rapp. Congrès internat. Phys., t. 1. Paris, 1900,
p. 1—29; Rev. gén. sci. pures et appl.,
1900, 11, 1163—1175; Rev. scient.,
4° sér., 1900, 14, 705—715.

452. Sur les principes de la Mécanique. —
Bibliotèque du Congrès internat.
Philosophie, Paris, 1900, t. 3. Paris,
1901, p. 457—494.

453* L'état actuel et l'avenir de la Physique mathématique. — Bull. Sci.
math., 2° sér., 1904, 28, 302—324;
Monist, 1905, 15, N 1.

454. L'avenir des Mathématiques. — Atti IV Congr. Internaz. Matematici, Roma, 11 Aprile 1908, p. 167—182; Bull. sci. math., 2° sér., 1908, 32, 1 partie, 168—190; Rendiconti Circolo mat. Palermo, 1908, 16, 162—168; Rev. gén. sci. pures et appl., 1908, 19, 930—939; Scientia (Rivista scienza), 1908, 2, N 3, 1—23.

455. L'évolution des lois. — Scientia, (Rivista scienza), 1911, 9, 275—292.

2. *Analysis of mathematical works.*

456. Notice sur les travaux scientifiques de M. Poincaré (rédigée par lui-même). Paris, Gauthier-Villars, 1884, 51 p.

457. Notice sur les travaux scientifiques de M. Poincaré. 2° éd. Paris, Gauthier-Villars, 1886, 75 p.

458*. Analyse des travaux scientifiques de Henri Poincaré faite par lui-même. — Acta math., 1921, 38, 36—135.

459. Savants et écrivains. Paris, Flammarion, 1910, XIV+281 p.

460. Notice sur la vie et les travaux de M. Laguerre. — C. r. Acad. sci., 1887, 104, 1643—1650; Préface des Oeuvres de Laguerre, t. 1. Paris, 1898, p. V—XV.

461. Notice sur Halphen. — J. École Polytechn., 1890, cah. 60°, 137—161.

462. Rapport sur un mémoire de M. Callerier intitulé: «Sur les variations des excentricités et des inclinaisons». — C. r. Acad. sci., 1890, 110, 942—944.

463. Rapport sur un mémoire presenté par M. Blondlot et relatif propagation des oscillations hertziennes. — C. r. Acad. sci., 1892, 114, 645—648.

464. Rapport sur le concours du prix Bordin. — C. r. Acad. sci., 1892, 115, 1126—1127.

465. Au Jubilé de M. Charles Hermite. — In: Jubilé de M. Hermite. Paris, Gauthier-Villars, 1893, p. 6—8; Rev. Questions scient., 2e sér., 1893, 3, 244—246.

466. Rapport verbal (concernant une démonstration du théorème de Fermat. adressée par M. G. Korneck). — C. r. Acad. sci., 1894, 118, 841.

467. Au Cinquantenaire de l'entrée de M. Joseph Bertrand dans l'Enseignement. — Rev. scient., 4° sér., 1894, 1, 685—686; Annuaire École polytechn., 1895, 107—108.

468. Rapport sur un mémoire de M. Stieltjes intitulé: «Recherches sur les fractions continues». — C. r. Acad. sci., 1894, 119, 630—632.

469. Rapport sur le Concours du prix Bordin (en commun avec M. M. Picard et Appell). — C. r. Acad. sci., 1894, 119, 1051—1056.

470. Prix Bordin. (Rapport sur le mémoire de M. Hadamard). — C. r. Acad. sci., 1896, 123, 1109—1111.

471. Sur la vie et les travaux de F. Tisserand. — Rev. gén. sci. pures et appl., 1896, 7, 1230—1233.

472. Discours prononcé aux funérailles de M. Tisserand. — Bull. Astron., 1896, 13, 430—432. Annuaire Bureau Longitudes, 1897, p. H. 15—H. 18.

473. Rapport sur un mémoire de M. Hadamard (Lignes géodésiques sur les surfaces à courbures opposées). — C. r. Acad. sci., 1897, 125, 589—591.

474. Rapport sur un mémoire de M. Le Roy (Sur l'intégrations des équations de la chaleur). — C. r. Acad. sci., 1897, 125, 847—849.

475. Grand prix des Sciences mathématiques (en commun avec M. Picard). — C. r. Acad. sci., 1898, 127, 1061—1065.

476. L'oeuvre mathématique de Weierstrass. — Acta math., 1898, 22, 1—18.

477. Analyse d'un ouvrage de Ch. André (intitulé «Traité d'Astronomie stellaire»). — Bull. astron., 1899, 16, 124—127.

478. Appréciation d'un ouvrage de M. V. Bjerknes (intitulé: «Vorlesungen über hydrodynamische Fernkräfte»). — C. r. Acad. sci., 1900, 130, 25.

479. La Géodésie Française (discours prononcé à la séance des Cinq Académies le 25 Octobre 1900). — Mémoires de l'Institut, 1900, 20, 13—25; Bull. Soc. astron. France, 1900, 14, 513—521.

480. Les Géométries non euclidiennes. — In: E. Rouché et Ch. de Comberousse. Traite de Géométrie, 11 partie. Paris. Gauthier-Villars, 1900, p. 581—583.

481. Rapport sur les papiers laissés par Halphen. — C. r. Acad. sci., 1901, 133, 722—724.

482. Les progrès de l'Astronomie en 1901. — Bull. Soc. astron. France, 1902, 16, 214—223.

483. Sur la vie et les travaux de M. Faye. — Bull. Soc. astron. France, 1902, 16, 496—501.

484. Analyse d'un mémoire de M. Zaremba. — Bull. sci. math., 2° sér., 1902, 26, 1 partie, 337—350.

485. Discours prononcé aux funérailles de M. A. Cornu (16.IV 1902). — Mémoires de l'Institut, 1902, 15—18; Bull. Soc. franç. phys., 1902, 186—188; Annuaire Bureau Longitudes, 1903, p. D. 7—D. 11.

486. Sur M. A. Cornu (lettre a M. C. M. Gariel, avr. 1902). — Bull. Soc. franç. phys., 1902, p. 32*—33*.

487. A. Cornu. — Éclairage électrique, 1902, 31, 81—82.

488. Les fondements de la Géométrie. — Bull. sci. math., 2° sér., 1902, 26, 249—272; J. Savants, 1902, 252—271; Oeuvres, t. XI, 92—113.

489. Les fondements de la Géométrie. — Bull. sci. math., 2° sér., 1903, 27, 115.

490. Sur les travaux de la Société Française du Physique. — Bull. Soc. franç. phys., 1903, p. 5—8.

491. Sur la Part des Polytechniciens dans l'Oeuvre scientifique du XIX° siècle. — Compte rendu. Paris, Gauthier-Villars, 1903, p. 11—17.

492. Rapport sur les travaux de M. Hilbert. — Proc. Phys-Mat. Soc. Kazan, 14 (104) 10—48. Concerns the work of Hilbert, Grundlagen der Geometrie, 1899, presented at the Third Lobachevsky Concourse.

493. Rapport sur le Concours du Prix Leconte. — C. r. Acad. sci., 1904, 139, 1120—1122.

494. Prix Damoiseau. — C. r. Acad. sci., 1905, 141, 1076—1077.

495. Rapport sur un Mémoire de M. Bachelier intitulé: "Les probabilités continues".—C. r. Acad. sci., 1905, 141, 647—648.

496. Preface. — In: G. W. Hill. Collected Mathematical Works, v. 1. Washington, 1905, p. V—XVIII.

497. Notice sur la vie et les oeuvres d'Alfred Cornu. — In: Alfred Cornu. Rennes, Francis Simon, 1904, p. 9—21; J. École polytechn., 2e sér., 1905, cah. 10e, 143—176.

498. A. Potier. — Éclairage électrique, 1905, 43, 281—282; In: A. Potier. Mémoires sur l'électricité et l'optique. Paris, Gauthier-Villars, 1912, p. V—X.

499. Sur de M. Langley, correspondant de l'Académie. — C. r. Acad. sci., 1906, 142, 925.

500. Sur de Membre de l'Académie M. Curie. — C. r. Acad. sci., 1906, 142, 939—941.

501. Sur de M. Bischoffsheim. — C. r. Acad. sci., 1906, 142, 1119.

502. Sur des Membres de l'Académie des Sciences et sur des Membres de la Mission géodesique à l'Équateur. — C. r. Acad. sci., 1906, 143, 989—998. Mémoirs de l'Institut, 1906, 23, p. 5—16.

503. Sur l'oeuvre de Marcelin Berthelot. — Le Matin, 25 mars 1907, p. 1.

504. Prix Vaillant. Rapport sur le Mémoire de M. Boggio et le Mémoire N 7 portant pour épigraphe «Barré de Saint-Venant», — C. r. Acad. sci., 1907, 145, 988—991.

505. Prix Monthyon. — C. r. Acad. sci., 1908, 147, 1199.

506. Préface. — In: Devaux-Charbonnel. État actuel de la science électrique. Paris, Dunod et E. Pinat, 1908, p. V—X.

507. Compte rendu d'ensemble des travaux du IV°. Congrès des Mathématiciens tenu à Rome en 1908. — Le Temps (Paris), 1908, 21 avril, p. 2—3.

508. Sur M. Maurice Loewy. — Annuaire Bureau Longitudes, 1908, p. D.1—D.18.

509. Lord Kelvin. — Lumière électrique, 2e sér., 1908, 1, 139—147.

510. Discours aux funérailles de M. Hippolyte Langlois. — Mémoires de l'Institut, 1909, 5 p.

511. Sur la vie et l'oeuvre poétique et philosophique (de Sully Prudhomme). — Mémoires de l'Institut, 1909, p. 3—37.

512. Sully Prudhomme, mathématicien. — Rev. gén. sci. pures et appl., 1909, 20, 657—662.

513. Rapport (sur le prix Bolyai, 18.X 1910). — Bull. sci. math., 1911, 31, 1re pt., 67—100; Acta math., 1912, 35, 1—28; Rendiconti Circolo mat. Palermo, 1912, 31, 109—132. Analysis of the work of D. Hilbert.

514. Préface de l'Oeuvrage de Jacques Lux. — In: J. Lux. Histoire de deux Revues francaises. Paris, 1911, p. 5—8.

515. Préface. — In: G. Lachapelle. La représentation proportionnelle en France et en Belgique. Paris, F. Alcan, 1911, p. III—XII.

516. Discours prononcé aux funérailles de M. Paul Gautier (9.XII 1909). — Annuaire Bureau Longitudes, 1911, p. D. 1—D. 11.

517. Notice nécrologique sur M. Bouquet de La Grye. — Annuaire Bureau Longitudes, 1911, p. C. 1—C. 13.

518. Discours prononcé aux funérailles de M. Rodolphe Radau (29.XII 1911). — Mémoires de l'Institut, 1911, 13—15; Bull. astron., 1912, 29, 88—89.

519. Discours au Jubilo de M. Gaston Darboux (21.I 1912). — Rev. internat. Enseignement, 1912, 59, 99—102.

520. Discours au Jubile de M. Camille Flammarion. — Bull. Soc. astron. France, 1912, 26, p. 101—103.

521. Rapport sur les travaux de M. Cartan (fait à la Faculté des Sciences de l'Université de Paris). — Acta math., 1921, 38, 137—145.

3. *Pedagogy.*

522. La notation différentielle et l'enseignement. — Enseignement Math., 1899, 1, 106—110; Oeuvres, t. XI, p. 125—128.

523. La logique et l'intuition dans la science mathématique et dans l'enseignement. — Enseignement Math., 1899, 1, 157—162; Oeuvres, t. XI, p. 129—133.

524. Les définitions générales en mathématiques. — Conférences du Musée pedagogique, 1904, 1—18; Enseignement Math., 1904, 6, 257—283.

525. L'Invention mathématique. - Enseignement Math., 1908, 10, 357–371; Bull. Institut gén. Psychol., 1908, 8, 175–187; Rev. du Mois, 1908, 6, 9–21; Rev. gén. sci. pures et appl., 1908, 19, 521–526; cf. also [400], Chap. III.

526. Les Astres. - In: H. Poincaré, E. Perrier, P. Painlevé. Ce que disent les choses. Paris, Hachett et Cie, 1912, p. 1–6.

527. En regardant tomber une pomme. - In: H. Poincaré, E, Perrier, P. Painlevé. Ce que disent les choses. Paris, Hachette et Cie, 1912, p. 1–6.

528. La Chaleur et l'Énergie. - In: H. Poincaré, E. Perrier, P. Painlevé. Ce que disent les choses. Paris, Hachette, et Cie, 1912, p. 11–14.

529. Les Mines. - In: H. Poincaré, E. Perrier, P. Painlevé. Ce que disent les choses. Paris, Hachette, et Cie, 1912, p. 69–74.

530. L'Industrie électrique. - In: H. Poincaré, E. Perrier, P. Painlevé. Ce que disent les choses. Paris, Hachette et Cie, 1912, p. 75–78.

4. *Other writings.*

531. Sur l'application du calcul des probabilités (lettre à M. P. Painlevé). In: Le Procès Dreyfus devant le Conseil du Guerre de Rennes, 7 aout–9 septembre 1899, t. 3. Paris, P.-V. Stock, 1900, p. 329–331.

532. Inauguration de la statue de F. Tisserand. — Annuaire Bureau Longitudes, 1900, E. 4—E. 12.

533. Sur la vérité scientifique et sur la vérité morale. — Université de Paris. Bull. officiel Assoc. gén. Etudiants de Paris, 1903, 18, 59—64.

534. Rapport relatif à la Fondation Jean Debrousse (1 avr. 1903). — Mémoires de l'Institut. Fondation Jean Debrousse, 1900—1905. Rapports, p. 45—67.

535. Rapport relatif à la Fondation Jean Debrousse (23 mars 1904).—Mémoires de l'Institut. Fondation Jean Debrousse, 1900—1905. Rapports, p. 69—86.

536. Rapport relatif à la Fondation Jean Debrousse (15 mars 1905).—Mémoires de l'Institut. Fondation Jean Debrousse, 1900—1905. Rapports, p. 87—101.

537. Rapport relatif à la Fondation Jean Debrousse. — Mémoires de l'Institut, 1906, p. 65—75.

538. Sur la participation des savants à la politique. — Rev. politique et littéraire (Revue bleue), 5e sér., 1904, 1, 708.

539. Sur la culture scientifique en Hongrie. — Magyar Szó (Budapest), 1906, N 303, suppl., p. 1—2.

540. Sur l'application du calcul des probabilités. - In: Affaire Dreyfus. La Revision du Procès de Rennes. Enquête de la Chambre criminalle de la Cour de Cassation, 5 mars–19 novembre 1904, t. 3. Paris, Ligue des Droits de l'Homme, 1909, p. 500–600 (with Darboux and Appell).

541. Discours au Banquet de la Société amicale des Lorrains de Meurtheet-Moselle (15.VI 1909). — Est Républicain (Nancy), 1909, N 8057, p. 2.

542. Discours a l'inauguration du monument élevé à la mémoire d'Octave Gréard (11.VII 1909). — Mémoires de l'Institut, 1909, p. 3—8; Le Temps (Paris), 1909, 12 juillet.

543. Discours.–A la réception en Sorbonne des Mèmbres de l'Expédition dans l'Antarctique, commandée par le Dr. J. Charcot (7.X II 1910). Paris, 1910, p. 4–6.

544. Sur la prépondérance politique du Midi.–L'Opinion (Paris), 25 mars 1911, p. 353–354.

545. Les sciences et les humanités. Paris, A. Fayard, 1911, 32 p.

546. Sciences et humanités. Published lecture, Vienna, May 22, 1912.

547. Preface to translation (of Science and Hypothesis). – In: H. Poincaré. The foundation of science. Lancaster, Science Press, 1913, p. 3–7.

Proceedings of Symposia in Pure Mathematics
Volume 39 (1983), Part 2

Books and Articles About Poincaré

(This compilation is almost certainly incomplete. A more detailed analysis of the literature involving Poincaré's views on the philosophy of mathematics can be found in the book of Mooij.)

R. d'Adhémar. Henri Poincaré. Paris, A. Hermann et Fils, 1912, 41 p.

P. S. Alexandroff. Poincaré and topology. Uspekhi Mat. Nauk (1973).

R. Apery. Conférence. — Livre du Centenaire [1], p. 148—153.

P. Appell. Henri Poincaré en mathématiques spéciales à Nancy. (Lettre à M. Mittag-Leffler. Paris, 22.XII 1912).— Acta math., 38, 1921, 189—195; Oeuvres de Henri Poincaré, t. XI, p. 139—145.

P. Appell. Henri Poincaré. Paris, 1925.

E. T. Bell. The last Universalist. Poincaré. — In: E. T. Bell. Men of mathematics. N. Y., Dover Publications, 1937, p. 526—554.

A. Bellivier. Henri Poincaré ou la vocation souveraine. 2ᵉ éd. Paris, Gallimard, 1956, 245 p.

R. Berthelot. Un romantisme utilitaire, v. 1. Le Progmatisme chez Nietzsche et chez Poincaré. Paris, Alcan, 1911.

E. W. Beth. Poincaré et la Philosophie. — Livre du Centenaire, p. 232—238.

E. Borel. Allocution. — Livre du Centenaire, p. 81—83.

E. Borel. Compte rendu de l'article de Poincaré: «La logique de l'infini». — Rev. du Mois, 1909, 8, 504.

E. Borel. La méthode de M. Poincaré. — Rev. du Mois, 1909, 7, 360—362.

A. Boulanger. Compte rendu et analyse de Calcul des probabilités. 2ᵉ éd. par H. Poincaré. — Bull. sci. math., 2ᵉ sér., 1912, 36, pt. 1, 169—184.

P. Boutroux. Henri Poincaré: L'oeuvre philosophique. — Rev. du Mois, 1913, 15, 155—183; Henri Poincaré. L'oeuvre scientifique. L'oeuvres philosophique. Paris, 1914, p. 205—250.

P. Boutroux. Lettre à M. Mittag-Leffler (Paris, 18.VI 1913). — Acta math., 1921, 38, 197—201; Oeuvres de Henri Poincaré, t. XI, p. 146—151.

M. Brillouin. Compte rendu et analyse de «Théorie mathématique de la Lumière» par H. Poincaré. — Bull. sci. math., 2ᵉ sér., 1889, 13, pt. 1, 173—198.

M. Brillouin. Extraits de l'Ouvrage de M. Henri Poincaré sur la theorie mathématique de la lumière. — In: Oeuvres de Henri Poincaré. t. X, p. 221–227.

L. de Broglie. Henri Poincaré et les theories de la physique. — In: L. de Broglie. Savants et découvertes. Paris, Albin Michel, 1951, p. 45–65.

L. de Broglie. Extraits de la conférence à la Société astronomique de France. — Livre du Centenaire, p. 140—146.

L. de Broglie. Henri Poincaré et les théories de la Physique. — Astronomie, 1954, 68, 217—229.

L. de Broglie. Henri Poincaré et les théories de la physique.* — Livre du Centenaire, p. 62—71.

L. de Broglie. Préface pour les tomes IX et X des Oeuvres de Henri Poincaré. – In: Oeuvres de Henri Poincaré, t. IX, p. VII–XIII.

M. de Broglie. Henri Poincaré et la philosophie. — Livre du Centenaire, p. 71—77.

L. Brunschvicg. L'oeuvre d'Henri Poincaré. Le philosophe. — Rev. métaphys. et morale, 1913, 21, N 5, 585—616.

S. G. Brush. Poincaré and cosmic evolution. Physics Today, 42 (1980), 42–49.

Many of the items in this compilation were reprinted directly from Selected Works of H. Poincaré (in Russian) edited by N. Bogoliubov and V. I. Arnold, 3 volumes, Moscow, 1972. (Bibliography compiled by E. I. Pogrebiskoi in Vol. 3.)

[1] Le Livre du Centenaire de la naissance de Henri Poincaré. 1854–1954. Paris, Gauthier-Villars, 1955 (Oeuvres de Henri Poincaré, t. XI).

P. G. Cath. Jules Henri Poincaré. — Euclides, 1954—1955, 30, 265—275.

A. Cecchini. Il concetto di convenzione matematica in Henri Poincaré. Torino, 1951.

L. Couturat. Etudes sur l'espace et le temps de M. M. Lechalas, Poincaré, Delboef, Bergson, L. Weber et Evellin. — Rev. métaphys. et morale, 1896, 4, 646—669.

C. Cuvaj. Henri Poincaré's mathematical contributions to relativity and the Poincaré stresses. — Amer. J. Phys., 1968, 36, N 12, 1102—1113.

T. Dantzig. Henri Poincaré. N. Y.—London, Charles Scribner's Sons, 1954, 149 p.

G. Darboux. Éloge historique d'Henri Poincaré membre de l'Académie lu dans la séance publique annuelle du 15 décembre 1913. — In: Oeuvres de Henri Poincaré, t. II, p. VII—LXXI.

G. Darmois. Répercussion des travaux d'Henri Poincaré dans le domaine du calcul des probabilités et de ses applications. — Livre du Centenaire, p. 127—132.

G. Darrieus. Contributions diverses d'Henri Poincaré à l'électrotechnique. — Livre du Centenaire, p. 132—139.

Dassault. Discours. — Livre du Centenaire, p. 97—106.

H. Dingle. Note on Mr. Keswani's article «Origin and Concept of Relativity». — Brit. J. Philos. Sci., 1965, 16, 242—246.

R. Dugas. Henri Poincaré devant les Principes de la Mécanique. — Rev. scient., 1951, 89, fasc. 2, N 3310, 75—82.

C. Eisele-Halpern. Poincaré's positivism in the light of C. S. Peirce's realism. — Actes IX Congrès internat. histoire sciences, t. 2. Paris, Hermann, 1960, p. 461—465.

E. Faguet. La philosophie de M. Henri Poincaré. — Rev. Latine, 1908, 7, 1—14.

G. Fornaro. Henri Poincaré e il valore della scienza. Napoli, 1924.

H. Freudenthal. Poincaré et les fonctions automorphes*. — Livre du Centenaire, p. 212—219.

R. Garnier. Les fonctions automorphes de Poincaré et la géométrie. — Livre du Centenaire, p. 29—48.

J. Giedymin, On the origin and significance of Poincaré's conventionalism. Studies in History and Philosophy of Science, 8 (4), (1977), 271–301.

S. Goldberg. Henri Poincaré and Einstein's theory of relativity. — Amer. J. Phys., 1967, 35, 934—944.

S. Goldberg. Poincaré's silence and Einstein's relativity. — Brit. J. History Sci., 1970, 5, N 17, 73—84.

E. Guillaume. Introduction. – In: H. Poincaré. La mécanique nouvelle. Paris, Gauthier-Villars et Cie, 1924, p. V—XVI.

J. Hadamard. Henri Poincaré et les mathématiques*.—Livre du Centenaire, p. 450—457.

J. Hadamard. Le problème des trois corps. – In: Henri Poincaré. L'oeuvre scientifique. L'oeuvre philosophique. Paris, 1914, p. 51—114.

J. Hadamard. L'oeuvre d'Henri Poincaré. Le mathématicien. — Rev. métaphys. et morale, 1913, 21, N 5, 617—658.

J. Hadamard. L'oeuvre mathématique de Poincaré. — Acta math., 38, 1921, 203—287; Oeuvres de Henri Poincaré, t. XI, p. 152—242.

G. B. Halsted. Henri Poincaré. – In: H. Poincaré. The foundation of science. Lancaster, Science Press, 1946, p. IX—XI.

Henri Poincaré (1854—1912). Nécrologie. — Rev. métaphys. et morale, 1912, Suppl., septembre, p. 1.

G. Holton. On the thematic analysis of science: the case of Poincaré and relativity. — Actes X Congrès internat. histoire sciences, t. 2. Paris, Hermann, 1962, p. 797—800.

G. Holton. On the origin of the special theory of relativity. — Amer. J. Phys., 1960, 28, 627—636.

G. Holton. On the thematic analysis of science: the case of Poincaré and relativity. — Mélanges Alexandre Koyré, 1964, 2, 257—268.

G. Humbert. Henri Poincaré. — La Nature, 1912, N 2044, 143—144.

D. M. Johnson. The problem of invariance of dimension in the growth of modern topology. Part II, Archive for History of Exact Sciences, 25 (1981),85—267 (spec. Chap. 5, pp. 85–112).

G. Julia. Henri Poincaré, sa vie et son oeuvre*. — Livre du Centenaire, p. 165—173.

G. H. Keswani. Origin and concept of relativity. — Brit. J. Philos. Sci., 1965, 15, 286—306.

G. H. Keswani. Origin and concept of relativity, II. — Brit. J. Philos. Sci., 1965, 16, 19—32.

F. Klein. Vorlesungen über die Entwicklung der Mathematik in 19. Jahrhundert, Teil 1. Berlin, J. Springer, 1926, S. 374—381.

G. Kropp. Poincaré. – In: G. Kropp. Geschichte der Mathematik. Heidelberg, 1969, S. 206–207.

F. Kuntze. Zum Gedächtnis an Henri Poincaré. – Kantstudien, 1912, 17, 337–348.

A. Lalande. Henri Poincaré: From *Science and hypothesis* to *Last thoughts*. – In: Roots of scientific thought. Ph. P. Wiener and A Noland (Eds). N. Y., Basic Books, 1957, p. 624–626.

C. Lanczos. Lorentz, Poincaré, Einstein, Minkowski. – In: C. Lanczos. Space through the ages. London a. N. Y., Acad. Press, 1970, p. 230–232.

P. Langevin. L'oeuvre d'Henri Poincaré. Le physicien. – Rev. métaphys. et morale, 1913, 21, N 5, 675–718.

P. Langevin. Le physicien. – In: Henri Poincaré. L'oeuvres scientifique. L'oeuvres philosophique. Paris, Félix Alcan, 1914, p. 115–202.

A. Lebeuf. L'oeuvre d'Henri Poincaré. L'astronome. – Rev. métaphys. et morale, 1913, 21, N 5, 659–674.

E. Lebon. Henri Poincaré. Paris, 1909 (Collection: Savants du Jour).

E. Lebon. Henri Poincaré. Biographie, bibliographie analytique des écrits. 2 éd. Paris, Gauthier-Villars, 1912, 111 p.

E. Lebon. Notice sur Henri Poincaré. – In: H. Poincaré, Lecons sur les hypothèses cosmogoniques. 2e éd. Paris, 1913, p. III–XLVIII.

J. Lévy. Poincaré et la Mécanique céleste. – Livre du Centenaire, p. 225–232.

G. Lippmann. Allocation (16.12.1912). – C. r. Acad. sci., 1912, 155, 1280–1283.

H. A. Lorentz. Deux mémoires de Henri Poincaré sur la physique mathématique. – Acta math., 38, 1921, 293–308.

F. L. Lot. Henri Poincaré et l'invention mathématique. – In: F. Lot. Visages des grands savants. Paris, Michel, 1963, p. 319–328.

A. Marie. Discours. – Livre du Centenaire, p. 84–89.

J. Marty. Compte rendu et analyse de «Sechs Vorträge über ausgewählte Gegenstände aus der reinen Mathematik und mathematischen Physik» par H. Poincaré. – Bull. sci. math., 2e sér., 1910, 34, partie 1, 100–104.

H. Russell McCormach. Henri Poincaré and the quantum theory. – Isis, 1967, 58, N 191 (1), 37–55.

G. Milhaud. La science et l'hypothèse, par M. H. Poincaré. – Rev. métaphys. et morale, 1903, 11, 773–791.

A. I. Miller. A Study of Henri Poincaré's «Sur la dynamique de l'électron.» – Arch. history exact sci. 1973, 10, 207–328.

N. Minorski. Influence d'Henri Poincaré sur l'évolution moderne de la théorie des oscillations non linéaires. – Livre du Centenaire, p. 120–126.

J. J. A. Mooij. La philosophie des mathématiques de Henri Poincaré (Collection de Logique mathématique. Sér. A20). Paris, Gauthier-Villars, 1966, 174 p.

Ch. Nordmann. Henri Poincaré. Son oeuvre scientifique – sa Philosophie. – Revue des deux Mondes, 1912, 82e ann., 11, 331–368.

P. Painlevé. Henri Poincaré. – Le Temps, 1912, N 18 642, p. 1.

P. Painlevé. Henri Poincaré. – Acta math., 38, 1921, 399–402.

F. Perrin. Henri Poincaré et Pierre Duhem. – In: F. Perrin. Histoire des sciences. Paris, Beaudart, 1956, p. 307–309.

E. Picard. L'oeuvre de Henri Poincaré. – Ann. Ecole Normale, 3e sér., 1913, 30, 463–482. Separately published, Paris, 1913, 22 p.

M. Planck. Henri Poincaré und die Quantentheorie. – Acta math., 1921, 38, 387–397; Oeuvres de Henri Poincaré, t. XI, p. 347–356.

R. Poirier. Henri Poincaré et le problème de la valeur de la science. – Rev. philos. France et etranger, 1954, N 10–12, 485–513; Livre du Centenaire, p. 176–202.

J. C. Pont. La Topologie Algebrique des Origines à Poincaré, Paris, 1974.

G. Rados. Rapport sur le Prix Bolyai, présenté à l'Academie Hongroise des Sciences. – Bull. sci. math., 2e sér., 1906, 30, pt. 1, 103–128.

G. Rados. Rapport sur le Prix Bolyai. – In: E. Lebon. Henri Poincaré. Paris, 1912, p. 21–26.

G. Rageot. La philosophie d'un géomètre: Henri Poincaré. – Revue de Paris, 1906, 13, 827–851.

De la Rive. Henri Poincaré le physicien. – Arch. sci. phys. et natur., 1914, 38, 159–163, 189–201.

L. Rougier. Henri Poincaré et la mort des vérités nécessaires. – La Phalange (Paris), 1913, 8, ann., 15, N 85, 1–20.

L. Rougier. La philosophie géométrique de Henri Poincaré. Thèse. Paris, F. Alcan, 1920.

J. *Royce.* Introduction (to Science and Hypothesis). – In: H. Poincaré. The foundation of science. Lancaster, Science Press, 1913, p. 9–25.

B. *Russell.* Compte rendu de Science et l'Hypothèse. — Mind, 1905, 14, 412–418.

T. *Sageret.* Henri Poincaré. Paris, Mercure de France, 1911, 80 p.

H. M. *Schwartz.* A note on Poincaré's contribution to relativity. — Amer. J. Phys., 1965, 33, 170.

H. M. *Schwartz.* Poincaré's Rendiconti paper on relativity. I—III. — Amer. J. Phys., 1972, 39, 1287—1294; 1972, 40, 862—872, 1282—1287.

L. *Schwartz.* L'oeuvre de Poincaré: Équations différentielles de la physique*. — Livre du Centenaire, p. 219—225.

Ch. *Scribner.* Henri Poincaré and the principle of relativity. — Amer. J. Phys., 1964, 32, N 9, 672—678.

M.-A. *Tonnelat.* Henri Poincaré et le principe de relativité. In: M.-A. Tonnelat. Histoire du principe de relativité. Paris, Flammarion, 1971, p. 123–129.

R. *Torretti,* Philosophy of Geometry from Riemann to Poincaré, Dordrecht, 1978.

Dr. *Toulouse.* Enquète médico-psychologique sur la supériorité intellectuelle: Henri Poincaré. Paris, 1910, 204 p.

H. *Villat.* Henri Poincaré et la mécanique. — Livre du Centenaire, p. 57—61.

V. *Volterra.* L'oeuvre mathématique (d'Henri Poincaré). – In: Henri Poincaré. L'oeuvre scientifique. L'oeuvre philosophique. Paris, 1914, p. 3—49.

K. *Weierstrass.* Über Poincarês Theorie der Fuchsschen Funktionen. — Acta math., 1923, 39, 240—245.

A. *Weil.* Poincaré et l'Arithmétique*. — Livre du Centenaire, p. 206—212.

R. M. *Wenley.* Scientific books: The value of Science by H. Poincaré. — Science, 1908, 27, 386—389.

E. *Whittaker.* The relativity theory of Poincaré and Lorentz. – In: E. Whittaker. A History of the theories of aether and electricity, v. 2. London–N. Y., 1953, p. 27–77.

W. *Wien.* Die Bedeutung Henri Poincaré's für die Physik. — Acta math., 38, 1921, 289—291; Oeuvres de Henri Poincaré, t. XI, p. 243—246.

E. B. *Wilson.* Compte rendu de Science et l'Hypothèse. — Bull. Amer. Math. Soc., 1906, 12, 187—193.

H. V. *Zeipel.* L'oeuvre astronomique d'Henri Poincaré. — Acta math., 38, 1921, 309—385; Oeuvres de Henri Poincaré, t. XI, p. 262—346.